# *The Mathematical Foundations of the Finite Element Method with Applications to Partial Differential Equations*

*ACADEMIC PRESS RAPID MANUSCRIPT REPRODUCTION*

*Proceedings of a Symposium*
*Held at the University of Maryland*
*Baltimore, Maryland*
*June 26–30, 1972*

Symposium on Mathematical Foundations of the Finite Element Method with Applications to Partial Differential Equations, University of Maryland, Baltimore County, 1972.

# The Mathematical Foundations of the Finite Element Method with Applications to Partial Differential Equations


Edited by

**A. K. Aziz**

*Division of Mathematics*
*University of Maryland*
*Baltimore, Maryland*

*and*

*Institute for Fluid Dynamics*
*and Applied Mathematics*
*University of Maryland*
*College Park, Maryland*



Academic Press
New York and London
1972

ACADEMIC PRESS, INC.
111 Fifth Avenue, New York, New York 10003

*United Kingdom Edition published by*
ACADEMIC PRESS, INC. (LONDON) LTD.
24/28 Oval Road, London NW1

LIBRARY OF CONGRESS CATALOG CARD NUMBER: 72-82668

PRINTED IN THE UNITED STATES OF AMERICA

# CONTENTS

## Part III. Short Communications

# CONTRIBUTORS

Numbers in parentheses indicate the pages on which authors' contributions begin.

**A. K. Aziz** (3), Division of Mathematics, University of Maryland, Baltimore, Maryland 21228

**Ivo Babuška** (3), Institute for Fluid Dynamics and Applied Mathematics, University of Maryland, College Park, Maryland 20742

**R. E. Barnhill** (749), Department of Mathematics, University of Utah, Salt Lake City, Utah 84112

**Alan E. Berger** (757), U. S. Naval Ordnance Laboratory, White Oak, Maryland 20910

**Garrett Birkhoff** (363), Department of Mathematics, Harvard University, Cambridge, Massachusetts 02138

**James H. Bramble** (387), Department of Mathematics, Cornell University, Ithaca, New York 14850

**P. G. Ciarlet** (409), Service de Mathématiques, Laboratoire Central des Ponts et Chaussées, 58, Boulevard Lefebvre, 75732 Paris Cedex 15, France

**J. E. Dendy, Jr.** (763), Department of Mathematics, University of Denver, Denver, Colorado 80220

**Jim Douglas, Jr.** (475), Department of Mathematics, University of Chicago, Chicago, Illinois 60637

**Todd Dupont** (491), Department of Mathematics, University of Chicago, Chicago, Illinois 60637

**S. C. Eisenstat** (505), Department of Computer Science, Yale University, New Haven, Connecticut 06520

**G. Fairweather** (767), Department of Mathematics, University of Kentucky, Lexington, Kentucky 40506

**George J. Fix** (525), Institute for Fluid Dynamics and Applied Mathematics, University of Maryland, College Park, Maryland 20742

**William J. Gordon** (769), Mathematics Department, General Motors Research Laboratories, Warren, Michigan 48090

**J. A. Gregory** (749), Mathematics Department, Brunel University, Uxbridge, England

**Charles A. Hall** (769), Department of Mathematics, University of Pittsburgh, Pittsburgh, Pennsylvania 15213

**Bruce M. Irons** (557), Department of Civil Engineering, University of Wales, Swansea SA2 8PP, United Kingdom.

**J. P. Johnson** (767), Department of Mathematics, University of Kentucky, Lexington, Kentucky 40506

**H. G. Kaper** (785), Applied Mathematics Division, Argonne National Laboratory, Argonne, Illinois 60439

**R. B. Kellogg** (589), Institute for Fluid Dynamics and Applied Mathematics, University of Maryland, College Park, Maryland 20742

**G. K. Leaf** (785), Applied Mathematics Division, Argonne National Laboratory, Argonne, Illinois 60439

**A. J. Lindeman** (785), Applied Mathematics Division, Argonne National Laboratory, Argonne, Illinois 60439

**Lois Mansfield** (791), Department of Computer Science, University of Kansas, Lawrence, Kansas 66044

**J. Nitsche** (603), Institute für Angewandte Mathematik der A-Ludwige-Universität, 78 Freiburg, West Germany

**J. T. Oden** (629), Department of Engineering Mechanics, University of Alabama, Huntsville, Alabama 35807

**John E. Osborn** (387), Department of Mathematics, University of Maryland, College Park, Maryland 20742

**Paul D. Patent** (793), Naval Undersea Research and Development Center, Pasadena, California 91107

**Theodore H. H. Pian** (671), Department of Aeronautics and Astronautics, Massachusetts Institute of Technology, Cambridge, Massachusetts 02139

**P.-A. Raviart** (409), Analyse Numérique, Tour 55, Université de Paris VI, 11, Quai Saint-Bernard, 75230 Paris Cedex 05, France

**Abdur Razzaque** (557), Engineering College, Rajsahi, Bangladesh

***A. H. Schatz**, Department of Mathematics, Cornell University, Ithaca, New York 14850

**M. H. Schultz** (505), Department of Computer Science, Yale University, New Haven, Connecticut 06520

**Ridgway Scott** (797), Massachusetts Institute of Technology, Cambridge, Massachusetts 02139

**Gilbert Strang** (689), Department of Mathematics, Massachusetts Institute of Technology, Cambridge, Massachusetts 02139

**Vidar Thomée** (711), Department of Mathematics, Chalmers Institute of Technology and the University of Göteborg, Göteborg, Sweden

**J. R. Whiteman** (749), Mathematics Department, Brunel University, Uxbridge, England

---

*Manuscript not received in time to be included.

# PREFACE

The Symposium on Mathematical Foundations of the Finite Element Method with Applications to Partial Differential Equations was held June 26-30, 1972, at the University of Maryland, Baltimore County Campus. Its purpose was to bring together a number of active numerical analysts currently involved in research in both theoretical and practical aspects of the finite element method. Among 250 participants were scientists from Canada, England, France, Germany, Ireland, Israel, Japan, Sweden, and Switzerland, thus providing the conference with a definite international flavor.

In recent years the scientific community, in particular the engineers, have focused considerable attention on the use of the finite element method. This is evidenced by the numerous national and international conferences held on this topic. In the last two years alone, 15 principal international conferences have been devoted to the finite element method, with the main emphasis on engineering applications.

As can be seen from the table of contents of these proceedings, the present symposium aims at bridging the gap between the mathematical and the practical aspects of the finite element method.

These proceedings consist of three parts. Part I gives the content of the 10 one-hour lectures given by Professor I. Babuška on the mathematical foundations of the field, while Part II contains all but one of the 16 one-hour lectures given by the invited speakers. These papers cover a large number of important results of both a theoretical and a practical nature. Part III contains the abstracts of 15-minute contributed talks.

The Division of Mathematics of the University of Maryland, Baltimore County Campus, and the U. S. Office of Naval Research were the joint sponsors of the symposium. The generous financial assistance of the U. S. Navy and the combined hard work of many members of the University of Maryland, faculty and staff, contributed immeasurably to the success of this meeting.

The editor wishes to express his sincere thanks to all these contributors. The advice and encouragement given by Professors I. Babuška, R. B. Kellogg, and G. J. Fix have been particularly helpful.

*The Mathematical Foundations of the Finite Element Method with Applications to Partial Differential Equations*

# PART I

# SURVEY LECTURES ON THE MATHEMATICAL FOUNDATIONS OF THE FINITE ELEMENT METHOD

*Ivo Babuška and A. K. Aziz*
with the collaboration of
*G. Fix and R. B. Kellogg*

# FOREWORD

As indicated in the preface, numerous meetings and conferences in the last few years have been devoted to recent developments in theory and applications of the finite element method. The symposium at the University of Maryland, Baltimore County Campus, differed from the others in its orientation which was exclusively directed toward the theoretical foundations of the method.

To this end the organizing committee, which consisted of A.K. Aziz (Chairman), I. Babuška, N.P. Bhatia, and R.B. Kellogg, included in the program a series of ten lectures dealing exclusively with basic theoretical concepts.

The notes which follow are an attempt to focus on some of the most important of these principles. In their preparation, some constraints were imposed by the nature and goals of the conference. Therefore, they should be considered as notes describing the content of the ten lectures and not as a monograph on the finite element method.

The notes were prepared by I. Babuška[1] and A.K. Aziz[2] with significant contributions and help provided by R.B. Kellogg[3] and G. Fix.[4] In particular, Chapter 3 on regularity of the solution was written by R.B. Kellogg; Chapter 11 and parts of Chapter 10 were written jointly with G. Fix.

*This work was supported in part by:*

[1]The Atomic Energy Commission, Contract No. AEC AT (40-1) 3443/5.

[2]The Naval Ordnance Laboratory Independent Research fund, and the Office of Naval Research, Contract No. NR-044-453.

[3]The National Science Foundation, Grant No. NSF GP 20555.

[4]The National Science Foundation, Grant No. NSF GP 18064.

We would like to express our gratitude to Professors Kellogg and Fix for their valuable assistance and understanding. We must not fail also to thank Mrs. L. Lau and Mrs. M. Small for their generous services, including the typing of the manuscript.

I. BABUŠKA and A. K. AZIZ

*College Park*
*September 1972*

# CHAPTER 1. PRELIMINARY REMARKS

## 1.1. Introduction.

The development of approximate methods for the numerical solution of partial differential equations has attracted attention of mathematicians, physicists and engineers for a long time. The methods, their mathematical foundation and their implementation have had a considerable impact on theoretical mathematics and on the level of sophistication of computational aids. Roughly speaking we may distinguish two stages in this development, namely the precomputer period and the computer era.

An excellent survey of the state of the art concerning the numerical solution of differential equations, up to the end of the precomputer period may be found in [1] and [2].

The application to partial differential equations is exemplified by the work of Southwell (see [3]) and others.

The advent of the computer served as a stimulus for new ideas in theory and in applications. Engineers, physicists and others have suggested many sophisticated, but often not theoretically founded, methods; nonetheless a great deal of experience has been gained. Mathematics has played a very central and essential role in this development, in particular in the understanding of the theoretical foundation of these methods, their applications and further very effective and new computer oriented methods have been devised by the numerical analysts. It can be stated without exaggeration that the present day numerical methods are essentially different from those of the precomputer period, and are primarily based on the achievements in the last 15 to 20 years.

The development of numerical methods for partial differential equations was particularly influenced by the innovations brought about by the computer era. One of the typical by-product is the present day's frequent use of the finite element method for the numerical solution of partial differential equations especially as far as the elliptic partial differential equations are concerned.

In recent years considerable attention has been focused on the finite element method primarily in the engineering literature. In the last few years an increasing number of papers on the finite element method has also appeared in the numerical analysis literature.

An extensive bibliography may be found in [4], [10] and [11]. For an excellent account of the interplay of mathematical and engineering ideas in the finite element method the reader is referred to [5]. [6] may be consulted for a mathematical formulation and [12] for a survay of the method.

The attention accorded to the finite element method by the scientific community is further evidenced by numerous national and international conferences held on this topic. In the last two years alone, 15 principal international conferences have been devoted to the finite element method, not counting a large number of small symposia, workshops and short courses in this field.

## 1.2. Numerical solution of partial differential equations.

In numerically solving a partial differential equation one first expresses approximately the solution by a finite number of parameters. Since, in general the solution is sought in a given class of functions, it is essential that one be able to express any function of this class in terms of a finite number of parameters, with a reasonable accuracy. Further details in this connection may be found in [7] and [8].

Secondly, we need to transform the given differential operator to expressions relating these parameters. For a general abstract approach the reader may consult chapter 14 of [9]. If the differential operator is linear then, in general, the relations among the parameters expressing the solution is also linear i.e., we are led to a linear system of algebraic equations. However, as supported by the general theory (see [7], [8]), in this process one cannot avoid dealing with a large number of parameters of order of at least a hundred or a thousand. Moreover, we need to determine the coefficients of the matrix, the number of which may be of order $10^4$ or $10^6$.

To avoid this complication it is necessary (but not sufficient) to choose the parameters in such a way that the resulting matrix is sparse. If we are seeking an approach which is applicable to the problems encountered in general practical cases, there are other restrictions too.

One of the most successful methods reflecting

these features and other important aspects is the finite element method.

## 1.3. Finite Element Method.

The name finite element method was invented by engineers (see [4]). We will be interested in a slightly more general version which we feel reflects all essential theoretical features of the classical finite element method. We shall call it the general finite element method. For the sake of brevity we will not in the sequel emphasize the word "general" nevertheless we shall always interpret it in this general context.

A cursory analysis of the so called finite element method reveals that there are two principles which appear to be essential from a mathematical point of view.

1) The choice of local parameters of the solution.
2) The use of various types of variational principles for transforming the given equations to relations among the parameters of the solution.

By the choice of local parameters it is understood that if the approximate solution $u$ is expressed in terms of the parameters $\alpha_i$ as $u = \sum \alpha_i \phi_i$, then the base functions $\phi_i$ have only small supports and if the number of the parameters is increased then the support of $\phi_i$ is decreased. A typical case is the introduction of the parameters which describe the set $S_h$ of the piecewise linear function (see Fig. 1.3.1).

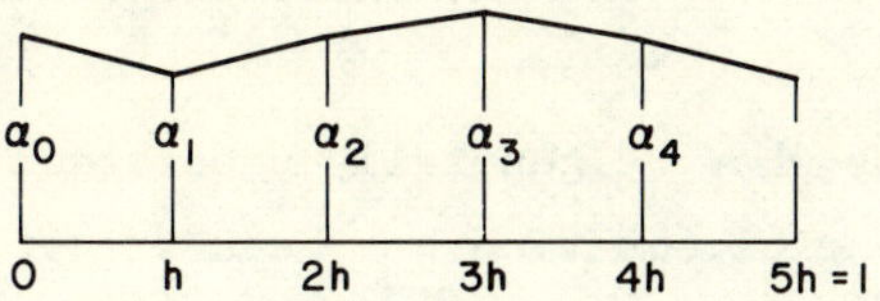


Fig. 1.3.1

It is clear that a change in one parameter, say $\alpha_3$, changes the function only in the interval $\langle 2h, 4h \rangle$. Obviously, for the set of piecewise linear functions, parameters may be introduced which are not local. There is a question as to whether one can find local parameters for every finite dimensional subspace. To answer this question consider the subspace $S_h = \sum_o^{1/h} \alpha_i \cos ix$ of all trigonometric polynomials. Indeed here we have quite a different situation. Thus, the assumption concerning the possibility of the introduction of local parameters for the solution means the possibility of a suitable selection of a finite dimensional subspace and a proper choice of the basis.

We remark that the term variational principle, in general, is understood in a broad sense. We interpret this term in a narrow sense. More precisely, we use the variational principle to mean that its application to a linear differential equation, transforms the given equation to a system of linear algebraic equations with the matrix $M$, provided the space $S_h$ of possible approximate solutions is linear. The elements of M are given by the bilinear form $B(\phi_i, \phi_j)$ , where $\phi_i$ and $\phi_j$ are basis functions. Moreover, we assume that the bilinear form is such that whenever $\phi_i$ and $\phi_j$ have disjoint support then $B(\phi_i, \phi_j) = 0$ .

This is what we generally understand when we say that the differential equation has been transformed to relations among the parameters of the solution, by the use of the variational principle. Under our assumptions the matrix M is sparse and the number of entries in the rows of the matrix is relatively very small, in fact is most frequently independent of the size of the matrix.

The introduction of the parameters and the construction of the matrix (variational principle) have very important physical significance in every stage in the application of the finite element method to engineering problems. This feature has greatly contributed to the widespread use of the finite element method. However, from the theoretical point of view, this aspect does not appear to be so significant.

### 1.4. The sources of the theory of the finite element method.

The theory of the finite element method in all its complexity is based on the fundamental knowledge in different fields. We feel that the heart of the theory of the finite element method lies in the following sources.

1) Functional analytic theory of partial differential equations.
2) Theory of approximation by piecewise

polynomial functions.

3) Computer science.

4) Concrete applications in different fields such as mechanics etc.

It is clear that all these sources have contributed greatly to the present day development of the finite element method. It would be impossible (and would not serve a useful purpose) if we tried to trace the different sources with important impact on the method, other than the advent of computers.

## 1.5. The mathematical foundations of the finite element method.

The main purpose of these notes is to assemble some basic results, and to treat them from a unified point of view.

These notes do not pretend to be complete, nevertheless they aim to develop the fundamentals of the method in such a way that it may be applied to a fairly large number of problems in different fields. For obvious limitations the discussion has been confined to some important preliminaries and basic ideas.

The aim of these notes is the development of a general and complete theory. Our goal is to underline the basic ideas and illustrate them by considering a selective number of examples. The chosen examples are very simple in appearance; however, they exhibit the typical characteristic of the problems encountered in general applications and at the same time, because of the simplicity of the form of the equations, many difficulties of a technical nature are avoided. For the above reason we have confined our discussion to two dimensional problems only. The setting of the problems is in a general framework and the approach utilized admits in all cases immediate and far reaching generalizations. In some cases a more special treatment could prove to be shorter and easier, but without a general character.

The model problems are selected not merely because of their mathematical interest, but also for their relevance from the point of view of applications. Therefore we have not restricted ourselves to the case of self-adjoint problems. In mechanics in the case of the second order equations it is natural to consider the self-adjoint form, but this is not the case in the diffusion problems e.g., diffusion in a moving medium etc.

References.

[1]. L. Collatz, "Numerische Behandlung von Differentialgleichungen, Springer-Verlag, 1951.

[2]. L. V. Kantorovich, V. J. Krylov, "Näherungsmethoden der Höheren Analysis, Berlin 1956.

[3]. F. S. Show, "An introduction to relaxation methods," New York, Dover Publ. Inc., 1953.

[4]. O. C. Zienkiewicz, "The finite element method: from intuition to generality," Applied Mechanics Reviews 23 (1970), pp. 249-256.

[5]. G. Strang, G. Fix, "An analysis of the finite element method," Prentice Hall, New York (to appear 1973).

[6]. J. P. Aubin, "Approximation of elliptic boundary-value problems," Wiley-Interscience, New York, London, Sydney, Toronto, 1972.

[7]. I. Babuška, S. L. Sobolev, "The Optimization of Numerical Processes," Apl. Mat. 10 (1965), pp. 96-129 (in Russian).

[8]. A. G. Vituskin, "Estimates from Tabulation Theory (in Russian), Moscow 1959.

[9]. L. V. Kantorovich, G. P. Akilov, "Functional analysis in Normed Spaces," The MacMillian Co., New York 1964.

[10]. J. E. Abin, D. L. Fenton, W. C. T. Stoddart, "The Finite Element Method -- A bibliography of its Theory and Applications." Report EM 72-1, Department of Engineering Mechanics, the University of Tennessee Knoxville, 1971.

[11]. J. R. Whiteman, "A Bibliography for Finite Element Method," Brunel University. Report TR/9, Department

of Mathematics, Brunel University, Uxbridge, 1972.

[12]. J. T. Oden, "Some Aspects of the Mathematical Theory of Finite Elements", pp. 3-38, Advances in Computation Computational Methods and Structural Mechanics and Design, edited by J. T. Oden, R. W. Clough, Y. Yamamoto, UAH Press, the University of Alabama in Huntsville, Huntsville, Alabama.

## CHAPTER 2. THE FUNDAMENTAL NOTIONS.

### 2.1. Introduction.

In these notes for the sake of technical simplicity we shall be concerned primarily with problems in two dimensions, but the approach used will be of quite general character. The properties of the domains considered play a fundamental role as far as the behavior of the solutions of partial differential equations are concerned. In the sequel we will be mainly interested in bounded domains, and of course in the entire plane $R_2$. First, we shall classify more precisely the type of domains which will be considered in this sequel.

### 2.2 Domains.

Let $R_2$ denote the two dimensional Euclidean space. For $x \equiv (x_1,x_2) \in R_2$ we use the notations:

$$\|x\| = \sqrt{x_1^2+x_2^2} \quad \text{and} \quad |x| = \max(|x_1|,|x_2|) \quad .$$

Further let $\Omega$ be a given bounded domain, with the boundary $\Omega^{\bullet}$ which fulfills the following assumptions. There exists a system of local coordinates $x_i^{[s]}$, $i = 1,2$, $s = 1,2,\cdots,\nu$ and open intervals $I_s \in R_1$ and $\bar{I}'_s \subset I_s$, and functions $\phi_s$ defined on $I_s$ which induce a mapping

$\chi_s$ of $I_s$ such that

$$\chi_s(I_s) = \Omega^{\bullet}_s = \left\{\left(x_1^{[s]}, \phi_s(x_1^{[s]})\right) \middle| x_1^{[s]} \in I_s\right\} \tag{2.2.1}$$

and

$$\bigcup_{s=1}^{\nu} \Omega^{\bullet\prime}_s = \Omega^{\bullet} ,$$

where

$$\Omega^{\bullet\prime}_s = \left\{\left(x_1^{[s]}, \phi_s(x_1^{[s]})\right) \middle| x_1^{[s]} \in I'_s\right\} . \tag{2.2.2}$$

with $\Omega$ locally on one side of $\Omega^{\bullet}$'s.
We shall say that the domain $\Omega$ is smooth if all functions $\phi_s$ have derivatives of all orders. The domain $\Omega$ is called Lipschitzian if every function $\phi_s$ satisfies a Lipschitz condition.

We remark that in many of our considerations we do not need that $\phi_s$ be infinitely differentiable, we merely require that they be sufficiently smooth. Nevertheless in order to avoid technical difficulties we shall always assume that $\phi_s$ are infinitely differentiable, whenever we use the term smooth domain. The Lipschitzian domain would be the most general domain with which we shall deal. In this setting inevitably we exclude some important domains such as those shown in Fig. 2.2.1.

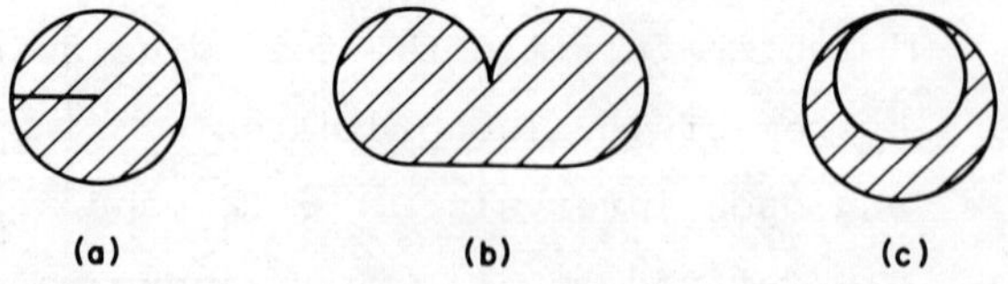


Fig. 2.2.1

or a more general domain.

We shall also deal with some special domains where the functions $\phi_s$ are merely piecewise smooth.

## 2.3. Sobolev spaces $H^\ell(\Omega)$ .

Consider a Lipschitz domain $\Omega$ and let $\mathcal{E}(\overline{\Omega})$ be the space of all [real] infinitely differentiable functions on $\Omega$ such that all the derivatives have continuous extensions to $\dot{\Omega}$ . Furthermore, denote by $\mathcal{D}(\Omega) \subset \mathcal{E}(\overline{\Omega})$ the subspace consisting of all functions with compact support in $\Omega$ . As usual let $L_2(\Omega)$ be the space of square integrable functions $u$ on $\Omega$ with the norm

(2.3.1) $$\|u\|^2_{L_2(\Omega)} = \int_\Omega u^2 \, dx \quad ,$$

where $dx = dx_1 dx_2$ . The scalar product will be denoted as $(u,v)_{L_2(\Omega)}$ . Sometimes we shall use the notation

$$L_2(\Omega) = H^\circ(\Omega) \, .$$

Suppose now that $\ell \geq 1$ is an integer. The Sobolev space $H^\ell(\Omega)$ (respectively $H^\ell_0(\Omega)$) will be defined as the closure of $\mathcal{E}(\overline{\Omega})$ (respectively $\mathcal{D}(\Omega)$) in the norm $\|\cdot\|_{H^\ell(\Omega)}$ , where

(2.3.2) $$\|u\|^2_{H^\ell(\Omega)} = \sum_{0 \leq |\alpha| \leq \ell} \|D^\alpha u\|^2_{L_2(\Omega)}$$

and

$$(2.3.3) \qquad D^{\alpha} = \frac{\partial^{\alpha_1+\alpha_2}}{\partial x_1^{\alpha_1} \partial x_2^{\alpha_2}}, \quad \alpha = (\alpha_1, \alpha_2)$$

$$|\alpha| = \sum_i \alpha_i \ .$$

($\alpha_i$ are non-negative integers.)

We now define the Sobolev spaces with fractional derivatives which were introduced by Aronszajn [1] and Slobodetskiĭ [2] (see also [3]). For $0 < \alpha = [\alpha] + \sigma$, $0 < \sigma < 1$ and $[\alpha]$ = integral part of $\alpha$, we define for $u \in \mathcal{E}(\bar{\Omega})$

$$(2.3.4) \quad \|u\|^2_{H^{\alpha}(\Omega)} = \|u\|^2_{H^{[\alpha]}(\Omega)} + \sum_{|K|=[\alpha]} \|D^K u\|^2_{H^{\sigma}(\Omega)} \ ,$$

where

$$(2.3.5) \qquad \|u\|^2_{H^{\sigma}(\Omega)} = \int_{\Omega}\int_{\Omega} \frac{[u(t)-u(\mathcal{T})]^2}{\|t-\mathcal{T}\|^{2+2\sigma}} \, dt d\mathcal{T} \ ,$$

and $H^{\alpha}(\Omega)$ is defined as the closure of $\mathcal{E}(\bar{\Omega})$ in the norm (2.3.4). Similary $H^{\alpha}_0(\Omega)$ will be defined as the closure of $\mathcal{D}(\Omega)$ in the norm defined by (2.3.4). In this definition of $H^{\alpha}_0(\Omega)$ the value $\alpha = n + \frac{1}{2}$, where $n$ is an integer is excluded. For this value of $\alpha$ the norm must be defined differently.

The spaces $H^{\alpha}(\Omega)$ respectively $H^{\alpha}_0(\Omega)$ are obviously Hilbert spaces. We may define the space $H^{\alpha}(\Omega)$ and $H_0^{\alpha}(\Omega)$ also for negative value of $\alpha$ as the closure of $E(\overline{\Omega})$ in the norm $\|\cdot\|_{H^{\alpha}(\Omega)}$ where for $\alpha < 0$

$$\|u\|_{H^{\alpha}(\Omega)} = \sup_{v \in H^{-\alpha}(\Omega)} \frac{\int_{\Omega} uv\,dx}{\|v\|_{H^{-\alpha}(\Omega)}} , \tag{2.3.6}$$

and the norm in $H_0^{\alpha}(\Omega)$ is similarly define. Obviously we may identify $H^{\alpha}(\Omega)$ and $H_0^{\alpha}(\Omega)$, $\alpha < 0$ with the dual spaces to $H^{-\alpha}(\Omega)$ and $H_0^{-\alpha}(\Omega)$ respectively.

We define the Sobolev spaces for a bounded domain $\Omega$, our definition is also valid for $\Omega = R_2$. In this case we may introduce the norm also by means of Fourier transform.

Let $u \in L_2(R_2)$ be given and

$$(Fu)(\sigma) = \hat{u}(\sigma) = \int_{R_2} e^{i\langle\sigma,x\rangle} u(x)\,dx ,$$

where $\sigma = (\sigma_1, \sigma_2)$ and $\langle\sigma,x\rangle = \sigma_1 x_1 + \sigma_2 x_2$. It is well known that

$$\|Fu\|^2_{L_2^c(R_2)} = (2\pi)^2 \|u\|^2_{L_2(R_2)} ,$$

where $L_2^c(R_2)$ is the usual space of complex square integrable functions.

Therefore for $\ell$ an integer

$$(2.3.7)\quad \|u\|^2_{H^\ell(\Omega)} = (2\pi)^{-2} \int_{R_2} |(Fu)(\sigma)|^2 \left[1+\mu^2+\cdots+\mu^{2\ell}\right] d\sigma ,$$

with $\mu^2 = \sigma_1^2 + \sigma_2^2$.

This norm is obviously equivalent to the norm $\|u\|_{H_1^\ell(\Omega)}$ given by

$$(2.3.8)\quad \|u\|^2_{H_1^\ell(\Omega)} = (2\pi)^{-2} \int_{R_2} |(Fu)(\sigma)|^2 \left[1+\mu^2\right]^\ell d\sigma .$$

We may define also the space $H_1^\ell(R_2)$ for non-integral values of $\ell$ by formula (2.3.8). By direct computation it is possible to verify that $\|u\|_{H^\ell(R_2)} \approx \|u\|_{H_1^\ell(R_2)}$ [1)], for all $\ell$. (We note that the

1) Norms $\|\cdot\|_{H_1}$ and $\|\cdot\|_{H_2}$ are said to be equivalent if there exist constants $0 < C_1 < C_2 < \infty$ such that

$$C_1\|u\|_{H_1} \leq \|u\|_{H_2} \leq C_2\|u\|_{H_1} .$$

We abreviate this by writing $\|\cdot\|_{H_1} \approx \|\cdot\|_{H_2}$. Throughout these notes $C$ denotes a generic constant with possibly different values in different contexts.

constants $C_1$ and $C_2$ appearing in the definition of the equivalent norms are not uniformly bounded in the case of the value $\ell$ lies between two integers.)

We shall list some basic well known properties of the Sobolev spaces. For a survey of the recent results in the theory of Sobolev spaces the reader is referred to [4].

Obviously for $\ell_1 \leq \ell_2$, $H^{\ell_1}(\Omega) \supset H^{\ell_2}(\Omega)$, with $\|\cdot\|_{H^{\ell_1}(\Omega)} \leq c\|\cdot\|_{H^{\ell_2}(\Omega)}$, i.e., the imbedding of $H^{\ell_2}(\Omega)$ into $H^{\ell_1}(\Omega)$ is continuous. Moreover for $\ell_1 < \ell_2$ and $\Omega$ bounded this imbedding is also compact, i.e., the unit sphere in $H^{\ell_2}(\Omega)$ is compact in $H^{\ell_1}(\Omega)$.

If $u \in H^{\ell}(\Omega)$, $\ell > 1$, then $u$ is continuous and $\|u\|_c \leq C\|u\|_{H^{\ell}(\Omega)}$ where $\|u\|_c = \max_{x\in\Omega}|u(x)|$ [1]. [In n dimensional case we need to require that $\ell > \frac{n}{2}$]. The continuity assertion of $u$ does not hold for $\ell = 1$.

The spaces $H^{\ell}(\Omega)$ as we have defined generate a Hilbert scale, more precisely they are equivalent to a Hilbert scale. For a detailed study of Hilbert scales the reader may consult [5].

---

1) Since the norm in $H^{\ell}$ is given by an integral in the sense of Lebesgue function u is continuous after possible change of its values on the set of measure zero.

In what follows we shall describe briefly the main ideas involved in a Hilbert scale which connects the space $H^{\ell_1}(\Omega)$ with the space $H^{\ell_2}(\Omega)$, $\ell_2 > \ell_1$.

For $v \in H^{\ell_1}(\Omega)$ we define the functional $\ell \in [H^{\ell_2}(\Omega)]'$ so that for $w \in H^{\ell_2}(\Omega)$ we have

$$\ell(w) = (w,v)_{H^{\ell_1}(\Omega)} , \tag{2.3.8}$$

and may write

$$\ell(w) = (w,z)_{H^{\ell_2}(\Omega)} = (w,Vv)_{H^{\ell_2}(\Omega)} \tag{2.3.9}$$

The operator $V$ may be considered as an operator in $H^{\ell_2}(\Omega)$ or in $H^{\ell_1}(\Omega)$. In both cases the operator is bounded, self-adjoint, positive, and in $H^{\ell_1}(\Omega)$, $V$ is a compact operator.

Denoting by $V^{\frac{1}{2}}$ the positive square root (which exists and it is uniquely determined), it is easy to show that $V^{\frac{1}{2}}$ induces an isometric isomorphism of the spaces $H^{\ell_1}(\Omega)$ and $H^{\ell_2}(\Omega)$, the operator $V^{\frac{1}{2}}$ maps $H^{\ell_1}(\Omega)$ into $H^{\ell_2}(\Omega)$, with $(u,v)_{H^{\ell_1}(\Omega)} = (V^{\frac{1}{2}}u,V^{\frac{1}{2}}v)_{H^{\ell_2}(\Omega)}$ and its inverse $V^{-\frac{1}{2}}$ maps $H^{\ell_2}(\Omega)$ onto $H^{\ell_1}(\Omega)$ with

$$(u,v)_{H^{\ell_2}(\Omega)} = (V^{-\frac{1}{2}}u, V^{-\frac{1}{2}}v)_{H^{\ell_1}(\Omega)} .$$

By denoting $T = V^{-\frac{1}{2}}$ we may write

$$V = T^{-2} = \sum_{k=1}^{\infty} \lambda_k^2 P_k , \tag{2.3.10}$$

where $\lambda_k$ are the eigenvalues and $P_k$ are the projection operators onto the subspaces of eigenfunctions $g_k$ of the equation

$$Vu = \lambda_k u .$$

We know that $\lambda_{k+1} \leq \lambda_k$ and $0 < \lambda_k \to 0$ as $k \to \infty$. Thus we have

$$T^2 = \sum_{k=1}^{\infty} \lambda_k^{-2} P_k . \tag{2.3.11}$$

We have

$$\|u\|_{H^{\ell_2}(\Omega)} = \|Tu\|_{H^{\ell_1}(\Omega)} , \tag{2.3.12}$$

and for $\ell_1 \leq \gamma \leq \ell_2$, we define

$$\|u\|_{H_\gamma(\Omega)} = \|T^{\frac{\gamma-\ell_1}{\ell_2-\ell_1}} u\|_{H^{\ell_1}(\Omega)} . \tag{2.3.13}$$

Now if

$$u = \sum_{k=1}^{\infty} a_k g_k \quad . \tag{2.3.14}$$

then

$$\|u\|^2_{H^{\ell_1}(\Omega)} = \sum_{k=1}^{\infty} |a_k|^2 \quad , \tag{2.3.15}$$

$$\|u\|^2_{H^{\ell_2}(\Omega)} = \sum_{k=1}^{\infty} \lambda_k^{-2}|a_k|^2 \quad , \tag{2.3.16}$$

$$\|u\|^2_{H_\gamma(\Omega)} = \sum_{k=1}^{\infty} \lambda_k^{-2\frac{\gamma-\ell_1}{\ell_2-\ell_1}}|a_k|^2 \quad , \tag{2.3.17}$$

and hence we see that

$$\|u\|_{H_\gamma(\Omega)} \leq \|u\|^{\frac{\ell_2-\gamma}{\ell_2-\ell_1}}_{H^{\ell_1}(\Omega)} \cdot \|u\|^{\frac{\gamma-\ell_1}{\ell_2-\ell_1}}_{H^{\ell_2}(\Omega)} \tag{2.3.18}$$

It is possible to show that $|\cdot|_{H_\gamma(\Omega)}$ is equivalent to the norm $|\cdot|_{H^\gamma(\Omega)}$ (see [5]) introduced earlier.

We notice that when connecting any two spaces $H^{\gamma_1}(\Omega)$ and $H^{\gamma_2}(\Omega)$, $\ell_1 < \gamma_1 < \gamma_2 < \ell_2$, by a Hilbert scale we obtain equivalent spaces to those described above.

In very much the same manner we can connect the space $H_0^{\ell_1}(\Omega)$ and $H_0^{\ell_2}(\Omega)$ by a Hilbert scale for integral values of $\ell_1$ and $\ell_2$. The spaces of the Hilbert scale are equivalent to the spaces $H_0^{\gamma}(\Omega)$ provided $\gamma \neq n + 1/2$, where n is an integer. In this case it is necessary to introduct $H_0^{\gamma}(\Omega)$ in a different form. Since we shall not make explicit use of this case in the sequel it will not be introduced.

We remark that due to the equivalence of the norms, we also have the inequality

(2.3.19) $$\|u\|_{H^{\gamma}(\Omega)} \leq C\|u\|_{H^{\ell_1}(\Omega)}^{\frac{\ell_2-\gamma}{\ell_2-\ell_1}} \|u\|_{H^{\ell_2}(\Omega)}^{\frac{\gamma-\ell_1}{\ell_2-\ell_1}}$$

The above inequality also holds for $H_0^{\gamma}(\Omega)$ provided $\gamma$, $\ell_1$, and $\ell_2$ are not of the form $n + 1/2$, with n an integer.

Let $H_\alpha$ and $H_\beta$ be two Hilbert scales with $\alpha_0 \leq \alpha \leq \alpha_1$ and $\beta_0 \leq \beta \leq \beta_1$. Moreover, assume that A is a given mapping which maps continuously $H_{\alpha_1}$ into $H_{\beta_1}$ and $H_{\alpha_0}$ into $H_{\beta_0}$. Then one can show that A maps continuously $H_{\alpha_t} = H_{\alpha_0 t+\alpha_1(1-t)}$ into $H_{\beta_t} = H_{\beta_0 t+\beta_1(1-t)}$, $0 \leq t \leq 1$, and

$$(2.3.20)\qquad \|A\|_{H_{\alpha_t}\to H_{\beta_t}} \leq C\|A\|^{t}_{H_{\alpha_0}\to H_{\beta_0}}\ \|A\|^{1-t}_{H_{\alpha_1}\to H_{\beta_1}} \ .$$

For more information on the theory and on the application of interpolation spaces the reader is referred to [5] and [6].

The spaces $H^{\alpha}(\Omega)$ are Hilbert spaces. Therefore in terms of the scalar product we may write

$$(2.3.21)\qquad \|u\|_{H^{\alpha}(\Omega)} = \sup_{v\in E(\bar{\Omega})} \frac{|(u,v)_{H^{\alpha}(\Omega)}|}{\|v\|_{H^{\alpha}(\Omega)}} \ .$$

on the other hand we defined for $u \in E(\bar{\Omega})$ and $\ell \leq 0$

$$(2.3.22)\qquad \|u\|_{H^{\ell}(\Omega)} = \sup_{v\in E(\bar{\Omega})} \frac{|(u,v)_{H^{0}(\Omega)}|}{\|v\|_{H^{-\ell}(\Omega)}} \ .$$

The connection between (2.3.21) and (2.3.22) is given by the following theorem.

<u>Theorem 2.3.1</u>. For every real $\gamma_1$ and $\gamma_2$ and $u \in H_{\gamma_1}(\Omega) \cap H_{\frac{\gamma_1+\gamma_2}{2}}(\Omega)$ we have

$$(2.3.23)\qquad \|u\|_{H^{\gamma_1}(\Omega)} \approx \sup_{v\in E(\bar{\Omega})} \frac{|(u.v)_{H^{\frac{\gamma_1+\gamma_2}{2}}(\Omega)}|}{\|v\|_{H^{\gamma_2}(\Omega)}} \ .$$

and $H_{\frac{\gamma_1+\gamma_2}{2}}(\Omega)$ is the Hilbert space obtained through the connection of the spaces $H^{\gamma_1}(\Omega)$ and $H^{\gamma_2}(\Omega)$.

<u>Proof</u>. Let $H^{\alpha}(\Omega)$, $H^{\beta}(\Omega)$, $\alpha < \beta$ be given and further let $H_{\gamma}(\Omega)$, $\alpha \leq \gamma \leq \beta$ be the Hilbert scale connecting the spaces $H^{\alpha}(\Omega)$ and $H^{\beta}(\Omega)$. We know that the space $H_{\gamma}(\Omega)$ and $H^{\gamma}(\Omega)$ are equivalent, $E(\bar{\Omega})$ is dense in $H^{\gamma}(\Omega)$. From (2.3.13) we have for $\alpha \leq \gamma_1 \leq \beta$

$$(2.3.24) \qquad (u,v)_{H_{\gamma_1}(\Omega)} = \left(T^{\frac{\gamma_1-\alpha}{\beta-\alpha}}u, T^{\frac{\gamma_1-\alpha}{\beta-\alpha}}v\right)_{H^{\alpha}(\Omega)},$$

and obviously

$$(2.3.25) \qquad \|u\|_{H_{\gamma_1}(\Omega)} = \sup_{\substack{v\in E(\bar{\Omega}) \\ v\neq 0}} \frac{|(u,v)_{H_{\gamma_1}(\Omega)}|}{\|v\|_{H_{\gamma_1}(\Omega)}}.$$

Let $\alpha \leq \gamma_2 \leq \beta$, then it follows that for every $v \in E(\bar{\Omega})$

$$(2.3.26) \qquad \|T^{\frac{\gamma_1-\alpha}{\beta-\alpha}}v\|_{H^{\alpha}(\Omega)} = \|T^{\frac{\gamma_2-\alpha}{\beta-\alpha}}T^{\frac{\gamma_1-\gamma_2}{\beta-\alpha}}v\|_{H^{\alpha}(\Omega)} =$$

$$= \|T^{\frac{\gamma_1-\gamma_2}{\beta-\alpha}}v\|_{H_{\gamma_2}(\Omega)}.$$

If we let

$$z = T^{\frac{\gamma_1-\gamma_2}{\beta-\alpha}} v \quad , \tag{2.3.27}$$

then

$$T^{\frac{\gamma_1-\alpha}{\beta-\alpha}} v = T^{\frac{\gamma_2-\alpha}{\beta-\alpha}} z \quad , \tag{2.3.28}$$

and therefore we have

$$\frac{(u,v)_{H_{\gamma_1}(\Omega)}}{\|v\|_{H_{\gamma_1}(\Omega)}} = \frac{\left(T^{\frac{\gamma_1-\alpha}{\beta-\alpha}} u, T^{\frac{\gamma_1-\alpha}{\beta-\alpha}} v\right)_{H^{\alpha}(\Omega)}}{\|T^{\frac{\gamma_1-\gamma_2}{\beta-\alpha}} v\|_{H_{\gamma_2}(\Omega)}} = \tag{2.3.29}$$

$$= \frac{\left(T^{\frac{\gamma_1-\alpha}{\beta-\alpha}} u, T^{\frac{\gamma_2-\alpha}{\beta-\alpha}} z\right)_{H^{\alpha}(\Omega)}}{\|z\|_{H_{\gamma_2}(\Omega)}}$$

$$= \frac{\left(T^{\frac{\left(\frac{\gamma_1+\gamma_2}{2}-\alpha\right)}{\beta-\alpha}} u, T^{\frac{\left(\frac{\gamma_1+\gamma_2}{2}-\alpha\right)}{\beta-\alpha}} z\right)_{H^{\alpha}(\Omega)}}{\|z\|_{H_{\gamma_2}(\Omega)}} = \frac{(u,z)_{H_{\frac{\gamma_1+\gamma_2}{2}}(\Omega)}}{\|z\|_{H_{\gamma_2}(\Omega)}}$$

Therefore, we have

$$\|u\|_{H_{\gamma_1}(\Omega)} = \sup_{z\in E(\bar{\Omega})} \frac{\left|(u,z)_{H_{\frac{\gamma_1+\gamma_1}{2}}(\Omega)}\right|}{\|z\|_{H_{\gamma_2}(\Omega)}} \quad . \tag{2.3.30}$$

Since $H^{\gamma_i}(\Omega) \approx H_{\gamma_i}(\Omega)$, $i = 1,2$, the theorem is proved.

There is a question as to whether or not our theorem is still true-up to an equivalence-if we change the scalar product

$$(u,z)_{H_{\frac{\gamma_1+\gamma_2}{2}}(\Omega)} \quad \text{to} \quad (u,z)_{H^{\frac{\gamma_1+\gamma_2}{2}}(\Omega)}$$

We know only that the spaces $H_{\frac{\gamma_1+\gamma_2}{2}}(\Omega)$ and $H^{\frac{\gamma_1+\gamma_2}{2}}(\Omega)$ are equivalent. Therefore, it is easy to show that there exists a bounded linear mapping $W$ of the space $H^{\frac{\gamma_1+\gamma_2}{2}}(\Omega)$ into itself with bounded inverse $W^{-1}$ such that

$$(u,v)_{H^{\frac{\gamma_1+\gamma_2}{2}}(\Omega)} = (Wu,v)_{H_{\frac{\gamma_1+\gamma_2}{2}}(\Omega)}$$

Therefore, we have only to write $\gamma_1 \leq \gamma_2$

$$\sup_{z\in \bar{E}(\bar{\Omega})} \frac{\left|(u,z)_{H^{\frac{\gamma_1+\gamma_2}{2}}}\right|}{\|z\|_{H^{\gamma_2}(\Omega)}} = \|Wu\|_{H^{\gamma_1}(\Omega)}$$

Theorem 2.3.1 holds for space $H_0^{\gamma_i}(\Omega)$ when $\gamma_i \neq$ integer + ½ . We remark also that if the domain $\Omega$ is smooth the theorem holds also for spaces $H^{\alpha}(\dot{\Omega})$ which will be introduced in the next section.

Now we state a well known extension theorem. As before let $\Omega$ be a bounded Lipschitz domain. Then we have the following theorem.

Theorem 2.3.2. There exists a mapping $S$ of $\mathcal{E}(\overline{\Omega})$ into $\mathcal{D}(R_2)$ such that for $\forall x \in \Omega$, we have $(Su)(x) = u(x)$, and

$$\|Su\|_{H^{\ell}(R_2)} \leq C(\ell)\|u\|_{H^{\ell}(\Omega)} \quad , \forall \ell \quad . \tag{2.3.31}$$

Since $\mathcal{E}(\overline{\Omega})$ is dense in $H^{\ell}(\Omega)$ , the extension holds for $u \in H^{\ell}(\Omega)$. For the construction of this extension which is called the universal prolongation, see [8]. For other constructions of the prolongation (not necessarily universal) see e.g. [9], [pp. 79-81].

The constant $C(\ell)$ in (2.3.31) depends on $\ell$ and the domain $\Omega$ . The boundary $\dot{\Omega}$ of $\Omega$ is described by $\nu$ functions $\phi_s$ which are Lipschitzian. Fixing $I_s$ and $I'_s$ , $s = 1,\cdots,\nu$ , there obviously exists a family of functions $\phi_s$ which characterize different domains. If these functions possess a uniform Lipschitz constant then there exists a uniform constant $C(\ell)$ in (2.3.31) for the entire family of such domains.

## 2.4. Sobolev spaces $H^{\alpha}(\dot{\Omega})$ .

In contrast to the previous section we shall

assume that the domain $\Omega$ is smooth.

Let $\psi_i(x) \in \mathcal{D}(R_2)$ , $i = 1,2,\cdots\nu$ satisfy the following properties:

$$\text{i) } \psi_i(x) \geq 0 \text{ ; ii) } \sum_{i=1}^{\nu} \psi_i(x) = 1 \text{ for } x \in \dot{\Omega} \text{ ;}$$

$$\text{iii) } \left\{x \in \dot{\Omega} \,|\, \psi_s(x) > 0\right\} \subset \dot{\Omega}_s' \text{ (for the definition}$$

of $\dot{\Omega}_s'$ see expression (2.2.2)).

Such functions $\psi_i(x)$ are easily constructed and they generate what is called a partition of unity on $\dot{\Omega}$ .

Consider a function $f$ defined on $\dot{\Omega}$ and let $f_s = f\psi_s$ , we see immediately that $f_s(x) = 0$ for $x \notin \dot{\Omega}_s'$ Therefore the function $f_s(\phi_s(x_1)) = g_s(x_1^{[s]})$ is a function defined on $I_s$ with support in $\bar{I}_s'$ . Thus we can define $g_s(x_1^{[s]}) = 0$ for $x_1^{[s]} \in R_1$ , $x_1 \notin I_s$ and

$$(2.4.1) \qquad \|f\|^2_{H^\ell(\Omega} = \sum_{s=1}^{\nu} \|g_s(x_1^{[s]})\|^2_{H^\ell(R_1)}$$

Clearly the above norm depends on the choice of $\psi_s$ and $\phi_s$ . Nonetheless different choices lead to equivalent norms. We note that $E(\dot{\Omega})$ is dense in $H^\ell(\dot{\Omega})$ , where $E(\dot{\Omega})$ has its obvious meaning. Moreover, $u \in E(\bar{\Omega})$ then the restriction of $u$ to $\dot{\Omega}$ belongs to $E(\dot{\Omega})$ .

We can easily compute $\dfrac{\partial^\ell u}{\partial n^\ell}$ on $\dot{\Omega}$ for $u \in E(\bar{\Omega})$

and formally denote $\frac{\partial^0 u}{\partial n^0} = u$ . Now we shall state a well known fundamental trace theorem (see [3],[4]).

Theorem 2.4.1. If $u \in E(\overline{\Omega})$ , then for $s > 0$ and $j \geq 0$ an integer we have

$$(2.4.2) \qquad \left\| \frac{\partial^j u}{\partial n^j} \right\|_{H^s(\Omega^\bullet)} \leq C \|u\|_{H^{s+\frac{1}{2}+j}(\Omega)} ,$$

where the constant $C$ is independent of $u$ . (The inequality (2.4.2) does not hold for $s \leq 0$ .)

Theorem 2.4.1 is formulated for $u \in E(\overline{\Omega})$ . If we interpret the expression $\frac{\partial^j u}{\partial n^j}$ as the limit of function in $E(\overline{\Omega})$ theorem 2.4.1 may be used for $u \in H^{s+1/2+j}(\Omega)$ .

The statement of theorem 2.4.1 is not true for $s \leq 0$ in general, nevertheless it is possible to prove the following special trace theorem using the results of chapter 3.

Theorem 2.4.2. Let $u \in E(\overline{\Omega})$ be such that $\Delta u = 0$ , then for any $s$ and every integer $j \geq 0$ we have

$$(2.4.3) \qquad \left\| \frac{\partial^j u}{\partial n^j} \right\|_{H^s(\Omega)} \leq C \|u\|_{H^{s+1/2+j}(\Omega)} ,$$

and the constant $C$ does not depend on $u$ .

Theorem 2.4.2 is a very special version of a general theorem concerning the regularity property of the solutions of elliptic partial differential equations. The operator $\Delta$ in theorem 2.4.2 may be replaced by any other elliptic operator with smooth coefficients.

By the use of the results of chapter 3 also one can easily prove the following theorem which is a converse of theorem 2.4.1.

<u>Theorem 2.4.3.</u> If $g \in H^s(\dot{\Omega})$, $s > 0$ is given, then there $u \in H^{s+1/2+j}(\Omega)$ such that $\frac{\partial^j u}{\partial n^j} = g$ and $\|u\|_{H^{s+1/2+j}(\Omega)} \leq C\|g\|_{H^s(\dot{\Omega})}$.

## 2.5. The n-width.

The concept of n-width was initially introduced by Kolmogorov. Its applications in the theory of numerical methods have received considerable attention in recent years, e.g., see [11], [12], [13], [14] and [15].

<u>Definition 2.5.1.</u> Let $H$ be a Hilbert space and $G \subset H$ a subset. Then the n-width of $G$ in $H$ is defined as

$$d_n(H,G) = \inf\left\{E(G,M) \mid M \subset H, \dim M = n\right\},$$

where

$$E(G,M) = \sup_{f \in G} \inf_{g \in M} \|f - g\|,$$

and $M$ is a linear manifold.

Obviously if $G_1 \subset G_2 \subset H$, then

(2.5.1) $$d_n(H,G_1) \leq d_n(H,G_2).$$

We introduce the n-width for some special spaces.

We define

$$(2.5.2) \qquad G(H^\beta(\Omega)) = \left\{u \in H^\beta(\Omega) \middle| \|u\|_{H^\beta(\Omega)} \leq 1\right\} .$$

$G(H_0^\beta(\Omega))$ and $G(H^\beta(\Omega^\bullet))$ are similarly defined. For $\beta \geq \alpha$ we define

$$D_n(H^\alpha(\Omega), H^\beta(\Omega)) = d_n(H^\alpha(\Omega), G(H^\beta(\Omega))) .$$

($D_n(H_0^\alpha(\Omega), H_0^\beta(\Omega))$, $D_n(H^\alpha(\Omega^\bullet)$, $H^\beta(\Omega^\bullet)$ etc. are similarly defined.)

Now we state and prove some theorems concerning the n-width.

<u>Theorem 2.5.1</u>. If for $\alpha \leq \beta$, $D_n(H^\alpha(\Omega), H^\beta(\Omega)) \geq |g(n)|$, $n = 1,2,\cdots$, then for every $\alpha_1 < \beta_1$

$$(2.5.3) \qquad D_n(H^{\alpha_1}(\Omega), H^{\beta_1}(\Omega)) \geq C|g(n)|^{\frac{(\beta_1-\alpha_1)}{\beta-\alpha}}$$

where $C$ does not depend on $n$ (generally $C$ depends on $(\alpha,\beta_1,\alpha_1,\beta_1)$).

<u>Proof</u>. Let $\alpha^* = \min(\alpha,\alpha_1)$, $\beta^* = \max(\beta,\beta_1)$ and consider the Hilbert scale $H_\gamma(\Omega)$ $\alpha^* \leq \gamma \leq \beta^*$ connecting the spaces $H^{\alpha^*}(\Omega)$ and $H^{\beta^*}(\Omega)$. (See section 2.3). Replacing $\ell_1$ and $\ell_2$ by $\alpha^*$ and $\beta^*$ respectively, then we immediately obtain from (2.3.15) - (2.3.17) (see also (2.3.11)) that

$$(2.5.4) \qquad D_n(H^{\alpha^*}(\Omega), H^{\beta^*}(\Omega)) = \lambda_{n+1} .$$

Also it is easily seen that

$$\text{(2.5.5)} \quad \|u\|_{H_\beta(\Omega)} = \|T^{\frac{(\beta-\alpha^*)}{(\beta^*-\alpha^*)}} u\|_{H^{\alpha^*}(\Omega)} = \|T^{\frac{(\beta-\alpha)}{(\beta^*-\alpha^*)}} u\|_{H_\alpha(\Omega)} \quad .$$

Therefore

$$\text{(2.5.6)} \qquad D_n(H_\alpha(\Omega)\ ,\ H_\beta(\Omega)) = \lambda_{n+1}^{\frac{(\beta-\alpha)}{(\beta^*-\alpha^*)}}$$

and

$$\text{(2.5.7)} \qquad D_n(H_{\alpha_1}(\Omega)\ ,\ H_{\beta_1}(\Omega)) = \lambda_{n+1}^{\frac{\beta_1-\alpha_1}{\beta^*-\alpha^*}} \quad .$$

By assumption we have $D_n(H^\alpha(\Omega)\ ,\ H^\beta(\Omega)) \gtrsim |g(n)|$. The spaces $H^\alpha(\Omega)$ and $H_\alpha(\Omega)$ are equivalent. The same being true of the spaces $H^\beta(\Omega)$ and $H_\beta(\Omega)$, we therefore have

$$\lambda_{n+1}^{\frac{\beta-\alpha}{\beta^*-\alpha^*}} \gtrsim C|g(n)| \quad .$$

thus

$$\lambda_{n+1}^{\frac{\beta_1-\alpha_1}{\beta^*-\alpha^*}} \gtrsim C|g(n)|^{\frac{\beta_1-\alpha_1}{\beta-\alpha}}$$

and the proof is complete.

The following theorem describes the asymptotic behavior of the n-width.

Theorem 2.5.2. $D_n(H^0(\Omega), H_0^1(\Omega)) = (\frac{C}{n})^{\frac{1}{2}} + o(n^{-\frac{1}{2}})$ 1) as $n \to \infty$, $C$ is independent of $n$. (For a proof see Corollary 3.5 in [13]). [The theorem is stated for two dimensional domains]. Since 2.5.1 holds for $\Omega \subset R_2$ we immediately obtain from (2.5.2) that

$$D_n(H^0(\Omega), H^1(\Omega)) \geq (\frac{C}{n})^{1/2}$$

and

$$D_n(H^0(\Omega^\bullet), H^1(\Omega^\bullet)) \geq \frac{C}{n} .$$

The above inequalities together with theorem 2.5.1 yields a lower bound for the n-widths $D_n(H^\alpha(\Omega), H^\beta(\Omega))$ and $D_n(H^\alpha(\Omega^\bullet), H^\beta(\Omega^\bullet))$.

2.6. The weighted Sobolev spaces.

In section 2.3 we discussed Sobolev spaces. Now we introduce the weighted Sobolev spaces. We confine ourselves only to Sobolev spaces with non-negative integral derivatives.

2.6.1. The weighted Sobolev spaces with weight given by the distance from the boundary.

Let $\Omega$ be a domain with smooth boundary. We denote by $\rho(x)$ the distance of the point $x \in \Omega$ from the boundary.

---

1) We write $o(n^{-1/2})$ if $n^{1/2} o(n^{-1/2}) \to 0$ for $n \to \infty$.

We defined the norm $\|\cdot\|_{H^{\ell}(\Omega)}$ for $\ell$ a non-negative integer by (2.3.2). Let us define

(2.6.1) $$\|u\|^2_{H^{\ell}_{\beta}(\Omega/\Omega^{\bullet})} = \sum_{0\leq|\alpha|\leq\ell} \|\rho^{\beta}D^{\alpha}u\|^2_{L_2(\Omega)} .$$

The following important imbedding results can be proved. For $0 \leq \beta \leq \ell$ we have

(2.6.2) $$\|u\|_{H^{\ell-\beta}(\Omega)} \leq C\|u\|_{H^{\ell}_{\beta}(\Omega/\Omega^{\bullet})} ,$$

and for $\beta - k + \frac{1}{2} > 0$ we have

(2.6.3) $$\|u\|_{H^{\ell-k}_{\beta-k}(\Omega)} \leq C\|u\|_{H^{\ell}_{\beta}(\Omega/\Omega^{\bullet})} .$$

For $\beta > \frac{1}{2}$, $r = \ell - \beta - \frac{1}{2} > 0$ the following trace theorem holds

(2.6.4) $$\|u\|_{H^{r}(\Omega^{\bullet})} \leq C\|u\|_{H^{\ell}_{\beta}(\Omega/\Omega^{\bullet})} .$$

There is a converse for the inequality (2.6.4), namely for $g \in H^{r}(\Omega^{\bullet})$ $r \geq 0$, there exists $u \in H^{\ell}_{\beta}(\Omega/\Omega^{\bullet})$ such that $u = g$ on $\Omega^{\bullet}$ and

(2.6.5) $$\|u\|_{H^{\ell}_{\beta}(\Omega/\Omega^{\bullet})} \leq C\|g\|_{H^{r}(\Omega^{\bullet})} .$$

For a survey of the results concerned with weighted Sobolev space [4] may consulted.

Before proving a decomposition theorem for weighted Sobolev spaces we introduce the definition $\Omega_h = \left\{ x \in \Omega | \rho(x) > h \right\}$. We observe that $\Omega_h$ is a domain with a smooth boundary if $h$ is sufficiently small. Now we state and prove the following result.

Theorem 2.6.1. If

I) $0 < \lambda < 1$ and $\xi > 0$ is sufficiently small,

II) $u \in H^{1+p}(\Omega)$ , $p \geq 0$ ,

III) (2.6.5) $\|u\|_{H^s(\Omega_h)} \leq C(s,p)h^{-(-s-1-p)}$

$\|u\|_{H^{1+p}(\Omega)}$ , with $s \geq 1 + p$

then

IV) $u$ may be written as

(2.6.6) $$u = \sum_{j=0}^{N} u_j \ ,$$

where $u_j$ have the following properties:

(2.6.7) $$u_j \in H^s(\Omega) \ , \ j = 0,1,\cdots , N - 1 \ ,$$

(2.6.8) $$u_j = 0 \quad \text{on} \quad \Omega_{\xi\lambda^j} \ , \ j = 1,2,\cdots , N \ ,$$

(2.6.9) $$\|u_j\|_{H^{1+p}(\Omega)} \leq C\|u\|_{H^{1+p}(\Omega)}$$

(2.6.10) $$\|u_j\|_{H^s(\Omega)} \leq C(\xi\lambda^j)^{-(s-1-p)} \|u\|_{H^{1+p}(\Omega)} ,$$

$$j = 0,1,\cdots , N - 1 ,$$

and $C$ does not depend on $N$.

<u>Proof</u>. For $\xi$ sufficiently small the domains $\Omega_{\xi\lambda^j}$, $j = 0,1,\cdots$, are such that i) $\Omega_{\xi\lambda^{j+1}} \supset \overline{\Omega}_{\xi\lambda^j}$ ; ii) the domains $\Omega_{\xi\lambda^j}$ have smooth boundaries, iii) The domains $\Omega_{\xi\lambda^j}$ have a uniform Lipschitz constant and consequently the constant $C$ in the prolongation theorem does not depend on $j$ (see section 2.3). By the prolongation theorem we can construct functions $w_j$, $j = 1,\cdots,N$ such that

(2.6.11) $$w_j = u \quad \text{on} \quad \Omega_{\xi\lambda^j} ,$$

with

(2.6.12) $$\|w_j\|_{H^s(R_2)} \leq C\|u\|_{H^s(\Omega_{\xi\lambda^j})} ,$$

where the constant $C$ does not depend on $j$.
Thus we have

(2.6.13) $$\|w_j\|_{H^s(\Omega)} \leq C(\xi\lambda^j)^{-(s-1-p)} \|u\|_{H^{1+p}(\Omega)} ,$$

and further

(2.6.14) $$\|w_j\|_{H^{1+p}(\Omega)} \leq C\|u\|_{H^{1+p}(\Omega)} .$$

Now we set

(2.6.15) $$u_0 = w_1$$

(2.6.16) $$u_j = w_{j+1} - w_j \ , \ j = 1,2,\cdots , N-1$$

(2.6.17) $$u_N = u - w_N \ ,$$

and the theorem easily follows.

Obviously the condition (2.6.5) is very much related to weighted Sobolev spaces.

We may remark that all the results concerning weighted spaces are valid if instead of the distance function $\rho$ we used any other positive function with similar characteristics.

2.6.2. Weighted Sobolev spaces with weight given by the distance from isolated points to the boundary.

Let $\Omega$ be a domain with either a smooth boundary or with a smooth boundary except at a vertex where the domain may have a corner with angle which is neither zero nor $2\pi$ .

Let $P \in \Omega$ and assume that $\Omega$ has a corner at $P$ . We denote by $\sigma(x)$ the distance of a point $x \in \Omega$ to point $P$ . Further we define

(2.6.18) $$\|u\|^2_{H^\ell_\beta(\Omega/P)} = \sum_{0\leq|\alpha|\leq\ell} \|\sigma^\beta D^\alpha u\|^2_{L_2(\Omega)} .$$

We have the following imbedding theorem.

If $\beta > 0$ , then

(2.6.19) $$\|u\|_{H_{\beta-1}(\Omega/P)} \leq C\|u\|_{H^1_\beta(\Omega/P)} ,$$

where $C$ is independent of $u$ . For the proof see e.g. [16].

As in section 2.6.1, we may prove a factorization theorem similar to theorem 2.6.1.

Let $\Omega_h = \{x \in \Omega | \sigma(x) > h\}$ , we have the following theorem.

Theorem 2.6.2. If

I) $0 < \lambda < 1$ , $\xi > 0$ is sufficiently small;

II) $u \in H^{1+p}(\Omega)$ $p > 0$

III) (2.6.20)

$$\|u\|_{H^s(\Omega_h)} \leq C(s,p)h^{-(s-1-p)} \|u\|_{H^{1+p}(\Omega)} , \quad s \geq 1 + p ,$$

then

IV) $u$ can be written as

(2.6.21) $$u = \sum_{j=0}^{N} u_j ,$$

where $u_j$ have the following properties.

(2.6.22) $$u_j \in H^s(\Omega) , \; j = 0,1,\cdots , N - 1 ,$$

(2.6.23) $$u_j = 0 \text{ on } \Omega_{\xi\lambda^j} , \; j = 1,2,\cdots , N ,$$

(2.6.24) $$\|u_j\|_{H^{1+p}(\Omega)} \leq C\|u\|_{H^{1+p}(\Omega)}$$

(2.6.25) $$\|u_j\|_{H^s(\Omega)} \leq C(\xi\lambda^j)^{-(s-1-p)} \|u\|_{H^{1+p}(\Omega)} .$$

The proof is very similar to the proof of theorem 2.6.1. (See [17]).

2.6.3. <u>The weighted Sobolev spaces on infinite domains</u>.

We introduced the Sobolev spaces $H^\ell(\Omega)$ and $H^\ell(R_n)$ . Now we introduce the space $H^\ell_\mu(R_n)$ as a weighted space for any real $\mu$ . The norm is defined by

$$\|u\|^2_{H^\ell_\mu(R_n)} = \sum_{0\leq|\alpha|\leq\ell} \|e^{\mu|x|} D^\alpha u\|^2_{L_2(\Omega)} .$$

where

$$|x| = |x_1| + |x_2| \quad .$$

The space $H^{\ell}(R_n)$ is defined as the closure of $\mathcal{D}(R_n)$ in the norm defined above.

References.

[1]. H. Aronszajn, "Boundary Value of Functions with Finite Dirichlet Integral". Conference on Partial Differential Equations No. 14, University of Kansas, Lawrence 1955.

[2]. M. I. Sloboditskii, "Generalized Sobolev Spaces and their Applications to Boundary Problems for Partial Differential Equations," Amer. Math. Soc., Tran . (2), 57 (1966), p. 207-275.

[3]. J. L. Lions, E. Magenes, "Problemes aux limites non homogenes et applications, Vol. 1, Dunod, Paris, 1968.

[4]. O. V. Besov, V. P. Il'in, L. D. Kudrjavcev, P. I. Lizorkin, S. M. Nikolskii, "Embedding Theorems of the Classes of Functions with Several Variables (in Russian), pp. 30-63, Partial Differential Equations, Editor A. V. Bicadze, Nauka, 1970.

[5]. S. G. Krein, Yu I. Petunin, "Scales of Banach Spaces Russian Math. Surveys, 21 (1966), No. 2, pp. 85-160.

[6]. P. L. Butzer, H. Berens, "Semi-Groups of Operators and Approximation, Springer, New York, 1967.

[7]. R. S. Varga, "Functional Analyses and Approximation Theory in Numerical Analyses, Conference Board of the Math. Sciences, No. 3, 1971, SIAM.

[8]. E. M. Stein, Singular Integrals and Differentiability Properties of Functions," Princeton University Press, Princeton, New Jersey, 1971.

[9]. J. Nečas, "Les methodes dirictes en théorie des equations elliptiques, Academia, Prague, 1967.

[10]. A. N. Kolmogorov, "Über die beste Annäherung von

Functionen einer gegebeven Funktionklasse," Math. Ann. 37, (1936), pp. 107-111.

[11]. J. Jerome, "On $L_2$ n-width of Certain Classes of Functions of Several Varriables, J. Math. Anal. Appl. 20 (1967), pp. 110-123.

[12]. J. Jerome, "Asymptotic Estimates of the $L_2$ n-width" Ibid 22 (1968), pp. 449-464.

[13]. J. Jerome, "On n-width in Sobolev Spaces and Applications to Elliptic Boundary Value Problems," Ibid 29 (1970), pp. 201-215.

[14]. I. Babuška, S. L. Sobolev, "The Optimization of Numerical Processes," (in Russian), Apl. Math. 10 (1965), pp. 96-129.

[15]. J. P. Aubin, "Approximation of Elliptic Boundary-Value Problems," Wiley-Interscience, New York, London, Sydney, Toronto, 1972.

[16]. A. Kufner, "Einige Eigenschaften der Sobolevschen Raume mit Belegungs function," Czechoslovak Mat., J. 15 (1961), pp. 597-620.

[17]. I. Babuška, "Finite Element Method for Domains with Corners." Computing 6, pp. 264-273, (1970).

[18]. I. Babuška, "Appriximation by Hill Functions," Comment. Math. Univ. Corolinae, 11, 4 (1970) pp. 787-811, [Tech. Note BN-648, (March 1970), The Institute for Fluid Dynamics and Applied Mathematics, University of Maryland].

## CHAPTER 3[1]. PROPERTIES OF SOLUTIONS OF ELLIPTIC BOUNDARY VALUE PROBLEMS

Introduction.

In this chapter we discuss some regularity properties of solutions of elliptic boundary value problems. No proofs are given, but the theorems are illustrated by various examples. The material is for the most part taken from references [1] and [2]. Everywhere in this section we assume that $\Omega$ is a smooth domain. (See section 2.2).

### 3.1. Strongly elliptic boundary value problems.

We consider linear differential operators $A$ of the form

$$(3.1.1) \qquad A(D,x)u(x) = \sum_{|\alpha|\leq 2m} a_\alpha(x) D^\alpha u(x)$$

where $\alpha = (\alpha_1,\alpha_2)$ $\alpha_i \geq 0$ integral, $i = 1,2$, and $|\alpha| = \alpha_1 + \alpha_2$. The operator is of order $2m$, and the coefficients $a_\alpha(x)$ are real functions. Everywhere in this section we assume that $a_\alpha(x)$ have all derivatives on $\Omega$ which are continuously prolongable on $\bar{\Omega}$. We let

1) This chapter was written by B. Kellogg.

$$A_0(D,x)u(x) = \sum_{|\alpha|=2m} a_\alpha(x) D^\alpha u(x)$$

denote the highest order part of $A$, also called the principal part of $A$. For fixed $x$ we regard $A_0(D,x)$ as a homogeneous polynomial of order $2m$ in the variables $\partial/\partial x_1$, and $\partial/\partial x_2$. We associate with $A_0(D,x)$ a polynomial in $\xi = (\xi_1, \xi_2)$ so that

$$A_0(\xi,x) = \sum_{|\alpha|=2m} A_\alpha(x) \xi^\alpha$$

where for $\alpha = (\alpha_1, \alpha_2)$ we write $\xi^\alpha = \xi_1^{\alpha_1} \xi_2^{\alpha_2}$. Regarding the ellipticity of $A$ we make the following definitions.

$A$ is <u>elliptic</u> at a point $x$ if $A_0(\xi,x) \neq 0$ for any real vector $\xi$, $\xi \neq 0$,

$A$ is <u>properly elliptic</u> at a point $x$ if it is elliptic, and if for any two real linearly independent nonzero vectors $\xi$ and $\eta$, the polynomial $A_0(\tau\xi+\eta,x)$ (in $\tau$ has exactly $m$ roots with positive imaginary part.

$A$ is <u>strongly elliptic</u> at a point $x$ if $A$ is properly elliptic at $x$ and if for some choice of signs and some $\gamma > 0$,

$$\pm A_0(\xi,x) \geq \gamma |\xi|^{2m}$$

for any real vector $\xi$ and $|\xi|^2 = \xi_1^2 + \xi_2^2$ . We will speak about elliptic, properly elliptic or strongly elliptic operators on $\bar{\Omega}$ if they are elliptic, properly elliptic or strongly elliptic at every point $x \in \bar{\Omega}$ . In these definitions $m$ is understood to be independent of $x$ .

Some remarks are in order. For more than two independent variables, ellipticity implies proper ellipticity. We remark that while the solutions of elliptic equations have the usual desirable regularity properties, the spectrum of an elliptic boundary value problem may be the entire complex plane. The spectrum of a strongly elliptic boundary value problem is discrete, and the resolvant operator is defined.

## 3.2. Examples of elliptic operators.

From the definition it is evident that the ellipticity of an operator $A$ depends only on the principal part, $A_0$ . To determine which second order operators[1] are

$$A_0(D)u = a_{11}u_{x_1x_1} + 2a_{12}u_{x_1x_2} + a_{22}u_{x_2x_2}$$

elliptic, we must consider the corresponding polynomial $A_0(\xi)$. We assume that the coefficients $a_{jk}$ are real. Evidently $A_0(\xi) \neq 0$ for real $\xi \neq 0$ if and only if $a_{11}a_{22} > a_{12}$ , which is the usual condition for ellipticity. To consider whether $A_0$ is properly elliptic it is convenient to introduct the 2 x 2 matrix

1) We are using the notation $u_{x_1x_1} = \frac{\partial^2 u}{\partial x_1^2}$ , $u_{x_1x_2} = \frac{\partial^2 u}{\partial x_1 \partial x_2}$

etc.

$$A = \begin{bmatrix} a_{11} & a_{12} \\ a_{12} & a_{22} \end{bmatrix} .$$

The requirement of ellipticity means that $A$ is positive definite. We see that for real $\xi$ and $\eta$,

$$A_0(\tau\xi+\eta) = \tau^2(A\xi,\xi) + 2\tau(A\xi,\eta) + (A\eta,\eta) \quad ,$$

and that, if $a_{11}a_{22} > a_{12}^2$ , this quadratic polynomial in $\tau$ has 2 complex conjugate roots with non-zero imaginary part. Thus $A_0$ is also properly elliptic. Similarly, it is easily verified that if $A_0$ is elliptic, then $A_0$ is strongly elliptic. In [1] there is given an example of a second order elliptic operator with complex coefficients which is properly elliptic but not strongly elliptic.

We next consider the biharmonic operator

$$\tilde{A}u = \tilde{A}_0 u = \Delta^2 u$$

and the associated form

$$\tilde{A}_0(\xi) = \xi_1^4 + 2\xi_1^2\xi_2^2 + \xi_2^4$$

$$= (\xi_1^2+\xi_2^2)^2 \quad .$$

Evidently $\tilde{A}$ is an elliptic operator. Since the roots of

$\tilde{A}_0(\tau\xi+\eta)$ are the roots of $A_0(\tau\xi+\eta)$, where $A_0(D)$ is the Laplace operator, $\tilde{A}$ is properly elliptic. Finally, we see by inspection that $\tilde{A}$ is strongly elliptic.

3.3. Boundary conditions.

Let there be given, in a domain $\Omega$, an elliptic operator $A$ of order $2m$. We suppose there is given on $\dot{\Omega}$, the boundary of $\Omega$, a set of $m$ differential operators $B_j$, of order $m_j$, of the form

$$(3.3.1)\quad B_j(D,x)u(x) = \sum_{|\alpha|\leq m_j} b_{j\alpha}(x)D^\alpha u(x),\ x\in\dot{\Omega}, 1\leq j\leq m$$

The functions $b_{j\alpha}(x)$ are then defined on the boundary $\dot{\Omega}$. We will always assume that the functions $b_{j\alpha}(x)$ are real functions with all continuous derivatives on $\dot{\Omega}$. If $m_j = 0$, then $B_j$ takes the form of a multiplication, $B_j u = b_j(x)u(x)$. As above, we consider the principal part, $B_{j0}$ of $B_j$, defined by

$$(3.3.2)\qquad B_{j0}(D,x)u(x) = \sum_{|\alpha|=m_j} b_{j\alpha}(x)D^\alpha u(x)\ ,$$

$$x\in\dot{\Omega}\ ,\ 1\leq j\leq m\ .$$

We define the system $B_j$ to be a normal system if the orders $m_j$ which are assumed independent of $x$, are all distinct: $m_j \neq m_k$ for $j\neq k$; if $0\leq m_j < 2m$; and if

$B_{j0}(\xi,x) \neq 0$ , $x$ $\dot{\Omega}$ , $\xi$ normal to $\dot{\Omega}$ at $x$ , $1 \leq j < m$.

One cannot hope to prescribe an arbitrary set of boundary conditions and obtain a regularity theory for the corresponding elliptic boundary value problem. The boundary conditions that are to be satisfied on $\dot{\Omega}$ must in some sense conform to the elliptic equation that is to be satisfied in $\Omega$ . For this purpose, we now define what is meant by a covering family of boundary conditions. Let $\eta \neq 0$ be a tangent vector to $\dot{\Omega}$ at the point $x$ , and let $\xi \neq 0$ be an outward pointing normal vector to $\dot{\Omega}$ at $x$ . Let $\phi(t)$ , $t \geq 0$ , be a function satisfying

$$A_0(\eta + \frac{1}{i}\,\xi\frac{d}{dt}, x)\phi(t) = 0 \ , \ t > 0 \tag{3.3.3}$$

$$B_{j0}(\eta + \frac{1}{i}\xi\frac{d}{dt}, x)\phi(t)\Big|_{t=0} = 0 \quad , \ 1 \leq j \leq m \ , \tag{3.3.4}$$

$$|\phi(t)| \leq c_1 e^{-c_2 t} \quad , \ t \geq 0 \ \text{ for some } \ c_1 > 0, c_2 > 0 \ . \tag{3.3.5}$$

If, for each $\xi$ and $\eta$ , the only function $\phi(t)$ satisfying (3.3.3), (3.3.4), and (3.3.5) is the function $\phi(t) \equiv 0$, then the boundary conditions $B_j$ are said to <u>cover</u> the operator $A$ at the point $x \in \dot{\Omega}$ . We will speak about normal resp. covering system at $\dot{\Omega}$ if it is normal resp. covering at every point $x \in \dot{\Omega}$ .

Some examples illustrating the use of this

condition are given in the next section. For the moment, one should note that (3.3.3) is an equation of order $2m$, for which one can give $2m$ initial conditions. Thus, (3.3.3) and (3.3.4) admit $m$ linearly independent solutions.

Using the Fourier transform, one may gain a heuristic understanding of the ellipticity condition and the covering conditions. To understand the ellipticity condition, imagine that the domain $\Omega$ is all of $R^2$, and that the differential operator $A = A_0$ and has constant coefficients. Then the problem $A(D)u(x) = f(x)$ can be formaly solved by taking the Fourier transform, $A(i\xi)\hat{u}(\xi) = f(\xi)$. To solve for $\hat{u}$ we must be able to divide by $A(i\xi) = (-1)^m A_0(\xi)$, and the nonvanishing of this quantity is the ellipticity condition. To understand the covering condition, imagine for the moment that the domain $\Omega$ is the half plane given by the inequality $x \cdot \xi < 0$ [1)] and that the differential operators $A$ and $B_j$ all have constant coefficients. Then the boundary value problem may be studied by taking the Fourier transform on the line $x \cdot \xi =$ constant. This leads ultimately to a consideration of an ordinary differential equation of the form (3.3.3) with boundary conditions (3.3.4).

## 3.4. Examples of elliptic boundary value problems.

In the first example we consider the boundary conditions associated with the Laplace operator,

$$A(D)u = u_{x_1x_1} + u_{x_2x_2} \quad .$$

1) We write $x \cdot \xi = x_1\xi_1 + x_2\xi_2$ .

We let $x \in \dot{\Omega}$ be a point with outer normal $\xi = (0,1)$. The boundary condition at $x$ is a linear differential operator of order at most 1,

$$Bu = \alpha u_{x_1} + \beta u_{x_2} + \gamma u \ .$$

If one of the coefficients $\alpha$, or $\beta$, is not zero, then the order of $B$ is 1, and $B$ has principal part

$$B_0(\zeta) = \alpha\zeta_1 + \beta\zeta_2 \ ,$$

The condition of normality gives $\beta \neq 0$. For the covering condition, we set $\eta = (\sigma,0)$, $\sigma \neq 0$. Then (3.3.3) us

$$\phi''(t) - \sigma^2\phi(t) = 0 \ ,$$

so

$$\phi(t) = c_1 e^{-\sigma t} + c_2 e^{\sigma t} \ .$$

From (3.3.5), $c_2 = 0$. The boundary condition, (3.3.4) is

$$\alpha\sigma\phi(0) - i\beta\phi'(0) = 0 \ ,$$

or,

$$(\alpha + i\beta)\sigma c_1 = 0 \ .$$

If $\alpha$ and $\beta$ are both real, we conclude $c_1 = 0$. Hence any boundary condition of the form

$$Bu = \alpha u_{x_1} + \beta u_{x_2} + \gamma u \ , \ \beta \neq 0 \ , \ \alpha \ , \ \beta \ \text{ real}$$

covers the Laplace operator at a point $x \in \dot{\Omega}$ with normal (0,1).

We next consider the biharmonic operator, $\tilde{A}(D)u = \Delta^2 u$, whose polynomial is $\tilde{A}(\zeta) = (\zeta_1^2+\zeta_2^2)^2$. As above, we let $\xi = (0,1)$ be a normal vector and $\eta = (\sigma,0)$, $\sigma \neq 0$, be a tangent vector to $\dot{\Omega}$ at a point $x$. We list in Table 1 all pairs of boundary operators, $B_1$, $B_2$, which are normal and cover $\tilde{A}$ at the point $x$. We list only the principal parts of the operators, and we assume that the principal parts have real coefficients. There are 6 cases, depending on the orders of $B_1$ and $B_2$.

We emphasize that in Table 1, the boundary is <u>not</u> assumed to be a straight line near the point in question, but just to be tangent to the horizontal at the point in question. For the covering condition at the other boundary points, we may use a rotation of axes to reduce it to the case of Table 1. As an example of a set of boundary conditions which are not covering conditions, but which nevertheless arise in a natural manner, we mention the conditions

$$B_2 u = \Delta u$$

$$B_3 u = \frac{\partial}{\partial n} \Delta u \quad .$$

| Order | | Covering Condition |
|---|---|---|
| 0&1 | $B_1u = u$ <br> $B_2u = \alpha u_{x_1} + u_{x_2}$ | all $\alpha$ |
| 0&2 | $B_1u = u$ <br> $B_2u = \alpha u_{x_1x_1} + \beta u_{x_1x_2} +$ <br> $+ u_{x_2x_2}$ | all $\alpha$ and $\beta$ |
| 0&3 | $B_1u = u$ <br> $B_2u = \alpha u_{x_1x_1x_1} + \beta u_{x_1x_1x_2} +$ <br> $+ \gamma u_{x_1x_2x_2} + u_{x_2x_2x_2}$ | $\|\gamma\|+\|\beta-3\|\neq 0$ |
| 1&2 | $B_1u = au_{x_1} + u_{x_2}$ <br> $B_2u = \alpha u_{x_1x_1} + \beta u_{x_1x_2} +$ <br> $+ u_{x_2x_2}$ | $\|a\|+\|\alpha+1\|\neq 0$ |
| 1&3 | $B_1u = au_{x_1} + u_{x_2}$ <br> $B_3u = \alpha u_{x_1x_1x_1} + \beta u_{x_1x_1x_2} +$ <br> $+ \gamma u_{x_1x_2x_2} + u_{x_2x_2x_2}$ | $\|a\gamma-1\|+\|\gamma+3a-a\beta\|$ $\neq 0$ |
| 2&3 | $B_1u = au_{x_1x_1} + bu_{x_1x_2} + u_{x_2x_2}$ <br> $B_2u = \alpha u_{x_1x_1x_1} + \beta u_{x_1x_1x_2} +$ <br> $+ \gamma u_{x_1x_2x_2} + u_{x_2x_2x_2}$ | $\|b-a\gamma\|+\|-a\beta+3a+b\gamma-$ $-\beta-1\|\neq 0$ |

Table 1
Covering operators for $\Delta^2$ at a point with normal (0,1) .

These arise from the problem of minimizing the integral

$$\int_{\Omega} [(\Delta u)^2 + u^2]dx$$

over all functions $u \in H^4(\Omega)$ .

3.5. The adjoint problem.

The formal adjoint to the differential operator $A$ given by (3.1.1) is defined by the formula

$$A^*u(x) = \sum_{|\alpha| \leq 2m} (-1)^{|\alpha|} D^\alpha (a_\alpha(x)u(x)) \quad . \tag{3.5.1}$$

Since $A^*$ has the same principal part as $A$ , if $A$ is elliptic, properly elliptic, or strongly elliptic, so is $A^*$ .

Let $A$ be a properly elliptic operator, and let $B_j$ , $1 \leq j \leq m$ be a family of boundary operators which are normal and which cover $A$ . Then there exist three families of operators, $C_j$ , $S_j$ , and $T_j$ , $1 \leq j \leq m$ , such that:

a) each of these is a normal family of operators;
b) if the order of $S_j$ is $\mu_j$ , the number $(m_j , \mu_j)$ comprise the set of numbers $(1,2,\cdots,2m)$. Also, $C_j$ has order $2m - 1 - \mu_j$ , and $T_j$ has order $2m - 1 - m_j$ ;

c) the family $C_j$ covers the operator $A^*$

d) a generalized Green's formula holds.

(3.5.2)
$$\int_\Omega (Au)v\,dx - \int_\Omega (A^*v)u\,dx = \sum_j{}' \oint_{\dot\Omega} (S_j u)(C_j v)\,ds - \sum_j \oint_{\dot\Omega} (B_j u)(T_j v)\,ds \quad .$$

The existence of these families is proved in [1, p. 127]. It is also shown there that the families $C_j$ and $T_j$ depend uniquely on the families $B_j$ and $S_j$. The family $C_j$ will be called a set of adjoint boundary conditions to the boundary conditions $B_j$.

We will be interested in solutions of the boundary value problem

(3.5.3)
$$Au = f \quad \text{in} \quad \Omega \; ; \; B_j u = g_j \quad \text{on} \quad \dot\Omega \; , \; 1 \leq j \leq m \quad .$$

The reason for our introducing the adjoint problem is to treat cases, like the Neumann problem, when the solution of (3.5.3) does not exist unless the data $f$ and $g$ are restricted. To see this, let

(3.5.4)
$$N^* = \left\{ v \,\middle|\, A^*v = 0 \quad \text{in} \quad \Omega \; , \; C_j v = 0 \quad \text{on} \quad \dot\Omega \right\} .$$

Then from (3.5.2), we see that if $u$ is a solution of (3.5.3), we have

(3.5.5) $$\int_{\Omega} fv\,dx + \sum_j \oint_{\dot{\Omega}} g_j T_j v\,ds = 0 \ , \ v \in N^* \ .$$

Thus if $N^*$ is not empty, (3.5.5) gives a necessary restriction on $f$ and $g$ in order that (3.5.3) have a solution. Furthermore, let

$$N = \left\{ v \,\middle|\, Av=0 \ \text{ in } \ \Omega \ , \ B_j v=0 \ \text{ on } \ \dot{\Omega} \right\} \ .$$

Then if $u$ is a solution of (3.5.3), and if $v \in N$, then $u + v$ is a solution of (3.5.3). Thus, if $N$ is not empty, the solution (3.5.3) is not unique.

The sets $N$ and $N^*$ are linear spaces of finite dimension, consisting of smooth functions. The quantity

$$\text{dimension } N - \text{dimension } N^*$$

is known as the index of the elliptic boundary value problem. For most problems of physical interest, the index is zero.

## 3.6. Solvability and regularity theorems.

We will be interested in the solvability and regularity of the problem $Au = f$ and $B_j u = g_j$ where $A$ is a differential operator of order $2m$ given by expression (3.1.1), which is properly elliptic on $\overline{\Omega}$ and $B_j, j = 1,2,\cdots,m$, are differential operators of the form (3.3.1) and order $m_j$ which create a normal and covering system on

$\dot{\Omega}$ . Such a problem will be called a <u>regularly elliptic boundary value problem of order 2m</u> . Let us remark that the adjoint operators $A^*$ , $C_j$ also give rise to regularly elliptic boundary value problems of order 2m .

In chapter 2 there was defined, for any real $s$ , the Sobolev spaces $H^s(\Omega)$ and $H^s(\dot{\Omega})$. In this section we state the basic theorems which determine when an elliptic boundary value problem has a solution $u \in H^s(\Omega)$ for $s \geq 2m$ .

If $s \geq 2m$ , and $u \in H^s(\Omega)$ , then evidently $Au$ has meaning and $Au \in H^{s-2m}(\Omega)$ . Furthermore, if $D_\nu^j$ denotes differentiation $j$ times in the outward normal direction, then for $0 \leq j \leq 2m - 1$ ,

$$D_\nu^j u\Big|_{\dot{\Omega}} \in H^{s-j-\frac{1}{2}}(\dot{\Omega}) \quad .$$

Hence, by using the imbedding theorem 2.4.1 if $m_j \leq 2m - 1$ , the boundary operator $B_j$ of order $m_j$ may be applied to $u$ , and

$$B_j u\Big|_{\dot{\Omega}} \in H^{s-m_j-\frac{1}{2}}(\dot{\Omega}) \quad .$$

Now if $A$ is a differential operator of order 2m , and $\left\{B_j\right\}$ , $1 \leq j \leq m$ is a normal family of boundary operators of order $m_j$ $m_j \leq 2m - 1$ , $1 \leq j \leq m$ , the map

$$T: u \rightarrow \left\{Au, B_j u\Big|_{\dot{\Omega}}, 1 \leq j \leq m\right\}, \ u \in H^s(\Omega) \ , \ s \geq 2m \ ,$$

gives rise to a bounded map

$$T:H^s(\Omega) \to H^{s-2m}(\Omega) \times H^{s-m_1-\frac{1}{2}}(\dot{\Omega}) \times \cdots \times H^{s-m_m-\frac{1}{2}}(\dot{\Omega}) \quad .$$

By a boundary value problem we mean the following. Given functions $f$ and $g_j$ , $1 \leq j \leq m$ ,

$$f \in H^{s-2m}(\Omega) \ , \ g_j \in H^{s-m_j-\frac{1}{2}}(\dot{\Omega})$$

to find a function $u \in H^s(\Omega)$ such that $Tu = (f, g_j)$ . Thus, the boundary value problem can always be solved if the map $T$ is onto, and the problem has a unique solution if the map $T$ is one to one. The following theorem give the main results on when this can happen.

Theorem 3.6.1. Let $\{A, B_j\}$ be operators of a regularily elliptic boundary value problem of order $2m$. Let $s \geq 2m$ . If $N = N^* = \{0\}$ , then $T$ is a one to one map of $H^s(\Omega)$ onto

$$H^{s-2m}(\Omega) \times H^{s-m_1-\frac{1}{2}}(\dot{\Omega}) \times \cdots \times H^{s-m_m-\frac{1}{2}}(\dot{\Omega}) \quad .$$

There is a constant $C > 0$ independent of $u$ such that for $f \in H^{s-2m}(\Omega)$ , $g_j \in H^{s-m_j-\frac{1}{2}}(\dot{\Omega})$ , $Tu = (f, g_j)$

(3.6.1)

$$\|u\|_{H^s(\Omega)} \leq C\left[\|f\|_{H^{s-2m}(\Omega)} + \sum_j \|g_j\|_{H^{s-m_j-\frac{1}{2}}(\dot{\Omega})}\right] .$$

In general, $(f,g_j)$ lies in the range of $T$ if and only if

(3.6.2)
$$(f,v)_{H^\circ(\Omega)} + \sum_1^m (g_j,C_j v)_{H^\circ(\dot{\Omega})} = 0 \ , \quad \text{all} \quad v \in N^* \ .$$

If (3.6.2) holds, there is a unique $u \in H^s(\Omega)$ such that $Tu = (f,g_j)$ , and such that $u$ satisfies the orthogonality condition

(3.6.3)
$$(u,v)_{H^\circ(\Omega)} = 0 \ , \quad \text{all} \quad v \in N \ .$$

The map $T$ is an isomorphism from the subspace of $H^s(\Omega)$ consisting of all $u$ satisfying (3.6.3) onto the subspace of

$$H^{s-2m}(\Omega) x H^{s-m_1-\frac{1}{2}}(\dot{\Omega}) x H^{s-m_2-\frac{1}{2}}(\dot{\Omega}) x \cdots x H^{s-m_m-\frac{1}{2}}(\dot{\Omega})$$

consisting of all $(f,g_j)$ satisfying (3.6.2), and with this understanding, there is a constant $C > 0$ such that (3.6.1) holds.

This theorem is proved in [1] or in [2]. We now illustrate the theorem with several examples. If $A = \Delta$ is the Laplace operator, and if $B_1 = u|_{\dot{\Omega}}$ , then we have the problem

(3.6.4)
$$\begin{cases} \Delta u = f & \text{in} \quad \Omega \\ u = g & \text{on} \quad \dot{\Omega} \ . \end{cases}$$

In this case, $N = N^* = \{0\}$, so if $f \in H^{s-2}(\Omega)$ and $g \in H^{s-\frac{1}{2}}(\dot{\Omega})$, then there is a unique $u \in H^s(\Omega)$ satisfying (3.6.4). If $B_1 = \partial u/\partial\nu|_{\dot{\Omega}}$, then our boundary value problem is

$$(6.3.5) \qquad \begin{cases} \Delta u = f & \text{in } \Omega \\ \partial u/\partial\nu = g & \text{on } \dot{\Omega} \end{cases}$$

In this case, $N = N^* = \{1\}$, the one dimensional space generated by the constant function. Then (3.6.5) has a solution $u \in H^s(\Omega)$ for any $f \in H^{s-2}(\Omega)$, $g \in H^{s-3/2}(\dot{\Omega})$ which satisfy

$$\int_\Omega f dx + \oint_{\dot{\Omega}} g ds = 0 \quad ;$$

The solution $u$ is unique if we prescribe the condition

$$\int_\Omega u dx = 0 \quad .$$

One can easily give examples of this theorem for the biharmonic operator. In these examples it should be noted that the regularity of the functions $g_j$ depend on the order of the operator $B_j$.

To consider data $f$, $g_j$ with weaker regularity properties, we must introduce some new spaces. This is done in the next section.

3.7. Another family of spaces.

In chapter 2 there was defined, for any real number $s$, the Sobolev spaces $H^s(\Omega)$ and $H^s(\Omega^{\bullet})$. For $s > 0$, the norm on $H^{-s}(\Omega)$ is given by

$$\|u\|_{H^{-s}(\Omega)} = \sup_{v \in \mathcal{E}(\overline{\Omega})} \frac{\int_\Omega uv\,dx}{\|v\|_{H^{+s}(\Omega)}} .$$

We identified $H^{-s}(\Omega)$ as the closure of $\mathcal{E}(\overline{\Omega})$ in this norm. Equivalently (see [3], $H^{-s}(\Omega)$ is the set of distributions $u$ in $H^{-s}(R^2)$ such that the support of $u$ lies in $\overline{\Omega}$. For the theory of distributions see e.g. [8]. Now if $v \in H^s(\Omega)$, then $v$ may be extended to a function $v \in H^s(R^2)$. We define

$$\langle u,v\rangle = \langle u,V\rangle ,$$

where the right side is interpreted in the sense that u is a distribution on $R^2$, acting on a function $V$ on $R^2$. If $V_1$ is another extension of $v$, then $V - V_1 = 0$ on $\overline{\Omega}$, so $\langle u,V\rangle = \langle u,V_1\rangle$. Thus, any such distribution $u$ gives a linear functional on $H^s(\Omega)$, and one can show that all elements of $H^{-s}(\Omega)$ can be obtained in this way.

To state a general theorem on the solutions of (3.5.3) it is convenient to introduce another family of spaces. For $k$ a positive integer and $s$ an integer we define a norm by

$$\|u\|^2_{\tilde{H}^{s,k}(\Omega)} = \|u\|^2_{H^s(\Omega)} + \sum_{j=0}^{k-1} \left\|\frac{\partial^j u}{\partial \nu^j}\right\|^2_{H^{s-j-\frac{1}{2}}(\Omega^\bullet)} ,$$

$$u \in E(\overline{\Omega}) .$$

We let $\tilde{H}^{s,k}(\Omega)$ be the closure of $E(\overline{\Omega})$ in this norm. Obviously $\tilde{H}^{s,k}(\Omega)$ is a Hilbert space and therefore we may use the theory of Hilbert scales for connecting $\tilde{H}^{s_1,k}(\Omega)$ and $\tilde{H}^{s_2,k}(\Omega)$ . For $s$ not an integer we define $\tilde{H}^{s,k}(\Omega)$ by interpolation.

To gain an understanding of these spaces we define a space $K^{s,k}(\Omega)$

$$K^{s,k}(\Omega) = H^s(\Omega) \times H^{s-\frac{1}{2}}(\Omega^\bullet) \times H^{s-1-\frac{1}{2}}(\Omega^\bullet) \times \cdots \times H^{s-(k-1)-\frac{1}{2}}(\Omega^\bullet)$$

We consider the map $S$ , which maps $\tilde{H}^{s,k}(\Omega)$ into $K^{s,k}(\Omega)$ ,

$$S : \tilde{H}^{s,k}(\Omega) \to K^{s,k}(\Omega) , \quad s \neq \text{integer} + \frac{1}{2} ,$$

defined, for $u \in E(\overline{\Omega})$ , by

$$Su = \left\{u, u_0, \cdots, u_{k-1}\right\} , \quad u_j = \left.\frac{\partial^j u}{\partial \nu^j}\right|_{\Omega^\bullet} ,$$

and defined by continuity for any $u \in H^{s,k}(\Omega)$ .

If $s < \frac{1}{2}$, then $S$ maps $\tilde{H}^{s,k}(\Omega)$ onto $K^{s,k}(\Omega)$ in a one to one manner, and the two spaces may be identified in this case. If $s > k - 1/2$, then

$$S\tilde{H}^{s,k}(\Omega)\left\{= \left\{u, u|_{\Omega^\cdot}, \cdots, \frac{\partial^{k-1} u}{\partial \nu^{k-1}}\Big|_{\Omega^\cdot}\right\} \Big| \; u \in H^s(\Omega)\right\},$$

so $\tilde{H}^{s,k}(\Omega)$ may be identified with $H^s(\Omega)$ in this case. In the intermediate case, $\frac{1}{2} < s < k - 1/2$, suppose that for some integer $\ell$,

$$\ell - 1/2 < s < \ell + 1/2 \quad .$$

Then

(3.7.1)

$$\begin{cases} S\tilde{H}^{s,k}(\Omega) = \text{the set of } \{u, u_0, \cdots, u_{k-1}\} \text{ such that} \\ u \in H^s(\Omega), \; u_j = \dfrac{\partial^j u}{\partial \nu^j}\Big|_{\Omega^\cdot} \text{ for } 0 \leq j < \ell, \text{ and} \\ u_j \in H^{s-j-\frac{1}{2}}(\Omega^\cdot) \text{ is arbitrary for } \ell \leq j \leq k-1 \quad . \end{cases}$$

The case when $s = \text{integer} + 1/2$ are, to some extent, pathological. For an understanding of these results the reader should refer also to Theorem 2.4.1.

Let $A$ be a linear differential operator of order $k$ with smooth coefficients. We will define $A$ on $u \in \tilde{H}^{s,k}(\Omega)$ so that $Au \in H^{s-k}(\Omega)$, and so that

(3.7.2) $\quad A: \tilde{H}^{s,k}(\Omega) \to H^{s-k}(\Omega)$ is bounded ,

For $s \geq k$ , $\tilde{H}^{s,k}(\Omega)$ may be identified with $H^s(\Omega)$, $s \neq$ integer $+ \frac{1}{2}$ , and the definition of $A$ is evident. For $s < k$ , the definition is more elaborate. We shall consider several cases.

(i) $0 \leq s < k$ , $s$ is an integer. From Gauss's theorem and some integrations be parts over the boundary, if necessary, we may write, for $u$ , $v \in E(\bar{\Omega})$ ,

$$(Au,v)_{H^\circ(\Omega)} = (A_s u, A_{k-s} v)_{H^\circ(\Omega)} + \sum_{j=0}^{k-s-1} \left( \frac{\partial^{k-j-1} u}{\partial \nu^{k-j-1}}, B_j v \right)_{H^\circ(\Omega^\bullet)} ,$$

where $A_s$ , $A_{k-s}$ , and $B_j$ are differential operators whose orders are given by the subscript. Taking absolute values and using the trace inequalities we get

$$\left| (Au,v)_{H^\circ(\Omega)} \right| \leq C \|u\|_{H^s(\Omega)} \|v\|_{H^{k-s}(\Omega)}$$

$$+ C \sum_{j=0}^{k-s-1} \left\| \frac{\partial^{k-j-1} u}{\partial \nu^{k-j-1}} \right\|_{H^{j+s+\frac{1}{2}-k}(\Omega^\bullet)} \|B_j v\|_{H^{k-j-s-\frac{1}{2}}(\Omega^\bullet)}$$

$$\leq C \|u\|_{\tilde{H}^{s,k}(\Omega)} \|v\|_{H^{k-s}(\Omega)} .$$

From this we see that for $u \in \mathcal{E}(\bar{\Omega})$, $(Au,v)$ may be extended to a bounded linear functional of $v \in H^{k-s}(\Omega)$, and the norm of this linear functional is bounded by $C\|u\|_{\tilde{H}^{s,k}(\Omega)}$. Since $\mathcal{E}(\bar{\Omega})$ is dense in $\tilde{H}^{s,k}(\Omega)$, we may use continuity to obtain (3.7.2).

(ii) $0 \leq s < k$, s not an integer. We define A on $\tilde{H}^{s,k}(\Omega)$ by a Hilbert scale connecting the adjacent integral values of s.

(iii) $s < 0$, s an integer. We start from the formula

(3.7.3)

$$(Au,v)_{H^\circ(\Omega)} = (u,A^*v)_{H^\circ(\Omega)} + \sum_{j=0}^{k-1} \left( \frac{\partial^{k-j-1}u}{\partial\nu^{k-j-1}} , B_j v \right)_{H^\circ(\dot{\Omega})}$$

where $A^*$ denotes the formal adjoint to A. We define a mapping T

$$T : H^{-s+k}(\Omega) \to H^{-s}(\Omega) \times H^{-s+k-\frac{1}{2}}(\dot{\Omega}) \times H^{-s+k-1-\frac{1}{2}}(\dot{\Omega})$$

$$\times \cdots \times H^{-s+k-(k-1)-\frac{1}{2}}(\dot{\Omega})$$

by

$$Tv = (A^*v, B_0 v|_{\dot{\Omega}}, \cdots B_{k-1}v|_{\dot{\Omega}}) \; .$$

From the trace theorem (2.4.1) we see that $T$ is a bounded linear transformation. Hence the adjoint map

$$T^*: H^s(\Omega) \times H^{s-k+\frac{1}{2}}(\dot{\Omega}) \times H^{s-k+3/2}(\dot{\Omega}) \times \cdots \times H^{s-\frac{1}{2}}(\dot{\Omega}) \to H^{s-k}(\Omega)$$

is a bounded linear map. Since, for $s < 0$,

$$\tilde{H}^{s,k}(\Omega) = H^s(\Omega) \times H^{s-k+\frac{1}{2}}(\dot{\Omega}) \times \cdots \times H^{s-\frac{1}{2}}(\dot{\Omega})$$

where the identification is made by means of the map $S$, we may regard $T^*$ as a bounded map from $\tilde{H}^{s,k}(\Omega)$ into $H^{s-k}(\Omega)$. Now let $u \in \mathcal{E}(\bar{\Omega})$, so $u \in \tilde{H}^{s,k}(\Omega)$, and

$$Su = \left\{ u, u\Big|_{\dot{\Omega}}, \cdots, \frac{\partial^{k-1} u}{\partial \nu^{k-1}}\Big|_{\dot{\Omega}} \right\}$$

$$\in H^s(\Omega) \times H^{s-\frac{1}{2}}(\dot{\Omega}) \times \cdots \times H^{s-k+\frac{1}{2}}(\dot{\Omega}) \quad .$$

By the definition of $T^*$, we have for $u \in \mathcal{E}(\bar{\Omega})$, $v \in \mathcal{E}(\bar{\Omega})$,

$$\langle T^*u, v \rangle = \langle u, Tv \rangle = (u, A^*v)_{H^\circ(\Omega)} + \left( \frac{\partial^{k-1} u}{\partial \nu^{k-1}}, B_0 v \right)_{H^\circ(\dot{\Omega})}$$

$$+ \cdots + (u, B_{k-1} u)_{H^\circ(\dot{\Omega})} \quad .$$

Using (3.7.3), we see that $(Au,v)_{H^\circ(\Omega)} = \langle T^*u,v\rangle$ and from the boundedness of $T^*$ ,

$$|(Au,v)_{H^\circ(\Omega)}| \leq C\|u\|_{\tilde{H}^{s,k}(\Omega)} \|v\|_{H^{k-s}(\Omega)} .$$

Hence $Au$ can be extended to be a linear functional on $H^{k-s}(\Omega)$ , so $Au \in H^{s-k}(\Omega)$ , and

$$\|Au\|_{H^{s-k}(\Omega)} \leq C\|u\|_{\tilde{H}^{s,k}(\Omega)} , \quad u \in E(\bar{\Omega}) .$$

Since $E(\bar{\Omega})$ is dense in $\tilde{H}^{s,k}(\Omega)$ , we may use continuity to obtain (3.7.2).

iv) $s < 0$ , $s$ not an integer. We define $A$ on $\tilde{H}^{s,k}$ by interpolating between the adjacent integral values of $s$ .

Now let $B$ be a linear differential operator of order $p \leq k - 1$ , with smooth coefficients. We define a bounded map

$$B:\tilde{H}^{s,k}(\Omega) \to H^{s-p-\frac{1}{2}}(\dot{\Omega})$$

as follows. We write, for $x \in \dot{\Omega}$ ,

$$B = \sum_{j=0}^{p} \tau_j \frac{\partial^j}{\partial\nu^j} ,$$

where $\tau_j$ is a differential operator of order $t_j \leq p - j$ consisting of tangential directional derivatives at the point $x \in \dot{\Omega}$ . Then using the isomorphism $S$ and (3.7.1), we define

$$Bu = \sum_{0}^{p} \tau_j u_j$$

Since $u_j \in H^{s-j-\frac{1}{2}}(\dot{\Omega})$, we have

$$\tau_j u_j \in H^{s-j-t_j-\frac{1}{2}}(\dot{\Omega})$$

$$\subset H^{s-p-\frac{1}{2}}(\dot{\Omega}) \quad .$$

3.8. Solvability and regularity theorems (continued).

With the help of the spaces $\tilde{H}^{s,2m}(\Omega)$ defined in the previous section we are now able to extend the statement of Theorem 3.6.1 to $s < 2m$ . Let $A$ be a properly elliptic operator of order $2m$ , and let $B_j$ be a normal covering family of boundary operators, the operator $B_j$ having order $m_j$ . We consider the map

$$T:\tilde{H}^{s,2m}(\Omega) \to H^{s-2m}(\Omega) \times H^{s-m_1-\frac{1}{2}}(\dot{\Omega}) \times \cdots \times H^{s-m_m-\frac{1}{2}}(\dot{\Omega})$$

defined by

$$Tu = \left\{Au, B_1 u, \cdots, B_m u\right\}, \quad u \in \tilde{H}^{s,2m}(\Omega) \quad .$$

Theorem 3.8.1. Let $\left\{A, B_j\right\}$ be a regularly elliptic boundary value problem of order $2m$. If $N = N^* = \{0\}$, then $T$ is a one to one map of $\tilde{H}^{s,2m}(\Omega)$ onto

$$H^{s-2m}(\Omega) \times H^{s-m_1-\frac{1}{2}}(\dot{\Omega}) \times \cdots$$

$$\cdots \times H^{s-m_m-\frac{1}{2}}(\dot{\Omega})$$

and there is a constant $C > 0$, independent of $u$, such that for $f \in H^{s-2m}(\Omega)$, $g_j \in H^{s-m_j-\frac{1}{2}}(\dot{\Omega})$,

(3.8.1)

$$\|u\|_{\tilde{H}^{s,2m}(\Omega)} \leq C\|f\|_{H^{s-2}(\Omega)} + C\sum_j \|g_j\|_{H^{s-m_j-\frac{1}{2}}(\dot{\Omega})}$$

In general, $\{f,g_j\}$ lies in the range of $T$ if and only if (3.6.2) is satisfied. If (3.6.2) holds, there is a unique $u \in \tilde{H}^{s,2m}(\Omega)$ such that $Tu = \{f,g_j\}$, and such that $u$ satisfies the orthogonality condition (3.6.3), where, if $s < 0$, $(u,v)_{H^0(\Omega)}$ is interpreted as the value of the linear functional $u \in H^s(\Omega)$ on the function $v \in H^{-s}(\Omega)$. For this $u$, (3.8.1) holds.

We give several examples of this theorem, connected with the problem

(3.8.2)
$$\begin{cases} \Delta u = f & \text{in } \Omega \\ u = g & \text{on } \dot{\Omega} \end{cases}.$$

Suppose $f \in H^s(\Omega)$ , $g \in H^{s-\frac{1}{2}}(\dot{\Omega})$ . By a solution of (3.8.2) is meant a <u>triple</u> of functions, $U = \left\{u, u_0, u_1\right\}$ with

$$u \in H^{s+2}(\Omega)\ ,\ u_0 \in H^{s+3/2}(\dot{\Omega})\ ,\ u_1 \in H^{s+\frac{1}{2}}(\dot{\Omega})\ ,$$

such that

(3.8.3)
$$(u,\Delta\phi)_{H^\circ(\Omega)} + (u_1,\phi)_{H^\circ(\dot{\Omega})} - (u_0,\frac{\partial\phi}{\partial n})_{H^\circ(\dot{\Omega})}$$

$$= (f,\phi)_{H^\circ(\Omega)}\ ,\ \phi \in E(\overline{\Omega})\ ,$$

(3.8.4)
$$u_0 = g\ .$$

We may interpret $u_0$ as being a kind of trace, on $\dot{\Omega}$ , of $u$ , and we may interpret $u_1$ as being a kind of trace, on $\dot{\Omega}$ , of $\partial u/\partial\nu$ . In particular, suppose $u \in H^1(\Omega)$ and suppose in addition $u$ is a harmonic function in $\Omega$ , $\Delta u = 0$ . Then if $g = u|_{\dot{\Omega}}$ , we have $g \in H^{\frac{1}{2}}$ . Then we may consider the problem (3.8.2) with $f = 0$ and $g$ as described. Setting $s = 1$ , there is a solution.

$$U = \left\{u, u_0, u_1\right\} \in H^1(\Omega)xH^{\frac{1}{2}}(\dot{\Omega})xH^{-\frac{1}{2}}(\dot{\Omega})$$

which satisfies (3.8.3). Interpreting $u_1$ as $\partial u/\partial n$ , we see that by this means a meaning may be given to such a normal derivative when $u \in H^1(\Omega)$ is a harmonic function. If we approximate $g$ by a sequence $g^{(k)}$ of smooth functions, the corresponding solutions $U^{(k)} = \left\{u^{(k)}, u_0^{(k)}, u_1^{(k)}\right\}$ of (3.8.2) satisfy

$$u_0^{(k)} = u^{(k)}\Big|_{\dot{\Omega}} ,$$

$$u_1^{(k)} = \frac{\partial u^{(k)}}{\partial n}\Big|_{\dot{\Omega}} .$$

The smooth functions $u_1^{(k)}$ are a Cauchy sequence in $H^{-\frac{1}{2}}(\dot{\Omega})$ whose limit is $u_1$. This construction of a normal derivative was also carried out in theorem 2.4.2.

In general, the weak solutions whose existence is asserted in Theorem 3.8.1 may be obtained as a limit of smooth solutions. The smooth solutions are obtained by selecting sequences $f^{(k)}$, $g_j^{(k)}$ of smooth data and solving the corresponding boundary value problems.

In the next series of examples we let $\varepsilon$ be a fixed number with $0 < \varepsilon < 1/2$. We let $u$ be a solution, in the classical sense, of the Dirichlet problem

$$(3.8.5) \qquad \begin{cases} \Delta u = 0 & \text{in } \Omega \\ u = g & \text{on } \dot{\Omega} \end{cases}$$

It is known that there are continuous functions $g$ such that $u \notin H^1(\Omega)$. Such a solution $u$ will, however, belong to $H^{1-\varepsilon}(\Omega)$. We let $g$ be such a continuous function, with $g \in H^{\frac{1}{2}-\varepsilon}(\dot{\Omega})$.

Now we consider the problem (3.8.5) in the sense of our Theorem 3.8.1. There is a triple

$$\tilde{U} = \left\{\tilde{u}, \tilde{u}_0, \tilde{u}_1\right\} \in \tilde{H}^{1-\varepsilon,2}(\Omega)$$

such that

(3.8.6)

$$\begin{cases} 0 = \int_\Omega \tilde{u}\Delta\phi dx + \oint_{\dot{\Omega}} \tilde{u}_1 \phi ds - \oint_{\dot{\Omega}} \tilde{u}_0 \frac{\partial\phi}{\partial n} ds \ , \ \phi \in E(\bar{\Omega}) \ , \\ \tilde{u}_0 = g \ . \end{cases}$$

By Weyl's theorem we see that $\tilde{u}$ is a harmonic function in $\Omega$ which is continuous on $\bar{\Omega}$ and such that $\tilde{u} = g$ on $\dot{\Omega}$. By the uniqueness of (3.8.5), we see that $u = \tilde{u}$, so we write our generalized solution as $U = \left\{u, u_0, u_1\right\}$. Thus, $U$ solves the problem

$$\begin{cases} \Delta u = 0 & \text{in} \quad \Omega \\ u_0 = g & \text{on} \quad \dot{\Omega} \\ u \in \tilde{H}^{1-\varepsilon,2}(\Omega) \ . \end{cases} \tag{3.8.7}$$

We may also interpret $u$ as a generalized solution of (3.8.2) with $g = 0$. We define a function $f \in H^{-3/2-\varepsilon}(\Omega)$ by

$$\int_{\Omega} f\phi dx = \oint_{\dot{\Omega}} g \frac{\partial\phi}{\partial n} ds \quad . \tag{3.8.8}$$

Then if $\phi \in H^{3/2+\varepsilon}(\Omega)$ , $\frac{\partial\phi}{\partial n} \in H^{\varepsilon}(\dot{\Omega})$ , and since $g \in H^{-\varepsilon}(\dot{\Omega})$, (3.8.8) defines $f$ as a linear functional on $H^{3/2+\varepsilon}(\Omega)$, so $f \in H^{-3/2-\varepsilon}(\Omega)$ . Now we consider the solution

$$V = \{v, v_0, v_1\} \in \tilde{H}^{\frac{1}{2}-\varepsilon,2}(\Omega)$$

of the problem

$$\begin{cases} \Delta V = f & \text{in } \Omega \\ V = 0 & \text{on } \dot{\Omega} \quad . \end{cases} \tag{3.8.9}$$

Then the solution satisfies

(3.8.10)

$$\begin{cases} \int_{\Omega} f\phi dx = \int_{\Omega} v\Delta\phi dx + \oint_{\dot{\Omega}} v_1 \phi ds \ , \quad \phi \in E(\overline{\Omega}) \ , \\ v_0 = 0 \quad . \end{cases}$$

Using (3.8.8), we see that (3.8.10) is the same as (3.8.6). Thus $v = u, v_1 = u_1$ . We have interpreted the same harmonic function $u$ , as a solution of the classical Dirichlet problem (3.8.5), the generalized Dirichlet problem (3.8.7), and the problem (3.8.9). We may note that in (3.8.9), the solution $V \notin \tilde{H}^{\frac{1}{2}+\varepsilon,2}(\Omega)$ . For the function $u = v$ is in fact, continuous on $\overline{\Omega}$ . But our solution V involves $v_0 = 0$ , so $v_0$ cannot be the trace of v . Hence the regularity of the problem must be such that v does not have a trace.

We give another example connected with our harmonic function u . For this, we select a function $\chi \in H^{-\frac{1}{2}-\varepsilon}(\dot{\Omega})$ , and we consider the function

$$W = \left\{w, w_0, w_1\right\} \in \tilde{H}^{1-\varepsilon,2}(\Omega)$$

defined by

$$\begin{aligned} w &= u \\ w_0 &= u_o \\ w_1 &= u_1 + \chi \quad . \end{aligned}$$

Then we may calculate $\Delta W = \tilde{f} \in H^{-1-\varepsilon}(\Omega)$ to be the distribution defined by

$$\int_\Omega \tilde{f}\phi dx = \oint_{\dot{\Omega}} \chi\phi ds \quad , \quad \phi \in E(\overline{\Omega}) \quad .$$

Thus, $\tilde{f}$ has its support on $\dot{\Omega}$ , and W solves the problem

(3.8.11)
$$\begin{cases} \Delta W = \tilde{f} & \text{in } \Omega \\ W = g & \text{on } \dot{\Omega} \\ W \in \tilde{H}^{1-\varepsilon,2}(\Omega) & . \end{cases}$$

In this way, we see that $u = w$ occurs as the solution of another problem, the problem (3.8.11).

3.9. Additional results.

We have stated and illustrated the basic existence and regularity theorem for solutions of boundary value problems associated with an elliptic operator of order $2m$. The solution of a elliptic boundary value problem has another important property. Let us assume that the function $u$ satisfies the equation $Au = 0$. Then the smoothness of $u$ is determined by the theorem 3.6.1 by the smoothness of boundary conditions. But this condition influences the smoothness only in the neighborhood of the boundary. Inside the domain $\Omega$ the solution $u$ has all derivatives. Denoting $\Omega_\delta = \left\{x \in \Omega / d(x, \dot{\Omega}) > \delta\right\}$ where $d(x, \dot{\Omega})$ is the distance of the point $x$ to the boundary we may estimate $\|u\|_{H^\ell(\Omega_\delta)}$ for any $\ell$. The form of such an estimate is for $\ell \geq s$

$$\|u\|_{H^\ell(\Omega_\delta)} \leq C(\delta) \|u\|_{H^s(\Omega)}$$

where $C(\delta)$ depends on $\delta$. The typical dependence is $C(\delta) \leq C\delta^{-(\ell-s)}$ provided that the domain is smooth. Also for the domains with boundary $\dot{\Omega} \in C^\ell$ there are such estimates. See e.g. [9], [10]. There is a similar theory for a system of elliptic operators. In this theory, the orders of the operators in the system need not all be the same, and this complicates the statement of the results. For a presentation of some of the results for systems, see [2,4].

We remark also that the results for equations and systems hold for more than two independent variables, and in norms other than the $L_2$ norm.

In case the boundary curve $\dot{\Omega}$ or the coefficients of the operator are not smooth, the theorems presented here are not valid. If the boundary $\dot{\Omega}$ and the coefficients are piecewise smooth, there are results on the regularity of the solution. For some aspects of this, [5], [6], [7].

In chapter 8 we will refer to some concrete results of this type.

References.

[1]. J. L. Lions and E. Magenes, Problèmes aux limites non homogènes et applications, volume 1, Dunod, Paris, 1968.

[2]. Ja. A. Roitberg and Z. G. Seftel', "A Theorem on Homeomorphisms for Elliptic Systems and its Applications," Mat. Sb., vol. 78 (120) (1969), translated in Math. USSR Sb., vol. 7 (1969), pp. 439-465.

[3]. J. Peetre, "Another Approach to Elliptic Boundary Problems," Comm. Pure Appl. Math., vol. 14 (1961), pp. 711-731.

[4]. S. Agmon, A. Douglis, and L. Nirenberg, "Estimates Near the Boundary for Solutions of Elliptic Partial Differential Equations Satisfying General Boundary Conditions II," Comm. Pure Appl. Math., vol. 17 (1964), pp. 35-92.

[5]. V. A. Kondrat'ev, "Boundary Problems for Elliptic Equations in Domains with Conical or Angular Points," Trans., Moscow Math. Soc., vol. 16 (1967), pp. 227-313.

[6]. R. B. Kellogg, "Singularities in Interface Problems," in Numerical Solution of Partial Differential Equations II, B. Hubbard, editor, Academic Press, 1971.

[7]. R. B. Kellogg, "Higher Order Singularities for Interface Problems," these proceedings.

[8]. I. M. Gelfand and G. E. Shilov, Generalized Functions, vol. I, Academic Press, 1964.

[9]. A. S. Foht, "Some Inequalities in the $L_2$ Metric for Solutions of Equations of Elliptic Type and for Their Derivatives Near the Boundary," Trudy Mat. Inst. Steklov, vol. 77 (1965), pp. 168-191.

[10]. A. S. Foht, "Boundary Estimates for Derivatives of Solutions of a Class of Hypoelliptic Equations in the $L_2$ Metric," Differencial' nye Uravnenija, vol. 6 (1970), pp. 1673-1682.

## CHAPTER 4. THEORY OF APPROXIMATION.

### 4.1. The (t,k) - system $S_h^{t,k}(\Omega)$ .

As mentioned earlier in chapter 1, in the finite element method one makes use of special functions in order to approximate the desired solution. We shall formulate in an axiomatic manner some of the properties of these functions and analyze them.

Let $\Omega \subset R_2$ be a given domain. At this point it is irrelevant whether $\Omega$ is smooth or Lipschitzian only or $\Omega = R_2$ .

We assume that for every $0 < h < 1$ , we are given a linear system of functions $\phi$ defined on $\Omega$ . This system will be called the (t,k)-system and will be denoted by $S_h^{t,k}(\Omega)$ $(t>k \geq 0)$ if the following conditions are satisfied.

(i) $S_h^{t,k}(\Omega) \subset H^k(\Omega)$ ,

(ii) for every $g \in H^\ell(\Omega)$ , $\ell \geq 0$ and $0 \leq s \leq \min(\ell,k)$ , there exists a $\phi \in S_h^{t,k}(\Omega)$ such that

$$(4.1.1) \qquad \|g-\phi\|_{H^s(\Omega)} \leq C h^\mu \|g\|_{H^\ell(\Omega)} ,$$

where

(4.1.2) $$\mu = \min(t-s, \ell-s) \quad ,$$

and the constant $C$ in (4.1.1) is independent of $g$ and $h$. The function $\phi$ depends on $g$ and may be different for different $s$.

A $(t,k)$ - system $S_h^{t,k}(\Omega), k > 0$ will be called a $(t,k)$-regular system if the function $\phi$ in (4.1.1) can be chosen independently of $s$ so that (4.1.2) holds for all $s$, with $0 \leq s \leq \min(\ell,k)$. In a similar way the system $S_h^{t,k}(\dot{\Omega})$ may be introduced, provided $\Omega$ is smooth.

In the above discussion we restricted ourselves to the case where the parameters $s$, $t$, $k$ and $\ell$ are non-negative. Due to the density of $\mathcal{E}(\bar{\Omega})$ in $H^{\alpha}(\Omega)$ for every $\alpha$, without loss of generality we may confine $g \in \mathcal{E}(\bar{\Omega})$ in (4.1.1). For $s < 0$ we have the following theorem.

<u>Theorem 4.1.1</u>. Let $S_h^{t,k}(\Omega)$ be a $(t,k)$-system. Then for arbitrary $s \leq \min(\ell,k)$ and every $g \in H^{\ell}(\Omega)$, $\ell \geq 0$, there exists $\phi \in S_h^{t,k}(\Omega)$ such that

(4.1.3) $$\|g-\phi\|_{H^s(\Omega)} \leq Ch^{\mu}\|g\|_{H^{\ell}(\Omega)} \quad ,$$

with

(4.1.4) $$\mu = \min(t-s, \ell-s) \quad .$$

where $C$ does not depend on $g$ and $h$, but depends on $s$ and $\ell$.

Proof. 1) Let $g \in H^{\ell}(\Omega), \ell \geq 0$ be given. Let $P_0$ be the orthogonal projection of $H^0(\Omega)$ onto $\overline{S_h^{t,k}}(\Omega)$. Now let $\phi_0 = P_0 g$. Then clearly for every $\omega \in S_h^{t,k}$ $(g-\phi_0,\omega)_{H^0(\Omega)} = 0$. For $\gamma \geq 0$ we have

$$(4.1.5) \qquad \|g-\phi_0\|_{H^{-\gamma}(\Omega)} \equiv \sup_{v\in E(\overline{\Omega})} \frac{|(g-\phi_0,v)_{H^0(\Omega)}|}{\|v\|_{H^{\gamma}(\Omega)}}$$

By the basic properties of the $(t,k)$-system $S_h^{t,k}(\Omega)$ for every $v$ there exists a $w \in S_h^{t,k}(\Omega)$ such that

$$(4.1.6) \qquad \|v-w\|_{H^0(\Omega)} \leq Ch^{\mu}\|v\|_{H^{\gamma}(\Omega)} ,$$

where

$$(4.1.7) \qquad \mu = \min(t,\gamma) .$$

Therefore, we have

$$(4.1.8) \qquad \|g-\phi_0\|_{H^{-\gamma}(\Omega)} = \sup_{v\in E(\overline{\Omega})} \frac{|(g-\phi_0,v-w)_{H^0(\Omega)}|}{\|v\|_{H^{\gamma}(\Omega)}} =$$

$$\leq \sup_{v\in E(\Omega)} \frac{\|g-\phi_0\|_{H^0(\Omega)}\|v-w\|_{H^0(\Omega)}}{\|v\|_{H^{\gamma}(\Omega)}}$$

Since $\phi_0 = P_0 g$ , then it follows that

(4.1.9) $$\|\phi_0 - g\|_{H^0(\Omega)} \leq Ch^{\overline{\mu}} \|g\|_{H^\ell(\Omega)} ,$$

with

(4.1.10) $$\overline{\mu} = \min(t,\ell) .$$

From (4.1.6)-(4.1.10), we obtain

(4.1.11) $$\|g - \phi_0\|_{H^{-\gamma}(\Omega)} \leq Ch^{\mu+\overline{\mu}} \|g\|_{H^\ell(\Omega)} ,$$

where

(4.1.12) $$\mu + \overline{\mu} = \min(2t, t+\gamma, \gamma+t, \ell+t)$$

Thus (4.1.4) holds for $-t \leq \gamma \leq s \leq \min(\ell,k)$[1].

2) Now let $-2t \leq \gamma \leq -t$ and let us connect the spaces $H^\gamma(\Omega)$ and $H^0(\Omega)$ by a Hilbert scale. Then the space $H_{\gamma/2}(\Omega)$ and $H^{\gamma/2}(\Omega)$ are equivalent as explained in section 2.3.

Let us denote $P_{\gamma/2}$ the orthogonal projection of $H_{\gamma/2}(\Omega)$ onto $\overline{S_h^{t,k}}(\Omega)$. By letting $\phi_{\gamma/2} = P_{\gamma/2} g$ and using theorem 2.3.1 we have

---

1) In general $\Phi_0 \in \overline{S_h^{t,k}}$ but $\Phi_0 \notin S_h^{t,k}$ . Nevertheless it is obvious that by changing $C$ in (4.1.11) we are getting (4.1.11) by $\Phi \in S_h^{t,k}$ also.

(4.1.13) $$\|g-\phi_{\gamma/2}\|_{H_\gamma(\Omega)} \approx \sup_{v\in H(\Omega)} \frac{|(g-\phi_{\gamma/2},v)_{H_{\gamma/2}(\Omega)}|}{\|v\|_{H^0(\Omega)}}$$

Since $0 < |\frac{\gamma}{2}| \leq t$ , the spaces $H^{\frac{\gamma}{2}}(\Omega)$ and $H_{\frac{\gamma}{2}}(\Omega)$ are equivalent, using the above results we have

$$\|g-\phi_{\gamma/2}\|_{H_{\gamma/2}(\Omega)} \leq Ch^{\mu}\|g\|_{H^\ell(\Omega)}$$

where

$$\mu = \min(t-\gamma/2,\ell-\gamma/2)$$

Furthermore, for every $w \in S_h^{t,k}(\Omega)$ we have obviously $(g-\phi_{\gamma/2},w)_{H_{\gamma/2}(\Omega)} = 0$ . On the other hand using the above results for every $v \in H^0(\Omega)$ there exists a $w \in S_h^{t,k}(\Omega)$ such that

$$\|v-w\|_{H_{\gamma/2}(\Omega)} \leq Ch^{-\gamma/2}\|v\|_{H^0(\Omega)} \quad .$$

By the arguments used in the first part of the proof we obtain

(4.1.14) $$\|g-\phi\|_{H^\gamma(\Omega)} \leq Ch^{\mu}\|g\|_{H^\ell(\Omega)}$$

where

$$\mu = \min(t+\gamma, \ell+\gamma) \tag{4.1.15}$$

By repeating these steps we obtain our theorem.

Let us prove now the general theorem.

<u>Theorem 4.1.2</u>. Let $S_h^{t,k}(\Omega)$ be a $(t,k)$-system, then for arbitrary $\ell$ and $s \leq \min(\ell,k)$ and $g \in H^{\ell}(\Omega)$, there exists a $\phi \in S_h^{t,k}(\Omega)$ such that (4.1.1) and (4.1.2) hold.

<u>Proof</u>. Because of theorem 4.1.1 we have to prove the case $\ell \leq 0$ only.

1) Let us prove the theorem for $g \in H^{-r}(\Omega)$ $r \geq 0$, and $-2r - t < \gamma \leq -2r$. Connect the space $H^{\gamma}(\Omega)$ and $H^{-2r-\gamma}(\Omega)$ by a Hilbert scale. Denote by $P_{-r}$ the orthogonal projection of $H_{-r}(\Omega)$ onto $S_h^{t,k}(\Omega)$. Realizing that $-2r - \gamma \geq 0$ and that $\|g-P_{-r}g\|_{H_{-r}(\Omega)} \leq \|g\|_{H_{-r}(\Omega)}$, and using theorem 2.3.1 and 4.1.1, by a similar argument as used in the proof of theorem 4.1.1 we obtain the theorem for given $r$ and $\gamma$.

2) The mapping $I - P_{-r}$ maps the space $H_{-r}(\Omega)$ into $H_{-r}(\Omega)$ and also $H_{-r}(\Omega)$ into $H_{\gamma}(\Omega)$. We have already proved that

$$\|I-P_{-r}\|_{H_{-r}\to H_{-r}} \leq 1 \text{ , } \|I-P_{-r}\|_{H_{-r}\to H_{\gamma}} \leq Ch^{\mu}$$

with $\mu = -r - \gamma$. Therefore, using (2.3.20) we

have for $\gamma \leq -s \leq -r$ $\|I-P_{-r}\|_{H_{-r}\to H_{-s}} \leq Ch^{\mu}$ with $\mu = -r-s$. Therefore the theorem holds for all $2\ell - t \leq s \leq \ell \leq 0$. By repeating these arguments we can easily complete the proof of the theorem.

Remark. The system $S_h^{t,k}(\Omega)$ was introduced so that the inequalities (4.1.1) and (4.1.2) hold for all $0 \leq s \leq \min(\ell,k)$. It is not hard to see that a system will be a $(t,k)$ system if (4.1.1) holds only for $k - \varepsilon \leq s \leq k, \ell \geq k$ $\varepsilon > 0$. We also note that theorem 4.1.2 holds for the spaces $H^{\alpha}(\Omega^{\bullet})$ and $S_h^{t,k}(\Omega^{\bullet})$.

We will say that the system $S_h^{t,k}(\Omega)$ satisfies the inverse assumption if there exists $k \geq \varepsilon > 0$ such that for every $k - \varepsilon \leq s \leq k$ and every $\phi \in S_h^{t,k}(\Omega)$ we have

$$\|\phi\|_{H^k(\Omega)} \leq Ch^{-(k-s)}\|\phi\|_{H^s(\Omega)}, \tag{4.1.16}$$

where $C$ does not depend on $h$ and $\phi$ (it may depend on $k$, $\varepsilon$). Now we prove the following result.

Theorem 4.1.3. Let $S_h^{t,k}(\Omega)$ be a $(t,k)$-system which satisfies the inverse assumption. Then for every $s_1 \leq s_2 \leq k$ and $\phi \in S_h^{t,k}(\Omega)$ we have

$$\|\phi\|_{H^{s_2}(\Omega)} \leq Ch^{s_1-s_2}\|\phi\|_{H^{s_1}(\Omega)}, \tag{4.1.17}$$

where $C$ does not depend on $h$ and $\phi$ (it depends in general on $s_1$ and $s_2$).

Proof. 1) Let $k - \varepsilon \leq s < k$, then from (2.3.19) we have

$$\|\phi\|^2_{H^s(\Omega)} \leq C\|\phi\|_{H^k(\Omega)} \|\phi\|_{H^{2s-k}(\Omega)} . \tag{4.1.18}$$

By hypotheses

$$\|\phi\|_{H^k(\Omega)} \leq Ch^{-(k-s)} \|\phi\|_{H^s(\Omega)} , \tag{4.1.19}$$

and therefore

$$\|\phi\|^2_{H^s(\Omega)} \leq C\|\phi\|_{H^s(\Omega)} h^{-(k-s)} \|\phi\|_{H^{2s-k}(\Omega)} ,$$

and thus

$$\|\phi\|_{H^s(\Omega)} \leq Ch^{-(k-s)} \|\phi\|_{H^{2s-k}(\Omega)} . \tag{4.1.20}$$

By using (4.1.19) once again, we obtain

$$\|\phi\|_{H^k(\Omega)} \leq Ch^{-(k-s)} \|\phi\|_{H^s(\Omega)} \leq \tag{4.1.21}$$

$$\leq Ch^{-2(k-s)} \|\phi\|_{H^{2s-k}(\Omega)} .$$

Therefore (4.1.16) holds for all $k - 2\varepsilon \leq s \leq k$. By induction it easily follows that (4.1.16) holds for all $s \leq k$.

2) Now let $s_1 \leqslant s_2 \leqslant k$ be given, then by (2.3.19) and other results we have

$$\|\phi\|_{H^{s_2}(\Omega)} \leqslant C\|\phi\|_{H^{s_1}(\Omega)}^{\frac{k-s_2}{k-s_1}} \|\phi\|_{H^k(\Omega)}^{\frac{s_2-s_1}{k-s_1}}, \tag{4.1.22}$$

By (4.1.19) we obtain

$$\|\phi\|_{H^{s_2}(\Omega)} \leqslant C\|\phi\|_{H^{s_1}(\Omega)}^{\frac{k-s_2}{k-s_1}} h^{-(k-s_1)\frac{(s_2-s_1)}{k-s_1}} \|\phi\|_{H^{s_1}(\Omega)}^{\frac{s_2-s_1}{k-s_1}}, \tag{4.1.23}$$

which reduces to

$$\|\phi\|_{H^{s_2}(\Omega)} \leqslant Ch^{s_1-s_2}\|\phi\|_{H^{s_1}(\Omega)}, \tag{4.1.24}$$

and the proof of our theorem is complete.

Analyzing the proof of the previous theorem we see that if

$$\|\phi\|_{H^{r+z}(\Omega)} \leqslant C(z)h^{-z}\|\phi\|_{H^r(\Omega)} \tag{4.1.25}$$

for every $0 \leqslant z \leqslant \varepsilon$ and every $\phi \in S_h^{t,k}(\Omega)$ then

$$(4.1.26) \qquad \|\phi\|_{H^r(\Omega)} \le C(z)h^{-z}\|\phi\|_{H^{r-z}(\Omega)}$$

i.e., the system $S_h^{t,k}(\Omega)$ $k \geq r + \varepsilon$ may be understood as a system $S_h^{t,r}(\Omega)$ which satisfies the inverse assumption.

Let us prove a theorem which is useful for checking the inverse assumption.

Theorem 4.1.4. Let a $(t,k)$-system $S_h^{t,k}(\Omega)$ be such that there exists $\varepsilon > 0$ so that for every $\phi \in S_h^{t,k}(\Omega)$

$$(4.1.27) \qquad \|\phi\|_{H^k(\Omega)} \le Ch^{-\varepsilon}\|\phi\|_{H^{k-\varepsilon}(\Omega)}$$

then for every $0 \le z \le \varepsilon$ we have

$$(4.1.28) \qquad \|\phi\|_{H^{k-\varepsilon+z}(\Omega)} \le C(z)h^{-z}\|\phi\|_{H^{k-\varepsilon}(\Omega)}$$

Proof. Let us connect the spaces $H^{k-\varepsilon}(\Omega)$ and $H^k(\Omega)$ by Hilbert scale. It was shown in section 2.3 that every $u \in H^{k-\varepsilon}(\Omega)$ could be written in a series (see (2.3.14))

$$(4.1.29) \qquad u = \sum_{k=1}^{\infty} a_k g_k$$

with

$$\|u\|^2_{H^{k-\varepsilon}(\Omega)} = \sum_{k=1}^{\infty} |a_k|^2 \tag{4.1.30}$$

and if $u \in H^k(\Omega)$ then

$$\|u\|^2_{H^k(\Omega)} = \sum_{k=1}^{\infty} |a_k|^2 \lambda_k^{-2} \tag{4.1.31}$$

with

$$0 < \lambda_{k+1} \leqslant \lambda_k$$

So let $\phi \in S_h^{t,k}(\Omega)$ be given. Then we may write

$$\phi = \sum_{k=1}^{\infty} b_k g_k \tag{4.1.32}$$

and by (4.1.27) we have

$$\sum_{k=1}^{\infty} |b_k|^2 \lambda_k^{-2} \leqslant Ch^{-2\varepsilon} \sum_{k=1}^{\infty} |b_k|^2 \tag{4.1.33}$$

or

$$\sum_{k=1}^{\infty} |b_k|^2 \left(h^{\varepsilon} \lambda_k^{-1}\right)^2 \leqslant C \sum_{k=1}^{\infty} |b_k|^2 \tag{4.1.34}$$

Using weighted means inequality[1)] we obtain for $0 \leq \theta \leq 1$

$$(4.1.35)\qquad \frac{\left[\sum_{k=1}^{\infty} |b_k|^2 \left(h^{\varepsilon}\lambda_k^{-1}\right)^{2\theta}\right]^{\frac{1}{\theta}}}{\left[\sum_{k=1}^{\infty} |b_k|^2\right]^{1/\theta}} \leq \frac{\sum_{k=1}^{\infty} |b_k|^2 \left(h^{\varepsilon}\lambda_k^{-1}\right)^2}{\sum_{k=1}^{\infty} |b_k|^2}$$

Using (4.1.33) we get

$$(4.1.36)\qquad \left[\sum_{k=1}^{\infty} |b_k|^2 \left(h^{\varepsilon}\lambda_k^{-1}\right)^{2\theta}\right]^{\frac{1}{\theta}} \leq$$

$$\leq C\left[\sum_{k=1}^{\infty} |b_k|^2\right]^{1/\theta}$$

and therefore we may write

$$(4.1.37)\qquad \sum_{k=1}^{\infty} |b_k|^2 \lambda_k^{-2\theta} \leq Ch^{-2\varepsilon\theta} \sum_{k=1}^{\infty} |b_k|^2$$

---

1) Let $p_\nu > 0$ , $a_\nu > 0$ , $\nu = 1,2,\cdots, r \leq 1$ . Then

$$\left(\frac{\sum pa^r}{\sum p}\right)^{1/r} \leq \frac{\sum pa}{\sum p}$$

For proof see theorem 16 in [16].

Using (2.3.17) the inequality (4.1.37) may be rewritten in the form

$$\|\phi\|_{H_{k-\varepsilon(1-\theta)}(\Omega)} \leq Ch^{-\varepsilon\theta}\|\phi\|_{H^{k-\varepsilon}(\Omega)}$$

The equivalency of the spaces $H_{k-\varepsilon(1-\theta)}(\Omega)$ and $H^{k-\varepsilon(1-\theta)}(\Omega)$ gives immediately the theorem.

The practical importance of theorem 4.1.4 is that we may check the inequality say for $\varepsilon = 1$ and $k$ integral only.

In theorem 4.1.2 we showed that for a $(t,k)$-system $S_h^{t,k}(\Omega)$ (4.1.1) and (4.1.2) hold for every $s \leq \min(\ell,k)$ . A question arises as to whether similar inequalities hold for a regular $(t,k)$-system $S_h^{t,k}(\Omega)$ . This problem appears to be unsolved. However, if a $(t,k)$-regular system satisfies the inverse assumption, then a theorem analogueous to theorem 4.1.2 holds.

Theorem 4.1.5. If

I) $S_n^{t,k}(\Omega)$ is a $(t,k)$-regular system which satisfies the inverse assumption,

II) a $\sigma < k$ is given

then

III) for every $g \in H^{\ell}(\Omega)$ , a $\phi \in S_h^{t,k}(\Omega)$ exists such that for every $\sigma \leq s \leq k$ , $s \leq \ell$

(4.1.38) $$\|g-\phi\|_{H^s(\Omega)} \leq Ch^{\mu}\|g\|_{H^{\ell}(\Omega)} ,$$

where

$$(4.1.39) \qquad \mu = \min(t-s, \ell-s) ,$$

and the constant $C$ in (4.1.38) is independent of $g$ and $h$.

The difference between this theorem and theorem 4.1.2 is that now the function $\phi$ is independent of $s$, where as in theorem 4.1.2 the function $\phi$ in general depends on $s$.

Proof. The proof is similar to that of theorem 4.1.2. We prove theorem 4.1.5 only for $-t \leq \sigma$ and $\ell \geq 0$. In the proof of theorem 4.1.1, we chose $\phi_0$ as the projection of $g \in H^0(\Omega)$ onto $S_h^{t,k}(\Omega)$ and $\phi_0$ was independent of $s$ for $-t \leq s \leq 0$. On the other hand since $S_h^{t,k}(\Omega)$ is $(t,k)$-regular, there exists $\psi \in S_h^{t,k}(\Omega)$ such that for $0 \leq s \leq \leq \min(k,\ell)$

$$(4.1.40) \qquad \|g-\psi\|_{H^s(\Omega)} \leq Ch^{\mu} \|g\|_{H^{\ell}(\Omega)} ,$$

with

$$(4.1.41) \qquad \mu = \min(t-s, \ell-s) ,$$

and $\psi$ is independent of $s$. Clearly we have

$$(4.1.42) \qquad \|\psi-\phi_0\|_{H^0(\Omega)} \leq Ch^{\mu_0} \|g\|_{H^{\ell}(s)} ,$$

where

$$\mu_0 = \min(t,\ell) \quad , \tag{4.1.43}$$

Moreover, for $0 \leq s \leq \min(\ell,k)$ , we have

$$\|g-\phi_0\|_{H^s(\Omega)} \leq \|g-\psi\|_{H^s(\Omega)} + \|\psi-\phi_0\|_{H^s(\Omega)} \quad . \tag{4.1.44}$$

Since the inverse assumption holds we have

$$\|\psi-\phi_0\|_{H^s(\Omega)} \leq Ch^{-s}\|\psi-\phi_0\|_{H^0(\Omega)} \tag{4.1.45}$$

By using (4.1.40), (4.1.42), (4.1.44) and (4.1.45) we obtain for $0 \leq s \leq \min(\ell,k)$

$$\|g-\phi_0\|_{H^s(\Omega)} \leq Ch^{\mu}\|g\|_{H^{\ell}(\Omega)} \quad , \tag{4.1.46}$$

with

$$\mu = \min(t-s,\ell-s) \quad . \tag{4.1.47}$$

Hence (4.1.40), (4.1.41) hold for all $-t \leq s \leq \min(\ell,k)$ , and the theorem is proved.

<u>Remark</u>. A close analysis of the proof of theorem 4.1.1 reveals that we have actually proved that if $S_h^{t,k}(\Omega)$ is a $(t,k)$- system then for every $u \in H^{\ell}(\Omega)$ $\ell \geq 0$ there

exists a $\phi \in S_h^{t,k}(\Omega)$ such that for all $-t < s \leq 0$, we have

$$\|u-\phi\|_{H^s(\Omega)} \leq Ch^{\mu}\|u\|_{H^s(\Omega)} ,$$

where

$$\mu = \min(t-s, \ell-s) \quad .$$

Thus the system $S_h^{t,k}(\Omega)$ has similar properties as a $(t,k)$ regular system provided we restrict ourselves in the interval $-t < s \leq 0$.

In the theorems proved in this section we dealt primarily with the spaces $H^{\alpha}(\Omega)$ and the systems $S_h^{t,k}(\Omega)$. Obviously, all these theorems hold as well for the spaces $H^{\alpha}(\Omega^{\bullet})$, $H^{\alpha}(R_2)$ and the systems $S_h^{t,k}(\Omega)$ and $S_h^{t,k}(R_2)$. Now we introduce an additional hypothesis, namely the local assumption which may be satisfied by a system $S_h^{t,k}(\Omega)$. A $(t,k)$-system $S_h^{t,k}(\Omega)$ is said to fulfill the local assumption if for $g \in H^{\ell}(\Omega)$ with support $S(\Omega)$ the function $\phi \in S_h^{t,k}(\Omega)$ may be so chosen that the support $S_h$ of the function $\phi$ in (4.1.1) has the property that $S_h \subset S^{\lambda h}$, where $S^{\lambda h} = \left\{x \in \Omega \,|\, d(x,S) \leq \lambda h\right\}$, and $d(x,S)$ denotes the distance of $x$ to $S$ and $\lambda$ is independent of $h$. Obviously, the above assumption is also meaningful for the systems $S_h^{t,k}(\Omega^{\bullet})$ and $S_h^{t,k}(R_2)$.

We have introduced several properties for a $(t,k)$-system $S_h^{t,k}(\Omega)$ such as regularity, inverse and local assumptions. Obviously, a (t,k)-system $S_h^{t,k}(\Omega)$ may possess all or any combination of these properties.

Now we state and prove a theorem which is very useful for the construction of (t,k)-systems.

Theorem 4.1.6. If

I) Given two domains $\Omega_1$ and $\Omega_2$ such that $\Omega_1 \subset \Omega_2$, $\Omega_1$ is a Lipschitz domain and $\Omega_2$ is either a Lipschitz domain or the entire plane,

II) $S_h^{t,k}(\Omega_2)$ is a (t,k)-(resp. (t,k)-regular, resp. system satisfying the local assumption) then

III) The restriction of the set of functions from $S_h^{t,k}(\Omega_2)$ to $\Omega_1$ generates a (t,k)-system $S_h^{t,k}(\Omega_1)$ (resp. (t,k)-regular, resp. system satisfying the local assumption).

Proof. For $g \in H^\ell(\Omega_1)$, by the use of the extension theorem (see section 2.3) we may construct a function $\bar{g} \in H^\ell(\Omega_2)$ such that

(4.1.48) $$\|\bar{g}\|_{H^\ell(\Omega_2)} \leq C\|g\|_{H^\ell(\Omega_1)} ,$$

and for $x \in \Omega_1$, we have $\bar{g}(x) = g(x)$. By hypothesis the system $S_h^{t,k}(\Omega_2)$ is (t,k)-system (resp. (t,k)-regular system, resp. system with local assumption). Therefore, there exists $\phi \in S_h^{t,k}(\Omega_2)$ such that (4.1.1) resp. (4.1.2) holds on $\Omega_2$. By restricting the approximation on $\Omega_1$ and making use of (4.1.48) we obtain our theorem, since the local assumption is obviously preserved.

We note that the local property as well as regularity property are preserved by the above restriction. However the inverse assumption on $S_h^{(t,k)}(\Omega_2)$ does not carry over to the system $S_h^{(t,k)}(\Omega_1)$ in general.

This is particularly important when we use regular elements say of square shape in $R_2$ and restrict them on $\Omega$ . If $\Omega$ is a square then the inverse assumption holds provided the square elements used have sides which are in relation with the square $\Omega$ . Unfortunately for a smooth domain it is impossible to construct a system which satisfies the inverse assumption using square elements restricted to the domain $\Omega$ . (See [1]).

The (t,k) systems $S_h^{(t,k)}(\Omega)$ have a long history in connection with the finite element method. They have been studied extensively, in particular, attention was focused on the verification of the conditions for a (t,k)-system. We mentioned some but not all of the papers which seem to be relevent especially in connection with the finite element method.

It was proved in [2] (and later referred to in [3]) that the triangular elements generate a (t,k) system. Some criteria for (t,k)-system may be found in [4]. In the case of piecewise polynomials (see also [5]) the (t,k)-systems have been introduced and analyzed in [6], [7] and [8]. Some necessary and sufficient conditiond for a special class of (t,k)-regular systems have been given and the concept of the local property has been introduced and used.

References [8], [9], [10] and [17] are also relvant to problems connected with (t,k)-systems. For an application of the Fourier transform in this connection see [11]. [12] comtains related important results also. In [13] the (t,k)-regular systems are extensively used and theorem 4.1.3 is in fact proved. It appears that for the first time the prolongation theorem (see theorem 4.1.6) has been used in [14].

4.2. Simultaneous approximations by (t,k)-systems. In section 4.1 we dealt with the approximation of a function $g \in H^{\ell}(\Omega)$ by functions in $S_h^{t,k}(\Omega)$ and for a Lipschitz domain.

In this section we restrict ourselves to the smooth domains and we consider first the problem of approximating a function $g \in H^{\ell}(\dot{\Omega})$ by the traces of a function $\phi \in S_h^{t,k}(\Omega)$.

Theorem 4.2.1. If

I) $g \in H^{\ell}(\dot{\Omega})$ is given,

II) $S_h^{t,k}(\Omega)$ is a given (t,k)- system

then

III) the traces of a function $\phi \in S_h^{t,k}(\Omega)$ generate a $(t-\frac{1}{2},k-\frac{1}{2})$ - system $S_h^{t-\frac{1}{2},k-\frac{1}{2}}(\dot{\Omega})$ on the boundary.

Proof. By the use of theorem 2.4.3 we can find a function $u \in H^{\ell+\frac{1}{2}}$, $\ell > 0$

(4.2.1) $$\|u\|_{H^{\ell+\frac{1}{2}}(\Omega)} < C\|g\|_{H^{\ell}(\dot{\Omega})} ,$$

and

$$u = g \text{ on } \dot{\Omega}$$

Therefore, there is a $\phi \in S_h^{t,k}(\Omega)$ such that

$$\|u-\phi\|_{H^{s+\frac{1}{2}}(\Omega)} \leq Ch^{\mu}\|u\|_{H^{\ell+\frac{1}{2}}(\Omega)} \quad , \tag{4.2.2}$$

where

$$\mu = \min(\ell-s, t-s-\frac{1}{2}) \quad , \tag{4.2.3.}$$

By theorem 2.4.1 and using (2.4.1), (2.4.2) and (2.4.3) we have for $s \geq \varepsilon > 0$

$$\|g-\dot{\phi}\|_{H^{s}(\dot{\Omega})} \leq Ch^{\mu}\|g\|_{H^{\ell}(\dot{\Omega})} \quad , \tag{4.2.4}$$

where

$$\mu = \min(\ell-s, t-s-\frac{1}{2}) \quad , \tag{4.2.5}$$

and $\dot{\phi}$ denotes the trace of the function $\phi$ on $\dot{\Omega}$. From theorem 4.1.2 and the related remark our theorem follows immediately.

We note in our theorem the traces of the system $S_h^{t,k}(\Omega)$ create the system $S_h^{t-\frac{1}{2},k-\frac{1}{2}}(\dot{\Omega})$. The question arises under what additional conditions we will obtain the system $S_h^{t,k-\frac{1}{2}}(\dot{\Omega})$. We state some results concerning the simultaneous approximations on the boundary and inside $\Omega$. We state a version of one theorem proved in [15], although a slightly more general version is also valid.

Theorem 4.2.2. If

I) (t,k)-system $S_h^{t,k}(\Omega)$ satisfies the local assumption,

II) $g \in H_0^{\ell}(\Omega)$ , $\ell$ is a non-negative integer,

then

III) for $0 \leq s \leq \min(\ell,k)$ , $s$ an integer, there exists a $\phi \in S_h^{t,k}(\Omega)$ such that $\phi$ has a compact support (and therefore $g = 0$ on $\dot{\Omega}$) and

(4.2.6) $$\|g-\phi\|_{H^s(\Omega)} \leq Ch^{\mu}\|g\|_{H^{\ell}(\Omega)}$$

holds, where

(4.2.7) $$\mu = \min(\ell-s,t-s) \quad ,$$

C does not depend on g and h, and $\phi$ generally depends on s . If the system $S_h^{t,k}(\Omega)$ is an addition (t-k) regular then $\phi$ is independent of s .

Theorem 4.2.3. If

I) $g \in H^{\ell}(\Omega)$ , $\ell > 0$ integral and $\frac{\partial^j g}{\partial n^j} = 0$ on $\dot{\Omega}$ for $j = 0,1,2,\cdots r \leq \ell - 1$ .

II) $S_h^{t,k}(\Omega)$ is a (t,k)-regular system satisfying the local assumption,

then

III) for $t \geq \ell$ , $k \geq \max(\tau+\frac{1}{2}+r,\rho)$ , $\rho, \sigma, \tau \geq 0$ , there exists a $\phi \in S_h^{t,k}(\Omega)$ such that

(4.2.8)

$$\|g-\phi\|_{H^{\rho}(\Omega)} + \sum_{j=0}^{r} h^{-\sigma+j} \|\frac{\partial^j \phi}{\partial n^j}\|_{H^{\tau}(\Omega^{\bullet})} \leq C(\varepsilon) h^{\mu-\varepsilon} \|g\|_{H^{\ell}(\Omega)} ,$$

where

(4.2.9)
$$\mu = \frac{t-\chi}{t-\rho} (\ell-\rho) , \bar{\chi} = \max(\rho, \tau + \frac{1}{2} + \sigma) .$$

and $\varepsilon > 0$ is arbitrary. $C(\varepsilon)$ does not depend on $h$ and $g$ .

The proof of theorem 4.2.3 is lengthy and will not be given here. We remark that theorem 4.2.2 and 4.2.3 can be generalizes and e.g. the containing $\varepsilon$ can be removed. We now apply the above theorem to prove a special result which shall play an important role in our later considerations.

<u>Theorem 4.2.4.</u> If

I) $g \in H^s(\Omega^{\bullet})$, $f \in H^{\ell}(\Omega)$ , $\ell$ is is a non-negative integer and $s > 3/2$ ,

II) $S_h^{t,k}(\Omega)$ is a $(t,k)$-regular system , $t \geq \ell + 2$ , $k \geq 2$ and satisfies a local assumption,

then

III) there exists a $\phi \in S_h^{t,k}(\Omega)$ such that

(4.2.10)

$$\|\Delta\phi - f\|_{H^0(\Omega)} + h^{-3/2}\|\phi - g\|_{H^0(\Omega)} \leq C(\varepsilon)\Big[h^{\mu_1-\varepsilon}\|f\|_{H^{\ell}(\Omega)} +$$

$$+ h^{\mu_2 - \epsilon} \|g\|_{H^s(\Omega^\bullet)} \Big] ,$$

where

$$\mu_1 = \ell \ , \ \mu_2 = \min(t-2, s-3/2) \ ,$$

$\epsilon > 0$ is arbitrary and $C$ does not depend on $f$, $g$ and $h$.

Proof. 1) Let $u_1$ denote the solution of the equation $\Delta u_1 = f$, $u_1 = 0$ on $\dot{\Omega}$. Then using the results introduced in chapter 3

(4.2.12) $$\|u_1\|_{H^{\ell+2}(\Omega)} \leq C\|f\|_{H^{\ell}(\Omega)} \ ,$$

and $u_1 = 0$ on $\dot{\Omega}$. Therefore using theorem 4.2.3 we can find a $\phi \in S_h^{t,k}(\Omega)$ such that

(4.2.13) $$\|\phi - u_1\|_{H^2(\Omega)} + h^{-3/2}\|\phi\|_{H^0(\dot{\Omega})} \leq$$

$$C(\epsilon) h^{\mu-\epsilon} \|u_1\|_{H^{\ell+2}(\Omega)} \ ,$$

where

(4.2.14) $$\mu = \frac{t-2}{t-2} \ell = \ell \ .$$

2) Let $u_2$ the solution of the equation $\Delta u_2 = 0$ , $u_2 = g$ on $\dot{\Omega}$ . Then using the results of chapter 3 we have

(4.2.15)
$$\|u_2\|_{H^{s+\frac{1}{2}}(\Omega)} \leq C\|g\|_{H^s(\dot{\Omega})}$$

with $s > \frac{3}{2}$ by assumption. Therefore there exists a $\phi \in S_h^{t,k}(\Omega)$ such that

(4.2.16)
$$\|\phi - u_2\|_{H^2(\Omega)} \leq Ch^{\mu_1}\|u_2\|_{H^{s+\frac{1}{2}}(\Omega)} ,$$

and

(4.2.17)
$$\|\phi - u_2\|_{H^{\frac{1}{2}+\epsilon}(\dot{\Omega})} \leq Ch^{\mu_2 - \epsilon}\|u_2\|_{H^{s+\frac{1}{2}}(\Omega)} ,$$

where

(4.2.18)
$$\mu_1 = \min(t-2, s-3/2)$$

(4.2.19)
$$\mu_2 = \min(t-\tfrac{1}{2}, s) .$$

Therefore using imbedding theorem 2.4.1 we have

(4.2.20)
$$\|\phi - u_2\|_{H^2(\Omega)} + h^{-3/2}\|\phi - g\|_{H^0(\Omega)} \leq Ch^{\mu-\varepsilon}\|u_2\|_{H^{s+\frac{1}{2}}(\Omega)}$$

where

$$\mu = \min(t-2, s-3/2) \quad . \tag{4.2.21}$$

Now the use of (4.2.12) in conjunction with (4.2.13) and (4.2.15) together with (4.2.20) yields the theorem.

Theorem 4.2.4 was proved in a general form in [13] and was used very successfully in the least square method applied to the finite element method. The general version given in [13] is slightly sharper (it is possible to have $\varepsilon = 0$).

Finally we note that in theorem 4.2.3 it is essential that the boundary of $\Omega$ is smooth. It is not clear whether theorem 4.2.4 holds for domains with piecewise smooth boundary such as a rectangle. In [13] also essential use is made in the proof of the smoothness of the boundary.

References.

[1]. I. Babuška, "A Remark to the Finite Element Method," Comment. Math. Univ. Carolinae, 12, 2 (1971), pp. 367-375. [Tech. Note BN-675 (Nov. 1970), Institute for Fluid Dynamics and Applied Mathematics, University of Maryland].

[2]. M. Zlámal, "On the Finite Element Method," Numer. Math., 12 (1968), pp. 394-409.

[3]. J. H. Bramble, M. Zlámal, "Triangular Elements in the Finite Element Method," Math. Comp. 24 (1970), pp. 809-821.

[4]. J. P. Aubin, "Evaluation des erreurs de troncature des approximations des espaces de Sobolev," J. Math. Anal. Appl., vol. 21 (1968), pp. 356-368.

[5]. J. P. Aubin, "Approximation of Elliptic Boundary-Value Problems," Wiley Interscience, New York, London, Sydney, Toronto, 1972, chapter 5.

[6]. I. Babuška, "Error-Bounds for Finite Element Method," Numer. Math., vol. 16 (1971), pp. 322-333. [Tech. Note BN-630 (Nov. 1969). The Institute For Fluid Dynamics and Applied Mathematics, University of Maryland].

[7]. I. Babuška, "Approximation by Hill Functions," Comment Math., Univ. Carolinae, 11, 4 (1970), pp. 787-811. [Tech. Note BN-648 (March 1970), The Institute for Fluid Dynamics and Applied Mathematics, University of Maryland].

[8]. G. Fix, "Fourier Analysis of the Finite Element Method in Ritz-Galerkin Theory Studies," Appl. Math. 48 (1969), pp. 265-273.

[9]. F. D. Guglielmo, "Construction d'approximations des espaces de Sobolev $H^m(R^n)$, m entier positit sur dés réseaux en simplexes". CR. Acad. de Sci. 268 (1969), pp. 314-317.

[10]. F. D. Guglielmo, "Méthode dés elements finis; une famille d'approximation des espaces de Sobolev par les translates de p functions," Calcolo 7 (1970), pp. 185-234.

[11]. G. Strang, G. Fix, "A Fourier Analysis of the Finite Element Method," Proceedings of CIME Conference Sicily 1971, Crimonese, Rome, to appear.

[12]. C de Boor, G. Fix, "Spline Approximation by Quasi-Interpolants". To appear in J. Approximation Theory.

[13]. J. H. Bramble, A. H. Schatz, "Least Squares Method for 2mth Order Elliptic Boundary-Value Problems," Math. Comp. 25, (1971), pp. 1-32.

[14]. J. Céa, "Approximation Variationnelle des problems," aux limites," Ann. Inst. Fourier 14 (1964), pp. 345-444.

[15]. I. Babuška, "Approximation by Hill Functions II," Comment. Math., Univ. Carolinae, 13, I (1972), pp. 1-22. [Tech. Note BN-708 (July 1971), Inst. for Fluid Dynamics and Appl. Math., Univ. of Maryland].

[16]. G. H. Hardy, J. E. Littlewood, G. Polya, "Inequalities," Cambridge at the University Press, 1964.

[17]. S. G. Mikhlin, "About ase unctions for the ariational- ifference ethod, (in Russian) Dokl. Akad. Nauk. SSSR, 200 (1971), pp. 526-529.

[18]. P. G. Ciarlet, P. A. Raviart, "The Combined Effects of Curved Boundaries and Numerical Integration in Isoparametric Finite Element Methods". This proceedings.

[19]. J. T. Oden, "Generalizes Conjugate Functions for Mixed Finite Element Approximations of Boundary-Value Problems. This proceedings.

[20]. R. S. Varga, "Functional Analyzis and Approximation Theory in Numerical Analysis," SIAM (Reg. Conference Series in Applied Mathematics), Philadelphia, Pennsylvania, 1971.

# CHAPTER 5. VARIATIONAL PRINCIPLES.

## 5.1. Introduction.

An important and widely used tool in the finite element method is the variational principle. In general it provides a means for discretization of the original problem. This variational principle is based on the existence of a stationary point and not on the minimization of quadratic functionals. In the sequel the note of a variational principle shall be used in this sense only.

In chapter 3 we discussed the properties of some basic boundary value problems. In general there are many different variational principles such that their stationary point is the desired solution of the given problem.

Often the variational principle is used advantageously as a tool for proving existence theorems. However, our principal aim is quite different. We shall use the variational principle primarily for the purpose of discretization of the original problem.

## 5.2 The variational principle.

We now formulate more precisely the notion of a variational principle which will play a fundamental role in all our forthcoming considerations.

Theorem 5.2.1. If

(I) $H_1$ and $H_2$ are two real Hilbert spaces with scalar product $(\cdot,\cdot)_{H_1}$ and $(\cdot,\cdot)_{H_2}$ respectively.

(II) $B(u,v)$ is a bilinear form on $H_1 \times H_2$, $u \in H_1$, $v \in H_2$ such that

$$|B(u,v)| \leq C_1 \|u\|_{H_1} \|v\|_{H_2} \quad , \tag{5.2.1}$$

$$\inf_{\substack{u \in H_1 \\ \|u\|_{H_1}=1}} \sup_{\substack{v \in H_2 \\ \|v\|_{H_2} \leq 1}} |B(u,v)| \geq C_2 > 0 \quad , \tag{5.2.2}$$

$$\sup_{u \in H_1} |B(u,v)| > 0 \ , \ v \neq 0 \tag{5.2.3}$$

where $C_1 < \infty$ .

(III) $f \in H_2'$, i.e., $f$ is a linear functional on $H_2$ . Then

(IV) There exists a unique element $u_0 \in H_1$ such that

$$B(u_0,v) = f(v) \quad , \ \forall v \in H_2 \quad ,^{1)} \tag{5.2.4}$$

$$\|u_0\|_{H_1} \leq \frac{\|f\|_{H_2'}}{C_2} \quad . \tag{5.2.5}$$

1) $\forall v$ means: for every $v$ .

Proof. The proof consists of four steps.

1) From (5.2.1) for every $u \in H_1$.

$$\Phi_u(v) = B(u,v)$$

is a linear functional on $H_2$ with the norm

$$\|\Phi_u(v)\|_{H_2} = \sup_{\substack{v \in H_2 \\ \|v\|_{H_2} \leq 1}} |B(u,v)| \leq C_1 \|u\|_{H_1} .$$

Thus we may write

(5.2.6) $$(z,v)_{H_2} = B(u,v) \quad , \quad z \in H_2 \quad ,$$

i.e., there exists a mapping $R(u)$ of $H_1$ into $H_2$ such that

(5.2.7) $$(R(u),v)_{H_2} = B(u,v) \quad ,$$

and

(5.2.8) $$\|R\|_{H_1 \to H_2} \leq C_1 \quad .$$

Clearly $R$ is linear and continuous.

2) Now we show that $R(H_1)$ is a closed set in $H_2$. We note that

$$\|R(u)\|_{H_2} = \sup_{\substack{v\in H_2\\ \|v\|_{H_2}\leq 1}} |(R(u),v)| = \sup_{\substack{v\in H_2\\ \|v\|_{H_2}\leq 1}} |B(u,v)| .$$

By (5.2.2) we have

$$\text{(5.2.9)} \qquad \|R(u)\|_{H_2} \geq C_2\|u\|_{H_1} .$$

Now let $\{R(u_n)\}$ be a Cauchy sequence in $H_2$. Then $\{u_n\}$ is a Cauchy sequence in $H_1$. In fact

$$\|R(u_n)-R(u_m)\|_{H_2} = \|R(u_n-u_m)\|_{H_2} \geq C_2\|u_n-u_m\|_{H_1} .$$

and therefore $R(H_1)$ is closed.

3) Now we show that $R(H_1) = H_2$. Suppose that this is not the case. Since $R(H_1)$ is a closed subspace of $H_2$ then there exists a $v_0 \in H_2$, $(v_0\neq 0)$ such that

$$(R(u),v_0)_{H_2} = 0 \quad , \forall v \in H_2$$

But from (5.2.3) it follows that there exists $u' \in H_1$ such that

$$(R(u'),v_0)_{H_2} = B(u',v_0) > 0 \quad ,$$

which is a contradiction. Hence $R(H_1) = H_2$.

4). We observe that the equation $R(u) = v$ has a solution for every $v \in H_2$ and (5.2.9) holds. Thus the inverse operator $R^{-1}$ exists and it is linear and continuous. Moreover,

$$\|R^{-1}\| \leq \frac{1}{C_2} \quad . \tag{5.2.10}$$

(see e.g. [1], p. 168). Now for $f \in H'_2$ we have

$$f(v) = (v_0, v)$$

with $\|v_0\|_{H_2} = \|f\|_{H'_2}$ . By choosing $u_0 = R^{-1}v_0$ , the existence assertion of the theorem follows and the uniqueness is obvious.

Now it is clear what we understand by a variational principle. It will consist of a pair of Hilbert spaces and a bilinear form with properties (5.2.1)-(5.2.3).

Conditions (5.2.2) and (5.2.3) are sometimes called the coerciveness conditions.

Our aim is to construct different variational principles corresponding to boundary value problems for partial differential equations (see chapter 3). This correspondence should be such that the element $u_0 \in H_1$ which satisfies (5.2.4) is the desired solution.

The variational principle characterizes the notion of a weak solution, together with the class of functions where uniqueness holds. Different variational principles define different kinds of weak solutions. In the case where the input data are such that the existence of the classical solution is guaranteed, we, indeed, require that the weak solution be the classical solution.

Let us indicate briefly the general procedure for checking if the assumptions of our theorem are satisfied. In applications (5.2.1) is easy to check. The main difficulty lies in checking (5.2.2). The general scheme is to construct a mapping $T_1$ which maps $H_1$ into $H_2$ and such that

$$|B(u,T_1u)| \geq C_1\|u\|_{H_1}\|T_1u\|_{H_2} ,$$

with $C_1 > 0$ . Similarly we construct a mapping $T_2$ which maps $H_2$ into $H_1$ and such that

$$|B(T_2v,v)| > 0 .$$

The above technique will be illustrated through examples in the following sections of this chapter.

Remarks. 1). Theorem 5.2.1 shall be modified later. (See chapter 6). This theorem is a generalization of the well known Lax-Milgram theorem. The theorem may easily be generalized to the case where $H_1$ and $H_2$ are reflexive Banach spaces. The method proof is an adaptation from [2] and [3] (see also [4], p. 294).

2). The formulation of the theorem is reminiscent of a Galerkin method when $H_1$ is the space of the trial functions and $H_2$ is the space of test functions (see chapter 6).

3). In most of our considerations the spaces

$H_i$ , i = 1,2 will be real spaces. However, the theorem 5.2.1 holds for complex Hilbert spaces too.

4). We would like to emphasize that we do not assume that $H_1 = H_2 = H$ , and $B(u,u) \geq C\|u\|^2$ , i.e., that the form $B(u,v)$ is positive definite. Such an assumption is very restrictive and would virtually exclude a large group of finite element methods from consideration within the present framework.

5). The condition (5.2.3) may be replaced by a more resetrive condition namely

$$(5.2.3) \qquad \inf_{\substack{v \in H_2 \\ \|v\|_{H_2}=1}} \ \sup_{\substack{u \in H_1 \\ \|u\|_{H_2} \leq 1}} |B(u,v)| \geq C\|u\|_{H_1} \quad , \quad C > 0 \ .$$

It could be shown that the existence of such a C follows from the conditions (5.2.1), (5.2.2) and (5.2.3).

## 5.3. Variational Principles for Second Order Differencial Equations.

### 5.3.1. The boundary value problem.

In chapter 3 we introduced boundary value problems for the general second order differential equations and we discussed the basic properties of their solutions. In the present chapter we shall be concerned with the study of the following problem.

$$(5.3.1) \qquad L(u) = f \quad \text{on} \quad \Omega$$

(5.3.2) $$\Gamma(u) = g \quad \text{on} \quad \dot{\Omega} \ ,$$

and its adjoint

(5.3.3) $$L^*(u) = f^* \quad \text{on} \quad \Omega$$

(5.3.4) $$\Gamma^*(u) = g^* \quad \text{on} \quad \dot{\Omega} \ ,$$

where $\Omega$ is a smooth domain and

(5.3.5)
$$L(u) = -\sum_{i,j} \frac{\partial u}{\partial x_i} a_{ij} \frac{\partial u}{\partial x_j} + \sum_i b_i \frac{\partial u}{\partial x_i} + cu$$

(5.3.6) $$\Gamma(u) = \frac{\partial u}{\partial n} + \alpha \frac{\partial u}{\partial x} + \beta u \ ,$$

(5.3.7)
$$L^*(u) = -\sum_{i,j} \frac{\partial}{\partial x_j} a_{ij} \frac{\partial u}{\partial x_i} - \sum \frac{\partial}{\partial x_i} (bu_i) + cu \ ,$$

(5.3.8) $$\Gamma^*(u) = \frac{\partial u}{\partial n} + \alpha^* \frac{\partial u}{\partial s} + \beta^* u \ ,$$

with $\alpha^*$ and $\beta^*$ given by (5.3.14) and (5.3.15) respectively. Further, we let

(5.3.9) $$p = \sum_{i,j} a_{ij} \cos(ni) \cos(nj)$$

$$p^* = \sum_{i,j} a_{ij} \cos(nj) \cos(ni) \tag{5.3.10}$$

$$q = - \sum_{i,j} a_{ij} \sin(nj) \cos(ni) \tag{5.3.11}$$

$$q^* = - \sum_{i,j} a_{ij} \sin(ni) \cos(nj) \tag{5.3.12}$$

$$\mu^* = \sum_{i,j} b_i \cos(ni) \quad , \tag{5.3.13}$$

where $\cos(ni)$ and $\sin(ni)$ denote the directional cosine and sine of the outernormal to $\dot{\Omega}$ respectively. Now we let

$$\alpha^* = \frac{q^* - \alpha_1}{p^*} \quad , \tag{5.3.14}$$

$$\beta^* = \frac{\mu^* + \beta_1 - \frac{d\alpha_1}{ds}}{p^*} \quad , \tag{5.3.15}$$

where

$$\alpha_1 = \alpha p - q \ , \ \beta_1 = p\beta \quad . \tag{5.3.16}$$

In what follows we assume the ellipticity condition

$$(5.3.17) \qquad 4\, a_{11}a_{22} - (a_{12}+a_{21})^2 \geq H > 0 \quad ,$$

on $\overline{\Omega}$ . From (5.3.17) it follows that

$$(5.3.18) \qquad \sum_{i,j} a_{ij}\xi_i\xi_j \geq H'(\xi_1^2+\xi_2^2)$$

and

$$(5.3.19) \qquad \sum_{i,j} a_{ij}\xi_j\xi_i \geq H'(\xi_1^2+\xi_2^2) \quad ,$$

with $H' > 0$ . Further, we assume that the coefficients $a_{ij}$, $b_i$, $c$, $\alpha$, $\beta$, are infinitely differentiable. Then it follows that $p$, $p^*$, $q$, $q^*$, $\mu^*$ are also infinitely differentiable and moreover,

$$(5.3.20) \qquad p \geq H' \ , \ p^* \geq H' \quad ,$$

and therefore $\alpha^*$ and $\beta^*$ are also infinitely differentiable. Problem (5.3.1) and (5.3.2), together with its adjoint (5.3.3) and (5.3.4) belong to the class of the problems we discussed in chapter 3.

We shall make the following additional assumptions.

1). The solution of (5.3.1)-(5.3.2) exists for every sufficiently smooth $f$ and $g$ (for more details see chapter 3).

2). The only smooth solution of our problem with $f = g = 0$ is the trivial solution. Then the adjoint problem has a solution for all $f^*$ and $g^*$.

### 5.3.2. The variational principle of Dirichlet type.

In this section we investigate a variational principle in the sense of section (5.2) for the problem (5.3.1) and (5.3.2) introduced in the preceding section.

Let $H_1 = H_2 = H^1(\Omega)$ and consider the bilinear form $B(u,v)$ defined by

(5.3.21)

$$B(u,v) = \int_\Omega \left[ \sum_{i,j} a_{ij} \frac{\partial u}{\partial x_j} \frac{\partial v}{\partial x_i} + \sum_i b_i \frac{\partial u}{\partial x_i} v + cuv \right] dx$$

$$+ \oint_{\dot{\Omega}} \left[ \alpha_1(s) \frac{\partial u}{\partial s} v + \beta_1(s) uv \right] ds \quad ,$$

where

(5.3.22)
$$\alpha_1 = p\alpha - q$$

(5.3.23)
$$\beta_1 = p\beta \quad ,$$

($\alpha$ and $\beta$ are functions appearing in (5.3.6); $p$ and $q$ are given by (5.3.9) and (5.3.11) respectively.) We introduce the functionals:

(5.3.24)

$$F(u) = \int_{\Omega^{\bullet}} fudx + \oint_{\Omega^{\bullet}} pguds \quad ,$$

$$(5.3.25) \qquad F^*(v) = \int_{\Omega} f^*vdx + \int_{\Omega^{\bullet}} p^*g^*vds \quad .$$

For $f,g$ and $f^*,g^*$ sufficiently smooth there exist functions $u_0$ and $v_0$ which are the solutions of problems (5.3.1)-(5.3.2) and (5.3.3)-(5.3.4) respectively (see chapter 3). Now we show that for every $v \in E(\overline{\Omega})$ , we have

$$(5.3.26) \qquad B(u_0,v) = F(v) \quad ,$$

and for every $u \in E(\overline{\Omega})$

$$(5.3.27) \qquad B(u,v_0) = F^*(v) \quad .$$

Actually (5.3.26) and (5.3.27) hold then for $v \in H^1(\Omega)$ and $u \in H^1(\Omega)$ respectively. This is an immediate consequence of the density of $E(\overline{\Omega})$ in $H_1(\Omega)$ and the continuity of $B(u,v)$ , $F$ and $F^*$ . We obtain by integrating by parts

(5.3.28)

$$B(u_0,v) = \int_{\Omega} \left[ -\sum_{i,j} \frac{\partial}{\partial x_i} \quad a_{ij} \frac{\partial u_0}{\partial x_j} + \sum_i b_i \frac{\partial u_0}{\partial x_i} + cu_0 \right] vdx$$

((5.3.28) continued)

$$+ \oint_{\dot{\Omega}} \left[ T(u_0) + \alpha_1 \frac{\partial u_0}{\partial s} + \beta_1 u_0 \; vds \right]$$

where $T(u_0)$ is given by

$$T(u_0) = \sum_{i,j} a_{ij} \frac{\partial u_0}{\partial x_j} \cos(ni) \; . \tag{5.3.29}$$

Since

$$\frac{\partial u_0}{\partial x_i} = \frac{\partial u_0}{\partial n} \cos(ni) - \frac{\partial u_0}{\partial s} \sin(ni) \tag{5.3.20}$$

(n is the outernormal to $\dot{\Omega}$), we may write

$$T(u_0) = \frac{\partial u_0}{\partial n} \sum_{i,j} a_{ij} \cos(ni) \cos(nj) \tag{5.3.31}$$

$$+ \frac{\partial u_0}{\partial s} \sum_{i,j} - a_{ij} \sin(nj) \cos(ni) = p \frac{\partial u_0}{\partial n} + q \frac{\partial u_0}{\partial s}$$

Hence,

(5.3.32)

$$T(u_0) + \alpha_1 \frac{\partial u_0}{\partial s} + \beta_1 u_o = p \frac{\partial u_0}{\partial n} + (\alpha_1 + q) \frac{\partial u_0}{\partial s} + \beta_1 u_0$$

((5.3.32) continued)

$$= p\left[\frac{\partial u_0}{\partial n} + \frac{\alpha_1+q}{p}\frac{\partial u_0}{\partial s} + \frac{\beta_1 u_0}{p}\right] = p\Gamma(u_0) \quad .$$

Consequently, we have

$$(5.3.33) \qquad B(u_0,v) = \int_\Omega fvdx + \oint pgvds = F(v) \quad .$$

Thus (5.3.26) is proved.

Now we consider the expression $B(u,v_0)$ which after integration by parts yields

(5.3.34)

$$( B(u,v_0) = \int_\Omega \left[ -\sum_{i,j} \frac{\partial}{\partial x_j} a_{ij} \frac{\partial v_0}{\partial x_i} - \sum_i \frac{\partial}{\partial x_i} (b_i v_0) + cv_0 \right] udx$$

$$+ \oint_{\dot{\Omega}} \left[ T^*(v_0) - \frac{\partial}{\partial s} (\alpha_1 v_0) + \beta_1 v_0 \right] uds \quad ,$$

where $T^*(v_0)$ is given by

(5.3.35)

$$T^*(v_0) = \sum_{i,j} a_{ij} \frac{\partial v_0}{\partial x_i} \cos(nj) + \sum_i b_i v_0 \cos(ni) \; .$$

From (5.3.30) it follows that

$$(3.3.36) \qquad T^*(v_0) = \frac{\partial v_0}{\partial n} p^* + q^* \frac{\partial v_0}{\partial s} + \mu^* v_0 \quad .$$

Thus we have:

(5.3.37)

$$\oint_{\dot\Omega} \left[ T^*(v_0) - \frac{\partial}{\partial s}(\alpha_1 v_0) + \beta_1 v_0 \right] u\,ds =$$

$$= \oint p^* \left[ \frac{\partial v_0}{\partial n} + \frac{q^* - \alpha_1}{p^*} \frac{\partial v_0}{\partial s} + \frac{\beta_1 + \mu^* \frac{d\alpha_1}{ds}}{p^*} v_0 \right] u\,ds$$

$$= \oint p^* \Gamma^*(v_0)\,ds \quad .$$

and therefore,

(5.3.38) $$B(u, v_0) = F^*(u) \quad , \forall u \in E(\bar\Omega) \quad .$$

Next we show that the bilinear form (5.3.21) satisfies the conditions (5.2.1), (5.2.2) and (5.2.3). The condition (5.2.1) is obviously satisfied, for the first term in (5.3.21). By Sobolev's imbedding theorem we have

$$\|u\|_{H^{\frac{1}{2}}(\dot\Omega)} \leqslant C \|u\|_{H^1(\Omega)} \quad ,$$

and therefore,

$$\left\| \frac{\partial u}{\partial s} \right\|_{H^{-\frac{1}{2}}(\dot\Omega)} \leqslant C \|u\|_{H^1(\Omega)} \quad .$$

Hence, the second term in (5.3.21) satisfies the continuity condition. Next we propose to show that the inequality (5.2.2) is satisfied. To this end we first show the existence of a (sufficiently large) positive constant $\gamma$ such that

(5.3.39)

$$\left|\int_{\Omega}\left[\sum_{i,j} a_{ij}\frac{\partial u}{\partial x_i}\frac{\partial u}{\partial x_j} + \sum_i b_i \frac{\partial u}{\partial x_i} u + \gamma u^2\right]dx + \right.$$

$$\left. + \int_{\dot{\Omega}}\left[\alpha_1 \frac{\partial u}{\partial s} u + \beta_1 u^2\right] ds \right| \geq C\|u\|_{H^1(\Omega)} .$$

By the use of well known arithmetic-geometric mean, for arbitrary $\varepsilon > 0$ we have

(5.3.40)

$$\int_{\Omega}\left[\sum_i b_i \frac{\partial u}{\partial x_i} u\right]dx \leq \varepsilon \|u\|_{H^1(\Omega)} + \Lambda_1(\varepsilon)\|u\|_{H^\circ(\Omega)} ,$$

where $\Lambda_1(\varepsilon)$ is a sufficiently large number depending on $\varepsilon$. We further note that

(5.3.41)

$$\oint_{\dot{\Omega}} \alpha_1 \frac{\partial u}{\partial s} u ds = \frac{1}{2}\oint_{\dot{\Omega}} \alpha_1 \frac{\partial}{\partial s}(u^2) ds = -\frac{1}{2}\oint_{\dot{\Omega}} \alpha_1' u^2 ds .$$

Therefore,

(5.3.42)

$$J = \oint_{\dot{\Omega}} \left[\alpha_1 \frac{\partial u}{\partial s} u + \beta_1 u^2\right] ds = \oint_{\dot{\Omega}} \left[\beta_1 - \frac{1}{2}\alpha_1'\right] u^2 ds \quad .$$

Thus,

(5.3.43) $$|J| \leq C\|u\|^2_{H^\circ(\dot{\Omega})} \leq C\|u\|^2_{H^{3/4}(\Omega)} \quad .$$

By Lions' lemma (see [13] and [3], p. 108)[1] for arbitrary $\varepsilon > 0$ we have

(5.3.44) $$|J| \leq \varepsilon\|u\|_{H^1(\Omega)} + \Lambda_2(\varepsilon)\|u\|_{H^\circ(\Omega)} \quad .$$

Hence,

---

1) The lemma is as follows. Let $B_i$ , $i = 1,2,3$ be Banach spaces and let $B_1 \subset B_2 \subset B_3$ hold in algebraic and topological sense. Further, let the imbedding of $B_1$ into $B_2$ be compact. Then for every $\varepsilon > 0$ there exists $\Lambda(\varepsilon)$ such that for $u \in B_1$ we have

$$\|u\|_{B_2} \leq \varepsilon\|u\|_{B_1} + \Lambda(\varepsilon)\|u\|_{B_3} \quad .$$

(5.3.45)

$$\int_\Omega \left[ \sum_{i,j} a_{ij} \frac{\partial u}{\partial x_i} \frac{\partial u}{\partial x_j} + \sum_i b_i \frac{\partial u}{\partial x_i} u + \gamma u^2 \right] dx +$$

$$\oint_{\dot\Omega} \left[ \alpha_1 \frac{\partial u}{\partial s} u + \beta_1 u^2 \; ds \right] \geq \int_\Omega \sum_{i,j} a_{ij} \frac{\partial u}{\partial x_i} \frac{\partial u}{\partial x_j} +$$

$$+ \gamma u^2 - 2\varepsilon \|u\|^2_{H^1(\Omega)} - \left[ \Lambda_1(\varepsilon) + \Lambda_2(\varepsilon) \right] \|u\|^2_{H^\circ(\Omega)} .$$

Therefore the choice of $\varepsilon \leq \frac{1}{4} H'$ yields

$$\gamma \geqq \left[ \Lambda_1(\varepsilon) + \Lambda_2(\varepsilon) \right] + \frac{1}{2} H' ,$$

where $H'$ is given by (5.3.18). Hence, (5.3.39) holds with $C = \frac{1}{2} H'$ . Let $u \in E(\overline{\Omega})$ be given. Now we construct a $v \in H'(\Omega)$ of the form $v = u + z$ such that

(5.3.46) $$|B(u,v)| \geq C\|u\|_{H^1(\Omega)} \|v\|_{H^1(\Omega)}$$

with $C > 0$ independent of $u$ and $v$ with $z$ determined later. We note that since $E(\overline{\Omega})$ is dense in $H^1(\Omega)$ it is sufficient to assume that $u \in E(\Omega)$ . We have

(5.3.47) $$|B(u,v)| = |Q(u,u) + B(u,z) + R(u,u)| ,$$

where

(5.3.48)

$$Q(u,u) = \int_{\Omega} \left[ \sum_{i,j} a_{ij} \frac{\partial u}{\partial x_i} \frac{\partial u}{\partial x_j} + \sum_i b_i \frac{\partial u}{\partial x_i} u + \gamma u^2 \right] dx$$

$$+ \oint_{\dot{\Omega}} \left[ \alpha_1 \frac{\partial u}{\partial s} u + \beta_1 u^2 \right] ds$$

(5.3.49)

$$R(u,u) = \int_{\Omega} (c-\gamma) u^2 dx \quad .$$

We have already shown that

(5.3.50)

$$Q(u,u) \geq C \|u\|^2_{H^1(\Omega)} \quad .$$

Now let $z$ be such that

$$L^*(z) = -(c-\gamma)u$$

(5.3.51)

$$\Gamma^*(z) = 0 \quad .$$

Such a $z$ exists (see chapter 3 and section 5.3.1), and we have

(5.3.52)

$$\|z\|_{H^3(\Omega)} \leq C \|u\|_{H^1(\Omega)} \quad .$$

On the other hand we have shown that

$$(5.3.53) \qquad B(w,z) = -\int_{\Omega} (c-\gamma)uw\,dx \ , \ \forall w \in H^1(\Omega) \ .$$

Now the choice $w = u$ yields

$$(5.3.54) \qquad B(u,z) = -R(u,u) \ .$$

Therefore,

$$(5.3.55) \qquad B(u,v) \geq Q(u,u) \geq C\|u\|^2_{H^1(\Omega)} \ .$$

From (5.3.51) we see that

$$(5.3.56) \qquad \|v\|_{H^1(\Omega)} \leq C\|u\|_{H^1(\Omega)} \ ,$$

and consequently (5.2.2) holds. By reversing the role of $u$ and $v$ in the above argument we obtain the inequality

$$\sup_{u \in H_1} |B(u,v)| \geq C\|u\|_{H_1} \|v\|_{H_1} \ , \ C > 0 \ ,$$

which implies (5.2.3). Thus we have proved the following theorem.

Theorem 5.3.2. The bilinear form (5.3.21) on $H^1(\Omega) \times H^1(\Omega)$ associated with problem (5.3.1)-(5.3.2) satisfies the condition (5.2.1)-(5.2.3) provided the assumptions in § 5.3.1 are fulfilled. Further, we observe that if $f$ and $g$ are such that the existence of the smooth solution $u_0$ is guaranteed, then this solution satisfies (5.3.26) with $F(u)$ defined by (5.3.24) for every $v \in H^1(\Omega)$.

The variational principle gives an extension of the classical solution to the generalized one. Furthermore, it characterizes the class of functions where we have uniqueness, i.e., given a $u \in H^1(\Omega)$ we may check whether $u$ is a solution by merely checking (5.3.26).

Obviously, since the continuous dependence on $f$ and $g$ is guaranteed by theorem 5.2.1, our generalized solution has the regularity described in chapter 3.

### 5.3.3. The variational principle of least square type.

For the sake of the clarity of exposition and to avoid purely technical difficulties we shall limit ourselves to a special case of the problem (5.3.1)-(5.3.2), namely we shall study the problem

$$L(u) = -\Delta u + cu = f \quad \text{on} \quad \Omega \tag{5.3.57}$$

$$\Gamma(u) = \frac{\partial u}{\partial n} + \beta u = g \quad \text{on} \quad \dot{\Omega} \tag{5.3.58}$$

Let us emphasize that this restriction is merely for technical convenience and the following considerations in a obvious analogous manner can be applied to the general case.

We do not assume that $c > 0$ and $\beta > 0$. However, we do assume the existence and uniqueness of the solution as in section 5.3.1. First we define the bilinear form

$$(5.3.59) \qquad B(u,v) = \int_{\Omega} (-\Delta u + cu)(-\Delta v + cv)\,dx$$

$$+ \oint_{\dot{\Omega}} \left(\frac{\partial u}{\partial n} + \beta u\right)\left(\frac{\partial v}{\partial n} + \beta v\right) ds \quad .$$

Now the main question is the choices of the spaces $H_1$ and $H_2$ in Theorem 5.2.1. Obviously the choice $H_1 = H^1(\Omega) = H_2$ will hurt the condition (5.2.1). We may try $H_1 = H_2 = H^2(\Omega)$ . Then obviously (5.2.1) holds. Unfortunately with this choice (5.2.2) is in trouble. To see this let $K^2(\Omega)$ be the space of all function $u \in H^2(\Omega)$ such that $-\Delta u + cu = 0$ . Obviously, for $u \in K^2(\Omega)$ we have

$$(5.3.60) \qquad B(u,v) = \oint \left(\frac{\partial u}{\partial n} + \beta u\right)\left(\frac{\partial v}{\partial n} + \beta v\right) ds \quad .$$

We show that

$$(5.3.61) \qquad \inf_{\|u\|_{H^2(\Omega)}=1} \ \sup_{\|v\|_{H^2(\Omega)}\leq 1} B(u,v) = 0 \quad .$$

Clearly from (5.3.60) we have

$$(5.3.62) \qquad |B(u,v)| \leq \left\|\frac{\partial u}{\partial n} + \beta u\right\|_{H^{-\frac{1}{2}}(\dot{\Omega})} \left\|\frac{\partial v}{\partial n} + \beta v\right\|_{H^{\frac{1}{2}}(\dot{\Omega})}$$

((5.3.62) continued)

$$\leq C\left\|\frac{\partial u}{\partial n} + \beta u\right\|_{H^{-\frac{1}{2}}(\Omega^{\bullet})} \|v\|_{H^2(\Omega)} ,$$

by the imbedding theorem.

It is very easy to see that there exists a sequence $u_i \in K^2(\Omega)$ such that

$$\|u_i\|_{H^2(\Omega)} = 1 \quad \text{and} \quad \|u_i\|_{H^1(\Omega)} \to 0 .$$

Since $u_i$ satisfies the differential equation $-\Delta u + cu = 0$, we can use theorem 2.4.2 (resp. its counterpart for the above differential equation) and obtain

$$\left\|\frac{\partial u_i}{\partial n} + \beta u_i\right\|_{H^{-\frac{1}{2}}(\Omega^{\bullet})} \leq C\|u_i\|_{H^1(\Omega)}$$

and therefore (5.3.61) is proved.

Now the main problem is to find an appropriate variational principle in the sense defined earlier. In this connection the question arises as to what is the right choice of the spaces under these circumstances.

We now explore one possibility. Let

$$H_1(\Omega) = H_2(\Omega) = \mathcal{H}(\Omega) ,$$

where for $u \in \mathcal{E}(\overline{\Omega})$ we define

(5.3.63)

$$\|u\|^2_{H(\Omega)} = \|-\Delta u+cu\|^2_{L_2(\Omega)} + \|u\|^2_{H^{3/2}(\Omega)} ,$$

and $H(\Omega)$ is the closure of $E(\bar{\Omega})$ with respect to (5.3.63). From the imbedding theoresm (see chapter 2) we know that the trace of $\frac{\partial u}{\partial n}$ exists and $\frac{\partial u}{\partial n} \in H^\circ(\dot{\Omega})$ [1], and therefore the bilinear form $B(u,v)$ satisfies (5.2.1).

We now check (5.2.2). Let $u \in H(\Omega)$. It is sufficient to prove (5.2.2) for $u \in H(\Omega) \cap E(\bar{\Omega})$.

(5.3.64) $$B(u,u) = \int_\Omega (-\Delta u+cu)^2 dx + \oint_{\dot{\Omega}} (\frac{\partial u}{\partial n} + \beta u)^2 ds .$$

We proceed to show that

(5.3.65) $$B(u,u) \geq C\|u\|^2_{H(\Omega)} , \quad C > 0 .$$

Assume to the contrary that (5.3.65) does not hold. Then there exists a sequence $u_n \in E(\bar{\Omega})$ such that $\|u_n\|_{H(\Omega)} = 1$ and $B(u_n,u_n) \to 0$ as $n \to \infty$. By the hypothesis we know that $u_n$ is a solution of the boundary value problem

---

1) We are using the fact that $u = u_1 + u_2$ where

$$u_1 \in H^2(\Omega) , \quad \|u_1\|_{H^2(\Omega)} \leq C\|u\|_{H(\Omega)}$$

and

$$- \Delta u_2 + u_2 = 0 \quad \text{with} \quad \|u_2\|_{H^{3/2}(\Omega)} \leq C\|u\|_{H(\Omega)}$$

$$-\Delta u_n + cu_n = f \quad ,$$

$$\frac{\partial u_n}{\partial n} + \beta u_n = g \quad ,$$

where by imbedding theorems as above we have

$$\|f\|_{H^\circ(\Omega)} = \|\Delta u_n + cu_n\|_{H^\circ(\Omega)} \leq \|u_n\|_{\mathcal{H}(\Omega)} \quad ,$$

and

$$\|g\|_{H^\circ(\dot{\Omega})} \leq \left\|\frac{\partial u_n}{\partial n} + \beta u_n\right\|_{H^\circ(\dot{\Omega})} \leq C\|u_n\|_{\mathcal{H}(\Omega)} \quad .$$

Since $u_n \in \mathcal{E}(\overline{\Omega})$ and $u_n$ is unique by assumption, we have

$$\|u_n\|_{H^{3/2}(\Omega)} \leq C\left[\|f\|_{L_2(\Omega)} + \|g\|_{H^0(\dot{\Omega})}\right] \quad ,$$

and consequently

$$\|u_n\|_{\mathcal{H}(\Omega)} \leq CB(u_n,u_n)$$

which is the desired contradiction. By reversing the role of $u$ and $v$ a similar argument yields (5.2.3). Thus we have proved the theorem.

<u>Theorem 5.3.3</u>. Let $H_1 = H_2 = \mathcal{H}(\Omega)$ and let $B(u,v)$ be defined by (5.3.58), then conditions (5.2.1)-(5.2.3) of theorem 5.2.1 are satisfied.

Now it is clear what is the variational principle. Intuitively we search for the minimum of the functional

$$\int_\Omega (-\Delta u+cu-f)^2 dx + \int_{\dot\Omega} \left(\frac{\partial u}{\partial n} +\beta u-g\right)^2 ds \quad .$$

The existence of the minimum is equivalent with the existence of $u_0$ such that

$$B(u_0,v) = \int_\Omega (-\Delta v+cv) f ds + \int_{\dot\Omega} \left(\frac{\partial v}{\partial n} +\beta v\right) g ds \quad .$$

Clearly, for $f \in L_2(\Omega)$ and $g \in L_2(\dot\Omega)$ , we have

(5.3.66)

$$F(v) = \int_\Omega (-\Delta v+cv) f dx + \int_{\dot\Omega} \left(\frac{\partial v}{\partial n} +\beta v\right) g ds \in H'(\Omega) \quad ,$$

and therefore it is clear how to apply theorem 5.2.1.

5.3.4. <u>The Problems with the boundary condition of the second type</u>.

In section 5.3.1 we considered the problem with boundary condition of the first type. The presence of $\frac{\partial u}{\partial n}$ in the boundary 5.3.6 was a characteristic of this case. If we do not have the term $\frac{\partial u}{\partial n}$ present in the boundary condition, we cannot have the term $\frac{\partial u}{\partial s}$ present in the boundary condition either. Otherwise the boundary value problem considered would lie outside the framework of our theory as described in chapter 3. Therefore we introduce the boundary value problem

(5.3.1) $$Lu = f \quad \text{on} \quad \Omega$$

(5.3.67) $$\Gamma_2(u) = u = g \quad \text{on} \quad \dot{\Omega} \quad ,$$

and its adjoint problem

(5.3.2) $$L^*(u) = f^* \quad \text{on} \quad \Omega \quad ,$$

(5.3.68) $$\Gamma_2^*(u) = \Gamma_2(u) = u = g^* \quad ,$$

where the operators $L$ and $L^*$ are defined by (5.3.5) and (5.3.7) respectively.

We will assume that the coefficients of the operator $L$ satisfy the same hypothesis as those in section (5.3.1). We will also assume that the problem (5.3.1), (5.3.67) has solutions for all sufficiently smooth $f$ and $g$ and that the only solution for $f = 0 = g$ is the trivial solution. Thus we know (see the results in chapter 3) that the problem (5.3.2),(5.3.67) has also smooth solutions for all $f^*$ and $g^*$ (sufficiently smooth).

5.3.5. <u>The variational principle of the Dirichlet type-boundary conditions of the second type</u>.

We suppose there exists a $\phi \in H^1(\Omega)$ such that $\phi = g$ on $\dot{\Omega}$ . A necessary and sufficient condition for the existence of such a $\phi$ is that $g \in H^{\frac{1}{2}}(\dot{\Omega})$.

Now let $H_1 = H_2 = H_0^1(\Omega)$ and define

(5.3.69)

$$B(u,v) = \int_\Omega \left[ \sum_{i,j} a_{ij} \frac{\partial u}{\partial x_j} \frac{\partial v}{\partial x_i} + \sum_i b_i \frac{\partial u}{\partial x_i} v + cuv \right] dx \quad .$$

Let

(5.3.70)

$$F(v) = \int_\Omega fvdx + \int_\Omega \left[ \sum_{i,j} a_{ij} \frac{\partial \phi}{\partial x_j} \frac{\partial v}{\partial x_i} \sum_i b_i \frac{\partial \phi}{\partial x_i} vc\phi v \right] dx \quad .$$

Obviously, for $f \in H^\circ(\Omega)$ and $\phi \in H_0^1(\Omega)$, the functional $F(v)$ is continuous on $H_0^1(\Omega)$. By employing the argument used in section (5.3.2) one concludes that the bilinear form (5.3.69) satisfies the conditions (5.2.1), (5.2.2) and (5.2.3). Therefore there exists a unique $u_0$ such that

(5.3.71) $$B(u_0,v) = F(v) \quad , \forall v \in H_0^1(\Omega) \quad .$$

If $f$, $g$ and $\phi$ are sufficiently smooth, then $u_0$ will be the classical smooth solution of

$$L(u_0) = f + L(\phi) \quad \text{on} \quad \Omega \quad ,$$

and thus $u_0 + \phi$ is the solution of our original problem.

We note that it is very important that $B(u,v)$ be defined on $H_0^1(\Omega) \times H_0^1(\Omega)$, i.e., the function $u$ and $v$

must satisfy the homogeneous boundary conditions. Such a restriction has serious practical disadvantages. In view of the above remark we search for another variational principle.

5.3.6. The variational principle of the Dirichlet type with Lagrange multipliers-Boundary conditions of of the second type.

If $b_i = 0$ and $c \geq 0$, it is well known that the condition

$$B(\bar{u}_0+\phi_1,v) = f(v) \quad , \forall v \in H_0^1(\Omega) \quad ,$$

where $B(u,v)$ is given by (5.3.69) $\phi = g$ on $\dot{\Omega}$ and $\bar{u}_0 \in H_0^1(\Omega)$ is equivalent with the condition that $u_0$ minimizes the quadratic functional $B(u,u) - 2f(u)$ over all functions $u \in H^1(\Omega)$ such that $u$ satisfies the boundary condition $u = g$ on $\dot{\Omega}$.

Formally we may interpret this as a minimization of the above quadratic form with the constraint $u = g$ on $\dot{\Omega}$. This interpretation suggests the possibility of approaching the problem by Lagrange multipliers technique.

In the foregoing discussion we shall closely follow the method described in [6] for a model problem.

Let

$$H_1 = H^1(\Omega) \times H^{-\frac{1}{2}}(\dot{\Omega}) = H_2 \quad . \tag{5.3.72}$$

For $(u,\lambda) \in H_1$, we define

$$\|(u,\lambda)\|^2_{H_1(\Omega)} = \|u\|^2_{H^1(\Omega)} + \|\lambda\|^2_{H^{-\frac{1}{2}}(\dot{\Omega})} ,$$

(similarly for $(v,\mu) \in H_2$). On $H_1 \times H_2$ we define the bilinear form

(5.3.73)

$$B((u,\lambda),(v,\mu)) = \int_\Omega \left[\sum_{i,j} a_{ij} \frac{\partial u}{\partial x_j} \frac{\partial u}{\partial x_j} + \sum_i b_i \frac{\partial u}{\partial x_i} v + cuv\right] dx$$

$$- \oint_{\dot{\Omega}} [u\mu + \lambda v] ds .$$

Define further

(5.3.74) $$F((v,\mu)) = \int_\Omega fv dx - \oint_{\dot{\Omega}} g\mu ds .$$

Let $u_0$ be the classical smooth solution of the problem (5.3.1), (5.3.67), (i.e., $u_0$ is twice continuously differentiable on $\overline{\Omega}$). Let us take $\lambda_0 = T(u_0)$ , where $T(u_0)$ is given by (5.3.29) and show that

(5.3.75) $$B((u_0,\lambda_0),(v,\mu)) = F(v,\mu) ,$$

for every $v \in E(\overline{\Omega})$ , $\mu \in E(\dot{\Omega})$ and $F(v,\mu)$ given by

(5.3.74). In fact by integrating by parts (5.3.73), we obtain (see (5.3.28))

$$(5.3.76)\quad B((u_0,\lambda_0),(\mu,v)) = \int_\Omega \Big[-\sum_{i,j} \frac{\partial}{\partial x_i} a_{ij} \frac{\partial u_0}{\partial x_j} + \sum_i b_i \frac{\partial u_0}{\partial x_i} + cu_0\Big] vdx + \oint_{\dot\Omega} \big[T(u_0) - \lambda_0\big] vdx - \oint_{\dot\Omega} u_0 \mu ds = \int fvdx - \oint_{\dot\Omega} g\mu ds = F((v,\mu)) \quad .$$

Next we show that the conditions (5.2.1), (5.2.2) and (5.2.3) are satisfied. First note that condition (5.2.1) is obviously satisfied. To show that (5.2.2) is fulfilled, we can find as in section (5.3.2) a positive constant $\gamma$ such that the auxiliary bilinear form

(5.3.77)

$$B_1(u,v) = \int_\Omega \Big[\sum_{i,j} a_{ij} \frac{\partial u}{\partial x_j} \frac{\partial v}{\partial x_i} + \sum_i b_i \frac{\partial u}{\partial x_i} + \gamma uv\Big] dx$$

defined on $H^1(\Omega) \times H^1(\Omega)$ satisfies the inequality

(5.3.78) $$B_1(u,u) \geq C\|u\|^2_{H^1(\Omega)} \quad , \quad C > 0 .$$

Clearly this bilinear form satisfies all the assumptions of the theorem 5.2.1. Let $g \in H^{-\frac{1}{2}}(\Omega)$ . Then obviously

(5.3.79) $$F(u) = \int_{\Omega^\bullet} guds \in (H^1(\Omega))' \equiv H^{-1}(\Omega) \quad .$$

with

$$\|F(u)\|_{(H^1(\Omega))'} \leq C\|g\|_{H^{-\frac{1}{2}}(\Omega^\bullet)} \quad .$$

Consequently there exists a $z_0 \in H^1(\Omega)$ such that

(5.3.80) $$B_1(u,z_0) = F(u) \, \forall u \in H^1(\Omega) \quad ,$$

and

(5.3.81) $$\|z_0\|_{H^1(\Omega)} \leq C\|g\|_{H^{-\frac{1}{2}}(\Omega^\bullet)} \quad .$$

Now we prove a lemma.

Lemma 5.3.1. Let $z_0$ the element satisfying the condition (5.3.80) then

(5.3.82) $$F(z_0) = \int_{\dot{\Omega}} z_0 g ds \geq C\|g\|^2_{H^{-\frac{1}{2}}(\dot{\Omega})} .$$

Proof. We have by definition

(5.3.83) $$\|g\|_{H^{-\frac{1}{2}}(\dot{\Omega})} = \sup_{v \in H^{\frac{1}{2}}(\dot{\Omega})} \frac{\left|\int_{\dot{\Omega}} gvds\right|}{\|v\|_{H^{\frac{1}{2}}(\dot{\Omega})}} .$$

By the inverse imbedding theorem (see theorem 2.4.3) there exists a bounded linear mapping $T$ of $H^{\frac{1}{2}}(\dot{\Omega})$ into $H^1(\Omega)$ with

$$v = T(v) \quad \text{on} \quad \dot{\Omega} .$$

Thus we have

$$\left|\int_{\dot{\Omega}} gvds\right| = |B_1(Tv, z_0)| \leq C\|v\|_{H^{\frac{1}{2}}(\dot{\Omega})} \|z_0\|_{H^1(\Omega)}$$

and therefore

(5.3.84) $$\frac{\left|\int_{\dot{\Omega}} gvds\right|}{\|v\|_{H^{\frac{1}{2}}(\dot{\Omega})}} \leq C\|z_0\|_{H^1(\Omega)} .$$

On the other hand we have

$$(5.3.85) \qquad \oint_{\dot{\Omega}} g z_0 ds = B_1(z_0, z_0) \geq C\|z_0\|^2_{H^1(\Omega)}$$

Therefore

$$(5.3.86) \qquad \frac{\left|\int_{\dot{\Omega}} g v ds\right|}{\|v\|_{H^{\frac{1}{2}}(\dot{\Omega})}} \leq C\left[\int_{\dot{\Omega}} g z_0 ds\right]^{\frac{1}{2}}$$

and (5.3.82) is proved. Now let $u \in H^1(\Omega)$ and $\lambda \in H^{-\frac{1}{2}}(\dot{\Omega})$ be given, then

$$(5.3.87) \qquad B((u,\lambda),(u-z_1+z_2,\mu)) =$$

$$B_1(u,u) - B_1(u,z_1) + \int_{\Omega}(c-\gamma)u(u-z_1)dx + B((u,0)(z_2,0))$$

$$- \oint_{\dot{\Omega}} \left[u\mu + u\lambda + z_2\lambda - \lambda z_1\right] ds \quad .$$

By selecting $z_1$ such that

$$(5.3.88) \qquad B_1(v,z_1) = \oint_{\dot{\Omega}} \lambda v ds \ \forall v \in H^1(\Omega)$$

we have

(5.3.89) $$\|z_1\|_{H^1(\Omega)} \leq C\|\lambda\|_{H^{-\frac{1}{2}}(\dot{\Omega})} \quad .$$

Further, let $z_2$ be the solution of the adjoint problem

$$L^*(z_2) = -\ (c-\gamma)(u-z_1) \quad ,$$

$$\Gamma^*(z_2) = z_2 = 0 \quad \text{on} \quad \dot{\Omega} \quad .$$

By the hypothesis of section 5.3.4, we have

(5.3.90)

$$\|z_2\|_{H^3(\Omega)} \leq C\left[\|u\|_{H_1} + \|z_1\|_{H_1}\right] \leq C\|(u,\lambda)\|_{H_1(\Omega)} \quad .$$

Hence,

(5.3.91) $$B((u,0),(z_2,0)) = -\int_\Omega u(u-z_1)(c-\gamma)dx + \oint_{\dot{\Omega}} T^*(z_2)u\,ds$$

and therefore,

(5.3.92)

$$B((u,\lambda),(u-z_1+z_2,\mu)) = B_1(u,u) + \oint_{\dot{\Omega}} \lambda z_1 ds -$$

$$- \oint_{\dot{\Omega}} \left[u\mu + 2u\lambda + T^*(z_2)u\right] ds \quad .$$

Now by taking $\mu = -2\lambda - T^*(z_2)$ and using lemma 5.3.1 we

$$(5.3.93) \qquad B((u,\lambda)(u-z_1+z_2,\mu)) \geq C\|(u,\lambda)\|^2_{H_1} \quad .$$

On the other hand from (5.3.89) and (5.3.90) it follows that

$$(5.3.94) \qquad \|u-z_1+z_2\|_{H^1(\Omega)} \leq C\|(u,\lambda)\|_{H_1(\Omega)} \quad .$$

Consequently,

$$\|(u-z_1+z_2,\mu)\|_{H_1} \leq C\|(u,\lambda)\|_{H_1} \quad .$$

and (5.2.2) holds for our bilinear form.

Analogously the inequality (5.2.3) may be proved. Our results may be summarized as

<u>Theorem 5.3.2</u>. Let the bilinear form $B((u,\lambda),(v,\mu))$ be defined by (5.3.73) on $H_1 \times H_2$ , where $H_1$ and $H_2$ are defined by (5.3.72). Then $B((u,\lambda),(v,\mu))$ satisfies the conditions (5.2.1)-(5.2.3) of the theorem 5.2.1. Moreover, if $f$ and $g$ are such that the classical smooth solution $u_0$ of our problem exists then $\lambda_0 = T(u_0)$ , with $T$ given by (5.3.29) and the pair $(u_0,\lambda_0)$ satisfies (5.3.75) for all $v \in H^1(\Omega)$ and $\mu \in H^{-\frac{1}{2}}(\dot{\Omega})$ .

## 5.3.7. Variational principle of Dirichlet type Revisited.

We consider once again the problem (5.3.1)-(5.3.2) introduced in section 5.3.1. For clarity of exposition and technical convenience we confine our discussion to the model problem

(5.3.95) $$-\Delta u + u = f \quad \text{in} \quad \Omega$$

(5.3.96) $$\frac{\partial u}{\partial n} = g \quad \text{on} \quad \dot{\Omega} \quad .$$

Our considerations for the above special case can be extended by a similar technique as used in section 5.3.2 to deal with the general case. The bilinear form (5.3.21) obviously assumes the form

(5.3.97)
$$B(u,v) = \int_{\Omega} \left[ \frac{\partial u}{\partial x_1} \frac{\partial v}{\partial x_1} + \frac{\partial u}{\partial x_2} \frac{\partial v}{\partial x_2} + uv \right] dx$$

and the functionals (5.3.24) and (5.3.25) take the form

(5.3.98) $$F(v) = \int_{\Omega} fvdx + \oint_{\dot{\Omega}} gvds \quad .$$

It was shown in the theorem 5.3.2 that the bilinear form (5.3.97) defined on $H_1 \times H_2$ with $H_1 = H_2 = H^1(\Omega)$ satisfies all the conditions of theorem 5.2.1.

It was remarked in section 5.2 that changing the

spaces $H_1$ and $H_2$ one changes also the variational principle. Let us change these spaces. We will distinguish two cases. First, we choose $H_1 = H^{1-\alpha}(\Omega)$, $H_2 = H^{1+\alpha}(\Omega)$ and let $0 < |\alpha| < \frac{1}{2}$. In the second case we study the choice $H_1 = H^0(\Omega)$ and $H_2 = H^2(\Omega)$. These cases cover, in fact, the general choice $H_1 = H^{1-\alpha}(\Omega)$, $H_2 = H^{1+\alpha}(\Omega)$ for $\alpha$ arbitrary. (We exclude always the case $\alpha =$ integer + 1/2.) For simplicity we analyze only these two cases, which are in some sense the most important ones. Furthermore, the generalization is very easy. The obvious problem is how to interpret the bilinear form (5.3.97). First, we shall analyze the case $|\alpha| < \frac{1}{2}$.

In chapter 3 we introduced the triples $(u,u_0,u_1)$ with $u \in H^{1-\alpha}(\Omega)$, $u_0 \in H^{\frac{1}{2}-\alpha}(\dot{\Omega})$ and $u_1 \in H^{-\frac{1}{2}-\alpha}(\dot{\Omega})$. This triple is isomorphic with the couple $(u,u_1)$ with $u_0$ the trace of $u$ (we note that $u \in H^{\beta}(\Omega)$, $\beta > \frac{1}{2}$). Denoting $\tilde{H}_1 = H^{1-\alpha}(\Omega) \times H^{-\frac{1}{2}-\alpha}(\Omega)$, and $\tilde{H}_2 = H^{-1-\alpha}(\Omega) \times H^{-\frac{1}{2}-\alpha}(\Omega)$ there exists by theorem 3.6.1 an isomorphisum $T$ of $\tilde{H}_1$ and $\tilde{H}_2$ such that $T(u,g) = (f,g)$, $(u,g) \in \tilde{H}_1$, $(f,g) \in \tilde{H}_2$ and for every $v \in E(\bar{\Omega})$ we have

$$\int_{\Omega} u(-\Delta v+v) + \oint_{\dot{\Omega}} u \frac{\partial v}{\partial n} ds - \oint_{\dot{\Omega}} gvds = \int_{\Omega} fvdx .$$

For the particular choice $g = 0$ we obtain

$$B(u,v) = \int_{\Omega} fvdx . \tag{5.3.99}$$

Therefore, we have

$$|B(u,v)| \leq \|f\|_{H^{-1-\alpha}(\Omega)} \|v\|_{H^{1+\alpha}(\Omega)}$$

with

$$\|f\|_{H^{-1-\alpha}(\Omega)} \leq C\|u\|_{H^{1-\alpha}(\Omega)} .$$

Therefore,

$$B(u,v) \leq C\|u\|_{H^{1-\alpha}(\Omega)} \|v\|_{H^{1+\alpha}(\Omega)} \tag{5.3.100}$$

for every $v \in E(\bar{\Omega})$ . By continuity we define $B(u,v)$ for $v \in H^{1+\alpha}(\Omega)$ .

Let us remark that for $u \in E(\bar{\Omega})$ we obtain $f = - \Delta u + u + \theta \frac{\partial u}{\partial n}$ where $\theta$ is the Dirac function concentrated on $\dot{\Omega}$ . Thus we have proved the condition (5.2.1) for $B(u,v)$ defined on $H^{1-\alpha}(\Omega) \times H^{1+\alpha}(\Omega)$ .

We shall prove the following theorem 5.3.3.

Theorem 5.3.3. Let $B(u,v)$ be defined by (5.3.97) on $H^{1-\alpha}(\Omega) \times H^{1+\alpha}(\Omega)$ with $|\alpha| < \frac{1}{2}$ then $B(u,v)$ satisfies conditions (5.2.1)-(5.2.3) of theorem 5.2.1.

Proof.

1) We have already shown that $B(u,v)$ defined for $u,v \in E(\bar{\Omega})$ may be continued to $H^{1-\alpha}(\Omega) \times H^{1+\alpha}(\Omega)$ and that the condition (5.2.1) is satisfied.

2) To show that (5.2.2) holds let $u \in E(\bar{\Omega})$ be given. Then there exists $f$ such that

$$\|f\|_{H^{-1+\alpha}(\Omega)} = \|u\|_{H^{1-\alpha}(\Omega)}$$

and

(5.3.101) $$\int_\Omega ufdx \geq \frac{1}{2}\|u\|^2_{H^{1-\alpha}(\Omega)} \quad .$$

Let $v$ be the solution of (5.3.95) and (5.3.96) with $f$ given as above and $g = 0$ . Then we have

(5.3.102) $$B(w,v) = \int_\Omega fwdx$$

for all $w \in E(\bar{\Omega})$ and using the results of chapter 3 we have $v \in H^{1+\alpha}(\Omega)$

(5.3.103) $$\|v\|_{H^{1+\alpha}(\Omega)} \leq C\|u\|_{H^{1-\alpha}(\Omega)} \quad .$$

Using $u$ for $w$ in (5.3.102), we obtain by using (5.3.101) and (5.3.103) the desired result. Changing the role of $u$ and $v$ we prove (5.2.3). Theorem 5.3.3 is completely proved.

Let us now analyze the second case introduced above, namely the case $H_1 = H^0(\Omega)$ and $H_2 = H^2(\Omega)$ . Recalling the definition of the solution in chapter 3, we know that the solution is a triple $(u_1, u_0, u_1)$ with $u \in H^0(\Omega)$ , $u_0 \in H^{-\frac{1}{2}}(\Omega)$ and $u_1 = g$ . In contrast to the case $|\alpha| < \frac{1}{2}$ studied earlier the space $H^0(\Omega) \times H^{-\frac{1}{2}}(\dot{\Omega})$

of couples $(u,u_0)$ is not isomorphic to $H^0(\Omega)$.
Therefore, let

$$H_1 = H^0(\Omega) \times H^{-\frac{1}{2}}(\dot{\Omega})$$

and

$$H_2 = H^2(\Omega)$$

The bilinear form $B(u,v)$ with $u = (u,u_0)$ is now defined on $H_1 \times H_2$ so that

$$B(u,v) = \int_\Omega u(-\Delta v+v)\,dx + \tag{5.3.104}$$

$$+ \int_{\dot{\Omega}} u_0 \frac{\partial v}{\partial n}\,ds$$

The following theorem holds.

Theorem 5.3.4. Let the bilinear form $B(u,v)$ be defined on $(H^0(\Omega) \times H^{-\frac{1}{2}}(\dot{\Omega})) \times H^2(\Omega)$ by expression (5.3.104). Then $B(u,v)$ satisfies the condition (5.2.1)-(5.2.3) of theorem 5.2.1.

Proof.

1) The continuity condition (5.2.1) is obvious because by the imbedding theorem we have $\frac{\partial v}{\partial n} \in H^{\frac{1}{2}}(\dot{\Omega})$.

2) To show that (5.2.2) holds let $u = (u,u_0) \in H_1$ be given. There exists a $w \in E(\dot{\Omega})$ with

$$\|w\|_{H^{\frac{1}{2}}(\Omega^{\bullet})} = \|u_0\|_{H^{-\frac{1}{2}}(\Omega^{\bullet})}$$

and such that

$$(5.3.105) \qquad \frac{1}{2}\|u_0\|^2_{H^{-\frac{1}{2}}(\Omega^{\bullet})} \leq \int_{\Omega^{\bullet}} u_0 w ds \quad .$$

Let $z$ be the solution of the problem

$$(5.3.106) \qquad -\Delta z + z = u \quad \text{on} \quad \Omega$$

$$\frac{\partial z}{\partial n} = w \quad \text{on} \quad \Omega^{\bullet}$$

Using theorem 3.6.1 we have

$$(5.3.107) \qquad \|z\|_{H^2(\Omega)} \leq C\|(u,u_0)\|_{H_1}$$

Putting $v = z$ in (5.3.104), using (5.3.106), (5.3.105), and (5.3.107) we obtain (5.2.2).

3) Let $v \in H^2(\Omega)$ be given. Then $\frac{\partial v}{\partial n} \in H^{\frac{1}{2}}(\Omega^{\bullet})$ and $-\Delta v + v \in H^0(\Omega)$ .

In an anologous way as before we may find $w \in E(\Omega^{\bullet})$ , $\|w\|_{H^{-\frac{1}{2}}(\Omega^{\bullet})} = \|\frac{\partial v}{\partial n}\|_{H^{\frac{1}{2}}(\Omega)}$ and such that

$$\frac{1}{2}\left\|\frac{\partial v}{\partial n}\right\|^2_{H^{\frac{1}{2}}(\Omega)} \leq \oint_{\dot{\Omega}} w \frac{\partial v}{\partial n} ds$$

and the choice $u = -\Delta v + v$ and $u_0 = w$ gives immediately the desired result and the theorem is completely proved.

We observe that every solution $u$ of (5.3.95) and (5.3.97) which is sufficiently smooth gives the pair $u = (u,u_0)$ with $u_0$ the trace of $u$ and

$$B((u,u_0),v) = \int_\Omega fvdx + \int_{\dot{\Omega}} gvds$$

for every $v \in H^2(\Omega)$ .

We discussed the problem with boundary conditions of the first type. In an anologous manner the cases of boundary conditions of the second type may be studied.

### 5.3.8. Variational Principle of Dirichlet type for Infinite Domains.

In the previous sections we discussed the problem on a bounded domain with smooth boundary. In the present section we study the problem on a domain $\Omega$ , where $\Omega$ is the entire plane. The theory of partial differential equation on unbounded domains has not been developed to the same extent as the theory in bounded domains. For this and other technical reasons we will confine our discussion to the prototype equation

(5.3.108) $$-\Delta u + cu = f \quad ,$$

with $0 < \alpha < c < K$, on $R^2$. For the uniqueness we do not prescribe any boundary conditions. The behavior at $\infty$ will be replaced by the requirement that the solution belongs to some Sobolev space.

In section 2.6.3 we introduced the space $H^{\ell}_{\mu}(R^n)$ Now we choose $H_1 = H^1_{\mu}(R^2)$ and $H_2 = H^1_{-\mu}(R^2)$ and let the bilinear form $B(u,v)$ on $H_1 \times H_2$ be defined by

(5.3.109)

$$B(u,v) = \int_{R_2} \left[\frac{\partial u}{\partial x_1}\frac{\partial v}{\partial x_1} + \frac{\partial u}{\partial x_2}\frac{\partial v}{\partial x_2} + cuv\right]dx \quad ,$$

with $\mu$ to be determined later. We prove the following theorem.

Theorem 5.3.5. For $|\mu| < \frac{\alpha}{4}$, the bilinear form (5.3.109) satisfies the conditions (5.2.1)-(5.2.3), where $\alpha$ is given in (5.3.108).

Proof.

1) To prove (5.2.1) we note that

(5.3.110)

$$|B(u,v)| = \left|\int_{R^2}\left[\sum_{i=1}^{2} \frac{\partial u}{\partial x_i} e^{\mu|x|} \frac{\partial v}{\partial x_i} e^{-\mu|x|} + c^{\frac{1}{2}} u e^{\mu|x|} v e^{-\mu|x|} c^{\frac{1}{2}}\right]dx\right.$$

$$\leq \sum_{i=1}^{2}\left[\int_{R^2}\left(e^{\mu|x|}\frac{\partial u}{\partial x_i}\right)^2 dx\right]^{\frac{1}{2}}\left[\int_{R^2}\left(e^{-\mu|x|}\frac{\partial v}{\partial x_i}\right)^2 dx\right]^{\frac{1}{2}} +$$

((5.3.110) continued)

$$+ \left[\int \left(e^{\mu|x|} c^{\frac{1}{2}} u\right)^2 dx\right]^{\frac{1}{2}} \left[\int \left(e^{-\mu|x|} c^{\frac{1}{2}} v\right)^2 dx\right]^{\frac{1}{2}}$$

$$\leqslant C\|u\|_{H^1_\mu(R_2)} \|u\|_{H^1_{-\mu}(R_2)}$$

2) To prove (5.2.2) let $u \in H^1_\mu(\Omega)$ and let $v = ue^{2\mu|x|}$. By direct computation we have

(5.3.111)

$$\frac{\partial}{\partial x_i}(ue^{2\mu|x|}) = e^{2\mu|x|} \frac{\partial u}{\partial x_i} + 2\mu e^{2\mu|x|} u \operatorname{sgn} x_i$$

Hence

(5.3.112)

$$\|v\|^2_{H^1_{-\mu}(R^2)} \leqslant C\left[\int_{R^2} \left[\sum_{i=1}^{2} e^{4\mu|x|}\left(\frac{\partial u}{\partial x_i}\right)^2 + 4\mu^2 2u^2 e^{4\mu|x|} + \right.\right.$$

$$\left.\left. + u^2 e^{4\mu|x|}\right] e^{-2\mu|x|}\right] dx \leqslant C\|u\|^2_{H^1_\mu(R^2)} \quad ,$$

which implies that $v \in H^1_{-\mu}(R^2)$ . Furthermore,

(5.3.113)

$$|B(u,ue^{2\mu|x|})| = \left|\int_{R^2} \sum_{i=1}^{2} \left[\frac{\partial u}{\partial x_i}\right]\left[\frac{\partial u}{\partial x_i} + 2\mu u \operatorname{sgn} x_i\right] e^{2\mu|x|} dx + \right.$$

((5.3.113)continued)

$$+\int_{R^2} cu^2 e^{2\mu|x|}dx| \geq$$

$$\geq \alpha\|u\|^2_{H^1_\mu(R^2)} - 2\mu\ 2\|u\|^2_{H^1_\mu(R^2)} \geq C\|u\|^2_{H^1_\mu(R^2)}$$

provided $|\mu| < \frac{\alpha}{4}$. By combining the inequalities (5.3.112) and (5.3.113) we obtain (5.2.2).

3) (5.2.3) is obtained by interchanging the role of $u$ and $v$ in the above argument.

Now if $f$ is such that $F(v) = \int_{R^2} fvdx \in H^1_{-\mu}(R^2)$, then there exists a unique $u_0 \in H^1_\mu(R^2)$ such that

$$B(u_0,v) = F(v)\ ,\ \forall v \in H^1_{-\mu}(R^2)\quad .$$

If $f$ is sufficiently smooth then $u_0$ satisfies equation (5.3.108) in the classical sense. Thus we have constructed a variational principle related to problem (5.3.108).

### 5.3.9. Variational principle of Dirichlet type for Neumann problem.

Throughout the previous sections the existence of a unique solution was assumed. However, often one has to solve problems where the solution is not unique. As a typical model of such problems we consider the problem:

(5.3.114) $$-\Delta u = f \quad \text{on} \quad \Omega$$

(5.3.115) $$\frac{\partial u}{\partial n} = g \quad \text{on} \quad \dot{\Omega} \quad .$$

The foregoing discussions concerning this particular problem hold in general and may be applied equally well to the more general problem considered in section 5.3.1.

Clearly, the solution (if it exists) of the problem (5.3.114) and (5.3.115) is not unique, it is determined up to a constant function. On the other hand a solution exists if and only if

(5.3.116) $$\int_{\Omega} f dx + \oint_{\dot{\Omega}} g ds = 0 \quad .$$

Let us assume apriori that (5.3.116) is satisfied. In practice in numerous cases it is impossible to check (5.3.116), since the only possibility for evaluation of the integrals in (5.3.116) is the use of the quadrature formulas. To avoid this difficulty we modify the problem (5.3.114), (5.3.115) in such a way that the modified problem has a unique solution whenever the original problem has a solution and moreover the solution of our modified problem is also a solution of the original problem. There are several possibilities for this modification. We shall point out one such possibility. To this end we consider the problem of finding the function $u$ and a constant $\lambda$ such that

(5.3.117) $$-\Delta u = f + \lambda \quad \text{on} \quad \Omega$$

$$\frac{\partial u}{\partial n} = g \quad \text{on} \quad \dot{\Omega}$$

(5.3.118) $$\int_\Omega u dx = 0$$

Obviously, the problem has the required properties (i.e., existence and uniqueness of a solution for all $f$ and $g$) mentioned above. Namely we have

(5.3.119) $$\int_\Omega (\lambda + f) dx + \oint g dx = 0 \quad .$$

In case (5.3.116) is satisfied, clearly we obtain $\lambda = 0$.

Let us formulate a variational principle using the Langrange multipliers technique for the problem (5.3.117)-(5.3.118). Let

$$H_1 = H^1(\Omega) \times R_1 = H_2 \ ,$$

where $R_1$ denote the set of all real numbers. We write $(u,\lambda) \in H_1$, for $u \in H^1(\Omega)$, $\lambda \in R_1$ and $\|(u,\lambda)\|^2_{H_1} = \|u\|^2_{H^1(\Omega)} + |\lambda|^2$.

Now we introduce the bilinear form

(5.3.120)
$$B((u,\lambda);(v,\mu)) = \int_\Omega \left( \frac{\partial u}{\partial x_1} \frac{\partial v}{\partial x_1} + \frac{\partial u}{\partial x_2} \frac{\partial v}{\partial x_2} \right) dx - \int_\Omega (u\mu + \lambda v) dx \quad .$$

Moreover let

(5.3.121)

$$F(v,\mu) = \int_{\Omega} fv\,dx + \oint_{\dot{\Omega}} gv\,ds + \frac{1}{\text{mes}(\Omega)}\left[\oint_{\dot{\Omega}} g\,ds + \int_{\Omega} f\,dx\right]\int_{\Omega} v\,ds \quad .$$

We assume that $f$ and $g$ are sufficiently smooth and that $\lambda_0$ is determined by (5.3.119). We denote by $(u_0,\lambda_0)$ the solution which satisfies (5.3.118). By integration by parts we have from (5.3.120).

(5.3.122)

$$B((u_0,\lambda_0),(u,v)) = \int_{\Omega} -\Delta u_o v\,dx + \oint_{\dot{\Omega}} \frac{\partial u_0}{\partial n} v\,ds - \mu\int_{\Omega} u_0\,dx - \lambda_0\int_{\Omega} v\,dx =$$

$$= \int_{\Omega} fv\,dx + \oint_{\dot{\Omega}} gv\,ds + \frac{1}{\text{mes}(\Omega)}\left[\oint_{\dot{\Omega}} g\,ds + \int_{\Omega} f\,dx\right]\int v\,dx$$

$$= F(v,\mu) \quad .$$

Consequently, by use of the proposed variational principle we can solve our modified problem.

Now we shall show that the conditions (5.2.1), (5.2.2) and (5.2.3) are fulfilled. Obviously, the condition (5.2.1) is satisfied. To show that (5.2.2) holds, we proceed as in section (5.3.6).

Lemma 5.3.2. Let $z_0 \in H^1(\Omega)$ be such that

(5.3.123)

$$B_1(u,z_0) = \int_{\Omega}\left(\frac{\partial u}{\partial x_1}\frac{\partial z_0}{\partial x_1} + \frac{\partial u}{\partial x_2}\frac{\partial z_2}{\partial x_2} + uz_0\right)dx = \mu\int_{\Omega} u\,dx$$

for all $u \in H^1(\Omega)$ . Then

$$\mu \int_\Omega z_0 dx \geq C|\mu|^2 \quad , \tag{5.3.124}$$

where $C > 0$ does not depend on $\mu$ .

Proof. First we observe that the bilinear form $B_1(u,z)$ satisfies all the hypotheses of Theorem 5.2.1 and therefore there exists a unique $z_0$ satisfying (5.3.123).

We note that $z_0$ satisfies (see, e.g., section 5.3.2)

$$- \Delta z_0 + z_0 = \mu \quad , \tag{5.3.125}$$

and the boundary condition

$$\frac{\partial z_0}{\partial n} = 0 \quad . \tag{5.3.126}$$

Therefore $z_0 = \mu \bar{z}_0$ , where $\bar{z}_0$ is the solution of (5.3.125) and (5.3.126) with $\mu = 1$ . From (5.3.123) it follows immediately that

$$\int_\Omega \bar{z}_0 dx = \|\bar{z}_0\|^2_{H^1(\Omega)} > 0 \quad .$$

Hence,

$$\mu \int_\Omega z_0 dx = \mu^2 \int_\Omega \bar{z}_0 dx = c|\mu|^2 \quad ,$$

and (5.3.124) is proved.

Let $u \in H^1(\Omega)$ , $\lambda \in R_1$ be given. Then we have

(5.3.127)

$$B((u,\lambda),(u-z_1+z_2,\mu)) = B_1(u,u) - B_1(u,z_1) - \int_\Omega u(u-z_1)dx + B((u,\lambda),(z_2,0)) - \int_\Omega (\mu u+\lambda u)dx + \int_\Omega \lambda z_1 dx .$$

We choose $z_1$ such that

(5.3.128)
$$B_1(w,z_1) = \lambda\int_\Omega w dx \quad \text{for every} \quad w \in H_1 .$$

Since $z_1$ is the solution of (5.3.125) and (5.3.126), we have

(5.3.129)
$$\|z_1\|_{H^1(\Omega)} \leq |\lambda|C .$$

Now we search for $\gamma \in R_1$ such that

(5.3.130)
$$\int_\Omega (\gamma+u-z_1)dx = 0 .$$

From (5.3.129) we have

(5.3.131)
$$|\gamma| \leq C\|(u,\lambda)\|_{H_1} .$$

Let $z_2$ be such that

(5.3.132)
$$-\Delta z_2 = u - z_1 + \gamma ,$$

(5.3.133)
$$\frac{\partial z_2}{\partial n} = 0 ,$$

(5.3.134)
$$\int_\Omega z_2 dx = 0$$

The existence of such a $z_2$ is assured by (5.3.130), and from (5.3.131) it follows that

(5.3.135)
$$\|z_2\|_{H^3(\Omega)} \leq C\|(u,\lambda)\|_{H_1} .$$

Thus, by (5.3.134) we obtain:

(5.3.136)
$$B((u,\lambda),(z_2,0)) = \int_\Omega u(u-z_1)dx + \gamma\int_\Omega udx ,$$

and hence

(5.3.137)
$$B((u,\lambda),(u-z_1+z_2,\mu)) = \|u\|^2_{H^1(\Omega)} - \int_\Omega(\mu+2\lambda+\gamma)udx +$$

((5.3.137) continued)

$$+ \int_\Omega \lambda z_1 dx \quad .$$

Now, by letting $\mu = - 2\lambda - \gamma$ and using lemma (5.3.2) we obtain

(5.3.138)

$$B((u,\lambda),(u-z_1+z_2,\mu)) \geq \|u\|^2_{H^1(\Omega)} + C|\lambda|^2 \geq C\|(u,\lambda)\|^2_{H_1} \quad .$$

On the other hand, from (5.3.129), (5.3.131) and (5.3.135) it follows that

$$\text{(5.3.139)} \qquad \|(u-z_1+z_2,\mu)\|_{H_1} \leq C\|(u,\lambda)\|_{H_1} \quad .$$

Now the condition (5.2.2) is an immediate consequence of (5.3.138) and (5.3.139). The reversal of the roles of $(u,\lambda)$ and $(v,\mu)$ in the preceeding argument yields condition (5.2.3).

Thus, we have proved the following theorem.

Theorem 5.3.6. If

I) (5.3.114) and (5.3.115) are given, where $f$ and $g$ satisfy (5.3.116);

II) $f$ and $g$ are sufficiently smooth (i.e., there exists a classical solution $u_0$);

Then

III) The bilinear form $B(u,\lambda;v,\mu)$ defined by (5.3.120) on $(H^1(\Omega)xR_1) \times (H^1(\Omega)xR_1)$ satisfies the conditions (5.2.1), (5.2.2) and (5.2.3);

IV) The solution $u_o$ with $\int_\Omega u_o dx = 0$ satisfies for every $v \in H^1(\Omega)$ and $\mu \in R_1$ the condition

$$B(u,\lambda;v,\mu) = \int_\Omega fvdx + \oint_{\dot\Omega} gvds \quad . \tag{5.3.140}$$

We note that in the expression (5.3.121) for $F(v,\mu)$ there is an additional term which does not appear in (5.3.140) due to the condition (5.3.116).

In the above theorem we dealt with a special case. However, the theorem may be proved in general. Let us emphasize that we do not need to know explicitly the eigenfunction of the adjoint problem. We merely need that $\int wds = 0$ , where $w$ is the solution of the homogeneous adjoint problem. This assumption may be avoided by a slight change in our arguments. Up to now we have assumed that there is only one linearly independent solution of the homogeneous adjoint problem. It is easy to generalize the problem to the case where there is a finite number of such solutions.

### 5.3.10. Remarks on the Dirichlet and least square type variational principles.

In the previous sections we derived a variety of variation principles. It is possible to construct many other variational principles by a similar technique. In this connection we mention the variational principle with

Lagrange multipliers and least square type principles with different spaces of trial functions and test functions. (This would be analogous to the generalization of Dirichlet principle discussed in section 5.3.7.)

In section 5.3.8 we extended our variational principle to the weighted spaces for infinite domains. It is possible to use the weighted space approach in the case of bounded domains where the weight is chosen to be a power of the distance to the boundary or a point on the boundary, etc. One can always combine the weighted space technique with the approach using spaces with different differentiability properties, etc.

Since we have already discussed, in the previous sections, several typical cases we shall not introduce any more principles of this type.

## 5.4. Variational principle for second order systems.

The principles introduced in section 5.3 may be extended in an analogous manner to strongly elliptic systems. Nonetheless, some new aspects may be observed. As a typical problem we shall analyse the equations of elasticity. In particular we confine our discussions to the case of plane elasticity. More precisely, consider the problem with zero displacements and nonzero prescribed body forces. We have chosen this typical problem to bring out some new aspects in the study of systems of equations.

### 5.4.1. The elasticity problem.

It is well known (see e.g., [11], [12], etc.) that the equations of plane elasticity, for homogeneous isotropic bodies are

$$\mu\Delta u_1 + (\lambda+\mu)\frac{\partial\theta}{\partial x_1} = g_1 \ , \tag{5.4.1}$$

$$\mu\Delta u_2 + (\lambda+\mu)\frac{\partial\theta}{\partial x_2} = g_2 \ ,$$

$$\theta = \frac{\partial u_1}{\partial x_1} + \frac{\partial u_2}{\partial x_2} \ ,$$

and $\Delta$ denotes as usual the Laplace operator. The constants $\lambda > 0$ and $\mu > 0$ are known as the Lamé and shear module respectively. They are related to Poisson's ratio $\sigma$ and Young's module $E$ by

$$\lambda = \frac{2\mu\sigma}{1-2\sigma} \ , \ \mu = \frac{E}{2(1+\sigma)}$$

The function $u_1$ and $u_2$ denote the displacements in $x_1$ and $x_2$ directions respectively and $g_1$ and $g_2$ are the body forces in these directions. For ordinary materials $\sigma$ lies in the interval $(0,1/2)$ and $E$ is positive, which in turn gives $0 < \lambda < \infty$ , $0 < \mu < \infty$ . However, there are materials which are incompressible (or nearly incompressible) for which $\sigma = \frac{1}{2}$ , i.e., $\lambda = \infty$ and obviously equation (5.4.1) degenerates. As we shall see this effect will introduce a new feature in the solution.

### 5.4.2. <u>The variational principle for equations of elasticity</u>.

5.4.2.1. Compressible materials.

We consider equation (5.4.1) with the boundary conditions

$$u_1 = u_2 = 0 \quad \text{on} \quad \dot{\Omega} \tag{5.4.2}$$

Let

$$H_1 = H_0^1(\Omega) \times H_0^1(\Omega) = H_2 \quad ,$$

with

$$(u_1,u_2) \in H_1 \quad \text{and} \quad \|(u_1,u_2)\|_{H_1}^2 = \|u_1\|_{H_0^1(\Omega)}^2 + \|u_2\|_{H_0^1(\Omega)}^2 \quad .$$

On $H_1 \times H_2$ we define a bilinear form which is related to the well known strain-energy density. To this end let

(5.4.3)

$$B((u_1,u_2),(v_1,v_2)) = \int_\Omega \left[ \lambda\theta\phi + 2\mu\left( \frac{\partial u_1}{\partial x_1}\frac{\partial v_1}{\partial x_1} + \frac{\partial u_2}{\partial x_2}\frac{\partial v_2}{\partial x_2} + \right.\right.$$

$$\left.\left. + \frac{1}{2}\left(\frac{\partial u_1}{\partial x_2} + \frac{\partial u_2}{\partial x_1}\right)\left(\frac{\partial v_1}{\partial x_2} + \frac{\partial v_2}{\partial x_1}\right)\right)\right] dx \quad ,$$

where $\theta = \frac{\partial u_1}{\partial x_1} + \frac{\partial u_2}{\partial x_2}$ and $\phi = \frac{\partial v_1}{\partial x_1} + \frac{\partial v_2}{\partial x_2}$ . First we show that the bilinear form (5.4.3) satisfies all the requirements of theorem 5.2.1. In fact, condition 5.2.1 is clearly satisfied with $C$ depending on $\mu$ and $\lambda$ . This fact plays an essential role, when the material is nearly incompressible and $\lambda$ is very large. We note that since the bilinear form (5.4.3) is symmetric condition (5.2.2) implies (5.2.3). Thus we need only to prove (5.2.2).

Obviously, to prove (5.2.2) it is sufficient to show that

(5.4.4) $$B((u_1,u_2),(u_1,u_2)) \geq C\|u\|^2_{H_1} .$$

From (5.4.3) we have

(5.4.5)

$$B((u_1,u_2),(u_1,u_2)) = \int_\Omega \left[\lambda\theta^2+2\mu\left(\left[\frac{\partial u_1}{\partial x_1}\right]^2+\left[\frac{\partial u_2}{\partial x_2}\right]^2+\frac{1}{2}\left[\frac{\partial u_1}{\partial x_2}+\frac{\partial u_2}{\partial x_1}\right]^2\right)\right]dx.$$

We now state a well known lemma, the Korn inequality (see [7], [8]).

<u>Lemma 5.4.1</u> Let $u_1 \in H^1_0(\Omega)$ , $u_2 \in H^1_0(\Omega)$ . Then

(5.4.6)

$$C_1\left[\|u_1\|^2_{H^1(\Omega)} + \|u_2\|^2_{H^1(\Omega)}\right] \geq \int_\Omega \left[\left[\frac{\partial u_1}{\partial x_1}\right]^2 + \left[\frac{\partial u_2}{\partial x_2}\right]^2 + \right.$$

$$\left.\left[\frac{\partial u_1}{\partial x_2} + \frac{\partial u_2}{\partial x_1}\right]^2\right] dx \geq C_2\left[\|u_1\|^2_{H^1(\Omega)} + \|u_2\|^2_{H^1(\Omega)}\right] ,$$

where $C_i > 0$ $(i=1,2)$ are independent of $u_1$ , $u_2$ .

The application of lemma 5.4.1 to (5.4.5) yields

the theorem immediately:

Theorem 5.4.1. The bilinear form (5.4.3) fulfills all the conditions of Theorem 5.2.1.

We introduce the functional

(5.4.7) $$F(v_1,v_2) = -\int_\Omega g_1 v_1 dx - \int_\Omega g_2 v_2 dx \quad .$$

For $g_1$ , $g_2 \in H^0(\Omega)$ , obviously $F \in H_1'$ . We assume that $g_1$ and $g_2$ are sufficiently smooth and $(u_{1,0},u_{2,0})$ is the classical solution of the problem (5.4.1) and (5.4.2). We show that

$$B((u_{1,0},u_{2,0}),(v_1,v_2)) = F(v_1,v_2) \ \forall v_1,v_2 \in \mathcal{D}(\Omega) \quad .$$

In fact, by integration by parts we obtain

$$B(u_{1,0},u_{2,0},v_1,v_2) = \int_\Omega - \left[\lambda\left[\frac{\partial^2 u_1}{\partial x_1^2} + \frac{\partial^2 u_2}{\partial x_1 \partial x_2}\right] - 2\mu\left[\frac{\partial^2 u_1}{\partial x_1^2} + \frac{1}{2}\left[\frac{\partial^2 u_1}{\partial x_2^2} + \frac{\partial u_2}{\partial x_1\ x_2}\right]\right]\right] v_1 dx + \int_\Omega - \left[\lambda\left(\frac{\partial^2 u_1}{\partial x_1 \partial x_2} + \frac{\partial^2 u_2}{\partial x_2^2}\right) - 2\mu\left(\frac{\partial^2 u_2}{\partial x_2^2} + \frac{1}{2}\left(\frac{\partial^2 u_1}{\partial x_1 \partial x_2} + \frac{\partial^2 u_2}{\partial x_2^2}\right)\right)\right] v_2 dx = -\int_\Omega \left[\left(\mu\Delta u_1 + (\lambda+\mu)\frac{\partial\theta}{\partial x_1}\right)v_1 + \left(\mu\Delta u_2 + (\lambda+\mu)\frac{\partial\theta}{\partial x_2}\right)v_2\right] dx = F(v_1,v_2) \quad .$$

Therefore we see that our variational principle solves our original problem. We note that the constant $\mu$ and $\lambda$ may be normalized so that $\mu = 1$ . The case $\lambda = 0$ does not create any difficulties. However, as it was pointed out earlier the case $\lambda \to \infty$ is quite different and causes some difficulties.

5.4.2.2. Nearly incompressible and incompressible materials

The results presented in the preceding section have a considerable disadvantage when $\lambda$ is very large, i.e., when the Poisson's ratio is close to $\frac{1}{2}$ . In this case the constant $C_1$ in (5.2.1) associated with the bilinear form (5.4.3) is very large and causes serious troubles in the applications.

To surmount this difficulty we introduce a new variational principle based on the use of the Lagrange multipliers.

To this end we note that the problem of finding $(u_{1,0}, u_{2,0})$ such that

$$B((u_{1,0}, u_{2,0}), (v_1, v_2)) = F(v_1, v_2) \quad ,$$

holds for all $v_1$ , $v_2 \in H_0^1(\Omega)$ , where B and F are defined by (5.4.3) and (5.4.7) respectively, is equivalent to finding the pair $(u_1, u_2)$ which minimizes the quadratic functional

$$B((u_1, u_2), (u_1, u_2)) - 2F(u_1, u_2) \quad .$$

The above observation allows us to modify formally this

minimization problem by introducing a new quadratic functional defined for a triple $(u_1,u_2;\xi)$ by

$$\int_\Omega \left[\lambda\gamma^2\xi^2+2\mu\left[\left(\frac{\partial u_1}{\partial x_1}\right)^2 + \left(\frac{\partial u_2}{\partial x_2}\right)^2 + \frac{1}{2}\left(\frac{\partial u_1}{\partial x_2} + \frac{\partial u_2}{\partial x_1}\right)^2\right]\right] dx -$$

$$- 2F(u_1,u_2) \quad ,$$

with the restriction

$$\gamma\xi - \left(\frac{\partial u_1}{\partial x_1} + \frac{\partial u_2}{\partial x_2}\right) = 0 \quad .$$

Now by using the technique of Lagrange multipliers, we define on $H_1 \times H_2$ , where

$$H_1 = H_2 = H_0^1(\Omega) \times H_0^1(\Omega) \times \tilde{H}^0(\Omega) \times \tilde{H}^0(\Omega) \quad ,$$

and $\tilde{H}^0(\Omega)$ is the subset of $H^0(\Omega)$ orthogonal to constants, the bilinear form

(5.4.8)

$$B\left((u_1,u_2,\xi,\mathcal{H}),(v_1,v_2,\eta,\tau)\right) = \int_\Omega \left\{\lambda\gamma^2\xi\eta+2\mu\left[\frac{\partial u_1}{\partial x_1}\frac{\partial v_1}{\partial x_1} + \frac{\partial u_2}{\partial x_2}\frac{\partial v_2}{\partial x_2}\right.\right.$$

$$\left. + \frac{1}{2}\left(\frac{\partial u_1}{\partial x_2} + \frac{\partial u_2}{\partial x_1}\right)\left(\frac{\partial v_1}{\partial x_2} + \frac{\partial v_2}{\partial x_1}\right)\right] - \left[\mathcal{H}\left(\gamma\eta - \frac{\partial v_1}{\partial x_1} - \frac{\partial v_2}{\partial x_2}\right) +\right.$$

((5.4.8) continued)

$$+ \tau\left(\gamma\xi - \frac{\partial u_1}{\partial x_1} - \frac{\partial u_2}{\partial x_2}\right)\Bigg]\Bigg\}dx \quad .$$

We note that $(u_1, u_2, \xi, H) \in H_1$ , $(v_1, v_2, \eta, \tau) \in H_2$ and let

$$\|(u_1, u_2, \xi, H)\|^2_{H_1} = \|u_1\|^2_{H^1_0(\Omega)} + \|u_2\|^2_{H^1_0(\Omega)} +$$

$$+ \lambda^{\frac{1}{2}}\gamma\|\xi\|^2_{L_2(\Omega)} + \|H\|^2_{L_2(\Omega)} \quad ,$$

$$\lambda , \gamma > 0 \quad .$$

It is obvious that (5.4.8) satisfies condition (5.2.1) of theorem 5.2.1. In fact $C_1 \leq C \max[\mu, \lambda^{\frac{1}{2}}\gamma, \gamma^{-\frac{1}{2}}\lambda^{-1/4}]$ , with $C$ independent of $\mu$, $\lambda$, $\gamma$. For large $\lambda$ we may choose $\gamma$ sufficiently small so that this constant is always of the same order of magnitude as $\mu$ . Thus the situation for large $\lambda$ in (5.4.9) is quite different from that in (5.4.3).

Before we investigate the condition (5.2.2), we first prove two lemmas which we shall need in the sequel.

Lemma 5.4.2. If $H \in L_2(\Omega)$ and $\int_\Omega H dx = 0$ , then there exist $v_1$ , $v_2 \in H^1_0(\Omega)$ such that

(5.4.9)
$$\frac{\partial v_1}{\partial x_1} + \frac{\partial v_2}{\partial x_2} = \mathcal{H} ,$$

(5.4.10)
$$\|v_1\|^2_{H^1_0(\Omega)} + \|v_2\|^2_{H^1_0(\Omega)} \leq C\|\mathcal{H}\|^2_{L_2(\Omega)} .$$

Proof. We first establish the existence of $u_1$ , $u_2 \in H^1(\Omega)$ such that (5.4.9) and (5.4.10) are satisfied.

Since $\int_\Omega \mathcal{H} dx = 0$ , there exists a $w \in H^2(\Omega)$ such that $\Delta w = \mathcal{H}$ and $\frac{dw}{dn} = 0$ , with $\|w\|_{H^2(\Omega)} \leq C\|\mathcal{H}\|_{L_2(\Omega)}$ . (See results of chapter 3.) Now if we choose $v_1 = \frac{\partial w}{\partial x_1}$ , $v_2 = \frac{\partial w}{\partial x_2}$ , obviously $u_1$ , $u_2 \in H^1(\Omega)$ , and we have

(5.4.11)
$$\|u_1\|^2_{H^1(\Omega)} + \|u_2\|^2_{H^1(\Omega)} \leq C\|\mathcal{H}\|^2_{L_2(\Omega)} ,$$

and furthermore

(5.4.12)
$$\frac{\partial u_1}{\partial x_1} + \frac{\partial u_2}{\partial x_2} = \Delta w = \mathcal{H} .$$

Obviously, $v_1$ , $v_2 \notin H^1_0(\Omega)$ . However, we shall show that there exists a function $U \in H^2(\Omega)$ such that if we let

$$v_1 = \frac{\partial w}{\partial x_1} + \frac{\partial U}{\partial x_2} , \quad v_2 = \frac{\partial w}{\partial x_2} - \frac{\partial U}{\partial x_1} ,$$

then $v_1$ , $v_2 \in H_0^1(\Omega)$ . To this end we note that for any $U \in H^2(\Omega)$ and $Z_1 = \frac{\partial U}{\partial x_1}$ , $Z_2 = -\frac{\partial U}{\partial x_2}$ , we have $\frac{\partial Z_1}{\partial x_1} + \frac{\partial Z_2}{\partial x_2} = 0$ . Thus we try to find $U \in H^2(\Omega)$ such that

$$\text{(5.4.13)} \qquad \frac{\partial w}{\partial x_1} + \frac{\partial U}{\partial x_2} = 0 \quad \text{on} \quad \dot{\Omega}$$

$$\frac{\partial w}{\partial x_2} - \frac{\partial U}{\partial x_1} = 0 \quad \text{on} \quad \dot{\Omega} \ .$$

Indeed such a function $U$ exists. In fact, since $\frac{\partial w}{\partial n} = 0$ , (5.4.13) is equivalent to $\frac{\partial U}{\partial s} = 0$ and $\frac{\partial U}{\partial n} = \frac{\partial w}{\partial x_2} \cos(n,x_1) - \frac{\partial w}{\partial x_1} \cos(n,x_2) = f$ . On the other hand, we have $f \in H^{\frac{1}{2}}(\dot{\Omega})$ and $\|f\|_{H^{\frac{1}{2}}(\dot{\Omega})} \leq C\|w\|_{H^2(\Omega)} \leq C\|H\|_{L_2}$ . We know that there exists a biharmonic function $U$ (see chapter 3) such that $U = 0$ on $\dot{\Omega}$ and $\frac{\partial U}{\partial n} = f$ , with $\|U\|_{H^2(\Omega)} \leq C\|f\|_{H^{\frac{1}{2}}(\dot{\Omega})} \leq C\|H\|_{L_2}$ . Therefore the functions $v_1 = \frac{\partial w}{\partial x_1} + \frac{\partial U}{\partial x_2}$ , $v_2 = \frac{\partial w}{\partial x_2} - \frac{\partial U}{\partial x_1}$ satisfy the conditions of lemma 4.2.

Lemma 5.4.3. If

I) $H \in L_2(\Omega)$ and $\int_\Omega H dx = 0$ ;

II) $u_1$ , $u_2 \in H_0^1(\Omega)$ such that for every $v_1$ ,

$v_2 \in H_0^1(\Omega)$

(5.4.14)

$$\int_\Omega \left[\frac{\partial u_1}{\partial x_1}\frac{\partial v_1}{\partial x_1} + \frac{\partial u_2}{\partial x_2}\frac{\partial v_2}{\partial x_2} + \frac{1}{2}\left(\frac{\partial u_1}{\partial x_2} + \frac{\partial u_2}{\partial x_1}\right)\left(\frac{\partial v_1}{\partial x_2} + \frac{\partial v_2}{\partial x_1}\right)\right] dx$$

$$= \int_\Omega H\left(\frac{\partial v_1}{\partial x_1} + \frac{\partial v_2}{\partial x_2}\right) dx \quad ;$$

Then

III)

(5.4.15)

$$\int_\Omega \left[H\left(\frac{\partial u_1}{\partial x_1} + \frac{\partial u_2}{\partial x_2}\right)\right] dx \geq C\|H\|^2_{L_2(\Omega)} \quad ,$$

and

(5.4.16) $$\|u_1\|^2_{H^1(\Omega)} + \|u_2\|^2_{H^1(\Omega)} \leq C\|H\|^2_{L_2(\Omega)} \quad .$$

Proof. We introduce the bilinear form

(5.4.17)

$$B_1(u_1,u_2;v_1,v_2) = \int_\Omega \left[\frac{\partial u_1}{\partial x_1}\frac{\partial v_1}{\partial x_1} + \frac{\partial u_2}{\partial x_2}\frac{\partial v_2}{\partial x_2}\right.$$

$$\left. + \frac{1}{2}\left(\frac{\partial u_1}{\partial x_2} + \frac{\partial u_2}{\partial x_1}\right)\left(\frac{\partial v_1}{\partial x_2} + \frac{\partial v_2}{\partial x_1}\right)\right] dx \quad ,$$

defined on $H_1 \times H_2 = H_0^1(\Omega) \times H_0^1(\Omega)$. From lemma 5.4.1 it follows that the bilinear form (5.4.17) satisfies the conditions of theorem 5.2.1 and consequently $u_1$ and $u_2$ are uniquely determined and (5.4.16) holds. Now by lemma 5.4.2 we have

$$\|H\|^2_{H^0(\Omega)} = \int_\Omega \left[ \left( \frac{\partial u_1}{\partial x_1} \frac{\partial v_1}{\partial x_1} + \frac{\partial u_2}{\partial x_2} \frac{\partial v_2}{\partial x_2} + \frac{1}{2}\left(\frac{\partial u_1}{\partial x_2} + \frac{\partial u_2}{\partial x_1}\right)\left(\frac{\partial v_1}{\partial x_2} + \frac{\partial v_2}{\partial x_1}\right)\right)\right] dx \quad .$$

where $v_1$ and $v_2$ are the functions appearing in lemma 5.4.2. From the Schwartz' inequality, lemmas 5.4.1 and 5.4.2 we obtain

$$\|H\|^2_{H^0(\Omega)} \leq C\left[\int_\Omega \left[\left(\frac{\partial u_1}{\partial x_1}\right)^2 + \left(\frac{\partial u_2}{\partial x_2}\right)^2 + \frac{1}{2}\left(\frac{\partial u_1}{\partial x_2} + \frac{\partial u_2}{\partial x_1}\right)^2\right] dx\right]^{\frac{1}{2}}$$

$$\left[\int_\Omega \left[\left(\frac{\partial v_1}{\partial x_1}\right)^2 + \left(\frac{\partial v_2}{\partial x_2}\right)^2 + \frac{1}{2}\left(\frac{\partial v_1}{\partial x_2} + \frac{\partial v_2}{\partial x_1}\right)^2\right] dx\right]^{\frac{1}{2}}$$

$$\leq C\left[\int_\Omega H\left(\frac{\partial u_1}{\partial x_1} + \frac{\partial u_2}{\partial x_2}\right) dx\right]^{\frac{1}{2}} \left[\|v_1\|^2_{H^1(\Omega)} + \|v_2\|^2_{H^1(\Omega)}\right]^{\frac{1}{2}}$$

$$\leq C\left[\int_\Omega H\left(\frac{\partial u_1}{\partial x_1} + \frac{\partial u_2}{\partial x_2}\right) dx\right]^{\frac{1}{2}} \|H\|_{H^0(\Omega)} \quad .$$

Thus the proof of the lemma is complete.

Now we return to bilinear form (5.4.8) and show that it satisfies condition 5.2.2 of theorem 5.2.1. Let $(u_1,u_2,\xi,H) \in H_1$ and choose $v_1 = u_1 + z_1$, $v_2 = u_2 + z_2$, then we have

(5.4.18)

$$B((u_1,u_2,\xi,H),(v_1,v_2,\eta,\tau)) = \int_\Omega \Bigg\{\lambda\gamma^2\xi\eta + 2\mu\left[\left(\frac{\partial u_1}{\partial x_1}\right)^2 + \left(\frac{\partial u_2}{\partial x_2}\right)^2 + \frac{1}{2}\left(\frac{\partial u_1}{\partial x_2} + \frac{\partial u_2}{\partial x_1}\right)^2\right] + 2\mu\left[\frac{\partial u_1}{\partial x_1}\frac{\partial z_1}{\partial x_1} + \frac{\partial u_2}{\partial x_2}\frac{\partial z_2}{\partial x_2} + \frac{1}{2}\left(\frac{\partial u_1}{\partial x_2} + \frac{\partial u_2}{\partial x_1}\right)\left(\frac{\partial z_1}{\partial x_2} + \frac{\partial z_2}{\partial x_1}\right)\right]$$

$$- \left[H\left(\gamma\eta - \frac{\partial u_1}{\partial x_1} - \frac{\partial u_2}{\partial x_2}\right) + \tau\left(\gamma\xi - \frac{\partial u_1}{\partial x_1} - \frac{\partial u_2}{\partial x_2}\right)\right]$$

$$+ H\left(\frac{\partial z_1}{\partial x_1} + \frac{\partial z_2}{\partial x_2}\right)\Bigg\}dx \quad .$$

By choosing $z_1$ and $z_2$ by the use of lemma 5.4.3, the expression (5.4.18) reduces to

$$B((u_1,u_2,\xi,H),(v_1,v_2,\eta,\tau)) = \int_\Omega \Bigg\{2\mu\left[\left(\frac{\partial u_1}{\partial x_1}\right)^2 + \left(\frac{\partial u_2}{\partial x_2}\right)^2 + \frac{1}{2}\left(\frac{\partial u_1}{\partial x_2} + \frac{\partial u_2}{\partial x_1}\right)^2\right]$$

$$+ \left(\frac{\partial u_1}{\partial x_1} + \frac{\partial u_2}{\partial x_2}\right)\left(\left(2\mu+1\right)H + \tau\right) - H\gamma\eta - \tau\gamma\xi +$$

$$+ \lambda\gamma^2\xi\eta + H\left(\frac{\partial z_1}{\partial x_1} + \frac{\partial z_2}{\partial x_2}\right)\Bigg\}dx \quad .$$

Let $\eta = (2\mu+1)\xi$ and $\tau = -(2\mu+1)H$. Now by the use of lemmas 5.4.1 and 5.4.3 we obtain

(5.4.19)

$$|B(u_1,u_2,\xi,H;v_1,v_2,\eta,\tau)| \geq C\Big[2\mu\Big[\|u_1\|^2_{H^1_0(\Omega)} + \|u_2\|^2_{H^1_0(\Omega)}\Big] + \|H\|^2_{L_2(\Omega)} +$$

$$+ \lambda\gamma^2\xi^2(2\mu+1)\Big] \geq CP\|(u_1,u_2,\xi,H)\|^2_{H_1} ,$$

where $P = \min(\mu,\lambda^{\frac{1}{2}}\gamma)$ and $C$ is independent of $\lambda$ and $\mu$. On the other hand

$$\|(v_1,v_2,\eta,\tau)\|_{H_1} \leq CQ\|(u_1,u_2,\xi,H)\|_{H_1} ,$$

where $Q = \max(\mu,1)$. Hence we have:

(5.4.20)

$$|B((u_1,u_2,\xi,H;v_1,v_2,\eta,\tau)| \geq \frac{CP}{Q}\|(u_1,u_2,\xi,H)\|_{H_1}\|(v_1,v_2,\eta,\tau)\|_{H_1} .$$

We observe that one can always normalize the problem so that $\mu = 1$, and $C$ is independent of $\lambda$. We point out that the functions $z_1$ and $z_2$ satisfy the differential equations

(5.4.21)

$$\frac{\partial^2 u_1}{\partial x_1^2} + \frac{1}{2}\left(\frac{\partial^2 u_1}{\partial x_2^2} + \frac{\partial^2 u_2}{\partial x_1 \partial x_2}\right) = \frac{\partial H}{\partial x_1}$$

((5.4.21) continued)

$$\frac{\partial^2 u_2}{\partial x_2^2} + \frac{1}{2}\left(\frac{\partial^2 u_1}{\partial x_2 \partial x_2} + \frac{\partial^2 u_2}{\partial x_1^2}\right) = \frac{\partial H}{\partial x_2} \quad ,$$

with zero boundary conditions. Equations (5.4.21) are precisely equations (5.4.1) with $\lambda = 0$ and $\mu = \frac{1}{2}$. We may state our results as

Theorem 5.4.2. The bilinear form (5.4.8) defined on $H_1 x H_2$ (as specified above), satisfies all the assumptions of theorem 5.2.1, where the constants $C_i$ , $i = 1,2,3$ are independent of $\lambda$ and $\gamma$ , provided $0 < \alpha < \lambda\gamma^2 < \beta < \infty$ . We may now use the above theorem to find a unique element $(u_{1,0}, u_{2,0}, \xi_0, H_0) \in H_1$ such that

(5.4.22)

$$B((u_{1,0}, u_{2,0}, \xi_0, H_0),(v_1, v_2, \eta, \tau)) = -\int g_1 v_1 dx - \int g_2 v_2 dx$$

for all $(v_1, v_2, \eta, \tau) \in H_2$ .

First we consider the case where $\lambda < \infty$ . It is easily seen that $u_{1,0}, u_{2,0}, \xi_0$ and $H_0$ satisfy the equations

(5.4.23)

$$2\mu\left[\frac{\partial^2 u_{1,0}}{\partial x_1^2} + \frac{1}{2}\left(\frac{\partial^2 u_{1,0}}{\partial x_2^2} + \frac{\partial^2 u_{2,0}}{\partial x_1 \partial x_2}\right)\right] = g_1 + \frac{\partial H_0}{\partial x_1} \quad ,$$

(5.4.24)

$$2\mu\left[\frac{\partial^2 u_{1,0}}{\partial x_2^2} + \frac{1}{2}\left(\frac{\partial^2 u_{1,0}}{\partial x_1 \partial x_2} + \frac{\partial u_{2,0}}{\partial x_2^2}\right)\right] = g_2 + \frac{\partial H_0}{\partial x_2} \quad ,$$

(5.4.25) $$\lambda\gamma^2\xi_0 - H_0\gamma = 0 \quad ,$$

(5.4.26) $$\gamma\xi_0 - \frac{\partial u_{1,0}}{\partial x_1} - \frac{\partial u_{2,0}}{\partial x_2} = 0 \quad .$$

Thus we have

(5.4.27) $$\xi_0 = \frac{1}{\gamma}\left[\frac{\partial u_{1,0}}{\partial x_1} + \frac{\partial u_{2,0}}{\partial x_2}\right]$$

(5.4.28) $$H_0 = \lambda\left[\frac{\partial u_{1,0}}{\partial x_1} + \frac{\partial u_{2,0}}{\partial x_2}\right]$$

Hence, $u_{1,0}$ and $u_{2,0}$ satisfy equations (5.4.1).

We now consider the case $\lambda \to \infty$ . Let $\gamma = \lambda^{-\frac{1}{2}}$ . In this case our bilinear form takes the form

$$B((u_1,u_2,\xi,H),(v_1,v_2,\eta,\tau)) = \int_\Omega \Bigg[\xi\eta + 2\mu\left[\frac{\partial u_1}{\partial x_1}\frac{\partial v_1}{\partial x_1} + \frac{\partial u_2}{\partial x_2}\frac{\partial v_2}{\partial x_2} + \right.$$

$$\left. + \frac{1}{2}\left(\frac{\partial u_1}{\partial x_2} + \frac{\partial u_2}{\partial x_1}\right)\left(\frac{\partial v_1}{\partial x_2} + \frac{\partial v_2}{\partial x_1}\right)\right] -$$

$$+ H\left(\frac{\partial v_1}{\partial x_1} + \frac{\partial v_2}{\partial x_2}\right) + \tau\left(\frac{\partial u_1}{\partial x_1} + \frac{\partial u_2}{\partial x_2}\right)\Bigg]dx \quad .$$

Consequently our variational principle applies to the case of incompressible materials as well. In this case we have $\xi = 0$ and $u_1, u_2$ and $H$ satisfy the equations:

$$2\mu\left[\frac{\partial^2 u_1}{\partial x_1^2} + \frac{1}{2}\left(\frac{\partial^2 u_1}{\partial x_2^2} + \frac{\partial^2 u_2}{\partial x_1 \partial x_2}\right)\right] + \frac{\partial H}{\partial x_1} = g_1 \quad ,$$

$$2\mu\left[\frac{\partial^2 u_2}{\partial x_2^2} + \frac{1}{2}\left(\frac{\partial^2 u_1}{\partial x_1 \partial x_2} + \frac{\partial^2 u_2}{\partial x_1^2}\right)\right] + \frac{\partial H}{\partial x_2} = g_2 \quad ,$$

$$\frac{\partial u_1}{\partial x_1} + \frac{\partial u_2}{\partial x_2} = 0$$

In view of the last equation the above equations may be re-written as

$$\mu \Delta u_1 + \frac{\partial H}{\partial x_1} = g_1$$

$$\mu \Delta u_2 + \frac{\partial H}{\partial x_2} = g_2$$

$$\frac{\partial u_1}{\partial x_1} + \frac{\partial u_2}{\partial x_2} = 0 \quad ,$$

which are the equations of elasticity for incompressible bodies (see [11], p. 79 and [9]). It follows immediately that if $\lambda \to \infty$ then $u_1$ and $u_2$ converge to the solution corresponding to the incompressible case. In this connection see [10, p. 48].

5.4.3. Comments on the variational principles for second order systems.

We considered in the previous section only a typical problem in the theory of elasticity. All the ideas and techniques used in section 5.3 may be applied easily to second order systems. Many other important systems in the applications such as the system of differential equations encountered in the study of nuclear reactors and the equations arising in the theory of optimal control may be treated in a similar fashion.

The technique of Lagrange multipliers may be applied to handle the differential equations with interface conditions and many other special problems. Here we shall not go into details.

5.5 Variational principles for higher order equations and systems.

The problems in applications which give rise to higher order differential equations, e.g., equations of plates, shells etc., may be dealt with in a very similar manner as indicated in the previous section. Therefore we shall not study these cases here.

The variational principles in connection with higher order equations and higher order systems may be introduced in a similar and obvious fashion as indicated for the second order equations and systems.

References.

[1]. L. V. Kantorovich, G. P. Akilov, "Functional analysis in normed spaces," (Russian translation), New York, McMillan Press, (1964).

[2]. J. Nečas, Sur une méthode pour résoudre les equations a dérivées partielles du type elliptique vosine de la variationnelle, Ann. Soc. Norm. Sup. Pisa, Ser III, 16, 4 (1962), pp. 305-326.

[3]. L. Nirenberg, "Remarks on strongly elliptic partial differential equations," Comm. Pure Appl. Math. 8 (1955), pp. 649-675.

[4]. J. Nečas, Les méthodes directes dans la théorie des equations elliptiques, Academia, Prague, (1967).

[5]. I. Babuška, Error-Bounds for Finite Element Method Num. Math. 16 (1971), pp. 322-333.

[6]. I. Babuška, The Finite Element Method with Lagrangian Multipliers, Tech. Note BN-724, (Jan. 1972). The Institute for Fluid Dynamics and Applied Mathematics, University of Maryland. To appear in Num. Math.

[7]. S. G. Mikhlin, The Problem of the Minimum of a Quadratic Functional, Holden-Day, Inc. San Francisco, London, Amsterdam, (1965).

[8]. I. Hlaváček, J. Nečas, On Inequality of Korn's type, Arch. Rat. Mech. Anal., 36 (1970), pp. 305-311, 312-334.

[9]. J. Fuka, Das zweite Problem der ebenen Elastizitätstheorie für incompressible Körper, Aplikace Matematiky 7 (1962), pp. 21-36.

[10]. R. J. Knops. L. E. Payne, "Uniqueness Theorems in Linear Elasticity, Springer Tracts in Nat. Phil. Vol. 19, Springer-Verlag, New York, Heidleberg, Berlin, (1971).

[11]. I. S. Sokolnikoff, "Mathematical Theory of Elasticity, McGraw-Hill Book Co., New York, Toronto, London, (1956).

[12]. I. Babuška, K. Rektorys, F. Vyčichlo, Mathematische Elastizitätstheorie der ebenen Probleme, Akademie-Verlag, Berlin, (1960).

[13]. J. L. Lions, Lectures on elliptic partial differential equations. Tata Institute, Bombay, (1957).

# CHAPTER 6. RATE OF CONVERGENCE OF THE FINITE ELEMENT METHOD.

## 6.1. Introduction.

In chapter 5 we introduced variational principles for solving boundary value problems for elliptic partial differential equations. In this chapter we shall deal with the problem of finding an approximate method for calculating the solution. Our main aim is to find the stationary point for the same bilinear form, where we restrict the spaces $H_1$ and $H_2$ to the finite dimensional subspaces $M_1$ and $M_2$ respectively, i.e., we seek the stationary point of our bilinear form on $M_1 \times M_2$ where $M_1 \subset H_1$ and $M_2 \subset H_2$. In the applications the spaces $M_1$ and $M_2$ are called the spaces of trial functions and the space of test functions respectively. By a suitable choice of the subspaces $M_1$ and $M_2$ we obtain the approximate solution which is near the actual solution. The approach which will be described in the next section, will give us the rate of convergence for the finite element method.

## 6.2. Approximation of the variational principle.

In this section we shall state and prove a simple theorem which together with theorem 5.2.1 forms the basis for the study of the rate of convergence in the finite element approximation.

Theorem 6.2.1. If

I) The hypotheses of theorem 5.2.1 holds,

II) $M_1$ and $M_2$ are given subspaces[†] of $H_1$ and $H_2$ respectively such that

(6.2.1)

$$\inf_{\substack{u \in M_1 \\ \|u\|_{H_1}=1}} \sup_{\substack{v \in M_2 \\ \|v\|_{H_2} \leq 1}} |B(u,v)| \geq d_2(M_1,M_2)\|u\|_{H_1} \ , \quad d_2(M_1,M_2) > 0 \ ,$$

and for every $v \in M_2$

(6.2.2) $$\sup_{u \in M_1} |B(u,v)| > 0 \ ,$$

III) For given $f \in H_2'$, $u_0 \in H_1$ denotes the unique element such that

(6.2.3) $$B(u_0,v) = f(v) \ ,$$

holds for every $v \in H_2$ [the existence of such $u_0$ is assured by theorem 5.2.1],

IV) There exists $\omega \in M_1$ such that

(6.2.4) $$\|u_0 - \omega\|_{H_1} \leq \delta \ ,$$

† We mean here a linear but not necessarily closed set of $H_1$.

V) $\hat{u}_0 \in M_1$, $M_1$ is the closure of $M_1$ in $H_1$ such that

$$B(\hat{u}_0,v) = f(v) \quad , \tag{6.2.5}$$

for all $v \in \overline{M}_2$, then

VI)

$$\|u_0-\hat{u}_0\|_{H_1} \leq 1 + \left[\frac{C_1}{d_2(M_1,M_2)}\right]\delta \quad . \tag{6.2.6}$$

Proof. Let $P$ denote the projection (orthogonal) operator which maps $H_2$ onto $\overline{M}_2$ and let $R$ be the mapping constructed in the proof of theorem 5.2.1, i.e.,

$$(R(u),v)_{H_2} = B(u,v) \quad . \tag{6.2.7}$$

Now consider the continuous mapping of $H_1$ into $\overline{M}_2$ given by $Q = PR$. Let $S$ be a mapping of $\overline{M}_1$ onto $\overline{M}_2$ such that

$$(Su,v)_{H_2} = B(u,v) \ , \forall u \in \overline{M}_1 \ , \forall v \in \overline{M}_2 \quad . \tag{6.2.8}$$

Such a mapping may be constructed in the same way as in the proof of theorem 5.2.1. We show that $S$ is the restriction of $Q$ to $\overline{M}_1$, i.e., $S = PR = Q$ on $\overline{M}_1$. Thus, we must verify

$$(PRu,v)_{H_2} = B(u,v) \ , \forall u \in \overline{M}_1 \ , \forall v \in \overline{M}_2 \ . \tag{6.2.9}$$

Clearly

$$(PRu,v)_{H_2} = (Ru,Pv)_{H_2} = (Ru,v_{H_2}) = B(u,v) \quad ,$$

where use is made of the symmetry of $P$ and the fact that $Pv = v$ for $v \in \overline{M}_2$ . Now let $v_0 \in \overline{M}_2$ be such that $(v_0,v)_{H_2} = f(v)$ , $v \in M_2$ . Obviously, $v_0 = PRu_0$ . Hence

$$\hat{u}_0 = S^{-1}PRu_0 \quad . \tag{6.2.10}$$

By putting $u_0 = \omega - z$ , we obtain

$$\hat{u}_0 - u_0 = S^{-1}PRu_0 - u_0 = S^{-1}PR\omega + S^{-1}PRz - \omega - z =$$

$$= - z + S^{-1}PRz \quad ,$$

and therefore,

$$\|u_0 - \hat{u}\|_{H_1} \leq \left[1 + \frac{C_1}{d_2(M_1,M_2)}\right]\delta \quad , \tag{6.2.11}$$

which was to be proved.

<u>Remarks</u>. 1. We often have $M_1 = M_2$ . The subspace $M_1$ and $M_2$ are referred to as the space of trial functions and the space of test functions respectively.

2. As in theorem 5.2.1 the condition (5.2.2) may be replaced by the condition

$$\sup_{\substack{u \in M_1 \\ |u|_{H_1} \leq 1}} |B(u,v)| \geq d_3(M_1,M_2)\|v\|_{H_1} \tag{6.2.12}$$

3. In both theorems 5.2.1 and 6.2.1 for the construction of the mapping $R$ we have made use of the condition (5.2.1). $R$ maps $H_1$ into $H_2$ and in particular it maps $\bar{M}_1$ into $\bar{M}_2$. We may study the angle between the spaces $R(M_1)$ and $M_2$ in $H_2$. The notion of the angle between spaces was introduced in [2] and [3]. The conditions in (6.2.1) and (6.2.12) are equivalent with the statement about the angle of the space $R(M_1)$ and $M_2$ in $H_2$.

4. As in the case of theorem 5.2.1, theorem 6.2.1 may be extended to reflexive Banach spaces.

5. The proof of theorem 6.2.1 is patterened after that in [4].

## 6.3. Finite element method for the second order differential equations

6.3.1. Introduction. In chapter 5 section 3 we analyzed selected boundary value problems for the second order elliptic partial differential equations. In the present chapter we shall use the various variational principles to discuss the finite element method.

6.3.2. <u>Finite element method based on the Dirichlet type variational principle-boundary conditions of the first type.</u>

We assume that a (t,k)-regular system $S_h^{(t,k)}$ is given with $k \geq 1$. Let $M_1 = S_h^{(t,k)} \subset H^1(\Omega) = M_2$. In the finite element method we make use of the condition (5.3.26) for $u_h \in \overline{M}_1$ and we require that (5.3.26) must hold for every $v \in M_2$ with $F(v)$ given by (5.3.24). Let $u_h \in \overline{M}_1$ be the element satisfying (5.3.26). Our objective is to study the magnitude of the error $u_0 - u_h$. Theorem 6.2.1 will play a central role in this study. First we investigate whether the condition 6.2.1 is satisfied.

To this end we recall the technique used in the proof of theorem 5.3.2, to prove the condition (5.2.2). For a given $u \in H^1(\Omega)$ we construct $v = u + z$ such that

$$\|z\|_{H^3(\Omega)} \leq C\|u\|_{H^1(\Omega)}, \quad \text{and}$$

$$|B(u,v+z)| \geq C\|u\|_{H^1(\Omega)} \|u+z\|_{H^1(\Omega)},$$

where $C$ is independent of $u$. We emphasize that $z$ was the solution of the adjoint problem (5.3.51) and therefore $z \notin M_2$ in general. However, we shall show that we can approximate $z$ by an element $z_h \in M_2$ which will serve our purpose. Consider $z_h \in S_h^{t,k}$ such that

$\|z-z_h\|_{H^1(\Omega)} \leq Ch^{\mu}\|z\|_{H^3(\Omega)}$, $\mu = \min(t-1,2) > 0$. The existence $z_h$ is assured by the assumed properties of $S_h^{t,k}$. Hence

$$|B(u,u+z_h)| = |B(u,u+z) + B(u,z_h-z)|$$

$$\geq |B(u,u+z)| - C\|u\|_{H_1} \|z_h-z\|_{H_1} \geq C\|u\|_{H_1}^2 (1-Ch^{\mu}) \quad .$$

Since $\|u+z_h\|_{H^1(\Omega)} \leq C\|u\|_{H^1(\Omega)}$, it follows that 6.2.1 is satisfied for sufficiently small $h$. Reversing the role of $u$ and $v$ we obtain (6.2.2).

In the finite element treatment of the boundary value problem (5.3.1), (5.3.2), we obtain (6.2.4) by using the regularity properties of the solution. Thus from theorem 6.2.1 we have the following result.

Theorem 6.3.1. If

I) Given the boundary value problem (5.3.1), (5.3.2), where all the assumptions of section 5.3.1 of chapter 5 are satisfied;

II) Given a $(t,k)$-system $S_h^{t,k}(\Omega)$, with $k \geq 1$, then

III) For sufficiently small $h$ there exists a unique solution $u_h \in S_h^{t,k}(\Omega)$ such that $B(u_h,v) = F(v)$ for every $v \in S_h^{t,k}(\Omega)$, where $B(u,v)$ and $F(v)$ are defined by (5.3.21) and (5.3.24) respectively.

IV) The estimate

(6.3.1)

$$\|u_0-u_h\|_{H^1(\Omega)} \leq C\, h^{\mu_1}\|f\|_{H^\ell(\Omega)} + h^{\mu_2}\|g\|_{H^s(\Omega^\bullet)}$$

holds, where $u_0$ is the exact solution of the problem (5.3.1), (5.3.2); $\mu_1$ and $\mu_2$ are given by

(6.3.2) $$\mu_1 = \min(t-1, \ell+1) \quad ,$$

(6.3.3) $$\mu_2 = \min(t-1, s+\frac{1}{2}) \quad ,$$

with $\ell \geq -1$ , $s \geq -\frac{1}{2}$ .

Remarks. 1) It is clear that theorem 6.3.1 may hold only for sufficiently small $h$ . However, if the operator is positive definite, then of course the solution $u_h$ exists for all $h$ . We emphasize that the finite element method applies, as could be expected, whenever the problem to be solved has a unique solution.

2) The version of the finite element method which we have introduced is based theoreticaly on the Galerkin method. It appears that the first detailed analysis of the Galerkin method for the second order equations was undertaken in [5]. Relevant results concerning the Galerkin method can also be found in [6]. For a general abstract treatment of the Galerkin method and other similar methods we refer the reader to [1] and the references contained therein.

3) We observe that if we seek instead $u \in \overline{S_h^{t,k}}$ satisfying (5.3.26), for every $v \in S_h^{t,k}$ , a $v \in \overline{S_h^{t,k}(\Omega)}$ which satisfies (5.3.27) for every $u \in S_h^{t,k}(\Omega)$ , we obtain an approximate solution of the adjoint problem (5.3.3), (5.3.4).

Theorem 6.2.1 together with the results of section (5.3.2) gives the error estimate in the norm of $H^1(\Omega)$ . It would be of interest to investigate the rate of convergence in the space $H^\alpha(\Omega)$ with $\alpha > 1$ and $\alpha < 1$ .

First we consider the case $\alpha > 1$ . In this connection we prove the following result.

<u>Theorem 6.3.2.</u> If

I) All the assumptions of theorem 6.3.1 hold;

II) the system $S_h^{t,k}(\Omega)$ is regular and satisfies the inverse assumption on $\Omega$ (see chapter 4);

then

III) the approximate solution $u_h$ is such that for $1 \leq \alpha \leq k$ ,

$$\|u_0-u_h\|_{H^\alpha(\Omega)} \leq C\left[h^{\mu_1'}\|f\|_{H^\ell(\Omega)} + h^{\mu_2'}\|g\|_{H^s(\Omega^\cdot)}\right] , \tag{6.3.4}$$

where

$$\mu_1' = \min(t-\alpha, \ell+2-\alpha) \tag{6.3.5}$$

$$\mu_2' = \min(t-\alpha, s+\frac{3}{2}-\alpha) , \tag{6.3.6}$$

provided that

$$\ell \geq -1 \ , \ s \geq -\frac{1}{2} \ , \ \mu_1, \ \mu_2 \geq 0 \ .$$

Proof. By the basic properties of (t,k)-regular system and the regularity properties of the exact solution $u_0$ it follows that there exist a $\omega_0 \in S_h^{t,k}(\Omega)$ such that

$$\|u_0-\omega_0\|_{H^\alpha(\Omega)} \leq C\left[h^{\mu_1'}\|f\|_{H^\ell(\Omega)} + h^{\mu_2'}\|g\|_{H^s(\Omega^\cdot)}\right] \tag{6.3.7}$$

holds for all $0 \leq \alpha \leq \min(t,k)$ , where $\mu_1'$ and $\mu_2'$ are given by (6.3.5) and (6.3.6) respectively. Therefore we may write

(6.3.8)

$$\|u_0-u_h\|_{H^\alpha(\Omega)} \leq \|u_0-\omega_0\|_{H^\alpha(\Omega)} + \|\omega_0-u_h\|_{H^\alpha(\Omega)} \ .$$

Since $\omega_0 - u_h \in S_h^{t,k}$ , we may apply the inverse assumption to obtain

(6.3.9)

$$\|\omega_0-u_h\|_{H^\alpha(\Omega)} \leq \frac{C}{h^{\alpha-1}} \|\omega_0-u_h\|_{H^1(\Omega)} \ .$$

Since (6.3.7) holds for $\alpha = 1$ , from inequality (6.3.1) we conclude that $\|\omega_0-u_h\|_{H^1(\Omega)}$ is bounded by the expression on the right hand sind of (6.3.1). By the use of this bound in (6.3.9), the assertion of the theorem follows.

We would like to emphasize that the additional inverse assumption on the system $S_h^{t,k}$ is necessary, generally without such a hypothesis theorem 6.3.2 does not hold.

On the other hand use is often made of the spaces $S_h^{t,k}(\Omega)$ without the inverse assumption. In this case we may compute higher derivatives in a different way. We may construct a function $v_h$ such that (6.3.4) holds for $v_h$ rather than for $u_h$. It is possible to construct such $v_h$ from $u_h$ for several choices of $S_h^{t,k}(\Omega)$. For more details on this point the reader may consult [7].

So far we have dealt with case $\alpha \geq 1$. Now we study the case $\alpha \leq 1$. In this connection we state and prove the following result.

Theorem 6.3.3. If

I) The hypotheses of theorem 6.3.1 holds,

II) $\alpha \leq 1$,

then

III)

(6.3.10)

$$\|u_0-u_h\|_{H^\alpha(\Omega)} \leq C\left[h^{\mu_1+\mu'}\|f\|_{H^\ell(\Omega)} + h^{\mu_2+\mu'}\|g\|_{H^s(\Omega^\bullet)}\right] ,$$

where $\mu_1$ and $\mu_2$ are given by (6.3.2) and (6.3.3) respectively and

(6.3.11) $$\mu' = \min(t-1, 1-\alpha) \quad .$$

Proof. Let $\varepsilon_h = u_h - u_0$. In theorem 6.3.1 we obtained an estimate for $\|\varepsilon_h\|_{H^1(\Omega)}$. We note that from the construction of $u_h$ and the fact that $u_0$ is the solution of the original problem it follows that

(6.3.12) $$B(\varepsilon_h, v) = 0 \quad \text{for } \forall v \in S_h^{t,k}(\Omega) \quad .$$

Suppose $\alpha \leq 1$ is given, then there exists $\phi \in H^{-\alpha}(\Omega)$ such that

(6.3.13) $$\|\phi\|_{H^{-\alpha}(\Omega)} = \|\varepsilon_h\|_{H^{\alpha}(\Omega)} \quad .$$

(6.3.14) $$\int_\Omega \varepsilon_h \phi \, d\Omega \geq \frac{1}{2}\|\varepsilon_h\|^2_{H^{\alpha}(\Omega)} \quad ,$$

From the results of section 5.3.2 and chapter 3 it follows that there exists a $V \in H^{2-\alpha}(\Omega)$ such that

(6.3.15) $$\|V\|_{H^{2-\alpha}(\Omega)} \leq C\|\phi\|_{H^{-\alpha}(\Omega)} \quad ,$$

(6.3.16) $$B(u, V) = \int_\Omega \phi u \, dx \quad , \forall u \in H^1(\Omega) \quad .$$

Hence since $\varepsilon_h \in H^1(\Omega)$ , we have

$$B(\varepsilon_h, V) = \int_\Omega \phi \varepsilon_h dx \geq \frac{1}{2}\|\varepsilon_h\|^2_{H^\alpha(\Omega)} \quad . \tag{6.3.17}$$

Moreover since $V \in H^{2-\alpha}(\Omega)$ there exists a $v \in S_h^{t,k}(\Omega)$ such that

$$\|V-v\|_{H^1(\Omega)} \leq Ch^{\mu'} \|V\|_{H^{2-\alpha}(\Omega)} \quad , \tag{6.3.18}$$

where

$$\mu' = \min(t-1, 1-\alpha) \quad . \tag{6.3.19}$$

From (6.3.13) and (6.3.15) it follows that

$$\|V-v\|_{H^1(\Omega)} \leq Ch^{\mu'} \|\varepsilon_h\|_{H^\alpha(\Omega)} \quad . \tag{6.3.20}$$

Now using (6.3.17) and the fact that the bilinear form $B$ satisfies (5.2.1) we obtain

(6.3.21)

$$\frac{1}{2}\|\varepsilon_h\|^2_{H^\alpha(\Omega)} \leq B(\varepsilon_h, V) = B(\varepsilon_h, V-v) \leq C\|\varepsilon_h\|_{H^1(\Omega)} h^{\mu'} \|\varepsilon_h\|_{H^\alpha(\Omega)} \quad ,$$

and therefore

$$(6.3.22) \qquad \|\varepsilon_h\|_{H^{\alpha}(\Omega)} \leq Ch^{\mu'} \|\varepsilon_h\|_{H^1(\Omega)} \quad .$$

(6.3.22) together with theorem 6.3.1 completes the proof of our theorem.

Remarks. 1) We observe that the error in the norms $\|\cdot\|_{H^{\alpha}(\Omega)}$, $\alpha < 1$ is in general smaller than in that norm $\|\cdot\|$. However the increase in the rate is limited by the number $t - 1$, i.e., the rate increases as $\alpha$ decreases to $\alpha = 2 - t$ and then the rate remains constant. Since in the applications we always deal with integer values of $t$ and since we assume $t > 1$, we have $t - 1 \geq 1$. Therefore for the second order differential equations the rate of convergence in $L_2(\Omega)$ is higher by a power of $h$ than the rate of convergence in $H^1(\Omega)$. For differential equations of order $2k$ the rate of convergence is increased by $\mu = \min(t-k, k-\alpha)$, which implies that the use of element with parameter $t$ which is not sufficiently large leads to loss of powers of $h$ in the rate of convergence. It appears that this fact was first observed and stated explicitly (in a slightly different way) in [9]. Relevant material on this topic may be found in [10].

2) We noticed that the rate of convergence in a "lower" norm is governed by the smoothness of the solution of the adjoint problem. Since in general (if the coefficients are not $C^{\infty}$) the solution of the adjoint problem has different smoothness, one does not have the expected gain in the rate of convergence. [see section 6.4 on the

optimality of error estimates.]

3) The result related to the improvement of the rate of convergence for "lower" spaces may be interpreted as an oscillatory character of the error and its derivatives, depending on the values of the parameter $t$ . We illustrate the above remark through a simple example. Consider the boundary value problem

$$- u'' + \frac{1}{4} u = - \sin 2(x - \frac{\Pi}{2}) \quad ,$$

with the boundary conditions

$$u'(0) = u'(\Pi) = 0 \ .$$

This problem is of the type we studied and the exact solution is easily computable. We solve this by the finite element method using two different systems $S_h^{t,k}$ , with $t = 2$ and $t = 3$ . Such systems are easy to construct. We construct the subspace $S_h^{t,k}$ for $h = \frac{\Pi}{n}$ . Thus every $v \in S_h^{t,k}$ may be expressed in the form

$$v = \sum c_i \phi \, (x - \frac{i}{h}) \quad .$$

We take $c_i = 0$ if $\phi(x - \frac{i}{h}) = 0$ , $x \in \langle 0, \Pi \rangle$ ,

$$\phi = \phi_1(\xi) = \begin{cases} 0 & \text{for } |\xi| \geq 1 \\ \xi + 1 & \text{for } -1 \leq \xi \leq 1 \\ 1 - \xi & \text{for } 0 \leq \xi \leq 1 \end{cases} .$$

Thus we obtain the $S_h^{2,1}(\Omega)$ system (see e.g. [25]). We construct $S_h^{3,2}(\Omega)$ system by letting

$$0 \quad \text{for} \quad |\xi| \geq \frac{3}{2}$$

$$\phi = \phi_2 = \begin{cases} \frac{1}{2}(\xi + \frac{3}{2})^2 & \text{for } \xi \in \langle -\frac{3}{2}, -\frac{1}{2} \rangle , \\ -\xi^2 + \frac{3}{4} & \text{for } \xi \in \langle -\frac{1}{2}, \frac{1}{2} \rangle , \\ \frac{1}{2}(\xi - \frac{3}{2})^2 & \text{for } \xi \in \langle \frac{1}{2}, \frac{3}{2} \rangle \end{cases} .$$

We choose $h = \frac{\Pi}{16}$ and study the behavior of the error. Fig. 6.1 gives the error for the case $\phi = \phi_1$ i.e., $t = 2$ (the points $ih$ are encircled). Fig. 6.2 gives the error for the case $\phi = \phi_2$ i.e., $t = 3$ . We notice the following typical differences: (i) The error is significnatly less in the case $t = 3$ , then in the case $t = 2$ .

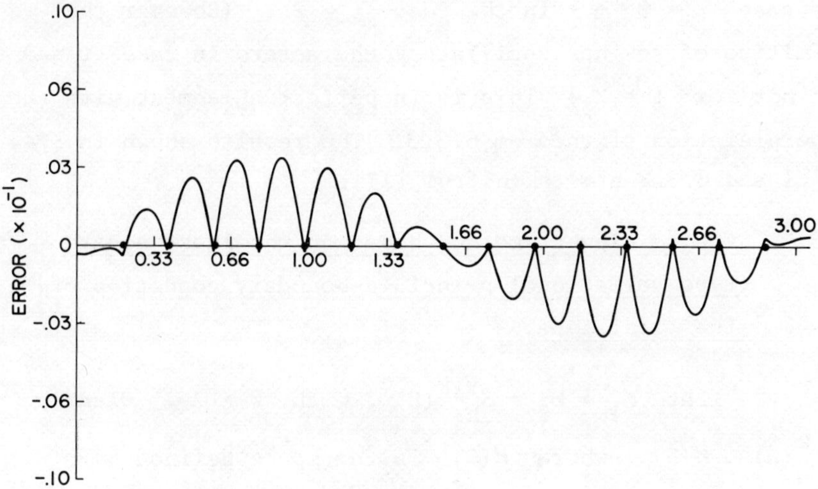


Fig. 6.3.1

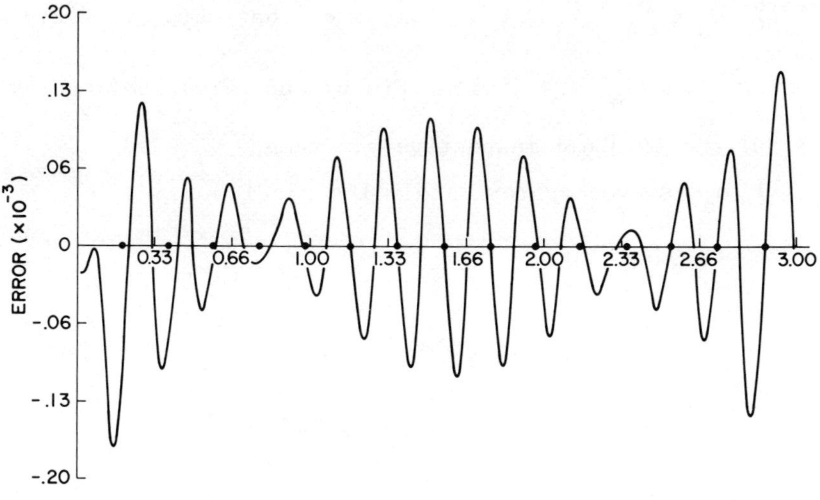


Fig. 6.3.2

(ii) In the interval of length $h$ there is oscillation in the case $t = 3$ and in the case $t = 2$ .. (However the primitive of $\varepsilon$ has oscillatory characters in case $t = 3$ but not for $t = 2$.) This is in perfect agreement with the inperpretation of theorem 6.3.3. The results shown in Fig. 6.3.1 and 6.3.2 are taken from [11].

6.3.3. Finite element method based on the least square type variational principle-boundary condition of the first type.

Let $M_1 = M_2 = S_h^{t,k}(\Omega)$ , with $k \geq 2$ . Clearly $S_h^{t,k}(\Omega) \subset H(\Omega)$ , where $H(\Omega)$ is the space defined by (5.3.63). In the same way as in section 5.3.3 we may easily show that all the assumptions of the theorem 6.2.1 are satisfied. We have the following result.

Theorem 6.3.4. If

I) $u_h \in \overline{S_h^{t,k}}(\Omega)$ , $k \geq 2$ is such that $B(u_h,v) = F(v)$ for every $v \in S_h^{t,k}(\Omega)$ , with $B(u,v)$ and $F(v)$ defined by (5.3.59) and (5.3.66) respectively, then

II) $u_h$ solves approximately the problem

$$-\Delta u + \alpha u = f \quad \text{in} \quad \Omega$$

$$\frac{\partial u}{\partial n} + \beta u = g \quad \text{in} \quad \dot{\Omega} \quad ,$$

II) (6.3.26)

$$\int_{\Omega} (-\Delta\varepsilon + c\varepsilon) U dx + \oint_{\dot{\Omega}} \left(\frac{\partial\varepsilon}{\partial n} + \beta\varepsilon\right) U ds = \int_{\Omega} \varepsilon^2 dx$$

Proof. Since $\mathcal{E}(\overline{\Omega})$ is dense in $\mathcal{H}(\Omega)$ it is sufficient to prove (6.3.26) for $\varepsilon \in \mathcal{E}(\overline{\Omega})$ and then $U \in E(\overline{\Omega})$ as well. Integrating the left hand side of (6.3.26) by parts and making use of the boundary conditions we obtain

$$\int_{\Omega} (-\Delta\varepsilon + c\varepsilon) U dx + \int_{\dot{\Omega}} \left(\frac{\partial\varepsilon}{\partial n} + \beta\varepsilon\right) U dx = \int_{\Omega} \varepsilon(-\Delta U + cU) dx - \int_{\dot{\Omega}} \frac{\partial\varepsilon}{\partial n} U dx$$

$$- \int_{\dot{\Omega}} \varepsilon\beta U ds + \int_{\dot{\Omega}} \left(\frac{\partial\varepsilon}{\partial n} U + \beta\varepsilon U\right) ds$$

$$= \int_{\Omega} \varepsilon^2 dx \quad .$$

The equality (6.3.26) is proved. From the results of chapter 3 using (5.3.63) we have

$$\|U\|_{H^{7/2}(\Omega)} \leq C \|\varepsilon\|_{\mathcal{H}(\Omega)} \quad ,$$

III) If $u_0$ is the exact solution of the problem then we have for $\ell \geq 0$ , the estimate

(6.3.23) $$\|u_0-u_h\|_{H} \leq C\left[h^{\mu_1}\|f\|_{H^{\ell}(\Omega)} + h^{\mu_2}\|g\|_{H^{s}(\dot{\Omega})}\right] ,$$

where

(6.3.24) $$\mu_1 = \min(t-2,\ell) \quad , \quad \ell \geq 0$$

(6.3.25) $$\mu_2 = \min(t-2,s-\frac{1}{2}) \quad , \quad s \geq \frac{1}{2}$$

The norm $\|\cdot\|_{H}$ is a very special norm. It would be therefore of interest to consider the error $\varepsilon = u_0 - u_h$ in the space $H^n(\Omega)$ , $n \leq 1$ . For technical convenience we treat the case $n = 0$ , but the appeoach used is quite general. We first prove a simple lemma.

Lemma 6.3.1. If

I) $\varepsilon \in H(\Omega)$ and $U$ is the solution of the boundary value problem

$$-\Delta U + cU = \varepsilon \quad \text{on} \quad \Omega$$

$$\frac{\partial U}{\partial n} + \beta U = 0 \quad \text{on} \quad \dot{\Omega}$$

then

and therefore

$$\|U\|_{H^3(\dot{\Omega})} \leq \|\varepsilon\|_{H(\Omega)} \quad .$$

As in section 6.3.2 let V be the solution of the problem

$$- \Delta V + c V = U \quad \text{on} \quad \Omega$$

$$\frac{\partial V}{\partial n} + \beta V = U \quad \text{on} \quad \dot{\Omega} \quad .$$

Then

$$\|V\|_{H^{9/2}(\Omega)} \leq C\|\varepsilon\|_{H(\Omega)} \quad ,$$

and by lemma 6.3.1

$$B(\varepsilon, V) = \|\varepsilon\|^2_{L_2(\Omega)} \quad .$$

Since there exists a $w \in S_h^{t,k}(\Omega)$ such that

$$\|V-w\|_{H(\Omega)} \leq Ch^{\mu'}\|V\|_{H^{9/2}(\Omega)} \leq h^{\mu'}C\|\varepsilon\|_{H(\Omega)} \quad ,$$

$(\mu' = \min(t-2,\frac{5}{2}))$ we have

$$\|\varepsilon\|^2_{L_2(\Omega)} \leqslant B(\varepsilon,V) = B(\varepsilon,V-w) \leqslant C\|\varepsilon\|_{H(\Omega)} h^{\mu'} \|\varepsilon\|_{H(\Omega)}$$

Hence

$$\|\varepsilon\|_{L_2(\Omega)} \leqslant Ch^{\frac{\mu'}{2}} \|\varepsilon\|_{H(\Omega)} \quad .$$

Thus we have proved the following theorem.

Theorem. 6.3.5. If

I) The assumptions of theorem 6.3.4 hold,

then

II) (6.3.27)

$$\|u_h - u_0\| \leqslant C\left[h^{\mu_1 + \frac{\mu'}{2}} \|f\|_{H^\ell(\Omega)} + h^{\mu_2 + \frac{\mu'}{2}} \|g\|_{H^s(\Omega)}\right] ,$$

where $\mu_1$ and $\mu_2$ are given by (6.3.24) and (6.3.25) respectively and

(6.3.28) $$\mu' = \min(t-2,\tfrac{5}{2}) \quad ,$$

Remarks. The comparison of the least square method as formulated above with the method based on the Dirichlet principle as described in section 6.3.2, reveals that the results obtained by the least square are not as good as those obtained by the method of section 6.3.2. The disadvantage of the least square method is apparent from the necessity of the choice $k = 2$, instead $k = 1$ for the method based on the Dirichlet principle. Further the rate of convergence is not the optimal one. Therefore it seems there is no advantage gained by the use of the least square method as described in the previous section. However, the method of least square could be suitabily modified to yield improved results, especially in connection with boundary conditions of Dirichlet type [see chapter 7].

We noticed in the previous section that the use of the lower norm improves the rate of convergence. As we saw this indicates the "ocilatory" character of the error.

### 6.3.4. The variational principle of the Dirichlet type Boundary condition of the second type.

In section 6.3.2 we analyzed the finite element method for solving problems with boundary conditions of the first type. The space used was $H^1(\Omega)$, i.e., a space with no restriction on the boundary $\dot{\Omega}$.

In section 5.3.5 it was shown that the variational principle is readily applicable to problems with homogeneous boundary conditions of the second type provided that the space $H_0^1(\Omega)$ is used in place of $H^1(\Omega)$. By the use of this argument it is easily shown that the finite element method converges and one obtains the rate of convergence in the same way provided that the subspace $M_1$ and $M_2$ satisfy the zero boundary condition. This in general is not easy to accomplish. It is done in practice by the use of curvlinear elements. [See e.g. [14] and others.]

6.3.5. Finite element method based on the variational principle of Dirichlet type with Lagrange multipliers-- boundary condition of the second type.

We recall that in using the classical Dirichlet principle for problems with the boundary conditions of the second type, one encounters serious difficulties in connection with fulfillment of the boundary conditions. One way to avoid these difficulties is to use the method based on the Dirichlet principle in conjunction with the technique of Lagrange multipliers.

In §5.3.6 we studied the bilinear form (5.3.73) defined on $H_1 \times H_2$, with $H_1 = H_2 = H^1(\Omega) \times H^{-1/2}(\Omega^\cdot)$. In theorem 5.3.2 we proved that this bilinear form satisfies all the hypotheses of theorem 5.2.1. Now we investigate the assumptions of the theorem 6.2.1.

Let $M_1 = M_{1,1} \times M_{1,2}$, with $M_{1,1} \subset H^1(\Omega)$, $M_{1,2} \subset H^{-1/2}(\Omega^\cdot)$ such that

$$M_{1,1} = S_{h_1}^{t,k}(\Omega) , \quad k \geq 1 , \quad t \geq 2 ,$$

$$M_{1,2} = S_{h_2}^{t,k}(\Omega^\cdot) , \quad k \geq 1 , \quad t \geq 2 .$$

Moreover, we suppose that the space $M_{1,2}$ satisfies the inverse assumption on $\Omega^\cdot$.

We proceed in the same way as in §5.3.6 to prove that the assumptions of theorem 6.2.1 are fulfilled. To this end, let $(u,\lambda)$ be given with $u \in M_{1,1}$ and $\lambda \in M_{1,2}$. Then there exists $z_1$ defined by (5.3.88) such that

(6.3.29) $$B_1(v,z_1) = \oint_{\Omega^\bullet} \lambda v ds \ , \forall \ v \in H^1(\Omega) \ .$$

Since $\lambda \in M_{1,2}$ and $M_{1,2}$ satisfies the inverse assumption, we have

(6.3.30) $$\|\lambda\|_{H^{1/2}(\Omega^\bullet)} \leq Ch_2^{-1}\|\lambda\|_{H^{-1/2}(\Omega^\bullet)} \ ,$$

and therefore

(6.3.31) $$\|z_1\|_{H^2(\Omega)} \leq Ch_2^{-1}\|\lambda\|_{H^{-1/2}(\Omega^\bullet)} \ .$$

We can construct a $z_2 \in H^1(\Omega)$ as in §5.3.6 such that

(6.3.32) $$\|z_2\|_{H^1(\Omega)} \leq C[\|u\|_{H^1(\Omega)} + \|z_1\|_{H^1(\Omega)}] \ .$$

Let $v = u - z_1 + z_2$ , $\mu = - 2\lambda - T^*(z_2)$ then as in (5.3.93) we have

(6.3.33) $$B\Big((u,\lambda), (u-z_1+z_2,\mu)\Big) \geq C\|(u,\lambda)\|^2_{H_1} \ .$$

We construct $\nu \in S^{t,k}_{h_2}(\Omega)$ , $w_1 \in S^{t,k}_{h_1}(\Omega)$ and $w_2 \in S^{t,k}_{h_1}(\Omega)$ such that

(6.3.34) $$\|T^*(z_2)-\nu\|_{H^{-1/2}(\Omega^\bullet)} \leq Ch_2\|(u,\lambda)\|_{H_1}$$

(6.3.35) $$\|z_1 - w_1\|_{H^1(\Omega)} \leq Ch_2^{-1}h_1\|\lambda\|_{H^{-1/2}(\Omega^\bullet)} \ ,$$

and

(6.3.36) $$\|z_2 - w_2\|_{H^1(\Omega)} \leq Ch_1\|(u,\lambda)\|_{H_1} \ .$$

Thus, from (6.3.34), (6.3.35) and (6.3.36) we obtain

$$|B((u,\lambda),(u-w_1+w_2,-2\lambda-\nu))| = |B((u,\lambda),(u-z_1+z_2,-2\lambda-T^*(z_2))$$
$$+ B((u,\lambda),(z_1-w_1+w_2-z_2,-\nu+T^*(z_2))| \geq C\|u,\lambda\|^2_{H_1}(1-Ch_2^{-1}h_1-Ch_1-Ch_2).$$

The constants appearing in (6.3.36) are independent of $u,\lambda$. Hence there exists a constant $K$ such that if $\frac{h_1}{h_2} \leq K$, and if $h_1$, $h_2$ is sufficiently small, then $(1 - Ch_2^{-1}h_1 - Ch_1 - Ch_2) \geq \eta > 0$. Now the proceeding in the same was as in § 5.3.6, we show that all the assumptions of theorem 6.2.1 are satisfied. Through the use of the properties of the $(t,k)$ system $S_h^{t,k}(\Omega)$, we may state our result as

<u>Theorem 6.3.6.</u> If

I) Given the boundary value problem (5.3.1), (5.3.67) where all the assumptions of §5.3.4 are satisfied,

II) Given $(t_1,k_1)$ system $S_{h_1}^{t_1,k_1}$, with $k_1 \geq 1$, $t_1 \geq 2$ and $(t_2,k_2)$ system $S_{h_2}^{t_2,k_2}(\Omega^{\cdot})$, $k_2 \geq \frac{1}{2}$, which satisfies the inverse assumption;

Then

III) There exists a constant $K$ such that if $\frac{h_1}{h_2} \leq K$ and if $h_1$, $h_2$ are sufficiently small, there is a unique $(u_h,\lambda_{h_2}), u_h \in S_{h_1}^{t_1,k_1}(\Omega)$, $\lambda_{h_2} \in S_{h_2}^{t_2 k_2}(\Omega^{\cdot})$ satisfying

$$(6.3.37) \qquad B((u_h,\lambda_{h_2})(v,\mu)) = F(v,\mu),$$

$$\forall\, v \in S_{h_1}^{t_1,k_1}(\Omega), \ \forall\, \mu \in S_{h_2}^{t_2,k_2}(\Omega^{\cdot}),$$

where $B\Big((u,\lambda),(v,\mu)\Big)$ and $F(v,\mu)$ are given by (5.3.73) and (5.3.74) respectively;

(6.3.38)

$$\text{IV)}\quad \|u_0 - u_{h_1}\|_{H^1(\Omega)} + \left\|\frac{\partial u_0}{\partial n} - \lambda_{h_2}\right\|_{H^{-1/2}(\Omega^\bullet)} \leq C\left[h_1^{\mu_1}\|f\|_{H^\ell(\Omega)} + h_2^{\mu_2}\|g\|_{H^s(\Omega^\bullet)}\right]$$

where $u_0$ denotes the exact solution of our problem, and for $t = \min(t_1,t_2) \geq 2$

$$\mu_1 = \min(t-1,\ell+1) \ , \ \ell \geq 0 \qquad (6.3.39)$$

$$\mu_2 = \min(t-1,s+\tfrac{1}{2}) \ , \ s \geq 0 \ . \qquad (6.3.40)$$

Remarks

1) The use of theorem 6.3.6 is restricted by the assumption that the elements on boundary $\Omega^\bullet$ must be "larger" than the elements in the domain $\Omega$. This is an important consideration in view of the fact that the constant K, $\left(\frac{h_1}{h_2} < K\right)$ in practice is not easy to determine. The inverse assumption for the elements on the boundary, particularly for two-dimensional domains does not appear to be too restrictive. The construction of such elements in practice is fairly easy.

2) The question arises as to whether or not the restriction $\frac{h_1}{h_2} < K$ is essential. We show the necessity of this condition by a simple example. To this end we consider the boundary value problem

$$-\Delta u + u = f \quad \text{on} \quad \Omega \ ,$$

$$u = 0 \quad \text{on} \quad \Omega^\bullet \ ,$$

where $\Omega$ is taken to be the unit circle. For $S_h^{t,k}(\Omega)$ , we use the regular triangulation with piecewise linear functions. For $S_h^{t,k}(\Omega^\bullet)$ choose the piecewise linear function (with respect to the length parameter) with equidistant spacing (see Fig. 6.3.3). In this case we have $t_1 = t_2 = 2$ , $k_1 = k_2 = 1$ , and the inverse assumption is satisfied. Obviously the number of triangles in the circle is of the order $\frac{C_1}{h_1^2}$ , and therefore the same number of unknown parameters. For the same reason there are $\frac{C_2}{h_2}$ unknown parameters of $\lambda$ .

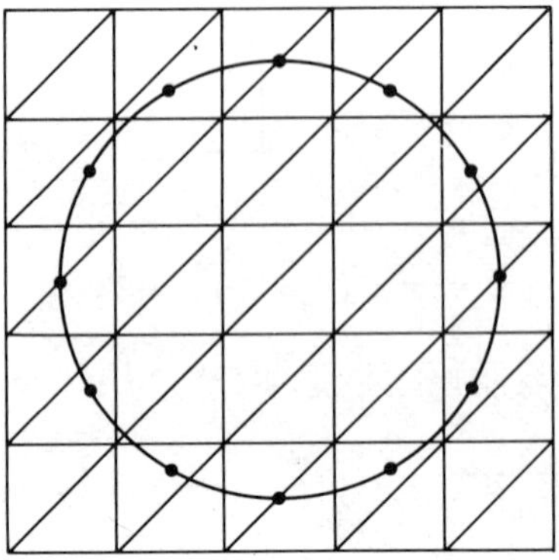

Fig. 6.3.3

On the other hand, we notice that in our case only $\frac{C}{h_1}$ of the functions $u \in S_{h_1}^{t_1,k_1}(\Omega)$ influence the behavior of $u$ on $\Omega^\bullet$ . Now let $h_2 = ph_1$ , for $p > 0$ sufficiently small. Now we can find

$0 \neq \lambda \in S_{h_2}^{t_2,k_2}(\Omega)$ such that $\oint_{\Omega^\bullet} \lambda v ds = 0$ for all $v \in S_{h_1}^{t,k}(\Omega)$ . Consequently, the matrix obtained from (6.3.37) for finding the approximate solution $(u_{h_1}, u_{h_2})$ which satisfies (6.3.37) is necessarily singular. This shows the necessity of the restriction that $h_1$ must be in relation to $h_2$ . In other words, it is not permissible to compute the Lagrange multipliers "too accurately."

3) Now we select for $S_{h_2}^{t,k_2}(\Omega)$ the space of the traces of the system $S_{h_1}^{t,k_1}(\Omega)$. Although this space does not satisfy our assumptions, nevertheless, it is of interest to analyze this case.

The bilinear form related to our problem is

$$B((u,\lambda),(v,\mu)) = \int_\Omega \left(\frac{\partial u}{\partial x_1}\frac{\partial v}{\partial x_1} + \frac{\partial u}{\partial x_2}\frac{\partial v}{\partial x_2} + uv\right)dx - \oint_{\Omega^\bullet} (\lambda v + \mu u)ds ,$$

and

$$F((v,\mu)) = \int_\Omega fvdx .$$

Let $(u^h,\lambda^h)$ be the solution which satisfies the condition (6.3. 37) for our choice of the subspaces. By letting $v = 0$ we obtain $\oint u^h\mu\, ds = 0$ for every $\mu$ which is the trace of the piecewise linear functions. Therefore, in this case if $\Omega$ is taken to be the circle shown in Fig. 6.3.3, then $u^h = 0$ on $\Omega^\bullet$. But this implies that $u^h$ is the solution to our problem with zero boundary condition on a subdomain of the domain $\Omega$ as shown in Fig. 6.3.4, which differs from the circle by the magnitude of order $h$. Therefore the error in $L_2(\Omega)$ norm is of the order $h$. On the other hand, using the spaces which satisfy the assumptions of theorem 6.3.6, we would obtain the error estimate of order $h^2$ when $f \in L_2(\Omega)$. This shows that the assumptions of theorem 6.2.1 cannot be satisfied with

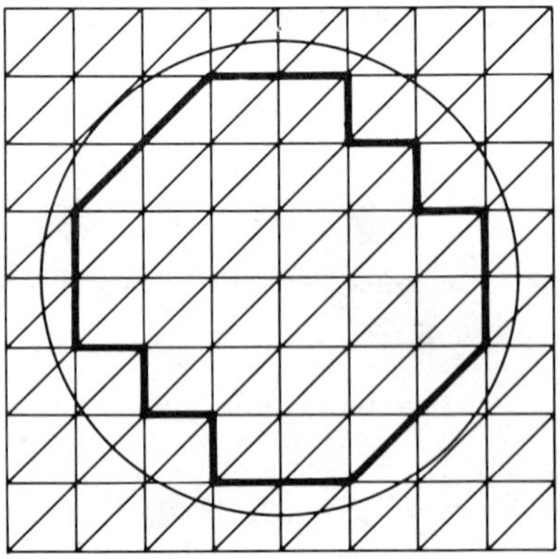

Fig. 6.3.4

constants independent of h .

In §6.3.2, we obtained error estimates in the space $H^1(\Omega) \times H^{-1/2}(\Omega)$. In exactly the same manner we may obtain error estimates in "lower" spaces. By the addition of the inverse assumption we can get error estimates in "higher spaces" as well.

Due to this straightforward analogy we shall not discuss these estimates any further. More details in this connection may be found in [12].

### 6.3.6. Finite element method based on variational principle of Dirichlet type. Revisited--boundary conditions of the first type.

In §6.3.2 we studied the finite element method for second order equations with boundary conditions of the first type. We now return to this method. For simplicity, instead of the general second order equation, we confine our discussion to the special case

$$-\Delta u + u = f \quad \text{on} \quad \Omega \ , \tag{6.3.41}$$

$$\frac{\partial u}{\partial n} = g \quad \text{on} \quad \Omega^{\bullet} \ . \tag{6.3.42}$$

As mentioned in §5.3.7, this specialization of the problem is not significant. The method used is easily adaptable to the general case. From the results of §6.3.2 we see that the finite element for our special problem consists of finding $u_h \in S_h^{t,k}(\Omega)$, $(k \geq 1)$ such that

$$(6.3.43) \qquad B(u_h,v) = F(v) \;, \forall\, v \in S_h^{t,k} \;,$$

where

$$(6.3.44) \qquad B(u,v) = \int_\Omega \left( \sum_{i=1,2} \frac{\partial u}{\partial x_i} \frac{\partial v}{\partial x_i} + uv \right) dx \;,$$

and

$$(6.3.45) \qquad F(v) = \int_\Omega fvdx + \oint_{\Omega^\bullet} gvds \;.$$

To find $u_h$ we need to solve a system of linear algebraic equations with the right-hand side given by (6.3.45). Clearly, for $f \in H^{-k}(\Omega)$ and $g \in H^{-k+\frac{1}{2}}(\Omega^\bullet)$, (6.3.45) formally make sense. Thus the ultimate question is that of convergence (and the sense of convergence) of the finite element method for $v \in S_h^{t,k}(\Omega)$ and general $f$ and $g$, e.g., when $f$ is a Dirac function or its derivatives, etc.

To answer this question, we recall the analysis which was made in §5.3.7. We begin to study the case when the bilinear form $B$ given by (6.3.44) is defined on $H^{1-\alpha}(\Omega) \times H^{1+\alpha}(\Omega)$ with $|\alpha| < \frac{1}{2}$. To answer the question stated above, we will study the case $\alpha > 0$. The case $\alpha < 0$ is covered by theorem 6.3.2. To this end, assume that $M_1 = S_h^{t,k}(\Omega) = M_2$, where $S_h^{t,k}(\Omega)$ is a $(t,k)$-regular system with $k \geq 1 + \alpha$ and satisfying the inverse assumption.

Under these assumptions we will show that the assumptions of theorem 6.2.1 are fulfilled. Let

$\omega \in S_h^{t,k}(\Omega)$ be given. As in (5.3.101), we can find $f \in E(\overline{\Omega})$ such that

$$\|f\|_{H^{-1+\alpha}(\Omega)} = \|\omega\|_{H^{1-\alpha}(\Omega)}$$

and

$$\int_\Omega \omega f dx \geq \frac{1}{2}\|\omega\|^2_{H^{1-\alpha}} . \tag{6.3.46}$$

The bilinear form $B(u,v)$ satisfies all the assumptions of theorem 5.2.1 on $H^1(\Omega) \times H^1(\Omega)$ and of theorem 6.2.1 on $S_h^{t,k}(\Omega) \times S_h^{t,k}(\Omega)$ .

Let $v \in H^1(\Omega)$ be such that

$$B(w,v) = \int_\Omega fwdx \tag{6.3.47}$$

for all $w \in E(\overline{\Omega})$ . Using the results of chapter 3, we have

$$v \in H^{1+\alpha}(\Omega) \text{ with } \|v\|_{H^{1+\alpha}(\Omega)} \leq C\|\omega\|_{H^{1-\alpha}(\Omega)} .$$

Further, let $v^h \in \overline{S_h^{t,k}(\Omega)}$† be such that

$$B(w^h,v^h) = \int_\Omega fw^h dx \tag{6.3.48}$$

for all $w^h \in S_h^{t,k}(\Omega)$ . Since $S_h^{t,k}(\Omega)$ is a (t,k)-regular system which satisfies the inverse assumption by the hypothesis, using theorem 6.3.2 we have

$$\|v^h\|_{H^{1+\alpha}(\Omega)} \leq C\|v\|_{H^{1+\alpha}(\Omega)} \leq C\|\omega\|_{H^{1-\alpha}(\Omega)} . \tag{6.3.49}$$

On the other hand, using $\omega$ for $w^h$ in (6.3.48), we obtain from (6.3.46)

† The closure of $\overline{S_h^{t,k}(\Omega)}$ is obviously taken in $H^1(\Omega)$ .

(6.3.50) $$B(\omega, v^h) = \int_\Omega f\omega dx \geq \frac{1}{2}\|\omega\|^2_{H^{1-\alpha}} .$$

Inequalities (6.3.44) and (6.3.50) give (6.2.1) with $d_2$ independent of $h$. By changing the role of $u$ and $v$ in the bilinear form, we may construct a $u^h \in S_h^{t,k}(\Omega)$ for any $\omega \in S_h^{t,k}(\Omega)$ such that

$$B(u^h, \omega) \geq \frac{1}{2}\|\omega\|^2_{H^{1+\alpha}} > 0 ,$$

and 6.2.2 is satisfied. Thus, we have proved the theorem.

Theorem 6.3.7. If

I) $\frac{1}{2} > \alpha \geq 0$ , $k \geq 1 + \alpha$

II) $M_1 = M_2 = S_h^{t,k}(\Omega)$ is a $(t,k)$-regular system with the inverse assumption;

then

III) (6.3.44) satisfies the conditions of theorem 5.2.1 on $H^{1-\alpha}(\Omega) \times H^{1+\alpha}(\Omega)$

IV) (6.3.44) satisfies the condition of theorem 6.2.1 with $d_2 > 0$ independent of $h$.

Theorem 6.3.7, together with theorems 6.2.1 and 5.2.1, provides us immediately with the answer to the question raised in the beginning for $0 \leq \alpha < \frac{1}{2}$.

Now we shall study the second case when $\alpha = 1$. In this case, as in §5.3.7, we take

$$H_1 = H^0(\Omega) \times H^{-1/2}(\Omega^\bullet)$$

and

$$H_2 = H^2(\Omega) .$$

Let $M_2 = S_h^{t,k}(\Omega)$ be a $(t,k)$-regular system with $k \geq 2$

which satisfies the inverse assumption. Further, let $M_1 \subset H_1$, $(\omega,\omega_0) \in M_1$ where $\omega \in S_h^{t,k}(\Omega)$ and $\omega_0$ is the trace of $\omega$ on $\Omega^\bullet$. Under these assumptions, (5.3.104) may be written for $u \equiv (\omega,\omega_0)$, and $v \in M_2$, as

$$(6.3.51)\quad B(u,v) = \int_\Omega \left(\frac{\partial\omega}{\partial x_1}\frac{\partial v}{\partial x_1} + \frac{\partial\omega}{\partial x_2}\frac{\partial v}{\partial x_2} + \omega v\right) dx .$$

Now we shall show that the assumptions of theorem 6.2.1 are satisfied. To this end, let $(\omega,\omega_0) \in M_1$ be given. Therefore there exists $w \in E(\Omega^\bullet)$ such that

$$\|w\|_{H^{1/2}(\Omega^\bullet)} = \|\omega_0\|_{H^{-1/2}(\Omega^\bullet)}$$

and

$$\oint \omega_0 w ds \geq \frac{1}{2}\|\omega_0\|^2_{H^{-1/2}(\Omega^\bullet)} .$$

Let $z$ (resp. $z^h \in S_h^{t,k}(\Omega)$) be such that

$$B(u,z) = \int_\Omega \omega u dx + \oint_{\Omega^\bullet} wu ds$$

for every $u \in H^1(\Omega)$ (resp. $S_h^{t,k}(\Omega)$), then $z$ is the solution of the problem

$$-\Delta z + z = \omega \quad \text{on} \quad \Omega$$

$$\frac{\partial z}{\partial n} = w \quad \text{on} \quad \Omega^\bullet ,$$

and therefore, using the results of chapter 3, we have

$$(6.3.52)\qquad \|z\|_{H^2(\Omega)} \leq C\|(\omega,\omega_0)\|_{H_1} .$$

Since $S_h^{t,k}(\Omega)$ is a (t,k)-regular system with the inverse assumption, we also have

$$(6.3.53)\qquad \|z^h\|_{H^2(\Omega)} \leq C\|(\omega,\omega_0)\|_{H_1} .$$

On the other hand, we have

$$B(\omega, z^h) = \int \omega^2 dx + \oint w\omega_0 ds \ ,$$

and therefore

(6.3.54) $$B(\omega, z^h) \geq C\|(\omega, \omega_0)\|^2_{H_1} \ .$$

From (6.3.53) and (6.3.54), theorem 6.2.1 follows immediately with $d_2 > 0$ and independent of $h$. Now we let $\omega \in S_h^{t,k}(\Omega)$ and take $u = (\omega, \omega_0)$. Then obviously $B(u, \omega) > 0$ and (6.2.2) is satisfied. Thus, the following theorem is proven.

Theorem 6.3.8. If

I) $M_1 = \{(\omega, \omega_0) \mid \omega \in S_h^{t,k}(\Omega)$ , $\omega_0$ is a trace of $\omega$ on $\Omega^{\bullet}\}$

$M_2 = S_h^{t,k}(\Omega)$ with $S_h^{t,k}(\Omega)$ a $(t,k)$-regular system, $k \geq 2$, and satisfying the inverse assumption,

II) $H_1 = H^0(\Omega) \times H^{-1/2}(\Omega^{\bullet}) \supset M_1$

$H_2 = H^2(\Omega) \supset M_2$ ;

then

III) the bilinear form (5.3.104) on $M_1 \times M_2$ may be written in the form

(6.3.55) $$B((\omega, \omega_0), v) = \int_\Omega \left( \frac{\partial \omega}{\partial x_1} \frac{\partial v}{\partial x_1} + \frac{\partial \omega}{\partial x_2} \frac{\partial v}{\partial x_2} + \omega v \right) dx \ ,$$

IV) the bilinear form (5.3.104) satisfies the conditions of theorem 5.2.1 on $H_1 \times H_2$,

V) the bilinear form (5.3.104) satisfies the conditions of theorem 6.2.1 with $d_2 > 0$ independent of $h$.

Theorems 6.3.7 and 6.3.8 answer the question

formulated at the beginning of this section completely. As stated in §5.3.7, we studied only two cases, namely $|\alpha| < \frac{1}{2}$ and $\alpha = 1$, but a similar analysis applies to the general case.

By using the theorems mentioned above, we may claim, roughly speaking, that using (t,k)-regular systems with the inverse assumption, the finite element method can be applied, as long as it makes sense computationally, and the error estimates are of the same structure as in theorem 6.3.1:

$$\|u_0 - u_h\|_{H^q(\Omega)} \leq C[h^{\mu_1}\|f\|_{H^\ell(\Omega)} + h^{\mu_2}\|g\|_{H^s(\Omega^\bullet)}]$$

with

$$\mu_1 = \min(t-q, \ell+2-q)$$

$$\mu_2 = \min(t-q, s+\tfrac{3}{2}-q) \ .$$

So far $\mu_1$ and $\mu_2$ are positive, and $k \geq \max(q+|q-1|, -\ell, -s+\frac{3}{2})$; the rate of convergence is optimal in the same sense as in theorem 6.3.1. As a very special case, convergence for the Green function, i.e., when $f$ is the Driac function, is guaranteed. For the analysis of the problem of the Green function, see also [20], [21] and [22].

In the foregoing discussions, we focused our attention primarily on problems with boundary conditions of the first type. In a very similar way, using the results of §5.3.8, one may obtain analogous theorems for problems with boundary conditions of the second type.

### 6.3.7. Finite element method based on the variational principle of Dirichlet type for infinite domains.

In the previous section we proved a number of theorems concerning the rate of convergence of the approximate solution. The rate of convergence was expressed in terms of powers of the parameter $h$ . This parameter is a good measure of the amount of work involved in finding the approximate solution. In fact, in two dimensions the number of unknowns to be determined is mostly of order $\text{mes}(\Omega)h^{-2}$. The situation is quite different in the case of an infinite domain. We may use the subspace $S_h^{t,k}(R_n)$ for our purposes. However, since the domain is infinite, the space $M_1 = S_h^{t,k}(R_2)$ can no longer be finite dimensional, and consequently we cannot solve this infinite system. In other words, the method based on the use of infinite dimensional $S_h^{t,k}(R_2)$ is not in fact an approximate method. Therefore, we have to measure the amount of work necessary for finding the approximate solution in terms of the power of the number of unknowns to be determined. In our discussion of this problem, we shall confine ourselves to the equation (5.3.108) and use the variational principle introduced in §5.3.8, where it was shown that for this variational principle, all the conditions of theorem 5.2.1 are fulfilled.

Now we shall investigate the fulfillment of the hypotheses of theorem 6.2.1. We will describe one possibility of applying the finite element to the case under consideration.

Let $p$ be a positive number and assume we are

given a (t,k)-system $S_h^{t,k}(R_2)$ , $k \geq 1$ . Further, let

(6.3.56) $M_1 = \{u \in S_h^{t,k}(R_2) \mid u(x) = 0 \text{ for } |x| \geq p\}$ ,

(6.3.57) $M_2 = \{v \mid v = e^{2\mu|x|}u, \ u \in M_1\}$ , $|\mu| < \frac{\alpha}{4}$ .

It is clear that the proof of theorem 5.3.5 can be used without any changes to obtain the following result.

Theorem 6.3.9. If

I) $\mu < \frac{\alpha}{4}$ , $M_1$ and $M_2$ are defined by (6.3.56) and (6.3.57),

then

II) the bilinear form (5.3.109) satisfies all the assumptions of theorem 6.2.1.

(Notice that the space of trial functions is different from the space of test functions.)

Now we consider the problem (5.3.108) for $f \in H^{\ell}(R_2)$ , with $\ell \geq 0$ . We know that there exists a unique solution $u \in H^1(\Omega)$ such that

(6.3.58) $$\|u\|_{H^{\ell+2}(R_2)} \leq C\|f\|_{H^{\ell}(R_2)} .$$

Obviously, if $f \in H^{\ell}(R_2)$ , then the functional

$$F(v) = \int_{\Omega} fv\,dx$$

is continuous on the space $H^1_{-\mu}(R_2)$ for $\mu \leq 0$ , i.e.,

(6.3.59) $$\|F\|_{\left(H^1_{-\mu}(R_2)\right)'} \leq C\|f\|_{H^{\ell}(R_2)} .$$

Hence, from theorems 5.3.5 and 5.2.1, we have that

$u \in H^1_\mu(R_2)$ , which is of course an immediate consequence of (6.3.58) as well.

Let $\psi(x)$ be an infinitely differentiable function such that $\psi(x) = 1$ for $|x_i| < 1$ , $i = 1,2$ , and $\psi(x) = 0$ for all $|x_i| > 2$, and for $q > 0$ , let $\psi_q(x) = \psi(\frac{x}{q})$ . Moreover, for $u \in H^{\ell+2}(R_2)$ let $u_q(x) = u(x)\psi_q(x)$ and $\varepsilon_q = u - u_q$ . Obviously, for $\mu < 0$ and $q > q_0 > 0$ , we have

(6.3.60) $$\|\varepsilon_q\|_{H^\ell_\mu(R_2)} \leqslant C(q_0)e^{q\mu}\|u\|_{H^\ell(R_2)} .$$

On the other hand, we also have

(6.3.61) $$\|u_q\|_{H^\ell(R_2)} \leqslant C(q_0)\|u\|_{H^\ell(R_2)} .$$

Now we assume that the system $S_h^{t,k}(R_2)$ is a (t,k)-system, $k \geqslant 1$ , and in addition satisfies the local assumption. Then for every $u \in H^\ell(R_2)$, $\ell \geqslant 1$ , we can find a $\omega \in S_h^{t,k}(R_2)$ with support in the cube $|x_i| \leqslant 2q + kh$ , $i = 1,2$ such that

(6.3.62) $$\|u_q - \omega\|_{H^1(R_2)} \leqslant Ch^\mu\|u\|_{H^{\ell+2}(R_2)} ,$$

where $u_q = \psi_q u$ and $\mu = \min(t-1,\ell+1)$ . Clearly, for $p = 3q$, $q \geqslant 1$ and $h$ sufficiently small, $\omega \in M_1$ , where $M_1$ is given by (6.3.56). From (6.3.60), (6.3.61) and (6.3.62), for $\mu < 0$ we obtain

(6.3.63) $$\|\omega - u\|_{H^1_\mu(R_2)} \leqslant C(e^{\mu q} + h^\sigma)\|f\|_{H^\ell(\Omega)} ,$$

with $\sigma = \min(t-1,\ell+1)$ . Now theorems 5.2.1, 5.3.5, 6.2.1 and 6.3.8 yield immediately the following result.

Theorem 6.3.10. If

I) In (5.3.108) $f \in H^{\ell}(R_2)$ , $\ell \geq 0$ and $-\frac{\alpha}{4} < \mu < 0$ ,

II) $M_1 \subset H^1(R_2)$ , $M_2 \subset H^1(R_2)$ given by (6.3.56) and (6.3.57) with $p = h^{-\varepsilon}$ , $\varepsilon > 0$ (arbitrary),

III) $S_h^{t,k}(\Omega)$ , $k \geq 1$ satisfies the local assumption,

IV) $u_0$ is the exact solution of (5.3.108) and $u_h \in \overline{M_1}$ satisfies

$$B(u_h,v) = \int_{\Omega} fv dx \ , \ \forall v \in M_2 \ ,$$

where $B(u,v)$ is defined by (5.3.109),

then

V) for every bounded domain $\Omega$ and sufficiently small $h$

(6.3.64) $$\|u_0 - u_h\|_{H^1(\Omega)} \leq K(\varepsilon,\Omega)h^{\sigma-\varepsilon}\|f\|_{H^{\ell}(R_2)} ,$$

where $\sigma = \min(t-1,\ell+1)$ .

With $N = \dim M_1 \leq \frac{p^2}{h^2}C$, then (6.3.64) may be written in the form

(6.3.65) $$\|u_0 - u_h\|_{H^1(\Omega)} \leq k(\varepsilon,\Omega)N^{-(\sigma+2\varepsilon)/2}\|f\|_{H^{\ell}(R_2)} ,$$

i.e., neglecting $\varepsilon > 0$ in the exponent, the rate of convergence is of the same order as if the problem were solved in $\Omega$ . The theorem requires only that $\mu < 0$ . We also note that $f$ may be chosen not only in $H^{\ell}(\Omega)$ , but in $H^{\ell}_{\mu}(\Omega)$ as well, with $\mu < \mu^* < 0$ . The method used employs different spaces for the trial and test functions.

The question remains as to whether theorem 6.3 still holds if $M_2$ is defined by (6.3.56) instead of (6.3.57), i.e., if $M_1 = M_2$ . The answer to the above question would be in the affirmative if we could show that (6.2.1) holds with the constant $d_2$ independent of p and h . This question is answered affirmatively in [13] for a special choice of $S_h^{t,k}(\Omega)$ .

For $f \in H_0^{\ell}(\Omega)$ , $\Omega$ bounded and fixed, our special problem may be solved directly. We briefly describe the main idea involved in this approach. Let $u_0$ denote the solution to our problem and let $u_{0,\rho}$ be the solution of the problem $-\Delta u + c = f$ on the cube $k(\rho)$ with the boundary condition $u = 0$ on the boundary of $k(\rho)$ . We may estimate the error $\|u_0 - u_{0,\rho}\|$ on every bounded domain $\Omega$ by the use of the results of §5.3.8. We note that the error decreases exponentially as $\rho \to \infty$ . Thus we can analyze the difference between the approximate solution and the solution $u_{0,\rho}$ . It is also possible to obtain error estimates in "higher" and lower norms (i.e., in the spaces $H^{\ell}(\Omega)$) .

6.3.8. Finite element method for the Neumann problem based on a Dirichlet type variational principle.

There are problems where the solution is not unique, and the solution exists only if the data satisfy certain special conditions. To treat these problems by the finite element method, some new approach is needed. In §5.3.9, we discussed a variational principle which may be used for this purpose. In the sequel we confine our discussion to the typical boundary value problem (5.3.114), (5.3.115). We shall suppose that a priori the necessary condition for the existence of a solution are satisfied, i.e., we write the functional $F(v,\mu)$ given by (5.3.121) in the form

$$F(v,\mu) = \int_{\Omega} fv\,dx + \oint_{\Omega^{\bullet}} gv\,ds \; . \tag{6.3.66}$$

It was shown in §5.3.9 that the bilinear form (5.3.120) defined on $H_1 \times H_2$, where $H_1 = H^1(\Omega) \times R_1$, satisfied all the conditions of theorem 5.2.1. In the following we shall show that for the case $M_1 = M_2 = S_h^{t,k}(\Omega) \times R_1$, with $k \geq 1$, where $S_h^{t,k}(\Omega)$ is a $(t,k)$-system, the assumptions of theorem 6.2.1 are fulfilled. In fact, we have the following theorem.

Theorem 6.3.11. If

I) $M_1 = M_2 = S_h^{t,k}(\Omega) \times R_1$, $k \geq 1$, where $S_h^{t,k}(\Omega)$ is a $(t,k)$-system;

then

II) the bilinear form (5.3.120) for sufficiently small $h$ satisfies all the hypotheses of theorem 6.2.1, with $d_2 > C > 0$, where $C$ is independent of $h$.

The proof of theorem 6.3.11 is based on the proof of theorem 5.3.6 (§5.3.9). Theorem 5.3.6 may be used in precisely the same way as it was used in the proofs of theorems 6.3.1 and 6.3.6. Since the space of Lagrange multipliers is one-dimensional, we do not face the difficulty which we encountered in the proof of theorem 6.3.6, where we needed some relationship between $h_1$ and $h_2$.

By combining theorems 5.2.1, 6.2.1, 6.3.11 and

the results of chapter 3, we obtain the following theorem.

Theorem 6.3.12. If

I) in the problem (5.3.114), (5.3.115), $f \in H^{\ell}(\Omega)$, $g \in H^{s}(\Omega)$, with $\ell > -1$, $s > \frac{-1}{2}$ and $\int_{\Omega} f dx + \oint_{\Omega^{\bullet}} g ds = 0$, $\int_{\Omega} u_0 dx = 0$, where $u_0$ denotes the exact solution of (5.3.114) and (5.3.115),

II) $u_h$ is the approximate solution obtained from the bilinear form (5.3.120) and the functions (5.3.121) using the subspaces $M_1 = M_2 = S_h^{t,k} \times R_1$, with $k \geq 1$; then

III) for $h < h_0 < 1$, $h_0$ sufficiently small (independent of $f$ and $g$) there exists a unique approximate solution $(u_h, \lambda_h)$ such that

$$\|u_0 - u_h\|_{H^1(\Omega)} + |\lambda_h| \leq C[h^{\mu_1} \|f\|_{H^{\ell}(\Omega)} + h^{\mu_2} \|g\|_{H^{s}(\Omega^{\bullet})}] , \tag{6.3.67}$$

where

$$\mu_1 = \min(t-1, \ell+1) \tag{6.3.68}$$

$$\mu_2 = \min(t-1, s+\tfrac{3}{2}) . \tag{6.3.69}$$

In exactly the same way as in §6.3.2, we derive error estimates in "lower" space, and if the $S_h^{t,k}(\Omega)$ satisfies the inverse assumption, then we may obtain error estimates in "higher" spaces as well.

### 6.3.9. Remarks on the finite element method.

In the previous section we showed that any of

the variational principles described in chapter 5 may be used for the construction of the finite element method. The rate of convergence is always obtained by the same procedure. First, one shows that the assumptions of theorem 6.2.1 are satisfied in very much the same way as we showed the main properties of the variational principles in chapter 5. The next step consists of the use of the regularity properties of the solution of the given problem to derive error estimates in the spaces on which the bilinear form is defined. The approach of §6.3.2 may be used to obtain error estimates in "lower" and "higher" spaces.

Although we discussed only a few variational principles associated with a few model problems, the extension to other variational principles and their use for the finite element method is obvious. We notice that in our discussions we have used the (t,k) system consistently in the construction of the finite element method. On several occasions, we required the (t,k) to be a regular system and to satisfy the inverse or the local assumption. We also had to restrict the ratio between the spacing of the elements in the case where the Lagrange multipliers technique was used. Although our discussions in the construction of the finite element have been confined to typical second-order equations, it is clear how the method can be extended in a more or less obvious way to higher order equations. We would like to emphasize that the regularity properties of the exact solution have played a very crucial role in all our preceding considerations.

Finally, a few remarks regarding the finite element method based on the results of §5.4.3 appear to be in order. The problem dealt with in §5.4.3 satisfied the Dirichlet boundary conditions. To treat this problem by the finite element method, we need to use elements which fulfill the boundary condition; as indicated earlier, the construction of such elements entails numerous difficulties. To avoid these difficulties, the Lagrange multipliers or some other techniques may be used.

In §5.4.3 we derived a variational principle for incompressible materials. The system (5.4.30) associated with this case is elliptic in a very general sense. The regularity properties of the solution are known, and the approach described in the previous section may be readily applied. We also derived a variational principle for nearly incompressible materials. As we observed earlier, the main advantage of this principle lies in the fact that the constants appearing in theorems 5.2.1 and 6.2.1 are independent of $\lambda$ , provided suitable spaces are used. If regularity theorems which are independent of $\lambda$ are known for the solution, then our approach is immediately applicable to the nearly compressible medium.

However, it appears that the regularity properties of equations of this type are not in general intensively studied, and needed deep results are not available. We emphasize that the approach in the treatment of the system of equations involved in the study of nearly compressible medium is based on the Lagrange multipliers method, and consequently it is necessary to assume some additional hypotheses about the elements such as the inverse assumption, and some restriction on the ratio of the spacing between elements, as indicated in §6.3.5.

## 6.4. Lower bounds for the finite element method.

In section 6.3 we obtained various error estimates. Typical of these estimates are those given in theorems 6.3.1, 6.3.2 and 6.3.3. We recall that (see theorem 6.3.1) if $u_0$ and $u_h$ denote the exact solution and the approximate solution of our problem respectively then

(6.4.1)
$$\|u_0-u_h\|_{H^1(\Omega)} \leq C\left[h^{\mu_1}\|f\|_{H^\ell(\Omega)} + h^{\mu_2}\|g\|_{H^s(\Omega)}\right] ,$$

where $\ell > -1$ , $s > -\frac{1}{2}$ , and

(6.4.2)
$$\mu_1 = \min(t-1,\ell+1)$$

(6.4.3)
$$\mu_2 = \min(t-1,s+\tfrac{1}{2})$$

A natural question arises as to whether the coefficients $\mu_1$ and $\mu_2$ could or could not be improved. The coefficients $\mu_i$ are determined as the minimum of two numbers the first of which is related to the structure of the (t,k)-system and in applications it is fairly easy to determine t so that $\mu_i$ cannot be large relative to the first term. Moreover by selecting the right (t,k)-system (i.e., t sufficiently large) we can always achieve to have $\mu_1 = \ell+1$ and $\mu_2 = s+1/2$ . To answer the question whether the rates $\mu_1 = \ell+1$ and $\mu_2 = s+1/2$ are optimal we need to know more about the (t,k)-system $S_h^{t,k}(\Omega)$ .

We introduce a special class of such (t,k)-systems. First we assume that for every $h > 0$ the system is finite dimensional and denote $N(h) = \dim S_h^{t,k}(\Omega)$ . We define (in two dimensions) the parameter $H(h) = [N(h)]^{-1/2}$ . In the majority of practical cases there exists constants $C_1$ and $C_2$ such that

$$C_1 h \leq H \leq C_2 h \quad . \tag{6.4.4}$$

We assume that (6.4.4) holds for the system $S_h^{t,k}(\Omega)$ . Now let $S \subset \Omega$ be given and let

$$S_h^{t,k}(\Omega/S) = \left\{ \phi \in S_h^{t,k}(\Omega) \mid \phi(x) = 0 \quad \text{for} \quad x \in \Omega - S \right\} \quad .$$

If there exists constants $C_1$ and $C_2$ such that

$$C_1(\text{meas } S) \leq h^2 N(S,h) \leq C_2(\text{meas } S) \quad ,$$

where meas $S$ is the measure of $S$ and $N(S,h) = \dim S_h^{t,k}(\Omega/S)$ then we refer to the (t,k)-system as an equidistant (t,k)-system. Such systems exist. In fact the majority of the systems used in practice are equidistant. The notion of an equidistant (t,k)-system has an intuitive meaning. To see this consider the commonly used systems of piecewise functions on triangles with sides of length of order $h$ . If the length of all the sides are the same in order, e.g., no refinement takes place, then we obtain an equidistant system. Clearly for such a system (6.4.4) holds. We note that there are systems which are not equidistant and yet satisfy (6.4.4).

Obviously the expression (6.4.1) can be written in the form

(6.4.5)
$$\|u_0-u_h\| \leqslant C\left[H^{\bar{\mu}_1}\|f\|_{H^{\ell}(S)} + H^{\bar{\mu}_2}\|g\|_{H^s(\Omega^{\bullet})}\right] ,$$

where $\bar{\mu}_1$ comes from $\mu_1$ and the relation between h and H . If (6.4.4) is satisfied then $\mu_i = \bar{\mu}_i$ , $i = 1,2$ . Now we can find lower bounds for $\bar{\mu}_i$ using the theory of n-widths (see chapter 2).

From theorems 2.5.1 and 2.5.2 we immediately obtain the following results:

Theorem 6.4.1. If

I) Given the boundary value problem (5.3.1), (5.3.2), where all the assumptions described in section 5.3.1 are satisfied,

II) $f \in H^{\ell}(\Omega)$ , $\ell > -1$ , $g \in H^s(\Omega)$ , $s > -1/2$ ,

III) $u_0$ is the exact solution and $u_h \in S_h^{t,k}(\Omega)$ , $(k \geq 1, t > 1)$ is the approximate solution constructed in section 6.3.2, then

IV) $\bar{\mu}_1 \leqslant \ell + 1$ and $\bar{\mu}_2 \leqslant 2(s+1/2)$ .

Proof. 1) By theorems 2.5.1 and 2.5.2 we see that the n-width d(n) of the space $H^{\ell+2}(\Omega)$ in the space $H^1(\Omega)$ is $d(n) \approx n^{-1/2(\ell+1)}$ . Further from chapter 3 we know that problem (5.3.1), (5.3.2) (with $g = 0$) induces an isomorphism between the spaces $H^{\ell+2}(\Omega)$ and the space $H^{\ell}(\Omega)$ . Therefore the best error estimate is given by the n-width of the space $H^{\ell+2}(\Omega)$ in the space $H^1(\Omega)$. Thus we obtain the first part of our statement concerning $\bar{\mu}_1$ .
2) Let $f = 0$ . We know that in this case the equations (5.3.1) and (5.3.2) induce an isomorphism between $H^1(\Omega)$ and $H^{-1/2}(\Omega^{\bullet})$ (see chapter 3 and theorem 2.4.2).

Therefore we can estimate $\overline{\mu}_2$ through the n-width $d(n)$ of the space $H^s(\Omega^\bullet)$ in $H^{-1/2}(\Omega^\bullet)$. From theorems (2.5.1) and (2.5.2) we obtain $\overline{\mu}_2 \leq 2(s+1/2)'$.

Now we compare the statement of our theorem with (6.4.1), (6.4.2), (6.4.3) and (6.4.4). If (6.4.4) holds we have $\overline{\mu}_1 = \overline{\mu}_2$ and we see that the rate (6.4.2) is optimal. By the optimality of the rate $\mu$ of convergence we understand $\mu$ cannot be increased under the assumption (6.4.4). The situation is quite different with respect to the rate $\mu_2$. We see that in (6.4.3) there is a great deal of loss in the rate of convergence. In this connection there are some obvious questions. Is the rate $\mu_2$ too pessimistic (under assumption (6.4.4))? Could it be improved?

In general, it can be shown that the rate cannot be improved for an equidistant system $S_h^{t,k}(\Omega)$. This gives rise to another question, namely whether there exist (t,k)-systems $S_h^{t,k}(\Omega)$ which satisfy (6.4.4) and such that (6.4.5) holds with $\overline{\mu}_2 = 2(s+1/2)$. We shall show that the answer to this question is in the affirmative. First we give an intuitive explanation of where the root of the problem lies. A characteristic property of the solution of a homogeneous elliptic equation is that its solution is very smooth inside the domain $\Omega$ and it is non-smooth only in the neighborhood of the boundary. Therefore it is intuitively clear that the optimal result cannot be obtained without some refinement of the elements around the boundary. Now we show that a suitable refinement of the elements around the boundary will bring about the desired effect. To this end we shall closely follow the ideas developed in [23].

For the sake of simplicity of exposition we restrict ourselves to the model problem

(6.4.6) $$-\Delta u + u = 0 \quad \text{on} \quad \Omega \ ,$$

(6.4.7) $$\frac{\partial u}{\partial n} = g \quad \text{on} \quad \dot{\Omega} \ ,$$

where $\Omega$ is a smooth domain. Let $H > 0$ denote by

$$\Omega_H = \left\{ x \in \Omega \,\middle|\, \rho(x) > H \right\} \ ,$$

where $\rho(x)$ is the distance of the point $x$ to the boundary $\dot{\Omega}$. $\Omega_H$ is a smooth domain for sufficiently small $H$.

First we state the theorem.

Theorem 6.4.2. Let $u \in H^\alpha(\Omega)$, $\alpha \geq 1$ such that

(6.4.8) $$B(u,v) = 0 \quad \text{for} \ \forall v \in \mathcal{D}(\Omega) \ ,$$

with

(6.4.9) $$B(u,v) = \int_\Omega \left( \frac{\partial u}{\partial x_1} \frac{\partial u}{\partial x_1} + \frac{\partial u}{\partial x_2} \frac{\partial u}{\partial x_2} \right) dx \ .$$

then for every $H > 0$, $u \in H^s(\Omega_H)$ for any $s \geq \alpha$, and

(6.4.10) $$\|u\|_{H^s(\Omega_H)} \leq C(\alpha, S) H^{-(s-\alpha)} \|u\|_{H^\alpha(\Omega)} \ .$$

For integral values of $\alpha$ and $s$ the theorem is proved in [24]. For arbitrary $\alpha$ and $s$ the theorem follows immediately by using the properties of Hilbert scales.

Now we assume that $S_h^{t,k}(\Omega)$, $(k \geq 1, t \geq 2(s+1/2)+1, s \geq 0)$, is an equidistant $(t,k)$-system satisfying the local assumption. For $g \in H^s(\Omega^\bullet)$ the solution $u$ of (6.4.6) and (6.4.7) satisfies the estimate (see chapter 3)

(6.4.11) $$\|u\|_{H^{s+3/2}(\Omega)} \leq C\|g\|_{H^s(\Omega^\bullet)} .$$

Let $\tau$ be a sufficiently small number and let $0 < \lambda < 1$ be given. Moreover for $h > 0$ sufficiently small let

$$M(h) = \left[\frac{|\log h|}{|\log \lambda|}\right] + 1 ,$$

where $[a]$ denotes the integral part of $a$. Then obviously

$$\lambda h < \lambda^{M(h)} \leq h .$$

By theorems 6.4.2 and 2.6.1 we can write the solution

(6.4.12) $$u = \sum_{j=0}^{M(h)} u_j ,$$

where $u_j = 0$ on $\Omega_{\tau\lambda^{2j}}$ and for $q \geq s + 3/2$ and $j = 0,1, \cdots, M(h) - 1$ we have

(6.4.13)
$$\|u_j\|_{H^q(H)} \leq C(q,s)(\tau\lambda^{2j})^{-(q-s-3/2)}\|g\|_{H^s(\Omega^\bullet)} .$$

Clearly the set $\Omega - \Omega_{\tau\lambda^{2j}} = S_{\tau\lambda^{2j}}$ is the support of $u_j$. By taking into accout the fact that the system $S_h^{t,k}(\Omega)$ satisfies the local property there exists $L > 0$ such that we can find $\omega_j \in S_{h_j}^{t,k}(\Omega/S_{\tau\lambda^{2j}+Lh_j})$ so that for $j \leq M(h)-1$

we have

(6.4.14)

$$\|u_j-\omega_j\|_{H^1(\Omega)} \leq Ch_j^{\mu}(\tau\lambda^{2j})^{-(q-s-3/2)}\|g\|_{H^s(\Omega^\bullet)} ,$$

with

$$\mu = \min(t-1,q-1) .$$

For $q = 2(s+1/2) + 1$ and $h_j = h\lambda^j$ we have

(6.4.15)

$$\|u_j-\omega_j\|_{H^1(\Omega)} \leq Ch_j^{2(s+1/2)}\lambda^{-2j(2(s+\frac{1}{2})+1-s-3/2)}|g\|_{H^s(\Omega^\bullet)} =$$

$$= Ch^{2(s+1/2)}\|g\|_{H^s(\Omega^\bullet)}$$

and for $j = M(h)$ we have

(6.4.16) $$\|u_M-\omega_M\|_{H^1(\Omega)} \leq Ch^{\mu_1}\lambda^{M\mu_1}\|g\|_{H^s(\Omega^\bullet)} ,$$

where $\mu_1 = s + \frac{1}{2}$. Therefore

(6.4.17) $$\|u_M-\omega_M\|_{H^1(\Omega)} \leq Ch^{2(s+1/2)}\|g\|_{H^s(\Omega^\bullet)} .$$

Now we construct a new $(t,k)$-system $\tilde{S}_h^{t,k}(\Omega)$.

$$\hat{S}_h^{t,k}(\Omega) = S_h^{t,k}(\Omega) \cup \bigcup_{j=1}^{M(h)} S_{h_j}^{t,k}\left(\Omega/S_{\tau\lambda^{2j}+L\lambda^j h}\right) ,$$

with $M(h)$ as given above.

From our discussion above it follows that there exists a $\omega \in \hat{S}_h^{t,k}(\Omega)$ such that

(6.4.18) $$\|u-\omega\|_{H^1(\Omega)} \leqslant Ch^{2(s+1/2)}\|g\|_{H^s(\Omega^\cdot)} \lg|h| .$$

Now we proceed to estimate the dimension of $S_h^{t,k}(\Omega)$. Since the system $S_h^{t,k}(\Omega)$ is assumed to be equidistant, we have

$$\dim S^{t,k}_{h\lambda^j}(\Omega/S_{\tau\lambda^{2j}+Lh\lambda^j}) \leqslant C(h\lambda^j)^{-2}(\tau\lambda^{2j}+Lh\lambda^j) \leqslant Ch^{-2}(\tau+Lh\lambda^{-j}).$$

Since $j \leqslant M(h)$ we have $h\lambda^{-j} \leqslant \lambda^{-1}$ and then

(6.4.19) $$\dim S^{t,k}_{h\lambda^j}(\Omega/S_{\tau\lambda^j+Lh\lambda^j}) \leqslant Ch^{-2}$$

for all $j$ such that $0 \leqslant j \leqslant M(h)$. Therefore

(6.4.20) $$\dim \tilde{S}_h^{t,k}(\Omega) \leqslant Ch^{-2}|\log h| ,$$

and hence in terms of the parameter $H$ we have

(6.4.21) $$H(h) \geqslant Ch|\log h|^{-1} .$$

By combining (6.4.18), (6.4.20) and (6.4.21) we obtain the following theorem.

<u>Theorem 6.4.3</u>. There exists a $S_h^{t,k}(\Omega)$ system such that for the error of the approximate solution of the problem (6.4.6) and (6.4.7) based on the method described in section 6.3.1 we have for $s \geqslant 0$ the estimate

$$\|u-u_h\|_{H^1(\Omega)} \leqslant C(\varepsilon)H^{2(s+1/2)-\varepsilon}\|g\|_{H^s(\Omega^\cdot)} ,$$

where $\varepsilon > 0$ is arbitrary.

We remark that by proving the above theorem we have effectively constructed a system with refinement around the boundary.

## REFERENCES


[1]. Krasnosel'skii, M. A. et al., Approximate solution of the operator equations (In Russian), Nauka, Moscow, 1969.

[2]. Krejn, M. G. and M. A. Krasnosel'skii, "Basic results about continuation of Hermit's operators and some applications to the theory of orthogonal polynomials and moment's problem" (In Russian), Uspehi Mat. Nauk. 2 (1947).

[3]. Nagy, B. Sz., "Perturbations des transformations autoadjointes dans l'espace de Hilbert," Comment Math. Helv. 19 (1947).

[4]. Babuška, I., "Error bounds for finite element method," Num. Math. 16 (1971), pp. 322-333. [Tech Note BN-630, (Nov. 1969) Institute for Fluid Dynamics and Applied Mathematics, University of Maryland.]

[5]. Michlin, S. G., "Some sufficient conditions for the convergence of the Galerkin method," (In Russian) Učen. Zap. Leningrad. G. Univ., Sir. Mat. 21 (1950).

[6]. Polskii, N. J., "Method of Galerkin and its foundation," Proceedings of 3rd Math. Congr. of USSR, Vol. 1 (1956).

[7]. Babuška, I., "Computation of derivatives in the finite element method," Comment Math. Univ. Carolinae 11, 3 (1970), pp. 545-558. [Tech Note BN-650 (April 1970), Institute for Fluid Dynamics and Applied Mathematics, University of Maryland.]

[8]. Nitsche, J., "Ein Kriterium für die Quasi-Optimalität des Ritichen Verfahrens," Num. Math. 11 (1968), pp. 346-348.

[9]. Strang, G. and G. Fix, "A Fourier analysis of the finite element method," Proceedings of CIME Conference Sicily 1971, Crimonese, Rome, To appear.

[10]. Schultz, M. H., "$L^2$ error bounds for the Rayleigh-Ritz-Galerkin method," SIAM J. Numer. Anal. 8 (1971), pp. 737-748.

[11]. Babuška, I., J. Segethová and K. Segeth, "Numerical experiments with finite element I," Tech Note BN-669 (August 1970), Institute for Fluid Dynamics

and Applied Mathematics, University of Maryland.

[12]. Babuška, I., "The finite element method with Lagrangian multipliers," Tech Note BN-724 (Jan. 1972), Institute for Fluid Dynamics and Applied Mathematics, University of Maryland, To appear in Numer. Math.

[13]. Babuška, I., "The finite element method for infinite domains," J. Math. Comp. 26 (1972), pp. 1-11. [Tech Note BN-670 (August 1970), Institute for Fluid Dynamics and Applied Mathematics, University of Maryland.]

[14]. Ciarlet, P. G. and P. A. Raviart, "The combined effects of curved boundaries and numerical integration in isoparametric finite element methods," These Proceedings.

[15]. Aubin, J. P., Approximation of Elliptic Boundary-Value Problems, Wiley Interscience, New York, London, Sydney, Toronto, 1972.

[16]. Schultz, M. H., Finite Element Analysis, Prentice-Hall, Inc., Englewood Cliffs, N. J., To appear, 1973.

[17]. Strang, G. and G. Fix, An Analysis of the Finite Element Method, Prentice-Hall, New York, To appear, 1973.

[18]. Ciarlet, P. G., M. H. Schultz and R. S. Varga, "Numerical methods of higher-order accuracy for nonlinear boundary-value problems in one-dimensional problems," Numer. Math. 9 (1967), pp. 394-430.

[19]. Ciarlet, P.G. and R. S. Varga, "Discrete variational Green's function," Numer. Math. 16 (1971), 115-128.

[20]. Ciarlet, P. G., "Discrete variational Green's function I," Aequationes Math. 4 (1970), pp. 74-82.

[21]. Bramble, J. H. and A. H. Schatz, "Least squares method for 2mth-order elliptic boundary-value problems," Math. Comp. 25 (1971), pp. 1-32.

[22]. Oganesjan, L. A. and P. A. Ruchovec, "Investigation of the convergence rate of variational-difference schemes for elliptic second-order equations in two-dimensional domains with a smooth boundary," (In Russian) Ž. Vyčisl. Mat. i Mat. Fiz. 9 (1969), pp. 1102-1120.

[23]. Babuška, I., "The rate of convergence for the finite element method," SIAM J. Numer. Anal. 8 (1971), pp. 309-315. [Tech Note BN-646 (March 1970), Institute for Fluid Dynamics and Applied Mathematics, University of Maryland.]

[24]. Focht, A. S., "Some inequalities for the solution and its derivatives of an equation of elliptic type in the $L_2$ norm," Proc. Steklov Inst. of Math., AMS Transl. 77 (1965), pp. 160-191.

[25]. I. Babuška, "Approximation by Hill Functions," Comment. Math., Univ. Carolinae 11 (1970), pp. 787-811.

# CHAPTER 7. ONE PARAMETER FAMILIES OF VARIATIONAL PRINCIPLES

## 7.1. Introduction.

We recall first the techniques used in chapters 5 and 6. Our starting point was a variational principle i.e., a pair of Hilbert spaces $H_1$ and $H_2$ and a bilinear form defined on $H_1 \times H_2$ . In theorem 5.2.1 we assumed that this bilinear form possessed certain basic properties. Similar assumptions were made in theorem 6.2.1. A close analysis of these theorems suggests a very natural generalization. Instead of fixed spaces $H_1$ and $H_2$ and the bilinear form $B$ , in general one deals with a family $H_1(\tau)$ , $H_2(\tau)$ of Hilbert spaces and a family $B(\tau)$ of bilinear forms defined on $H_1(\tau) \times H_2(\tau)$ , $0 < \tau < 1$ . If it is assumed that (5.2.1) and (5.2.2) hold with $C_1(\tau)$ , $C_2(\tau) > 0$, and (5.2.3) holds for all $\tau$ and if moreover there is a functional $f(\tau)$ defined on $H_2(\tau)$ for all $\tau$ , then we may obviously construct $u_o(\tau) \in H_1(\tau)$ such that

$$(7.1.1) \qquad B(\tau, u_o(\tau), v) = f(\tau, v) \ , \forall v \in H_2(\tau) \ ,$$

and

$$(7.1.2) \qquad \|u_o(\tau)\|_{H_1(\tau)} \leq \frac{\|f\|_{H_2'(\tau)}}{C_2(\tau)}$$

In a quite similar manner, using theorem 6.2.1, we may construct $u_h(\tau) \in \overline{M_1(\tau)} \subset H_1(\tau)$ such that

$$(7.1.3) \quad \|u_o(\tau)-u_h(\tau)\|_{H_1(\tau)} \leq \left[1 + \frac{C_1(\tau)}{d_2(\tau,M_1(\tau),M_2(\tau))}\right]\delta ,$$

where $\delta$ depends on the parameter $h$. Naturally, one would try to select the values of $\tau$ and $h$ simultaneously in an optimal manner.

There exist a very vaste field of applications for this procedure. We illustrate the scope of application of this idea by considering certain model problems.

This procedure has been used under different names in different contexts, to mention a few, the perturbed approximation, perturbed variational principle, penalty method etc. Perturbation methods in the study of differential problems were systematically used by Lions and other authors. In particular, in connection with numerical solution of partial differential equations the references [1, chapter 9] and [2]-[9] may be consulted.

7.2. The Penalty Method.

We shall describe the forgoing ideas by considering the Poisson problem on a smooth domain $\Omega$,

$$(7.2.1) \quad -\Delta u = f \text{ on } \Omega , \; f \in L_2(\Omega)$$

and

$$(7.2.2) \quad u = 0 \text{ on } \dot{\Omega} .$$

As we mentioned earlier in chapter 6, the main technical difficulty in solving problems (7.2.1), (7.2.2) by the finite element method arises in connection with the fulfillment of the boundary condition. Now we introduce a method which avoides this difficulty.

We consider a perturbed variational principle. To this end let for $0 < \tau \leq 1$, $\psi(\tau)$ be a given function such that $0 < \psi_0 < \psi(\tau) < \infty$ Moreover let $H_2(\tau) = H_1(\tau)$, where $H_1(\tau)$ is defined by the norm

(7.2.3)

$$\|u\|^2_{H_1(\tau)} = \iint_\Omega \left[\left(\frac{\partial u}{\partial x_1}\right)^2 + \left(\frac{\partial x}{\partial x_2}\right)^2\right] dx + \psi(\tau) \int_{\dot{\Omega}} u^2 ds \ .$$

Obviously the $H_1(\tau)$ is equivalent to the space $H^1(\Omega)$ (the constants appearing in the definition of equivalent norms depend on $\tau$). Now we define on $H_1(\tau) \times H_2(\tau)$ the bilinear form

$$B(\tau,u,v) = \iint_\Omega \left[\frac{\partial u}{\partial x_1}\frac{\partial u}{\partial x_1} + \frac{\partial u}{\partial x_2}\frac{\partial v}{\partial x_2}\right] dx + \psi(\tau)\int_{\dot{\Omega}} uv \, ds \ . \tag{7.2.4}$$

We have the following result.

<u>Theorem 7.2.1</u>. The bilinear form (7.2.4) satisfies the conditions (5.2.1), (5.2.2) and (5.2.3), with constants $C_1$ and $C_2$ independent of $\tau$.

Proof. The proof is in fact a mere repetition of the arguments used in section (5.3.2). In our special case the proof is very elementary. By Schwarz's inequality we see immediately that (5.2.1) holds with $C$ independent of $\tau$. Similarly by letting $v = u$, (5.2.2) follows at once with $C_2$ independent of $\tau$. It is obvious that (5.2.3) holds as well.

Now we introduce a functional $F$ on $H_2(\tau)$ defined by

$$F(v) = \int_{\Omega} fv dx \quad , \tag{7.2.5}$$

where $f$ is given in (7.2.1). Clearly for $f \in L_2(\Omega)$ the norm of the functional (7.2.5) is independent of $\tau$. Therefore by theorem 5.2.1 for every $0 < \tau < 1$, there exists a $u_o(\tau)$ such that

$$B(\tau, u_o(\tau), v) = F(v) \quad , \tag{7.2.6}$$

for every $v \in H_2(\tau)$. It is easily seen that $u_o(\tau)$ is a weak solution of (7.2.1) with the boundary condition

$$\frac{\partial u_0}{\partial n} + \psi(\tau) u_0 = 0 \quad . \tag{7.2.7}$$

We notice that $u_0(\tau)$ is not the solution of the original problems (7.2.1), (7.2.2) for any $\tau > 0$. We also note that the necessary condition that $u_o(\tau) \to u_o$ ($u_o$ being the solution of (7.2.1), (7.2.2)) is that $\lim_{\tau \to 0} \psi(\tau) = \infty$.

Let $M_1 = M_2 = S_h^{t,k}(\Omega)$, where $S_h^{t,k}(\Omega)$ is a (t,k)-system with $k \geq 1$. In a similar way as in theorem 7.2.1, we immediately see that the hypotheses of theorem 6.2.1 are satisfied with the constant $d_2$ independent of $\tau$ and $h$. Consequently in the estimate (6.2.6) the terms inside the parenthisis is independent of $\tau$ and $h$. Let $u_h(\tau) \in \overline{S_h^{t,k}}(\Omega) \subset H_1(\tau)$ be the function which satisfies (7.2.6) for every $v \in S_h^{t,k}(\Omega)$. The existence and uniqueness of such a function are assured by theorem (6.2.1).

We now proceed to estimate the norm of the error

$$E(h) = u_h(\tau) - u_0 \quad .$$

In this direction we have the following result.

Theorem 7.2.2. If

I) in problem (7.2.1), (7.2.2) $f \in H^{\ell}(\Omega)$, $\ell \geq 0$

II) $S_h^{t,k}(\Omega)$, $k \geq 1$, $t \geq \ell + 2$ is a (t,k)-regular system with the local property and let $\sigma > 0$.

III)

(7.2.8) $\quad B(\tau,u_h(\tau),v) = F(v) \ , \quad v \in S_h^{t,k}(\Omega) \ , \quad u_h \in \overline{S_h^{t,k}}(\Omega)$

where $B(\tau,u,v)$ and $f(v)$ are given by (7.2.4) and (7.2.5) respectively then

IV) the condition (7.2.8) determines $u_h(\tau)$ uniquely and for every $\varepsilon > 0$ we have

(7.2.9)

$$\|u_o - u_h(\tau)\|_{H_1(\tau)} \leq C(\varepsilon)\|f\|_{H^{\ell}(\Omega)} \Big[\psi^{-\frac{1}{2}} + \psi^{-1} + h^{\mu-\varepsilon}(\max[1,\psi h^{\sigma}])^{\frac{1}{2}}$$

$$+ \psi^{-1}h^{\ell} + h^{\ell+\frac{1}{2}-\varepsilon}\psi^{-\frac{1}{2}}\Big] ,$$

and

(7.2.10)

$$\|u_0-u_h(\tau)\|_{H^1(\Omega)} \leq C(E)\|f\|_{H^\ell(\Omega)}\Big[\psi^{-1}+h^{\mu-\varepsilon}(\max[1,\psi h^\sigma]^{\frac{1}{2}} + \psi^{-1}h^\ell+h^{\ell+\frac{1}{2}-\varepsilon}\psi^{-\frac{1}{2}}\Big] ,$$

where $u_o$ denotes the exact solution of (7.2.1), (7.2.2) and

$$\mu = \frac{t-\chi}{t-1}(\ell+1) \quad \text{and} \quad \chi = \max(1, \frac{\sigma+1}{2})) .$$

Proof. 1) Since by hypothesis $f \in H^\ell(\Omega)$ , $\ell \geq 0$ (see results of chapter 3) the solution $u_0$ of (7.2.1), (7.2.2) satisfies the inequality

$$\|u_0\|_{H^{\ell+2}(\Omega)} \leq C\|f\|_{H^\ell(\Omega)} , \tag{7.2.11}$$

and hence

$$\Big\|\frac{\partial u_0}{\partial n}\Big\|_{H^{\ell+\frac{1}{2}}(\Omega^\bullet)} \leq C\|f\|_{H^\ell(\Omega)} . \tag{7.2.12}$$

Let $\xi$ denote the solution of the problem

$$-\Delta\xi + \xi = 0 \quad \text{on} \quad \Omega , \tag{7.2.13}$$

$$\xi = \frac{\partial u_0}{\partial n} \quad \text{on} \quad \Omega^\bullet . \tag{7.2.14}$$

Then we have

(7.2.15) $$\|\xi\|_{H^{\ell+1}(\Omega)} \leq C\|f\|_{H^{\ell}(\Omega)} \quad .$$

Now we may write $u_o(\tau)$ as

(7.2.16) $$u_o(\tau) = u_0 - \psi^{-1}\xi + \zeta \quad ,$$

and (7.2.6) takes the form

(7.2.17)
$$B(\tau,u_0(\tau),v) = \iint_{\Omega}\left(\frac{\partial u_0}{\partial x_1}\frac{\partial v}{\partial x_1} + \frac{\partial u_o}{\partial x_2}\frac{\partial v}{\partial x_2}\right)dx + \psi\int_{\dot{\Omega}} u_o v ds +$$

$$- \psi^{-1}\iint_{\Omega}\left(\frac{\partial \xi}{\partial x_1}\frac{\partial v}{\partial x_1} + \frac{\partial \xi}{\partial x_2}\frac{\partial v}{\partial x_2}\right)dx - \int_{\dot{\Omega}} \xi v ds + B(\tau,\zeta,v) = F(v) \quad .$$

Since

(7.2.18) $$\iint_{\Omega}\left(\frac{\partial u_0}{\partial x_1}\frac{\partial v}{\partial x_1} + \frac{\partial u_0}{\partial x_2}\frac{\partial v}{\partial x_2}\right)dx = F(v) + \int_{\dot{\Omega}} v\frac{\partial u_0}{\partial n} ds \quad ,$$

and $u_0 = 0$ on $\dot{\Omega}$, we have

(7.2.19)
$$B(\tau,\zeta,v) = \psi^{-1}\iint_{\Omega}\left(\frac{\partial \xi}{\partial x_1}\frac{\partial v}{\partial x_1} + \frac{\partial \xi}{\partial x_2}\frac{\partial v}{\partial x_2}\right)dx \quad , \forall v \in H_2(\tau) \quad ,$$

on the other hand

$$\left|\iint_{\Omega}\left(\frac{\partial\xi}{\partial x_1}\frac{\partial v}{\partial x_1}+\frac{\partial\xi}{\partial x_2}\frac{\partial v}{\partial x_2}\right)\right| \leq \|\xi\|_{H^1(\Omega)}\|v\|_{H_1(\tau)} \quad .$$

Hence from (7.2.15) and (7.2.19) we obtain

$$\|\zeta\|_{H_1(\tau)} \leq C\psi^{-1}(\tau)\|f\|_{H^\ell(\Omega)} \quad . \tag{7.2.20}$$

2) We now approximate $u_0(\tau)$ in $H_1(\tau)$ by functions in $S_h^{t,k}(\Omega)$ . Let $w_1 \in S_h^{t,k}(\Omega)$ be such that

$$\|\xi-w_1\|_{H^s(\Omega)} \leq Ch^{\ell+1-s}\|f\|_{H^\ell(\Omega)} \quad , \tag{7.2.21}$$

with $\frac{1}{2}+\varepsilon \leq s \leq 1, \varepsilon > 0$. Moreover let $w_2 \in S_h^{t,k}(\Omega)$ be such that (see theorem 2.4.3).

(7.2.22)

$$\|u_o-w_2\|_{H^1(\Omega)} + h^{-\frac{\sigma}{2}}\|w_2\|_{H^\circ(\Omega^\bullet)} \leq C(\varepsilon)h^{\mu-\varepsilon}\|f\|_{H^\ell(\Omega)} \quad ,$$

with

$$\mu = \frac{t-\chi}{t-1}(\ell+1) \quad , \quad \chi = \max(1, \frac{1+\sigma}{2}) \quad .$$

By letting

$$w = w_2 - w_1\psi^{-1} \quad , \tag{7.2.23}$$

we obtain

(7.2.24)

$$\|u_o(\tau)-w\|_{H_1(\tau)} \leq \|\zeta\|_{H_1(\tau)} + \|u_o-w_2\|_{H_1(\tau)} + \psi^{-1}\|\xi-w_1\|_{H_1(\tau)}$$

$$\leq C(\varepsilon)\|f\|_{H^\ell(\Omega)}\left[\psi^{-1} + h^{\mu-\varepsilon}\left[\max[1,\psi h^\sigma]\right]^{\frac{1}{2}} + \psi^{-1}h^\ell + h^{\ell+\frac{1}{2}-\varepsilon}\psi^{-\frac{1}{2}}\right]$$

Consequently,

(7.2.25)

$$\|u_0(\tau)-u_h(\tau)\|_{H_1(\tau)} \leq C(\varepsilon)\|f\|_{H^\ell(\Omega)}\left[\psi^{-1}(\tau) + h^{\mu-\varepsilon}\left[\max[1,\psi h^\sigma]\right]^{\frac{1}{2}} + \psi^{-1}h^\ell + h^{\ell+\frac{1}{2}-\varepsilon}\psi^{-\frac{1}{2}}\right] .$$

3) We have the following inequalities

(7.2.26)

$$\|u_0-u_0(\tau)\|_{H_1(\tau)} \leq \psi^{-1}\|\xi\|_{H_1(\tau)} + \|\zeta\|_{H_1(\tau)} \leq C\psi^{-\frac{1}{2}}\|f\|_{H^\ell(\Omega^\bullet)} .$$

(7.2.27)

$$\|u_0-u_0(\tau)\|_{H^1(\Omega)} \leq \psi^{-1}\|\xi\|_{H^1(\Omega)} + \|\zeta\|_{H_1(\tau)} \leq C\psi^{-1}\|f\|_{H^\ell(\Omega)} .$$

Now our theorem follows as a consequence of the above inequalities.

From the inequalities (7.2.9) and (7.2.10) it is possible to select $\psi(\tau)$ and $\sigma$ in an optimal way, e.g., let $\psi(\tau) = h^{-\sigma}$, then (7.2.9) may be written in the form

(7.2.28) $$\|u_0-u_h(\tau)\|_{H_1(\tau)} \leq C(\varepsilon)\|f\|_{H^\ell(\Omega)} h^{\mu_1-\varepsilon} ,$$

where

(7.2.29) $$\mu_1 = \min\left[\frac{\sigma}{2} , \frac{t-\chi}{t-1} (\ell+1)\right] ,$$

and

(7.2.30) $$\|u_0-u_h(\tau)\|_{H^1(\Omega)} \leq C(\varepsilon)\|f\|_{H^\ell(\Omega)} h^{\mu_2-\varepsilon} ,$$

and

(7.2.31) $$\mu_2 = \min\left[\sigma , \frac{t-\chi}{t-1} (\ell+1) , \frac{\sigma+1}{2} + \ell\right]$$

and

(7.2.33) $$\chi = \max[1 , \frac{\sigma+1}{2}] .$$

We note that the relation between $\sigma$ and $t$ guarantees convergence (independent of $\ell$).

The error estimate obtained is in the norm $H^1(\Omega)$ only. By a similar technique as used in chapter 6 it is possible to derive an error estimate in $H^0(\Omega)$, e.g., for $\psi(\tau) = h^{-\sigma}$, we obtain

(7.2.33) $$\|u_0-u_h(\tau)\|_{H^0(\Omega)} \leq C(\varepsilon)h^{J-\varepsilon}\|f\|_{H^\ell(\Omega)} ,$$

where

(7.2.34) $$J = \min\left[H + \mu_2 , H + \frac{\sigma}{2} , \mu_2 + \frac{\sigma}{2} , \sigma\right] ,$$

with $\mu_2$ given by (7.2.31) and

$$H = \frac{t-\chi}{t-1} ,$$

We remark that the main term appearing in the error estimate comes from (7.2.22), which is based on theorem 2.4.3. Thus obviously by improving (7.2.22) we would sharpen our error estimate. This leads to the question whether one can generate (t,k)-systems $\mathcal{S}_h^{t,k}(\Omega)$ which will yield better approximation than the result contained in theorem 2.4.3. The answer is the affirmative, e.g., by using piecewise linear functions on triangles with nodes on the boundary we will obtain a (t,k)-regular system $\mathcal{S}_h^{t,k}(\Omega)$ with local property, $k = 1$, $t = 2$ and for $f \in H^2(\Omega)$, we can find $\overline{w} \in \mathcal{S}_h^{t,k}(\Omega)$ such that

(7.3.35) $$\|u_0-\overline{w}\|_{H^1(\Omega)} + h^{-1}\|\overline{w}\|_{H^0(\Omega)} \leqslant h\|f\|_{H^2(\Omega)} \quad .$$

The comparison of (7.2.35) with (7.2.22) reveals that under this assumption we obtain better results than before and the use of (7.2.35) with $\sigma = 1$ in (7.2.10) yields

(7.3.36) $$\|u_0 - u_h(\tau)\|_{H^1(\Omega)} \leqslant Ch\|f\|_{L_2(\Omega)} \quad ,$$

which is an optimal estimate (see section 6.4). For smoother $f$ we may proceed similarly, provided we choose our system $\mathcal{S}_h^{t,k}(\Omega)$ appropriately.

It is worth noting that the system of piecewise linear functions with nodes on the boundary has an additional feature that makes it possible to choose the function $w$ in (7.2.35) so that $w = 0$ at the nodal points and on the boundary. Therefore these points may be eliminated in the computation. We remark that in this case a further computational simplification is possible. In fact we may eliminate the term $\psi(\tau)\int_{\Omega} u^2 ds$ from (7.2.23). This may be treated as the error in the quadrature formulae, which will be discussed in chapter 9.

## 7.3. The weighted least square method.

In section 5.3.3 and section 6.3.3 a variational principle of least square type was derived for the problem

$$(7.3.1) \qquad -\Delta u + cu = f$$

$$(7.3.2) \qquad \frac{\partial u}{\partial n} + \beta u = g .$$

It was shown that this method did not yield optimal results, and we also remarked that a similar situation would occur in connection with the Dirichlet boundary conditions. We shall therefore discuss in this section the Dirichlet boundary conditions. For the sake of simplicity we restrict ourselves to the problems

$$(7.3.3) \qquad -\Delta u = f \quad \text{on} \quad \Omega ,$$

$$(7.3.4) \qquad u = 0 \quad \text{on} \quad \dot{\Omega} .$$

Let $\psi(\tau) > \psi_0 > 0$ be given for $0 < \tau \lesssim 1$ let the space $H_1(\tau)$ be the completion of $E(\bar{\Omega})$ with respect to norm

$$(7.3.5) \qquad \|u\|^2_{H_1(\tau)} = \int_{\Omega} (\Delta u)^2 dx + \psi \int_{\dot{\Omega}} u^2 ds , \quad u \in E(\bar{\Omega}) .$$

We define on $H_1(\tau) \times H_2(\tau)$ , $H_2(\tau) = H_1(\tau)$ , the bilinear form

(7.3.6) $$B(\tau,u,v) = \int_{\Omega} \Delta u \Delta v dx + \psi \int_{\dot{\Omega}} uvds$$

It is easily shown that the bilinear form $B(\tau,u,v)$ given by (7.3.6) satisfies all the assumptions of theorem 5.2.1 and constants involved are independent of $\tau$ .

We now introduce the functional $(f \in H^0(\Omega))$

(7.3.7) $$F(v) = -\int_{\Omega} f \, \Delta v \, dx \quad .$$

Since this functional is continuous on $H_2'(\tau)$ , there exists a unique $u_0(\tau)$ such that

(7.3.8) $$B(\tau,u_0(\tau),v) = F(v) \ , \forall v \in H_2(\tau) \quad .$$

It is easy to see that $u_0(\tau) = u_0$ , where $u_0$ is the solution of (7.3.3) and (7.3.4), i.e., $u_o(\tau)$ does not depend on $\tau$ . This the main difference between the present situation and that described in section 7.2. Therefore we have to differentiate between the penalty method and weighted method.

Now consider a (t,k)-regular system $S_h^{t,k}(\Omega)$ with local property and $k \geq 2$ . It is easily seen that the bilinear form $B(\tau,u,v)$ satisfies all the conditions of theorem 6.2.1, with the constants involved independent of $\tau$ .

Let $f \in H^{\ell}(\Omega)$ , $\ell \geq 0$ and $u_h(\tau) \in \overline{S_h^{t,k}}(\Omega)$ be such that

(7.3.9)
$$B(\tau,u_h(\tau),v) = F(v) \quad .$$

Using the results of chapter 3 we have $u_0 \in H^{\ell+2}(\Omega)$ and

(7.3.10)
$$\|u_0\|_{H^{\ell+2}(\Omega)} \leq C\|f\|_{H^{\ell}(\Omega)}$$

For $\sigma > 0$ , $t \geq \ell + 2$ and $k \geq 2$ , there exists a $w \in S_h^{t,k}(\Omega)$ such that (see theorem 2.4.3).

(7.3.11)
$$\|u_o - w\|_{H^2(\Omega)} + h^{-\frac{\sigma}{2}}\|w\|_{H^\circ(\Omega^\bullet)} \leq C(\varepsilon)h^{\mu-\varepsilon}\|f\|_{H^{\ell}(\Omega)} \quad ,$$

where

$$\mu = \frac{t-\chi}{t-2}\,\ell \quad , \chi = \max(2, \frac{\sigma+1}{2})$$

Hence

(7.3.12)
$$\|u_0-u_h\|^2_{H_1(\tau)} \leq C(\varepsilon)h^{2\mu-\varepsilon}\|f\|^2_{H^{\ell}(\Omega)} \cdot \max(1,h^{\sigma}\psi(\tau)) \quad .$$

Thus we proved the following result.

<u>Theorem 7.3.1.</u> If

I) For $\ell \geq 0$ , $t \geq \ell + 2$ , $S_h^{t,k}(\Omega)$ is a (t,k)-regular system with the local property,

then

II) (7.3.13)
$$\|u_0-u_h(\tau)\|_{H_1(\tau)} \leq C(\varepsilon)h^{\mu-\varepsilon}\|f\|_{H^{\ell}(\Omega)}\left[\max(1,h^{\sigma}\psi(\tau))\right]^{\frac{1}{2}} \quad ,$$

where

$$\mu = \frac{t-\chi}{t-2} \ell \ , \ \chi = \max(2, \frac{\sigma+1}{2}) \quad \text{and} \quad \epsilon > 0 \quad \text{arbitrary} \ .$$

From the results of section 6.4, it follows, that the rate of convergence in $h$ cannot exceed $\ell$. Hence $\chi = 2$ will not decrease the rate of convergence (up to $\varepsilon$) and therefore we may take $\sigma = 3$ and the function $\psi$ and $h$ related by $\psi(\tau) = h^{-3}$.

Next we study the error estimate in $L_2(\Omega)$. To this end denote $\eta = u_0 - u_h(\tau)$ and let

(7.3.4) $$\Delta U = \eta \quad \text{on} \quad \Omega$$

(7.3.15) $$U = 0 \quad \text{on} \quad \Omega \ .$$

Further let

(7.3.16) $$\Delta V_1 = U \quad \text{on} \quad \Omega \ ,$$

(7.3.17) $$V_1 = 0 \quad \text{on} \quad \dot{\Omega} \ ,$$

(7.3.18) $$\Delta V_2 = 0 \quad \text{on} \quad \Omega$$

(7.3.19) $$V_2 = \psi^{-1} \frac{\partial}{\partial n} (\Lambda_h U) \quad \text{on} \quad \dot{\Omega} \ ,$$

where $\Lambda_h$ is a smoothening operator such that $\Lambda_h$ mapps $H^0(\Omega)$ into $H^s(\Omega)$ for all $s > 0$ and $\|\Lambda_h u\|_{H^s(\Omega)} \leq Ch^{-(s-\ell)} \|u\|_{H^\ell(\Omega)}$ for $s \geq \ell$, and $\|U - \Lambda_h U\|_{H^s(\Omega)} \leq Ch^{\ell-s} \|U\|_{H^\ell(\Omega)}$ for $s \leq \ell$. From the results of chapter 3 it follows that

(7.3.20) $$\|U\|_{H^2(\Omega)} \leq C\|\eta\|_{L_2(\Omega)} \quad ,$$

and consequently

(7.3.21) $$\|V_1\|_{H^4(\Omega)} \leq C\|\eta\|_{L_2(\Omega)}$$

Moreover

(7.3.22) $$\|\Lambda_h U\|_{H^5(\Omega)} \leq Ch^{-3}\|U\|_{H^2(\Omega)} \leq Ch^{-3}\|\eta\|_{L_2(\Omega)} \quad ,$$

and thus

(7.3.23) $$\left\|\frac{\partial}{\partial n}(\Lambda_h U)\right\|_{H^{7/2}(\Omega^{\bullet})} \leq Ch^{-3}\|\eta\|_{L_2(\Omega)} \quad .$$

and therefore

(7.3.24) $$\|V_2\|_{H^4(\Omega)} \leq C\psi^{-1}h^{-3}\|\eta\|_{L_2(\Omega)} \quad .$$

Furthermore

(7.3.25)
$$\|U-\Lambda_h U\|_{H^{3/2+\varepsilon}(\Omega)} \leq C(\varepsilon)h^{\frac{1}{2}-\varepsilon}\|U\|_{H^2(\Omega)} \leq C(\varepsilon)h^{\frac{1}{2}-\varepsilon}\|\eta\|_{L_2(\Omega)} \quad ,$$

and by the imbedding theorem 2.4.1, we obtain

(7.3.26) $$\left\|\frac{\partial U}{\partial n} - \frac{\partial}{\partial n}(\Lambda_h U)\right\|_{H^0(\Omega^{\bullet})} \leq Ch^{\frac{1}{2}-\varepsilon}\|\eta\|_{L_2(\Omega)} \quad .$$

Now we have

(7.3.27)

$$B(\tau,\eta,V_1+V_2) = B(\tau,\eta,V_1) + B(\tau,\eta,V_2) =$$

$$\iint_\Omega (\Delta\eta\cdot U)\,dx + \psi \oint_{\Omega^\bullet} \eta\cdot\psi^{-1} \frac{\partial}{\partial n} (\Lambda_h U)\,ds =$$

$$= \|\eta\|^2_{L_2(\Omega)} - \oint_{\Omega^\bullet} \eta\left[\frac{\partial U}{\partial n} - \frac{\partial}{\partial n} (\Lambda_h U)\right] ds \quad .$$

By (7.3.26) we get

(7.3.28)

$$\|\eta\|^2_{L^2(\Omega)} \leq |B(\tau,\eta,V_1+V_2)| + C\|\eta\|_{L_2(\Omega^\bullet)} h^{\frac{1}{2}-\varepsilon} \|\eta\|_{L_2(\Omega)}$$

$$\leq |B(\tau,\eta,V_1+V_2)| + Ch^{\frac{1}{2}-\varepsilon} \|\eta\|_{L_2(\Omega)} \psi^{-\frac{1}{2}} \|\eta\|_{H_1(\tau)} \quad .$$

Moreover

(7.3.29) $$B(\tau,\eta,V_1+V_2) = B(\tau,\eta,V_1+V_2-\omega) \quad ,$$

for any $w \in S_h^{t,k}(\Omega)$ . By theorem 2.4.3 and inequality (7.3.26), for $t \geq 4$ there exists a $w_1 \in S_h^{t,k}(\Omega)$ such

(7.3.30)

$$\|V_1-w_1\|_{H_1(\tau)} \leq Ch^{\mu_1-\varepsilon}\left[\max\,(1,h^\sigma\psi(\tau)\right]^{\frac{1}{2}} \|\eta\|_{L_2(\Omega)} \quad ,$$

where

(7.3.31) $$\mu_1 = 2\frac{t-\chi}{t-2} \text{ , } \chi = \max(2, \frac{\sigma+1}{2}) \quad .$$

From the basic properties of the (t,k)-regular systems it follows that there exists a $w_2 \in S_h^{t,k}(\Omega)$ such that

(7.3.32)

$$\|V_2-w_2\|_{H_1(\tau)} \leq C\psi^{-1}h^{-3}\cdot h^2\|\eta\|_{L_2(\Omega)} + $$

$$+ C(\varepsilon)\psi^{\frac{1}{2}}\cdot\psi^{-1}h^{-3}h^{7/2-\varepsilon}\|\eta\|_{L_2(\Omega)} \leq$$

$$\leq C(\varepsilon)\|\eta\|_{L_2(\Omega)}h^{2-\varepsilon}\left[\psi^{-1}h^{-3} + \psi^{-\frac{1}{2}}h^{-3/2}\right] .$$

Hence we have
(7.3.33)

$$\|\eta\|^2_{L_2(\Omega)} \leq C(\varepsilon)\|\eta\|_{H_1(\tau)}\|\eta\|_{L_2(\Omega)}\ h^{\mu_1-\varepsilon}\Big[(\max[1,h^\sigma\psi(\tau)])^{\frac{1}{2}} +$$

$$+ h^{2-\varepsilon}\left[\psi^{-1}h^{-3} + \psi^{-\frac{1}{2}}h^{-3/2}\right]\Big] .$$

Now we state our result as:

<u>Theorem 7.3.2.</u> If

I) $S_h^{t,k}(\Omega)$ is a (t,k)-regular system with the local property and $k \geq 2$ , $t \geq \max(4,\ell+2)$ , $\ell \geq 0$ ,

II) $u_h(\tau) \in \overline{S_h^{t,k}}(\Omega)$ satisfies (7.3.9) for all $v \in S_h^{t,k}(\Omega)$ , then

III) for arbitrary $\varepsilon > 0$ .

(7.3.34)

$$\|\eta\|_{L_2(\Omega)} \leq C(\varepsilon)\|\eta\|_{H_1(\tau)}h^{2-\varepsilon}\left[h^{\mu_1-2}\max(1,h^\sigma\psi(\tau))^{\frac{1}{2}} + \right.$$

$$\left. + \psi^{-1}h^{-3} + \psi^{\frac{1}{2}}h^{-3/2}\right] ,$$

where $\eta = u_0 - u_h(\tau)$ , with $u_0$ denoting the solution of (7.3.3) and (7.3.4), $\|\eta\|_{H_1(\tau)}$ satisfies inequality (7.3.13) and $\mu_1$ is given by (7.3.32).

We remarked above that a particularly good choice was $\psi(\tau) = h^{-3}$ . In this case $\sigma = 3$ may be used as well and (7.3.34) takes the simple form

$$\text{(7.3.35)} \qquad \|\eta\|_{L_2(\Omega)} \leq C(\varepsilon)\|\eta\|_{H_1(\tau)} h^{2-\epsilon} .$$

Remarks. 1) We note that for $t \geq \max(4, \ell+2)$ (neglecting $\varepsilon$ in the estimate) we obtain the optimal error in $L_2(\Omega)$ (see section 6.4). It is also fairly obvious that the weight $\psi$ plays a central role for obtaining the proper estimates and numerical experiments show that in general the weight $\psi = h^{-\sigma}$ , $\sigma \neq 3$ yields worse results.

2) The results portaining to the weighted least square method were obtained first in [7] and [8].

3) A comparison of the methods described in sections 7.2 and 7.3 reveals similarities as well as significant differences between the methods. The main difference being that in the penalty we have $u_o(\tau) \neq u_0$ , where as in the weighted least square method $u_o(\tau) = u_0$ for every $\tau$ . Moreover the optimal penalty in general depends on the smoothness of the solution. In contract the optimal weight in the least square method is independent of the smoothness of the solution.

4) The concept of a one parameter family of variational principles could be used very successfully in many different situations. As an illustration we considered two simple model problems of special interest.

References.

[1]. J. P. Aubin, "Approximation of elliptic boundary-value problems," Wiley Interscience (J. Wiley), New York, London, Sydney, Toronto 1972.

[2]. J. P. Aubin, "Approximation des problèmes aux limites nonhomogènes et régularité de la convergence, Calcolo 6 (1969), pp. 117-140.

[3]. I. Babuška, "Numerical Solution of Boundary-Value Problems by Perturbed Variational Principle," Tech. Note BN-624 (October 1969), The Institute for Fluid Dynamics and Applied Mathematics, University of Maryland.

[4]. I. Babuška, "The finite element method for elliptic equations with discontinuous coefficients," Computing 5 (1970), pp. 207-213.

[5]. I. Babuška, "The Finite Element Method With Penalty," Tech. Note BN-710 (August 1971), The Institute for Fluid Dynamics and Applied Mathematics, University of Maryland.

[6]. I. Babuška, M. Zlámal, "Nonconforming Elements in the Finite Element Method," Tech. Note BN-729 (March 1972) The Institute for Fluid Dynamics and Applied Mathematics, University of Maryland.

[7]. J. H. Bramble, A. H. Schatz, "Rayleigh-Ritz-Galerkin Methods for Dirichlet's Problem Using Subspaces without Boundary Conditions," Comm. Pure Appl. Math. XIII (1970), pp. 653-675.

[8]. J. H. Bramble, A. H. Schatz, "Least Squares Methods for 2mth Order Elliptic Boundary-Value Problems," Math. Comp. 25 (1971), pp. 1-32.

[9]. J. L. Lions, "On Inequalities in Partial derivatives," (in Russian) Uspehi Mat. Nauk 158 (1971), pp. 205-263.

# CHAPTER 8. FINITE ELEMENT METHOD FOR NON-SMOOTH DOMAINS AND COEFFICIENTS.

## 8.1. Introduction.

In the previous sections we assumed that the domain $\Omega$ , the coefficients of the equations and the boundary conditions were smooth. In this case, as it was seen in chapter 3, the smoothness of the solution is determined by the smoothness of the right-hand side and the boundary conditions only. If the domain or the coefficients are not sufficiently smooth, the picture changes completely and the theory of the behavior of the solution is far from being satisfactory. In the following we shall deal separately with the case of a non-smooth domain and the case of non-smooth coefficients. In the case of a non-smooth domain, it is reasonable to distinguish between the case where the boundary is only Lipschitzian and the case where it is piecewise smooth. It is also essential to differentiate between two-dimensional domains and other multi-dimensional domains. One of the obvious reasons is that a piecewise smooth, two-dimensional domain has singular points at the corners only, whereas in a three-dimensional domain the singularities occur at the vertices and on the edges. The analysis of the three-dimensional case is very complicated and will not be dealt with here. We shall restrict ourselves to the analysis of the two-dimensional case only. The essential behavior is that the solution has singularities in the neighborhood of the corners of the domain, where the character of the singularity is known. In practice it is impossible to avoid the presence of such singularities. It is possible to use the finite element method in these cases. However, without the special treatment of the singularities, the rate of convergence is slowed down significantly. Special techniques for the treatment of the corners can be used to avoid this difficulty and to obtain the optimal rate of convergence. It is necessary to emphasize that the singularities affect the accuracy throughout the entire domain, and not only in the neighborhood of the corners. Singularities of similar character occur if the boundary conditions are abruptly changing from one type to another.

In the case of non-smooth coefficients, the most important case from the practical point of view is the case when the coefficients are piecewise smooth; when the interfaces are present it is advantageous to distinguish as before between the smooth interfaces and the interfaces with corners. At the interfaces, the solution exhibits similar behavior as at the corners, namely the occurance of singularities, treatment of which requires special techniques in order for the method to be effective computationally. In the next section we shall show how to deal with these problems.

### 8.2. Problems with Lipschitzian domain.

As a model problem we study

(8.2.1) $$-\Delta u = f \quad \text{on} \quad \Omega ,$$

(8.2.2) $$u = 0 \quad \text{on} \quad \Omega^{\bullet} ,$$

where $\Omega$ is a general Lipschitzian domain. We assume for the time being that $\Omega$ is a smooth domain. Then, as shown in chapter 3, we have

(8.2.3) $$\|u\|_{H^{\alpha+2}(\Omega)} \leq C(\alpha) \|f\|_{H^{\alpha}(\Omega)} ,$$

for all $\alpha > -\frac{3}{2}$. (In this case the triple $(u, u_0, u_1)$ (see chapter 3) reduces to the pair $(u, u_1)$ with $u_0$ as the trace of $u$.) Now there is a question as to whether (8.2.3) holds for general Lipschitzian domain. The following theorem provides the answer.

<u>Theorem 8.2.1</u>. Let $\Omega$ be a Lipschitzian domain, $u$ the solution of (8.2.1) and (8.2.2), then (8.2.3) holds for $-\frac{3}{2} < \alpha < -\frac{1}{2}$. For proof see [1].

For any $\alpha > -\frac{1}{2}$ one can easily construct a domain $\Omega$ such that (8.2.3) does not hold (the case $\alpha = \frac{1}{2}$ seems to be an open problem). However, if $\Omega$ is convex, then the theorem holds for $-\frac{3}{2} < \alpha \leq 0$ (see [2]). We shall restrict ourselves to the Dirichlet boundary conditions. There is an open question as to whether similar results are valid for other boundary conditions, say for the equation $-\Delta u + u = f$ and the boundary condition $\frac{\partial u}{\partial n} = 0$. It is very likely that a similar theorem holds in this case, but it seems that no proof is available so far. [In the case when the domain is piecewise smooth, the conjecture is proven.]

We remark that theorem 8.2.1 holds in the general

n-dimensional case and more general differential equations. The situation which occurs here is very similar to that described in chapters 5 and 6. Analogous results concerning the finite element method are obtained in a similar way.

Now let $S_h^{t,k}(\Omega)$ be a (t,k)-regular system with $k \geq 1$ and satisfying the local assumption. Further, let $\mathring{S}_h^{t,k}(\Omega) \subset S_h^{t,k}(\Omega)$ be the subset of all functions with compact support in $\Omega$. We now have the following lemma.

Lemma 8.2.1. Let $g \in H_0^{\ell}(\Omega)$, $\frac{1}{2} < \ell < \frac{3}{2}$. Then for every $0 \leq s \leq \min(\ell,k)$, there exists $\phi \in \mathring{S}_h^{t,k}(\Omega)$ such that

$$\|g - \phi\|_{H^s(\Omega)} \leq Ch^{\mu} \|g\|_{H^{\ell}(\Omega)} , \tag{8.2.4}$$

with

$$\mu = \min(1,\ell-s,t-s) . \tag{8.2.5}$$

The proof is given in [3]. The main part of the proof consists of showing that for any $h > 0$ and $g \in H_0^{\ell}(\Omega)$, there exists a $\bar{g} \in H_0^{\ell}(\Omega)$, with compact support such that $\|g - \bar{g}\|_{H^s(\Omega)} \leq Ch^{\mu} \|g\|_{H^{\ell}(\Omega)}$, with $\mu = \min(1,\ell-s)$.

Consider the bilinear form

$$B(u,v) = \int_{\Omega} \left( \sum_i \frac{\partial u}{\partial x_i} \frac{\partial v}{\partial x_i} \right) dx \tag{8.2.6}$$

defined on $H_1 \times H_2$, $H_1 = H_2 = H_0^1(\Omega)$. Obviously $B(u,v)$ satisfies all the assumptions of theorem 5.2.1. It is also possible to define $B(u,v)$ on $H_0^{1-\alpha}(\Omega) \times H_0^{1+\alpha}(\Omega)$,

$|\alpha| < \frac{1}{2}$ and to show that it satisfies the assumptions of theorem 5.2.1 (see [1]). Moreover, let

(8.2.7) $$F(v) = \int_\Omega fv dx ,$$

which is continuous for $f \in H_0^{-(1+\alpha)}(\Omega)$ . Thus there exists $u_0 \in H_0^{1-\alpha}(\Omega)$ such that

(8.2.8) $$B(u_0,v) = F(v) \ , \forall\ v \in H_0^{1+\alpha}(\Omega)$$

is the (weak) solution of the problem (8.2.1) and (8.2.2). As in section 6.3.6, we may prove the following theorem.

<u>Theorem 8.2.2</u>. Let $u_h \in \bar{\mathring{S}}_h^{t,k}(\Omega)$ , $k \geq 1$ be such that

$$B(u_h,v) = F(v) \ , \forall\ v \in \bar{\mathring{S}}_h^{t,k}(\Omega) ,$$

then

(8.2.9) $$\|u_0 - u_h\|_{H^1(\Omega)} \leq Ch^\mu \|f\|_{H^\ell(\Omega)} ,$$

$-1 < \ell < -\frac{1}{2}$ , with

(8.2.10) $$\mu = \min(t-1, \ell+1) ,$$

and $u_0$ the solution of (8.2.1) and (8.2.2). Further, if the (t,k) system $S_h^{t,k}(\Omega)$ is regular and it satisfies the inverse and local assumptions, then for every $|\alpha| < \frac{1}{2}$ , $|\alpha| \leq k - 1$ , we have

(8.2.11) $$\|u_0 - u_h\|_{H^{1-\alpha}(\Omega)} \leq Ch^\mu \|f\|_{H^\ell(\Omega)}$$ *

with $-\alpha - 1 < \ell < -\frac{1}{2}$ and

---

* This theorem improves the results of [3].

(8.2.12) $\mu = \min(t - (1-\alpha), \ell + 1 + \alpha)$ .

8.3. Problems with piecewise smooth domain.

In §8.2, we assumed that the domain was Lipschitzian. Now we shall suppose that the domain has a piecewise smooth boundary. Here, in contrast to the case considered in §8.2, it is essential that the domain be two-dimensional. We shall confine our discussion to a model problem, namely

(8.3.1) $$-\Delta u + u = f \quad \text{on} \quad \Omega ,$$

(8.3.2) $$\frac{\partial u}{\partial n} = g \quad \text{on} \quad \Omega^{\bullet} .$$

Furthermore, let the domain $\Omega$ be a polygon, i.e., $\Omega^{\bullet}$ consists of a finite number of straight-line segments. We emphasize that our model problem is not restrictive; in fact, a similar situation prevails in the case when the domain is piecewise smooth and the equation and the boundary condition have smooth coefficients. For the moment, assume that $\Omega$ is smooth. Then from chapter 3 we have the following a priori estimate for the solution of (8.3.1)

(8.3.3) $$\|u\|_{H^{s+2}(\Omega)} \leq C\left[\|f\|_{H^{s}(\Omega)} + \|g\|_{H^{s+1/2}(\Omega)}\right] \quad s \geq 0 .$$

In chapter 2 we introduced the space $H^{\alpha}(\Omega^{\bullet})$ for a smooth domain $\Omega$ . The smoothness of the domain was essential. In order to obtain the analogue of (8.3.3) for a polygonal domain $\Omega$ , we first need to clarify the meaning of the space $H^{\alpha}(\Omega^{\bullet})$ when $\Omega$ is a polygon. We shall use the ideas introduced in [4] here.

By hypothesis, the boundary $\Omega^{\bullet}$ consists of a finite number of straight-line segments. We denote these segments by $I_j$, $j = 1, \cdots, \nu$. These segments may be viewed as one-dimensional domains, and we can introduce the norms $\|g\|_{H_j^s(I_j)}$, $s \geq 0$ of $g$ after restriction to $I_j$, and then

$$(8.3.4) \qquad \|g\|^2_{H^s(\Omega^{\bullet})} = \sum_j \|g\|^2_{H_j^s(I_j)} .$$

We note that the norm $\|\cdot\|_{H^s(\Omega^{\bullet})}$ in chapter 2 was introduced in a similar way. The main difference lies here in the treatment of the endpoints of the intervals $I_j$. (8.3.4) does not say anything about the smoothness at these points. To see this, we may think of a "formal" vertex of the polygon in the middle of the straight-line segment. If $g \in H^s(\Omega^{\bullet})$ by removing this "formal" vertex $g \notin H^s(\Omega^{\bullet})$, obviously $g$ may have jumped there. Therefore different partitions of the boundary would not lead to the equivalent norms as it was the case in chapter 2. Thus, we have to assume the division is always given by the vertices of the polygon without excluding the vertex angles of value $\pi$ (i.e., the "formal" vertex). Now we formulate the analogue of the estimate (8.3.3)

<u>Theorem 8.3.1</u>. Let $u \in H^1(\Omega)$ be the solution of 8.3.1; then if $s \notin \mathcal{Q}$, where $\mathcal{Q}$ is an isolated set of exceptional indices, we have

(8.3.5) $$u = \sum_{j=1}^{\kappa} C_j v_j + w ,$$

where

a) $w \in H^{s+2}(\Omega)$ and

(8.3.6) $$\|w\|_{H^{s+2}(\Omega)} \leq C\left[\|f\|_{H^s(\Omega)} + \|g\|_{H^{s+1/2}(\Omega^\bullet)}\right] ;$$

b) the functions $v_j$ called the "singular functions" are independent of f, g ; each $v_j$ can be taken so that it is nonvanishing only in a neighborhood of a vertex of the domain $\Omega$ , and it depends only on the differential equation and the domain;

c) the coefficients $C_j$ are bounded linear functional on $H^s(\Omega) \times H^{s+1/2}(\Omega^\bullet)$ ;

d) the exceptional set $\mathcal{Q}$ does not depend on f and g ; it consists of a countable set of positive numbers whose only limit point is $+\infty$ .

For theorem 8.3.1, see [4].

As mentioned earlier, a similar situation occurs in the more general case (including the case of higher order differential equations). The domain we have considered is two-dimensional. I n three dimensions a similar theorem holds if the domain has no edges, only conical points. The singular functions $v_j$ in the theorem can be regarded as known. I n some cases it is easy to compute them. Sometimes, especially in the general case of piecewise smooth domains, it is hard to determine these functions from the practical point of view. I f $s \in \mathcal{Q}$, then the situation is more complicated. For further details see [4].

Under the assumption that the singular functions $v_j$ are known, the following implementation of the finite element method suggests itself.

Let $S_h^{t,k}(\Omega)$ be a (t,k) system with $k \geq 1$. Moreover, define on $H^1(\Omega) \times H^1(\Omega)$ the bilinear form

$$(8.3.6) \qquad B(u,v) = \int_\Omega \left( \sum_i \frac{\partial u}{\partial x_i} \frac{\partial v}{\partial x_i} + uv \right) dx \ ,$$

and the functional

$$(8.3.7) \qquad F(v) = \int_\Omega fvdx + \oint_{\Omega^\bullet} gvds \ .$$

Let $s \geq 0$ be given such that $s \notin \mathcal{Q}$ and let $v_j$ be the singular functions introduced in theorem 8.3.3. We denote the direct sum of $S_h^{t,k}(\Omega)$ and the set of the singular functions $v_j$ by $\tilde{S}_h^{t,k}(\Omega) \subset H^1(\Omega)$. The bilinear form $B(u,v)$ obviously satisfies all the assumptions of theorems 5.2.1 and 6.2.1 on $H_1^1(\Omega) \times H^1(\Omega)$ and on $\tilde{S}_h^{t,k}(\Omega) \times \tilde{S}_h^{t,k}(\Omega)$ respectively.

Let $\tilde{u}_h \in \bar{\tilde{S}}_h^{t,k}(\Omega)$ denote the approximate solution of the problem with the property

$$(8.3.8) \qquad B(\tilde{u}_h, v) = F(v) \ \forall \, v \in \tilde{S}_h^{t,k}(\Omega) \ .$$

Now a straightforward application of theorems 6.2.1 and 8.3.1 yields the following theorem.

<u>Theorem 8.3.2</u>. Let $s \notin \mathcal{Q}$, $s \geq 0$ and $\tilde{u}_h \in \bar{\tilde{S}}_h^{t,k}(\Omega)$, $k \geq 1$ be the approximate solution satisfying (8.3.8). Moreover, let $u_0$ be the solution of (8.3.1) and (8.3.2), then

(8.3.9) $$\|u_0 - \tilde{u}_h\|_{H^1(\Omega)} \leq Ch^{\mu} \|f\|_{H^s(\Omega)} + \|g\|_{H^{s+1/2}(\Omega)} ,$$

where

(8.3.10) $$\mu = \min(t-1,s+1) .$$

If $s \in \mathcal{Q}$, then (8.3.9) holds with $\mu^* = \mu - \varepsilon$, $\varepsilon > 0$ and $C$ depending on $\varepsilon$. We note that theorem 8.3.2 gives the same type of results as we obtained in the case of a smooth domain. We emphasize that the use of the subspaces $\tilde{S}_h^{t,k}(\Omega)$, i.e., the space enriched by the singular functions $v_j$, is essential. There remain some essential questions in this connection. First, we shall discuss the problem of what will happen if the singular functions are not added to the space $S_h^{t,k}(\Omega)$. Let us consider a particular case which can be dealt with by techniques similar to those used in [5].

Let $S_h^{t,k}(\Omega)$ be a (t,k) system generated by piecewise linear functions, linear on triangles with sides of length $d$ such that $C_2 h < d < C_1 h$. In this case we have $k = 1$ and $t = 2$. Assume further that the angles of the boundary of the polygon $\Omega$ at the vertices are $\alpha_i$, $i = 1,2,\cdots,\nu$ ; $g = 0$ and $f \in L_2(\Omega)$ in (8.3.1) and (8.3.2). Let $u_h \in S_h^{t,k}(\Omega)$ be the approximate solution, then we have

(8.3.11) $$\|u_0 - u_h\|_{H^1(\Omega)} \leq Ch^{\mu^*} \|f\|_{H^0(\Omega)} ,$$

(8.3.12) $$\begin{cases} \mu^* = 1 & \text{for } \alpha = \max \alpha_i \leq \pi \\ \mu^* = \dfrac{\pi}{\alpha} & \text{for } \alpha \geq \pi , \end{cases}$$

and the estimate is optimal (i.e., $\mu^*$ cannot be larger). As noted earlier, we can add singular functions to $S_h^{t,k}(\Omega)$. In the present case it is sufficient to add one function $v_i$ for every vertex with angle $\alpha_i \geqslant \pi$ (see [6]) and for $\alpha_i < \pi$ no $v_i$ is needed. We know that the corners of the domain create singularities around the corners only. This will give rise to a problem which we call the pollution problem. This problem may be described as follows. Let $O_i$ denote some of the neighborhoods of the vertices of the domain $\Omega$. Consider $\Omega' = \Omega - \sum_i O_i$. Let us assume

$$(8.3.13) \qquad \|u_0 - u_h\|_{H^1(\Omega')} \leqslant Ch^{\bar{\mu}}\|f\|_{H^0(\Omega)},$$

where the error is taken only over $\Omega'$ instead of the entire domain $\Omega$, i.e., the critical corners are excluded, then if $\bar{\mu} = \min(t-1,1)$ (in our particular case we have $\bar{\mu} = 1$), we say that no pollution is taking place in $H^1(\Omega)$. In [7] it is shown that under some (quite general) assumptions this is the case.

The next natural question is whether something can be said about the characterization of the growth of the error in the very neighborhood of the vertices by the use of weighted Sobolev spaces. In connection with this, there are still many open questions. (For some special cases, see the analysis in [5].) So far we have been concerned with the error estimate in the space $H^1(\Omega)$. There are analogous problems concerning the error estimate in the space $H^0(\Omega)$. We mention briefly the main results which can be easily obtained in a similar way as in chapter

6. By the use of the subspace $\bar{\tilde{S}}_h^{t,k}(\Omega)$, we obtain for the approximate solution $\tilde{u}_h \in \bar{\tilde{S}}_h^{t,k}(\Omega)$ the estimate

$$(8.3.14) \qquad \|u_0 - \tilde{u}_h\|_{H_h^0(\Omega)} \leq Ch^{\mu_1}[\|f\|_{H^s(\Omega)} + \|g\|_{H^{s+1/2}(\Omega')}] ,$$

with

$$(8.3.15) \qquad \mu_1 = \min(t,s+2) .$$

The comparison of the above result with the result of chapter 6 reveals that the error estimate is optimal.

Returning to the use of the subspace $S_h^{t,k}(\Omega)$, we face the problem of pollution mentioned earlier. As an example, we take, as before, the subspace of piecewise linear functions for $S_h^{t,k}(\Omega)$. In this case we will obtain

$$(8.3.16) \qquad \|u_0 - u_h\|_{H^0(\Omega)} \leq Ch^{\mu_1^*}\|f\|_{H^0(\Omega)} ,$$

where

$$(8.3.17) \qquad \begin{cases} \mu_1^* = 2 & \text{for} \quad \alpha = \max \alpha_i \leq \pi \\ \mu_1^* = \dfrac{2\pi}{\alpha} & \text{for} \quad \alpha \geq \pi . \end{cases}$$

In contrast with the estimates in $H^1(\Omega)$ (see (8.3.11) and (8.3.12)), it is not clear whether (8.3.17) is optimal, although numerical experiments are hinting at its optimality. It is possible to obtain lower bounds for $\mu^*$. It is not too difficult to show that $\mu_1^* \leq \mu^* + 1$.

Now we turn our attention to the study of the error outside the corners, i.e., the pollution problem. We shall be interested in the estimate

$$(8.3.18) \qquad \|u_0 - u_h\|_{\overset{0}{H}(\Omega')} \lesssim Ch^{\bar{\mu}_1^*} \|f\|_{\overset{0}{H}(\Omega)} .$$

We will say that the pollution is not taking place in $\overset{0}{H}(\Omega)$ if $\bar{\mu}_1^* = 2$ (in our case). It can be shown that the pollution is taking place, namely that $\mu_1^* \lesssim \frac{\pi}{\alpha} + 1$. The pollution problem in its full generality is far from being solved.

Finally, we remark that the behavior around the corners (singularities) plays a significant role in applications. As a typical case, we mention the problem of computation of the stress intensity factors in the theory of elasticity (see, e.g., [8], [9]). The use of the singular functions is a combination of the "analytical" and the approximate methods such as the finite difference, finite element methods, etc., which sometimes is referred to as the method of removing the singularities. The idea is very old, and it has been used successfully by many authors, e.g., [14], [18], and others. In a precise mathematical way this method has been analyzed in [16] where numerical experiments confirm theoretically predicted behavior. For the application of this idea to the eigenvalue problem, see [17].

## 8.4. The problem with a piecewise smooth domain--Continuation.

In §8.3 we studied the model problem (8.3.1) and (8.3.2) and the method of singular functions. The use of this method required a knowledge of these functions. It is often difficult to compute these functions. There is another approach which can be used to deal with the problem of singularities around the corners. It is the technique of refinement of the elements. Here we shall follow the procedure analyzed in [10]. For further details the reader may consult this reference. In §8.3 we generalized the error estimate (8.3.3) for non-smooth domain by using theorem 8.3.1. Now we will generalize (8.3.3) by the use of weighted Sobolev spaces.

Let $P = (P_1,\cdots,P_\nu)$, where $P_i$ are the vertices of the domain $\Omega$. Moreover, let $\beta = (\beta_1,\cdots,\beta_\Gamma)$, $\beta_i \geq 0$ and $k \geq 0$ is an integer. We denote the distance from the point $x$ to the vertex $P_i$ by $\sigma_i(x)$. Now we introduce the space $H^{k,s}_\beta(\frac{\Omega}{P})$ by

$$\|u\|^2_{H^{k,s}_\beta(\Omega/P)} = \|u\|^2_{H^s(\Omega)} + \sum_{|\alpha|=k}\int_\Omega [\min\{1,\prod_i \sigma_i^{\beta_i}(x)\}]^2[D^\alpha u]^2 dx.$$

For $k \geq 2$ and $s > 1$ we introduce the space $\mathcal{H}^{k,s}_\beta(\frac{\Omega}{P})$ by

$$\|u\|^2_{\mathcal{H}^{k,s}_\beta(\Omega/P)} = \sum_{j=2}^{k} \|u\|^2_{H^{j,s}_{\beta+j-2}(\Omega/P)},$$

where

$$\beta + j - 2 = \{\beta_i + j - 2\}.$$

Now we are in a position to generalize (8.3.3) for our domain.

<u>Theorem 8.4.1</u>. There exists $\varepsilon > 0$ (depending on $\Omega$) such that if $u \in H^1(\Omega)$ is the solution of problem (8.3.1) and (8.3.2), then

$$\|u\|_{H_{\beta}^{k+2,s}(\Omega/p)} \leq C\left[\|f\|_{H^{k}(\Omega)} + \|g\|_{H^{k+1/2}(\Omega^{\cdot})}\right],$$

where $s > \frac{3}{2} + \varepsilon$ and $\beta_i > \frac{1}{2} + \varepsilon$ .

Theorem 8.4.1 follows from theorem 8.3.1 and a closer analysis of the singular functions $v_i$ . Theorem 8.4.1, together with theorem 2.6.1, indicate a method for refining the elements optimally. Merely by changing $h$ to the parameter $H$ introduced in §6.4, we obtain estimates of the form (8.3.9) and (8.3.14), and the analysis is analogous to that in section 6.4 the reader is sent to reference [10] for more details regarding this optimal refinement. A comparison of the method of singular functions with the method of optimal refinements reveals that both methods yield asymptotically the same error estimate. Nonetheless, there are important differences between the two approaches. The method of singular functions usually guarantees the right order of convergence without a large increase in the size of the systems to be solved; however, some computational problems arise. The main problem in connection with this is the computation of the singular functions, which is generally a difficult problem. On the other hand, the method of refinement leads to a substantial increase in the size of the systems to be solved (i.e., the dimension of $S_h^{t,k}(\Omega)$), but the knowledge of the behavior of these functions is not necessary. Numerical experiments indicate that this approach gives very good results (see e.g. [19] and [20]).

## 8.5. The interface problem.

Another important problem in the applications is the interface problem. Once again we focus our attention on the model problem

$$(8.5.1) \qquad -\sum \frac{\partial}{\partial x_i} a(x) \frac{\partial u}{\partial x_i} + u = f \quad \text{on} \quad \Omega ,$$

$$(8.5.2) \qquad \frac{\partial u}{\partial n} = g \quad \text{on} \quad \Omega^{\bullet} ,$$

where $a(x) > a_0 > 0$ is a piecewise constant function. We distinguish several cases.

First we assume that $\Omega$ is smooth and the subdomains $\Omega_i \subset \Omega$ where $a(x)$ is constant, are also smooth. We define the space $\tilde{H}^s(\Omega)$ for $s \geq 1$ by

$$\|u\|^2_{\tilde{H}^s(\Omega)} = \|u\|^2_{H^1(\Omega)} + \sum_i \|u\|^2_{H^s(\Omega_i)} .$$

Now we can prove the following result.

Theorem 8.5.1. Let $u \in H^1(\Omega)$ be the solution of problem (8.5.1) and (8.5.2). Then for every $s \geq 0$ we have

$$\|u\|_{\tilde{H}^{s+2}(\Omega)} \leq e\left[ \|f\|_{H^s(\Omega)} + \|g\|_{H^{s+1/2}(\Omega^{\bullet})} \right] .$$

(For the proof see [11].)

From the above theorem we see that the solution is smooth but has jumps in the derivatives on the interfaces. By the use of finite element methods analogous results can be obtained as before. One may use the technique of Lagrange multipliers, the penalty method [12], or the weighted least square method [13]. The second case is the case when $\Omega$ and $\Omega_i$ are piecewise smooth. Then the singularities of the solution will occur in the neighborhood of the corners of $\Omega$ and $\Omega_i$. [4] contains a theorem similar to theorem 8.3.1. These singularities can be treated in the same way as the singularities around the corners discussed in sections 8.3 and 8.4. In this case it is necessary to deal with the singularities also, which jump around the smooth parts of the boundary of $\Omega_i$ as if $\Omega_i$ were smooth.

## 8.6. Abrupt changes of the boundary conditions.

An abrupt change of the boundary condition creates singularities which are very similar in character to those brought about by the presence of the interfaces and corners in the solution. All the results of the previous sections, including theorem 8.3.1, hold in this case (see [4]).

## REFERENCES


[1]. Nečas, J., "Sur la coercivité des formes sesquilinéares elliptiques," Rev. Roumaine de Meth. Pure et Appl. IX (1964), pp. 47-69.

[2]. Kadlec, J., "The regularity of the solution of the Passou problem in a domain whose boundary is similar to that of a convex domain," Czechoslovak Math. J. 14 (89) (1964), pp. 386-393.

[3]. Babuška, I., "Error bounds for finite element method," Num. Math. 16 (1971), pp. 322-333. [Tech Note BN-630 (Nov. 1969). Institute for Fluid Dynamics and Applied Mathematics, University of Maryland.]

[4]. Kellogg, R. B., "Higher order singularities for interface problems," These proceedings.

[5]. Babuška, I. and M. Rosenzweig, "A finite element scheme for domains with corners," To appear in Num.

Math. 1972. [Tech Note BN-720 (Nov. 1971), Institute for Fluid Dynamics and Applied Mathematics, University of Maryland.]

[6]. Babuška, I. and R. B. Kellogg, "Numerical solution of the neutron diffusion equation in the presence of corners and interfaces," To appear. [Tech Note BN-720 (Dec. 1971), Institute for Fluid Dynamics and Applied Mathematics, University of Maryland.]

[7]. Schatz, A., "Interior estimates for Ritz-Galerkin methods," These Proceedings.

[8]. Irwin, G. R., "Fracture," Handbuch der Physik, Vol. VI, Springer, Berlin (1958), p. 55

[9]. Irwin, G. R., "Analysis of stresses and shams near the end of a crack tronnersing plate," J. of Appl. Math. 24 (1957), p. 361.

[10]. Babuška, I., "Finite element method for domains with corners," Computing 6 (1970), pp. 264-273. [Tech Note BN-636 (Jan. 1970), Institute for Fluid Dynamics and Applied Mathematics, University of Maryland. ]

[11]. Seftel, Z. G., "Energy inequalities and general boundary problems for elliptic equations with discontinuous coefficients," Sibirsk. Math. Z. 6, 636-668.

[12]. Babuška, I., "The finite element method for elliptic equations with discontinuous coefficients," Computing 5 (1970), 207-213. [Tech Note BN-631 (Dec. 1969), Institute for Fluid Dynamics and Applied Mathematics, University of Maryland.]

[13]. Baker, G. A., "Finite element method for elliptic equations with discontinuous coefficients," Thesis at Cornell University, To appear.

[14]. Babuška, I., L. Mejzlik, E. Vitasek, "Effects of artificial cooling of concrete in a dam during its hardening," VII Congress des Grands Barrages, Rome, 1961, pp. 1-13.

[15]. Fix, G., "Higher-order Rayleigh-Ritz approximation," J. Math. Mech. 18 (1969), pp. 645-658.

[16]. Fix, G., S. Gulati, G. I. Wahoff, "On the use of singular functions with the finite element method," To appear in J. of Comp. Physics.

[17]. Fox, L., P. Henrici, C. Moler, "Approximation and bounds for eigenvalues of elliptic operators," J. SIAM Num. Anal. 4 (1967), pp. 89-102.

[18]. Yamamoto, Y. and N. Tokuda, "Determination of stress intensity factors in cracked plates by the finite element method," NAUT Report No. 4003 (1972), Dept. of Naval Architecture, University of Tokyo, Hongo, Bunkyo-ku, Tokyo.

[19]. Eisenstat, S. C. and M. H. Schultz, "Computational aspects of the finite element method," These Proceedings.

[20]. Suhara, J. and J. Fukuda, Automatic Mesh Generation for Finite Element Analysis Advances in Computational Methods in Structural Mechanics and Design, edited by J. T. Oden, R. W. Clough, Y. Yamamoto, UAH Press, University of Alabama in Huntsville, Huntsville, Alabama, 1972, pp. 607-624.

# CHAPTER 9. THE PROBLEMS OF PERTURBATIONS IN THE FINITE ELEMENT METHOD

## 9.1. Introduction.

In the previous chapters we studied the finite element method for numerical solution of boundary value problems for elliptic partial differential equations. We recall that the approach used consisted of the following four steps.

1) The selection of finite dimensional subspaces $M_1 \subset H_1$ (resp. $M_2 \subset H_2$) and the choice of the basis functions $\phi_i, \psi_i$ , $i = 1,\ldots,n$ . ($M_1$ and $M_2$ generally depend on the parameter $h$ .)

2) The computation of the coefficients $B(\phi_i,\psi_j) = a_{i,j}$ , where $B(\phi_i,\psi_j)$ is the bilinear form associated with the variational principle. The computation of the values $b_j = F(\psi_j)$ , where $F(\psi)$ is the functional on the right-hand side of the equation expressing the variational principle.

3) The solution of the linear system of algebraic equations

$$\sum_{i=1}^{n} a_{ij}\xi_i = b_j \quad , \qquad j = 1,\ldots,n \quad . \tag{9.1.1}$$

4) The evaluation of the solution in the form

$$u_h = \sum_{i=1}^{n} \xi_i \phi_i \quad . \tag{9.1.2}$$

We have always assumed that we know the coefficients $a_{ij}$ and $b_j$ exactly. However, this is not the case. The coefficients $a_{ij}$ and $b_j$ are defined as integrals and often the only possibility for their computation is by numerical integration.

In practice numerical integration is often used even if there is a possibility of computing the integral in a closed form.

Finally, the round-off errors are unavoidable in the computation of the coefficients $a_{ij}$, in solving the algebraic system (9.1.1), and in the evaluation (9.1.2).

Thus the situation is the following. In step 2 instead of the coefficients $a_{ij}$ we obtain the coefficients

$$\tilde{a}_{ij} = a_{ij} + \gamma_{ij} \quad , \tag{9.1.3}$$

and in place of $b_j$ we have

$$\tilde{b}_j = b_j + \gamma_j \quad . \tag{9.1.4}$$

As mentioned earlier the main reason for the appearance of the perturbations $\gamma_{ij}$ is the numerical quadrature and the round-off error.

We may formally write

$$\tilde{a}_{ij} = B(\phi_i, \psi_j) + \Gamma_1(\phi_i, \psi_j) + \Gamma_2(\phi_i, \psi_j) \quad , \tag{9.1.5}$$

where $\Gamma_1(\phi, \psi)$ is a bilinear form defined on $M_1 \times M_2$

$(M_1 \subset H_1 , M_2 \subset H_2)$ and $\Gamma_2$ is generally not a bilinear form, the structure of which will be discussed later. The term $\Gamma_1(\phi,\psi)$ is due to the error in the numerical quadrature. If the quadrature formula is assumed to be linear then obviously $\Gamma_1(\phi,\psi)$ is a bilinear form, with an additional feature which will be discussed later. The form $\Gamma_2(\phi,\psi)$ is due to round-off errors and obviously is not bilinear.

Similarly we may write

$$\tilde{b}_j = F(\psi_j) + \gamma_1(\psi_j) + \gamma_2(\psi_j) \quad , \tag{9.1.6}$$

where $\gamma_1$ is a linear form due to the quadrature formulas and the nonlinear expression $\gamma_2$ is due to round-off errors. Now instead of finding the vector $\{\xi_i\}$ which solves (9.1.1), we seek a vector $\{\tilde{\xi}_i\}$ which solves the system

$$\sum_i \tilde{a}_{ij}\tilde{\xi}_i = \tilde{b}_j \quad , \quad j = 1,\cdots,n \tag{9.1.7}$$

and this vector will be used for the computation of

$$\tilde{u}_h = \sum \tilde{\xi}_i \phi_i \quad . \tag{9.1.8}$$

In fact, due to round-off errors, we are not able to find the solution (9.1.7). On the basis of this fact we are actually seeking a vector $\{\tilde{\tilde{\xi}}_i\}$ which solves exactly the system

$$\sum \tilde{\tilde{a}}_{ij}\tilde{\tilde{\xi}}_i = \tilde{\tilde{b}}_j \quad , \tag{9.1.9}$$

where

(9.1.10) $$\tilde{\tilde{a}}_{ij} = \tilde{a}_{ij} + \bar{\gamma}_{i,j} \quad ,$$

(9.1.11) $$\tilde{\tilde{b}}_j = \tilde{b}_j + \bar{\gamma}_j \quad ,$$

where $\bar{\gamma}_{ij}$ and $\bar{\gamma}_j$ depend on the algorithms we use for solving the system (9.1.7) (see [1] and [2]). The structure now is more complicated, and as the structure of $\Gamma_2(\phi,\psi)$ and $\gamma_2(\psi)$, it depends on the solution $\xi_i$, etc. We treat these two perturbations together, assuming that the system (9.1.7) with (9.1.5) and (9.1.6) is solved exactly and that $\Gamma_2$ and $\gamma_2$ are zero, using the computer with infinite length of word or more practically if the length is sufficiently large in comparison with $\Gamma_1$ and $\gamma_1$, which is often the case. We said that we evaluate (9.1.8). Actually this is impossible to accomplish. We can evaluate the value of $\tilde{u}_h$ for some $x$, i.e., to evaluate $\tilde{u}_h(x) = \sum \tilde{\xi}_i \phi_i(x)$, or evaluate the derivatives, etc. Once again due to the presence of round-off errors we compute

$$\tilde{\tilde{u}}_h(x) = \sum \tilde{\xi}_i \tilde{\phi}_i(x) \quad ,$$

where

$$\tilde{\phi}_i(x) = \phi_i(x) + \sigma_i(x) \quad ,$$

$\sigma_i(x)$ has a complicated structure. However, this operation in the finite element method is harmless and it can be assumed that $\sigma_i(x) = 0$, since the other terms are much larger.

Now we emphasize some very important features of

the procedure just discussed.

1) We are not interested in the accuracy of the value $\xi_i$, i.e., not in the difference $\tau_i = \xi_i - \tilde{\xi}_i$, but only in the difference $\tilde{u}_h - u_h$ (in some norm, i.e., in the norm of the function $\varepsilon = \sum \tau_i \phi_i$. This may give a very different result. If $\phi_i(x)$ is small then large $\tau_i$ makes only a small contribution to $\varepsilon$, etc).

2) Our main interest does not lie in the solution of the algebraic problem in general. We are interested only in the case when $u_h$ given by (9.1.2) approximates well the exact solution $u$, which is the solution of an elliptic partial differential equation; i.e., it is a smooth function. In other words, we are interested in analyzing only the case where $b_j$ and $\xi_i$ have very special qualitative properties. In the case of the finite element method one of the most important of these properties is that the values of $\xi_i$ (resp. $\xi_j$) which converge to the values of the solution (or its derivatives) in some point $x_i$ (resp. $x_j$) are very close together if the points $x_i$ and $x_j$ are near and further as $h \to 0$ and the distance between $x_i$ and $x_j$ is of order $h$ then $\xi_i \to \xi_j$. The general theory of perturbations in variational methods has been deeply studied in [3], where several references are included. In this connection [4, §17] may also be consulted. The general problems of round-off errors concerned with the mathematical analysis have also been discussed in [5], [6].

## 9.2. The problem of linear perturbation operators.

In this section we consider the case when the

perturbations $\Gamma(\phi,\psi)$ and $\gamma(\psi)$ are a bilinear form and a linear function respectively. A special and most important application of this case is the use of quadrature formulae instead of bilinear forms of integral type. Problems of this type are thoroughly analyzed in [7], [8]. See also [9], [10]. Since these results are contained in these proceedings, we shall only give a brief description of the main ideas.

Let $u \in H_1$, $u_h \in \overline{M}_1 \subset H_1$ and $\tilde{u}_h \in \overline{M}_1$ be such that

(9.2.1) $$B(u,v) = F(v) \quad , \quad \forall v \in H_2 \quad ,$$

(9.2.2) $$B(u_h,v) = F(v) \quad , \quad \forall v \in M_2 \subset H_2 \quad ,$$

(9.2.3) $$B(\tilde{u}_h,v) + \Gamma(\tilde{u}_h,v) = F(v) + \gamma(v) \quad , \quad \forall v \in M_2 \subset H_2 \quad .$$

In what follows we assume that the above bilinear form $B(u,v)$ possesses all the properties required in theorems 5.2.1 and 6.2.1. Under these assumptions we know that if

$$\sup_{v \in H_2} \frac{|F(v)|}{\|v\|_{H_2}} \leq C$$

then

(9.2.4) $$\|u_h\|_{H_1} \leq C \qquad \text{(see theorem 5.2.1)} \quad .$$

By denoting $\tilde{u}_h - u_h = J$, we have

(9.2.5) $$B(J,v) + \Gamma(J,v) = -\Gamma(u_h,v) + \gamma(v) \quad .$$

By assumption $\Gamma(\phi,\psi)$ is a bilinear form defined on $M_1 \times M_2$. Obviously this form depends on $h$. In this connection we introduce the additional hypotheses:

1)

(9.2.6) $$|\Gamma(\phi,\psi)| \leq \alpha_0(h)\|\phi\|_{H_1}\ \|\psi\|_{H_2} \quad ,$$

$$\forall\phi \in M_1 \quad \text{and} \quad \forall\psi \in M_2 \quad ,$$

with $\alpha_0(h) \to 0$ as $h \to 0$ .

2)

(9.2.7) $$|\Gamma(u_h,\psi)| \leq \alpha(h)\|\psi\|_{H_2} \quad ,$$

where $\alpha(h)$ generally depends on the functional $F(v)$ on the right-hand side of (9.2.1). We remark that in applications $\alpha_0(h)$ and $\alpha(h)$ are in general different functions and $(\alpha_0(h))/(\alpha(h)) \to \infty$ as $h \to 0$ .

3)

(9.2.8) $$\|\gamma(v)\| \leq \beta(h)\|v\|_{H_2} \quad ,$$

where $\beta(h)$ depends on $F$ as does $\alpha(h)$ .

The condition (9.2.6) is equivalent to the "stability condition" and the conditions (9.2.7) and (9.2.8) to the accuracy conditions.

Now we may write

(9.2.9) $$J = J_1 + J_2 + J_3 + \cdots$$

so that for every $v \in M_2 \subset H_2$

(9.2.10) $$B(J_1,v) = -\Gamma(u_h,v) + \gamma(v)$$

(9.2.11) $$B(J_{i+1},v) = -\Gamma(J_i,v) \quad .$$

Hence we have from (9.2.4) and (9.2.6)

$$\|J_{i+1}\|_{H_1} \leq C\alpha_0(h)\|J_i\| \quad , \quad i = 1,2,\ldots \quad ,$$

where $C$ is independent of $h$ and $i$, and

$$\|J_1\| \leq C[\alpha(h) + \beta(h)] \quad .$$

Therefore for $h$ sufficiently small we have

$$|C\alpha_0(h)| < 1 - \varepsilon \quad , \quad \varepsilon > 0$$

and consequently

$$\|J\|_{H_1} \leq C[\alpha(h) + \beta(h)] \quad . \tag{9.2.12}$$

Thus we see that in fact $J \approx J_1$ and therefore in general we do not improve the error by resorting to the lower spaces.

## 9.3. Penalty method with linear perturbations.

As we pointed out in Section 7.2, spaces $S_h^{t,k}(\Omega)$ with additional properties may be used in order to obtain improved results in the penalty method. In this section, by combining the penalty method with the approach of linear perturbations, we obtain new results. We restrict ourselves as in Section 7.2 to the model problem

$$-\Delta u = f \quad \text{on} \quad \Omega \quad , \tag{9.3.1}$$

$$u = 0 \quad \text{on} \quad \Omega \quad , \tag{9.3.2}$$

where $\Omega$ is assumed to be a smooth domain. We suppose

that the system $S_h^{t,k}(\Omega)$ with $k \geq 1$ satisfies the additional hypotheses.

1) There exists a $p > 0$ such that for every $\omega \in S_h^{t,k}(\Omega)$ we have

$$\|\omega\|_{H^0(\Omega^\bullet)} \leq Ch^p \|\omega\|_{H^1(\Omega)} , \tag{9.3.3}$$

where $C$ is independent of $h$.

2) For $f \in H^{\ell+2}(\Omega)$, $f = 0$ on $\Omega^\bullet$, there are positive constants $\mu_1$ and $\mu_2$ depending on $\ell$ and $t$ such that there exists a $\omega \in S_h^{t,k}(\Omega)$ with the property

$$\|f - \omega\|_{H^1(\Omega)} + h^{-\mu_1}\|\omega\|_{H^0(\Omega^\bullet)} \leq Ch^{\mu_2}\|f\|_{H^\ell(\Omega)} , \tag{9.3.4}$$

where $C$ is independent of $f$ and $h$. We know that systems $S_h^{t,k}(\Omega)$ with the above properties exist, e.g., the system of piecewise linear functions (linear elements) with nodal points on $\Omega^\bullet$. In this case we have $t = 2$, $p = \frac{3}{2}$ and $\mu_1 = \mu_2 = 1$ for $\ell \geq 0$.

Now we consider the bilinear form

$$B(\tau,u,v) = \iint_\Omega \left[\frac{\partial u}{\partial x_1}\frac{\partial v}{\partial x_1} + \frac{\partial u}{\partial x_2}\frac{\partial v}{\partial x_2}\right]dx \tag{9.3.5}$$

instead of the form (7.2.4). Thus we have the perturbation operator

$$\Gamma(u,v) = \psi(\tau)\int_{\Omega^\bullet} uv\,ds , \tag{9.3.6}$$

and the functional $\gamma(v) = 0$. Hence for $u \in S_h^{t,k}(\Omega)$, $v \in S_h^{t,k}(\Omega)$ we have

$$|\Gamma(u,v)| \leq \psi(\tau)\|u\|_{H^0(\Omega^\bullet)}\|v\|_{H^0(\Omega^\bullet)}$$

(9.3.7)

$$\leq C\psi(\tau)h^{2p}\|u\|_{H^1(\Omega)}\|v\|_{H^1(\Omega)} ,$$

and since

$$\|u\|_{H_1(\tau)} \geq C\|u\|_{H^1(\Omega)} ,$$

we also have

$$|\Gamma(u,v)| \leq C\psi(\tau)h^{2p}\|u\|_{H_1(\tau)}\|v\|_{H_1(\tau)} . \tag{9.3.8}$$

Therefore if $\psi(\tau)h^{2p} \to 0$ as $h \to 0$ , the assumption (9.2.6) is satisfied. By returning to §7.2, taking $\omega_1 = 0$ in (7.2.2), and using (9.3.4), we obtain

$$\|u_0 - u_h(\tau)\|_{H_1(\tau)} \leq C\|f\|_{H^\ell(\Omega)}[\psi^{-\frac{1}{2}}$$

(9.3.9)

$$+ h^{\mu_2}[\max(1,\psi h^{2\mu_1})]^{\frac{1}{2}} .$$

Hence

$$|\Gamma(u_h(\tau),\omega| \leq |\Gamma(u_0,\omega)| + |\Gamma(u_0-u_h(\tau),\omega)|$$

$$\leq \psi^{-\frac{1}{2}}C\|f\|_{H^\ell(\Omega)}[\psi^{-\frac{1}{2}} + h^{\mu_2}[\max(1,\psi h^{2\mu_1})]]^{\frac{1}{2}}h^p\|\omega\|_{H_1(\tau)} .$$

Thus we have (see (9.2.7))

$$\alpha(h) = h^p C\|f\|_{H^\ell(\Omega)}[\psi^{-1} + \psi^{-\frac{1}{2}}h^{\mu_2}[\max(1,\psi h^{2\mu_1})]]^{\frac{1}{2}} . \tag{9.3.10}$$

Now we take $\psi = h^{2\mu_1}$ , then from (9.3.9) we have

$$\|u_0 - u_h(\tau)\|_{H^1(\Omega)} \leq C\|f\|_{H^\ell(\Omega)}[h^{\mu_1} + h^{\mu_2}] , \tag{9.3.11}$$

From (9.3.10) and (9.2.11) we obtain

(9.3.12)
$$\begin{aligned}\|\tilde{u}_h(\tau) - u_0\|_{H^1(\Omega)} &\leq C\,\|f\|_{H^\ell(\Omega)}\, h^p[h^{2\mu_1} + h^{\mu_1+\mu_2}] \\ &\leq C\,\|f\|_{H^\ell(\Omega)}\,[h^{\mu_1} + h^{\mu_2}] ,\end{aligned}$$

provided $p > \mu_1$ .

We notice that in the example of piecewise linear functions we have $\mu_1 = \mu_2 = 1$ , $p = \frac{3}{2} > 1$ and therefore for $\ell \geq 0$ we obtain

$$\|u_h(\tau) - u_0\|_{H^1(\Omega)} \leq C\|f\|_{H^\ell(\Omega)}\, h ,$$

which is the optimal rate of accuracy for piecewise linear functions. This implies that if $S_h^{t,k}(\Omega)$ consists of piecewise linear functions with nodal points on the boundary of triangular elements with sides of order $h$ angle $\theta_0 > 0$ , then the penalty method degenerates to the method where the boundary conditions are nearly satisfied (see also [9]). We would like to emphasize that the assumption that $u = 0$ on $\Omega^\cdot$ is not important and that the approach applies equally well to nonhomogeneous boundary conditions.

There are many other possibilities to use a similar approach. Let us briefly mention another one. We shall be interested in the solution of a model problem

(9.3.13) $$-\Delta u = f \quad \text{on} \quad \Omega$$

and

(9.3.14) $$u = 0 \quad \text{on} \quad \Omega^\cdot .$$

Solving this problem by the finite element technique, we select $M_1 = M_2 = H_0^1(\Omega)$ and we find $u \in M_1$ such that

$$B(u,v) = \int_\Omega \left(\frac{\partial u}{\partial x}\frac{\partial v}{\partial x} + \frac{\partial u}{\partial y}\frac{\partial v}{\partial y}\right)dx = \int_\Omega fv dx$$

for every $v \in M_2$ .

Let us assume that we have $\tilde{M}_1$ and a one-to-one mapping $Q$ of $M_1$ onto $\tilde{M}_1$ . It may be $\tilde{M}_1 \not\subset H_0^1$ ; nevertheless, the norm of the operator $I-Q$ could be small. Then we may find an approximate solution $\tilde{u}$ $\tilde{M}_1$ such that $B(\tilde{u},\tilde{v}) = \int f\tilde{v}dx$ for every $\tilde{v} \in \tilde{M}_1$ . Based on the structure of the operator $Q$ , the element $\tilde{u}$ is also a good approximation of the desired solution.

## 9.4. Problems of nonlinear perturbations.

The main application of nonlinear perturbations occurs in the analysis of the round-off errors. As mentioned earlier, the round-off errors appear in two stages. A thorough analysis of this phenomen depends on various lengthy details. Due to the limitation of space, we confine our attention to the main ideas only and illustrate them by considering a model problem. In connection with this, [6], [10] and [11] may be consulted. For an interesting idea of how to influence the conditioning of the matrix $\{a_{ij}\}$ , see [12].

We shall suppose that the round-off errors are very small and that the vector $\tilde{\xi}_i$ in (9.1.7) may be written as a solution of

(9.4.1) $$\sum_i a_{ij}\tilde{\xi}_i = b_j + \gamma_j ,$$

where

(9.4.2) $$|\gamma_j| \leq D \max_i |\xi_i| \varepsilon ,$$

where $\xi_i$ is the solution of (9.1.1) and $\varepsilon > 0$ is small. Further, we assume that the base functions are normalized, so that if $\xi_i$ is the exact solution of (9.1.1), then $\max|\xi_i| \leq C$, where $C$ does not depend on $h$.

Let us emphasize that in the finite element method we have a very special matrix $\{a_{ij}\}$. One of the typical and important features of this matrix is that for the majority of the rows we have

$$\sum_i a_{ij} = (h^{2\ell}) , \tag{9.4.3}$$

where $2\ell$ is the order of the equation. Moreover, we may write for the solution

$$\xi_i = \xi_j + O(h) , \tag{9.4.4}$$

where $i$ ranges over all $i$ such that $a_{ij} \neq 0$. This behavior is in fact a corollary of the local properties of the finite approximations mentioned in chapter 1 and a proper normalization. Let us remark that it is not true for all choices of base functions. As an example, consider the one-dimensional case on $\langle -1,+1 \rangle$, with base functions $\phi_j = x^j$.

Our last assumption justifies (9.4.2) since the number of the entries in a row of the matrix $\{a_{ij}\}$ is independent of $h$. Therefore, instead of (9.4.2) we may write

$$|\gamma_j| \leq C(h)\varepsilon \leq C_1\varepsilon , \tag{9.4.5}$$

where $C_1$ does not depend on $h$. In chapter 6 we saw that the coefficients in (9.1.1) are such that

(9.4.6) $$b_j = F(\phi_j) ,$$

where $F$ is a functional. In the case of the problem

(9.4.7) $$-\Delta u + u = f$$

(9.4.8) $$\frac{\partial u}{\partial n} = 0 ,$$

we have

(9.4.9) $$F(\phi_j) = \int_\Omega f\phi_j dx .$$

In §6 it was shown that if the system $S_h^{t,k}(\Omega)$ satisfied the inverse assumption, then the finite element method converged not only for $f \in L_2$ but for more general spaces, too. Therefore we may try to estimate

(9.4.10) $$\sup_{|\gamma_j| \leq \varepsilon} \quad \inf_{\int f\phi_j = \gamma_j} |f|_{H^\ell(\Omega)} = \varepsilon\Lambda\ (h,\ell) .$$

The influence of $\gamma_j$ on the solution of the problem (9.4.7) and (9.4.8) may be determined when $\Lambda(h,\ell)$ is known. The right-hand side $f$ is estimated by $\Lambda$. It is important to note that by using smaller $\ell$, we systematically decrease $\Lambda$, in particular for smaller $h$ and larger $t$ in $S_h^{t,k}(\Omega)$. We cannot analyze these problems in more detail here. Instead we illustrate them by considering an example.

Let us try to solve the problem

$$-y'' + y = -1$$

$$y(-1) = y(1) = 0$$

by the method based on the use of the $S_h^{t,k}(\Omega)$ described at the end of §6.3.2. We carry out the entire computation in single and double precision. We assume that the double

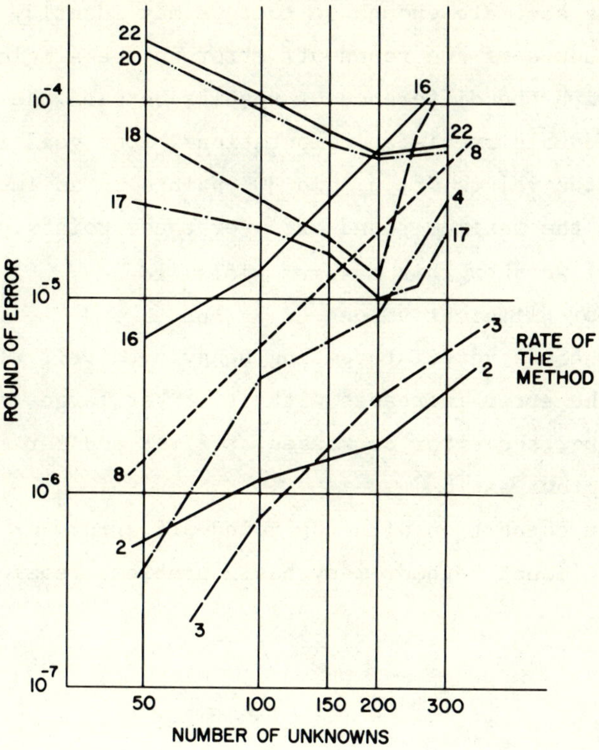
22
20
16
22
8
18
17
4
17
16
3
RATE OF THE METHOD
2
8
2
3
$10^{-4}$
$10^{-5}$
$10^{-6}$
$10^{-7}$
ROUND OF ERROR
50
100
150
200
300
NUMBER OF UNKNOWNS

Fig. 9.4.1

precision is accurate enough so that we may identify the global influence of the round-off error in the single precision with the difference of results between the single and double precision computations. Our goal is to compute the values of $u_h$ in 300 points in an interval and to take the maximum round off over these points. In Fig. (9.4.1) we show this maximum global round-off error for different values of t and h . We observe here a very interesting behavior, i.e., with large h the error increases with t . For large t and h decreasing, the error decreases at first and for h very small grows as $h^{-2}$ .

In connection with the round-off error in the finite element method, many basic problems remain to be solved.

## REFERENCES

[1]. Wilkinson, J. H., The Algebraic Eigenvalue Problem, Clarendon Press, Oxford, 1965.

[2]. Babuška, I., "Numerical stability in problems of linear algebra," SIAM J. Numer. Anal. 9 (1972), pp. 53-77.

[3]. Mikhlin, S. G., Numerical Realizations of the Variational Methods, Moscow, 1966 (In Russian).

[4]. Krasnoselskii, M. A. et al., Approximate Solution of Operator Equations, Moscow, 1969 (In Russian).

[5]. Babuška, I., M. Prager and E. Vitasek, Numerical Processes in Differential Equations, Wiley-Interscience, London, New York, Sydney, 1966.

[6]. Babuška, I., "Numerical stability in mathematical analysis," Information Processing 68 (1969), North-Holland Publ. Comp., Amsterdam.

[7]. Fix, G., "On the effects of quadrature error in the finite element method," Advances in Computational

Methods in Structural Mechanics and Design, edited by J. T. Oden, R. W. Clough and Y. Yamamoto, UAH Press, University of Alabama in Huntsville, Huntsville, Alabama, pp. 55-68.

[8]. Fix, G., "Effects of quadrature error in finite element approximation of eigenvalue and parabolic problems," This symposium.

[9]. Ciarlet, P. G. and R. A. Raviart, "The combined effect of curved boundaries and numerical integration in isoparametric finite element methods."

[10]. Herbold, R. J. and R. S. Varga, "The effect of quadrature errors in the numerical solution of two-dimensional boundary value problems by variational techniques," Aequationes Math. 7 (1972), pp. 36-58.

[11]. Yamamoto, Y., "Some considerations on round-off errors of the finite element method," Advances in Computational Methods in Structural Mechanics and Design, edited by J. T. Oden, R. W. Clough and Y. Yamamoto, UAH Press, University of Alabama in Huntsville, Huntsville, Alabama, 1972.

[12]. Ruchovec, L. A., "To the problem of the construction of variation-difference schemes for elliptic equations," (In Russian) Ž. Vyčisl. Mat. i. Mat. Fiz. 12 (1972), pp. 781-785.

## CHAPTER 10. THE EIGENVALUE PROBLEM.

### 10.1 Introduction.

In the previous chapters we studied the problem of finding a $u \in H_1$ so that $B(u,v) = F(v)$ for all $v \in H_2$, where $B(u,v)$ was a bilinear form defined on $H_1 \times H_2$ and $F(v)$ was a linear functional on $H_2$. An essential assumption was that the only $u \in H_1$ which satisfies the equation

$$B(u,v) = 0 \quad \text{for all} \quad u \in H_2$$

is $u = 0$. We saw that the general problem for the elliptic equations is covered in this framework.

In what follows we assume that our bilinear form depends on parameter (complex) $\lambda$, i.e. we are considering the complex Hilbert spaces". Now our problem in general is to find the value $\lambda = \lambda_0$ such that there exists $u_0 \in H_1$, $u_0 \neq 0$, with $B(\lambda,_0,u_0,v) = 0$ for all $v \in H_2$.

The technique of the finite element method consisted of fixing the finite dimensional subspaces $M_1(h) \subset H_1$ and $M_2(h) \subset H_2$ and finding of $u_h \in \overline{M}_1$ such that

1) As we remarked in chapter 5, theorem 5.2.1 obviously holds for complex spaces $H_1$ and $H_2$, similarly theorem **6.2.1.** In the applications we also have complex Sobolev spaces instead of real spaces. The definition of those is anologous to the definition of real spaces.

$$B(u_h,v) = F(v) \quad \text{for all} \quad v \in M_2(h) \ .$$

We showed that under suitable assumptions $u_h$ is very close to the exact solution $u$ if $h$ is sufficiently small. This provides us immediately with the hint of how to solve approximately the eigenvalue problem. We need to find $\lambda_0^h$ and $u_{0,h} \neq 0$ such that

$$B(\lambda_0^h,u_{0,h},v) = 0 \quad \text{for all} \quad v \in M_2(h) \ .$$

We can expect that for sufficiently small $h, \lambda_0^h$ and $u_{0,h}$ will be close to $\lambda_0$ and $u_0$ respectively.

We assume that the bilinear form has a special form, namely,

$$B(u,v) - D(\lambda,u,v) \ .$$

We shall speak of the general eigenvalue problem if we want to find $\lambda_0$ and $u_0 \in H_1$ such that

$$B(u_0,v) = D(\lambda_0,u_0,v) \ , \forall v \in H_2 \ .$$

The case where the operator $D(\lambda,u,v)$ has the special form

$$D(\lambda,u,v) = \lambda D(u,v) \quad ,$$

will be refered to as the linear eigenvalue problem.

The finite element method in connection with the linear eigenvalue problem has been studied by many authors in the past decade. We cannot quote here all the papers dealing with this problem. We mention here only a series of papers written by Vainikko (see [1], [2], [3]) summarized in [4]. An important result in this field has been obtained independently in [5]. The theory of self-adjoint linear problems where $H_1 = H_2$ has been studied by various authors (in particular see [6]). The most general case of the linear eigenvalue problem has been studied recently in [7], [8] and [9]. The main results in [7] and [9] are similar, but the method of proofs is very different.

In this chapter we shall present results which are similar to those in [9], generalized to nonlinear problems. The main ideas discussed here are taken from [9][2). Although the linear eigenvalue problem is the most important, nonetheless the nonlinear problems have many interesting physical applications.

In studying this general nonself-adjoint case we show that the method is applicable with almost no complications to this general situation. It is of interest to note that the assumption of self-adjointness does not actually lead to sharper results or simpler proofs.

10.2. <u>The eigenvalue problem</u>.

Let $H_1$ and $H_2$, $K_1$ and $K_2$ be given complex Hilbert spaces, with $H_1 \subset K_1$ and $H_2 \subset K_2$. We assume that the identity imbedding $H_1$ into $K_1$ and $H_2$ into $K_2$ are compact and $\|u\|_{K_i} \leq C\|u\|_{H_i}$, $i = 1,2$. Further let $B(u,v)$ and $D(\lambda,u,v)$ be bilinear forms defined on $H_1 \times H_2$ and $K_1 \times K_2$ respectively. We assume that the

2) We are indebted to G. Fix for very valuable discussions and help in the preparation of this chapter.

the bilinear form $D(\lambda,u,v)$ depends on a complex variable $\lambda$, i.e. is defined on $K \times K_1 \times K_2$, where $K$ denotes the space of complex numbers. Throughout this chapter we shall assume that the form $B(u,v)$ satisfies all the assumptions of theorem 5.2.1, i.e. the inequalities (5.2.1)-(5.2.3).

The form $D(\lambda,u,v)$ is assumed to satisfy the assumptions:

I.

(10.2.1) $$|D(\lambda,u,v)| \leq K(\lambda) \|u\|_{K_1} \|v\|_{K_2} ,$$

with $K(\lambda)$ bounded on every bounded set of $\lambda$.

II. For every fixed $u,v$ the function $D(\lambda,u,v)$ is an entire holomorphic function.

We first approve the lemma.

<u>Lemma 10.2.1</u>. For $\lambda_n \to \lambda$ we have

(10.2.2) $$\sup_{\substack{u \in H_1 \\ v \in H_2}} \frac{|D(\lambda_n,u,v) - D(\lambda,u,v)|}{\|u\|_{H_1} \|v\|_{H_2}} \to 0$$

<u>Proof</u>. Suppose the conclusion of the lemma is false. Then there exists a sequence $(u_n,v_n)$, $u_n \in H_1$, $v_n \in H_2$ with $\|u_n\|_{H_1} = 1$ and $\|v_n\|_{H_2} = 1$, and an $\varepsilon > 0$ such that

(10.2.3) $$|D(\lambda_n,u_n,v_n) - D(\lambda,u_n,v_n)| \geq \epsilon > 0 .$$

Since $H_1$ and $H_2$ are compact in $K_1$ and $K_2$ respectively, there exists a subsequence $(u_n,v_n)$ (denoted with the index n again) such that

$$\|u_n - u_0\|_{K_1} \to 0 \ , \ \|v_n - v_0\|_{K_2} \to 0$$

and (10.2.3) holds. We may write

$$\begin{aligned}
|D(\lambda_n,u_n,v_n) - D(\lambda,u_n,v_n)| &= |D(\lambda_n,u_n,v_n) - D(\lambda_n,u_n,v_0) + \\
&\quad D(\lambda_n,u_n,v_0) - D(\lambda_n,u_0,v_0) - D(\lambda,u_n,v_n) + D(\lambda,u_n,v_0) - \\
&\quad - D(\lambda,u_n,v_0) + D(\lambda,u_0,v_0) + D(\lambda_n,u_0,v_0) - D(\lambda,u_0,v_0)| \leq \\
&\leq K(\lambda_n)\|u_n\|_{K_1}\|v_n - v_0\|_{K_2} + K(\lambda_n)\|u_n - u_0\|_{K_1}\|v_0\|_{K_2} + \\
&\quad + K(\lambda)\|u_n\|_{K_1}\|v_n - v_0\|_{K_2} + K(\lambda)\|u_n - u_0\|_{K_1}\|v_0\|_{K_2} + \\
&\quad + |D(\lambda_n,u_0,v_0) - D(\lambda,u_0,v_0)| \ .
\end{aligned}$$

Our lemma follows immediately as a consequence of the assumptions on the form $D(\lambda,u,v)$ .

Let $\omega(\lambda)$ and $\omega^*(\lambda)$ be defined on $K$ as follows:

$$\omega(\lambda) = \inf_{\substack{u \in H_1 \\ \|u\|_{H_1}=1}} \ \sup_{\substack{u \in H_2 \\ \|v\|_{H_2}=1}} |B(u,v) - D(\lambda,u,v)| \tag{10.2.4}$$

$$\omega^*(\lambda) = \inf_{\substack{v \in H_2 \\ \|v\|_{H_2}=1}} \ \sup_{\substack{u \in H_1 \\ \|u\|_{H_1}=1}} |B(u,v) - D(\lambda,u,v)| \ . \tag{10.2.5}$$

Now we prove the following lemma.

Lemma 10.2.2. The functions $\omega(\lambda)$ and $\omega^*(\lambda)$ are continuous.

Proof. It is sufficient to prove the lemma for $\omega(\lambda)$ only.

1) as in the proof of theorem 5.2.1 we may write

$$B(u,v) - D(\lambda,u,v) = (L(\lambda)u,v)_{H_2} \quad , \tag{10.2.6}$$

where $L(\lambda)$ is a linear mapping of $H_1$ into $H_2$ which in general depends on $\lambda$ . Letting

$$S(\lambda_1,\lambda_2)(u) = L(\lambda_1)u - L(\lambda_2)u \quad ,$$

we obviously have

$$(S(\lambda_1,\lambda_2)u,v)_{H_2} = - D(\lambda_1,u,v) + D(\lambda_2,u,v)$$

and therefore from lemma 10.1.1 we have

$$\|S(\lambda_0,\lambda_n)\|_{H_1\to H_2} \to 0 \quad \text{as} \quad \lambda_n \to \lambda_0 \quad .$$

2) Clearly we have

$$\omega(\lambda) = \inf_{\substack{u\in H_1 \\ \|u\|_{H_1}=1}} \|L(\lambda)u\|_{H_2} \quad .$$

Now assume on the contrary that $\omega(\lambda)$ is not continuous at

$\lambda_0$ . Then there exists $\lambda_n \to \lambda_0$ and $\epsilon > 0$ such that

$$|\omega(\lambda_0)-\omega(\lambda'_n)| > \epsilon \quad .$$

Let $\varepsilon_n = \omega(\lambda_0) - \omega(\lambda_n)$ . Obviously there exists a subsequence, which is denoted again by the same index, so that either all $\varepsilon_n > \varepsilon$ or $-\varepsilon_n > \varepsilon$ . We analyze both cases.

I. Case $\varepsilon_n > \varepsilon > 0$ . In this case we have

$$\omega(\lambda_0) \geq \omega(\lambda_n) + \varepsilon \quad .$$

We construct now $u_n$ , $\|u_n\|_{H_1} = 1$ and such that

$$\|L(\lambda_n)u_n\|_{H_2} \leq \omega(\lambda_n) + \frac{\varepsilon}{5} \quad .$$

Obviously

$$\|L(\lambda_0)u_n\|_{H_2} \geq \omega(\lambda_0) \quad .$$

Therefore we have

$$\|L(\lambda_0)u_n\|_{H_2} - \|L(\lambda_n)u_n\|_{H_2} \geq \frac{\varepsilon}{2}$$

But

$$L(\lambda_0)u_n = L(\lambda_n)u_n + (L(\lambda_0) - L(\lambda_n))u_n \quad .$$

Therefore

$$\frac{\varepsilon}{2} \leq \|(L(\lambda_0)-L(\lambda_n))u_n\|_{H_2} \quad ,$$

which contradicts the results in the first part of the proof.

II. Case $-\varepsilon_n > \epsilon > 0$. In this case we have $\omega(\lambda_n) \geq \omega(\lambda_0) + \varepsilon$. Thus we can find $u_0$, $\|u_0\|_{H_2} = 1$ such that

$$\|L(\lambda_0)u_0\| \leq \omega(\lambda_0) + \frac{\varepsilon}{5} ,$$

and

$$\|L(\lambda_n)u_0\|_{H_2} \geq \|L(\lambda_0)u_0\|_{H_2} + \frac{\varepsilon}{5} .$$

By arguments similar to those used above we complete the proof of our lemma.

Now let $\Pi$ and $\Pi^*$ denote the set of zeros of $\omega$ and $\omega^*$ respectively. Throughout this chapter we shall assume that $\Pi$ and $\Pi^*$ are sets of isolated points. We would like to emphasize the importance of the above assumption. In the linear eigenvalue problem this assumption comes from the general theory. In the nonlinear case under certain hypotheses one obtains this result also from the general theory (see [10]). It appears that in the most general case the problem is unsolved. Nonetheless in the physically interesting cases one may assume this assumption on physical grounds.

Now we prove the following theorem.

<u>Theorem 10.2.1</u>. If $\lambda_0$ is a zero of $\omega(\lambda)$, i.e. $\omega(\lambda_0) = 0$, then there exists a $u_0 \neq 0$, $u_0 \in H_1$ such that

(10.2.6) $$B(u_0,v) = D(\lambda_0,u_0,v) , \forall v \in H_2 .$$

Proof. By hypotheses there exists a sequence $u_n \in H_1$, $\|u_n\|_{H_1} = 1$ and such that

(10.2.7) $$\sup_{\|v\|_{H_2}=1} |B(u_n,v)-D(\lambda_0,u_n,v)| \leq \frac{1}{n} .$$

Thus there exists a subsequence (which we denote by the same index n) and $u_0 \in H_1$ such that

$$B(u_n,v) \to B(u_0,v) \quad ,$$

for every fixed $v \in H_2$. Since $H_1$ is compact in $K_1$, $u_n \to u_0$ in $K_1$ and therefore

(10.2.8)

$$|D(\lambda_0,u_n,v)-D(\lambda_0,u_0,v)| \leq K(\lambda_0) \|u_n-u_0\|_{K_1} \|v\|_{K_2} .$$

Therefore by noticing that $\|v\|_{K_2} \leq C\|v\|_{H_2}$, we have

(10.2.9)

$$|B(u_n,v)-D(\lambda_0,u_0,v)| \leq \|v\|_{H_2}\left[\frac{1}{n} + \|u_n-u_0\|_{K_1}\right] .$$

Our assumption about B implies that there exists a $z \in H_1$ such that

$$B(z,v) = D(\lambda_0,u_0,v) \quad , \forall v \in H_2 .$$

Therefore (10.2.9) may be written in the form

$$|B(u_n,v)-D(\lambda_0,u_0,v)| = |B(u_n-z,v)| \leq \|v\|_{H_2}\left[\frac{1}{n} + \|u_n-u_0\|_{K_1}\right].$$

The assumption on the form B (see 5.2.2) implies that

$$u_n \to z \quad \text{in} \quad H_1 \quad \text{and hence} \quad z = u_0 \quad .$$

Therefore we have

$$B(u_0,v) = D(\lambda_0,u_0,v) \quad .$$

Since $\|u_n\|_{H_1} = 1$ we have $z \neq 0$ our theorem is completely proved.

Let $N(\lambda_0)$ denote the linear space of all functions $u \in H_1$ such that

$$B(u_0,v) = D(\lambda_0,u_0,v) \ , \ v \in H_2$$

We have shown that $N(\lambda_0)$ is not empty. Similarly we denote by $N^*(\lambda_0)$ the space associated with the zeros of the function $\omega^*(\lambda)$ .

Now we show that $\Pi = \Pi^*$ .

<u>Theorem 10.2.2.</u> We have $\Pi = \Pi^*$ .

<u>Proof.</u> Without any loss of generality we may assume that there exists $\lambda_0$ such that $\lambda_0 \in \Pi^*$ but $\lambda_0 \notin \Pi$ .

1) By hypothesis $\omega(\lambda_0) \neq 0$ . Therefore using lemma 10.2.2 there exists $\varepsilon > 0$ such that $\omega(\lambda) \geq \sigma > 0$ for all $|\lambda-\lambda_0| < \varepsilon$ and $\omega^*(\lambda) > 0$ for all $\lambda \neq \lambda_0$ and $|\lambda-\lambda_0| < \varepsilon$ . From theorem 5.2.1 we have that for any $F \in H_2'$ , $\lambda \neq \lambda_0$ and $|\lambda-\lambda_0| < \varepsilon$ there exists a unique $u_0(\lambda) \in H_1$ such that

(10.2.10) $$B(u_0(\lambda),v) - D(\lambda,u_0(\lambda),v) = F(v)$$

holds for every $v \in H_2$ and

(10.2.11) $$\|u_0(\lambda)\|_{H_1} \leq \frac{C}{\omega(\lambda)} \|F\|'_{H_2} \quad ,$$

with $C$ independent of $\lambda$ .

2. From the proof of theorem 5.2.1 it follows that there exists a mapping $L(\lambda)$ of $H_1$ into $H_2$ such that $|L(\lambda)| < C$ , where $C$ is independent of $\lambda$ . Further we have

(10.2.12) $$B(u,v) - D(\lambda,u,v) = (L(\lambda)u,v)_{H_2} \quad .$$

By repeating the arguments used in the proof of theorem 5.2.1, we see that $L(\lambda)(H_1)$, $\lambda \neq \lambda_0$ is closed in $H_2$ , $L(\lambda)(H_1) = H_2$ for $\lambda \neq \lambda_0$ and $L(\lambda_0)(H_1) \neq H_2$ . In fact for $v_0 \in N^*(\lambda_0)$ we have

$$0 = B(u,v_0) - D(\lambda_0,u,v_0) = (L(\lambda_0)u,v_0)_{H_2} \quad , \quad u \in H_1 \quad .$$

3. Let $F_0(v) = (v,v_0)_{H_2}$ and let $\lambda_n \neq \lambda_0$ , $\lambda_n \to \lambda_0$ , $|\lambda_n - \lambda_0| < \epsilon$ . Suppose $\hat{u}_0(\lambda_n)$ is the function satisfying (10.2.10) with $F_0(v)$. Now (10.2.11) implies that there exists a subsequence (using the same index n) $u_0(\lambda_n)$ converging weakly to $u_0$ and strongly in $K_1$. Therefore we have

(10.2.13) $$B(u(\lambda_n),v_0) \to B(u_0,v_0) \quad , \quad v \in H_2 \quad ,$$

(10.2.14)

$$D(\lambda_n,u_0(\lambda_n),v_0) = D(\lambda_n,u_0,v_0) + D(\lambda_n,u_0(\lambda_n)-u_0,v_0)$$

Since by hypothesis $D(\lambda,u_0,v_0)$ is holomorphic we have

$$D(\lambda_n,u_0,v_0) \to D(\lambda_0,u_0,v_0) \quad , \tag{10.2.15}$$

and

(10.2.16)

$$D(\lambda_n,u_0(\lambda_n)-u_0,v_0) \leq K\|u_0(\lambda_n)-u_0\|_{K_1}\|v_0\|_{H_2} \to 0 \quad .$$

Thus we have

$$B(u_0,v_0) - D(\lambda_0,u_0,v_0) = (v_0,v_0)_{H_2} \quad , \tag{10.2.17}$$

which is a contradiction, since $v_0 \in N^*(\lambda_0)$ . Our theorem is proved.

For every $\lambda$ such that $\omega(\lambda) \neq 0$ using theorem 2.5.1 we can construct a continuous mapping $R(\lambda)$ of $H_2'$ into $H_1$ such that

$$B(R(\lambda)F,v) - D(\lambda,R(\lambda)F,v) = F(v)$$

We prove the following theorem.

<u>Theorem 10.2.3</u>. If $\lambda_0$ is such that $\omega(\lambda_0) \neq 0$ , then

$$R(\lambda) = \sum_{n=0}^{\infty} (\lambda-\lambda_0)^n R_n \quad , \tag{10.2.18}$$

and the series converges absolutely in a neighborhood of $\lambda_0$.

Proof. The proof is quite similar to the proof in the usual spectral theory.

1. By hypothesis we have

$$D(\lambda,u,v) = \sum_{n=0}^{\infty} (\lambda-\lambda_0)^n D_n(u,v) \quad , \tag{10.2.19}$$

where

$$D_0(u,v) = D(\lambda_0,u,v) \quad , \tag{10.2.20}$$

$$D_n(u,v) = \frac{1}{2\pi i} \int \frac{D(\lambda,u,v)d\lambda}{(\lambda-\lambda_0)^n} \tag{10.2.21}$$

From our assumption it easily follows that

$$|D_n(u,v)| \leq K\|u\|_{H_1} \|v\|_{H_2} \quad ,$$

and therefore the series (10.2.19) converges absolutely for $|\lambda-\lambda_0| < 1$ .

2. Let $F \in H_2'$ be given and suppose that $u(\lambda) \in H_1$ is such that

$$B(u(\lambda),v) - D(\lambda,u(\lambda),v) = F(v) \quad . \tag{10.2.22}$$

Such a $u(\lambda)$ exists in some neighborhood of the point $\lambda_0$. We write

$$u(\lambda) = \sum_{0}^{\infty} (\lambda-\lambda_0)^n u_n \quad , \tag{10.2.23}$$

and hence

(10.2.24)

$$\sum_{n=0} (\lambda-\lambda_0)^n B(u_n,v) + \sum_{\substack{m,n \\ m,n\geq 0}} (\lambda-\lambda_0)^{m+n} D_m(u_n,v) = F(v) \quad .$$

Now let $R$ be the mapping of $H_2'$ into $H_1$ such that (10.2.22) holds with $\lambda = \lambda_0$ and $u(\lambda) = R(F)$. Thus we have

$$u_0 = RF \quad ,$$
$$u_1 = - RD_1(u_0,v)$$
$$u_2 = - R\left[D_1(u_1,v)+D_2(u_0,v)\right]$$
$$u_n = - R\sum_{m=1}^{n} D_m(u_{n-m},v) \quad ,$$

and therefore

$$\|u_n\|_{H_1} \leq K \sum_{j=0}^{n-1} \|u_j\|_{H_1} \quad .$$

Consequently

$$\|u_n\|_{H_1} \leq CK^n$$

From this we see immediately that the series (10.2.19) converges in some neighborhood of the point $\lambda_0$ and the proof is complete.

Theorem 10.2.4. If $\lambda_0$ is such that to $\omega(\lambda_0) = 0$ , then in some neighborhood of $\lambda_0$ we can write

$$(10.2.25) \qquad R(\lambda) = \sum_{n=-\infty}^{+\infty} (\lambda-\lambda_0)^n R_n \ ,$$

and the series converges absolutely (i.e., in the norm).

Proof. The proof is exactly the same as the proof of Laurent expansion.

Theorem 10.2.5. If $\lambda_0$ is such that $\omega(\lambda_0) = 0$ , then the spaces $N(\lambda_0)$ and $N^*(\lambda_0)$ are finite dimensional.

Proof. The proof easily follows from the observation that the set $N(\lambda_0) \cap \Sigma$ , where $\Sigma$ is the unit ball in $H_1$ is compact.

It is easy to see

Theorem 10.2.6. If $\omega(\lambda) \geq C|(\lambda-\lambda_0)|^\ell$ , $\ell \geq 0$ , then

$$(10.2.26) \qquad R(\lambda) = \sum_{n=-\ell}^{\infty} (\lambda-\lambda_0)^n R_n \ .$$

Let $\xi(\lambda) = \frac{1}{\omega(\lambda)}$ . We note that this function is defines everywhere except if the set $\Pi$ i.e., the domain of the definition of $\xi(\lambda)$ is $K - \Pi$ . Now we show that $\xi(\lambda)$ is subharmonic. By definition we have

$$\omega(\lambda) = \inf_{\|u\|_{H_1}=1} \sup_{\|v\|_{H_2}=1} |B(u,v)-D(\lambda,u,v)| \ .$$

Clearly

(10.2.27) $$\omega(\lambda) = \inf_{|u|_{H_1}=1} \|B(u)-D(\lambda)(u)\|_{H_2'} ,$$

where $B - D(\lambda)$ is understood as a mapping from $H_1$ into $H_2'$. Now

(10.2.28) $$\xi(\lambda) = \sup_{\|u\|_{H_1}=1} \frac{1}{\|(B-D(\lambda))u\|_{H_2'}} .$$

For $u = R(\lambda)v$, from (10.2.18), we have

(10.2.29)
$$\xi(\lambda) = \sup_{u\in H_1} \frac{\|u\|_{H_1}}{\|(B-D(\lambda))u\|_{H_2'}} = \sup_{v\in H_2'} \frac{\|R(\lambda)v\|_{H_1}}{\|v\|_{H_2'}} =$$

$$= \|R(\lambda)\|_{H_2'\to H_1} .$$

For every $v \in H_2'$ and $\lambda_0$ such that $\omega(\lambda_0) \neq 0$ we can write (see (10.2.18))

(10.2.30) $$R(\lambda)v = \sum_{n=0}^{\infty} (\lambda-\lambda_0)^n u_n ,$$

where $u_n \in H_1$ and the series is absolutely convergent. Now it follows immediately that $\|R(\lambda)v\|_{H_1}$ is a subharmonic function in the circle where the series (10.2.30) is convergent. Supremum of a set of bounded subharmonic functions is also subharmonic, which gives the desired result.

## 10.3. The linear eigenvalue problem.

In this section we shall study a special case of the eigenvalue problem. Namely we shall analyze in linear case

(10.3.1) $$D(\lambda,u,v) = \lambda G(u,v) \quad .$$

Thus we have

$$B(u,v) - \lambda G(u,v) \quad ,$$

and we note that $\lambda = 0$ cannot be an eigenvalue. The form $B$ under our hypothesis defines a mapping from $H_1$ into $H_2'$. Let $\Sigma$ denote a mapping from $H_1$ into $H_2$ such that

(10.3.2) $$B(u,v) = (\Sigma u,v)_{H_2} \quad .$$

We shall also write $(\Sigma u,v)_{H_2} = B(u) \in H_2'$. We can also write

(10.3.3) $$B(u,v) = (u,\Sigma^* v)_{H_1} \quad .$$

In theorem 5.2.1 it was shown that the mapping $\Sigma$ (and $\Sigma^*$) maps $H_1$ onto $H_2$ with continuous inverse. Further let

(10.3.4) $$G(u,v) = (\Gamma u,v)_{H_2} = (u,\Gamma^* v)_{H_2} \quad ,$$

and similarly let $G(u) \in H_2'$, i.e. $\Gamma$ is a mapping of $H_1$ into $H_2'$. Now we show that $\Gamma$ is a compact mapping. In fact $\Gamma$ may be considered as a mapping of $K_1$ into $H_2$. Since $H_1$ is compact in $K_1$, $\Gamma$ is compact from $H_1$ into

$H_2$. Similarly $\Gamma^*$ is a compact mapping of $H_2$ into $H_1$. We define

$$T = \Sigma^{-1}\Gamma \tag{10.3.5}$$

$$T^* = \Sigma^{*-1}\Gamma^* \quad . \tag{10.3.6}$$

Obviously $T$ and $T^*$ are compact mappings of $H_1$ and $H_2$ into themselves respectively. $T^*$ is the adjoint of $T$. Moreover we have

(10.3.7)

$$B(u,v) - \lambda G(u,v) = ((\Sigma-\lambda\Gamma)u,v)_{H_2} = B((I-\lambda T)u,v)$$

$$= B(u,(I-\bar{\lambda}T^*)v) \quad .$$

Now we use the well known theory of eigenvalues

$$I - \lambda T \quad \text{and} \quad I - \bar{\lambda}T^* \quad .$$

The set of eigenvalues is isolated and in the neighborhood of the eigenvalue $\lambda_0$ we can write

$$(\mu I-T)^{-1} = \sum_{j=-\ell}^{\infty} (\mu-\lambda_0^{-1})^j Q_j \quad , \tag{10.3.8}$$

where $Q_j$ are bounded operators which commute with $T$ and which satisfy

$$Q_{-1}^2 = Q_{-1} \quad , \tag{10.3.9}$$

$$(\lambda_0^{-1}-T)^{j+1} Q_j = I - Q_{-1} \quad , \tag{10.3.10}$$

$$(\lambda_0^{-1}-T)^j Q_{-1} = (-1)^j Q_{-j-1} \quad , \tag{10.3.11}$$

for $j \geq 0$ and $Q_j = 0$ for $j \geq \ell$ (see e.g. [11], p. 229). Throughout our discussion we shall assume that we have a simple eigenvalue problem, i.e. the algebraic and geometric multiplicity is the same. In this case $\ell = 1$. The general case will be considered in section 10.6. The expression (10.3.9) implies that $Q_{-1}$ is a projection and in our case ($\ell$=1) $Q_1$ project $H_1$ onto the space of eigenfunctions. For the sake of convenience we write $Q_{-1} = Q$. We defined the function $\omega(\lambda)$ (resp. $\omega^*(\lambda)$) as

$$\omega(\lambda) = \inf_{\substack{u \in H_1 \\ \|u\|_{H_1}=1}} \sup_{\substack{v \in H_2 \\ \|v\|_{H_2} \leq 1}} |B(u,v) - \lambda G(u,v)|$$

$$= \inf_{\substack{u \in H_1 \\ \|u\|_{H_1}=1}} \sup_{\substack{v \in H_2 \\ \|v\|_{H_2} \leq 1}} |B((I-\lambda T)u,v)| ,$$

and $\omega(\lambda_0) = 0$. Now we show that for $w = (I-Q)u$, $u \in H_1$ arbitrary we have

(10.3.12) $$\sup_{\|v\|_{H_2}=1} |B(w,v) - \lambda_0 G(w,v)| \geq C\|w\|_{H_1} .$$

In fact we have from (10.3.10).

$$(I-\lambda_0 T)Q_0 = \lambda_0 (I-Q) ,$$

and therefore using (10.3.9) we obtain

(10.3.13) $$\|w\|_{H_1} = \|(I-Q)u\|_{H_1} = \|(I-Q)^2u\|_{H_1} \leq$$

$$\leq \frac{1}{\lambda_0} \|Q_0(I-\lambda_0T)(I-Q)u\|_{H_1} \leq \frac{C}{\lambda_0} \|(I-\lambda_0T)(I-Q)u\|_{H_1} =$$

$$\leq \frac{C}{\lambda_0} \|\Sigma(I-\lambda_0T)w\|_{H_2} \leq \frac{C}{\lambda_0} \sup_{\|v\|_{H_2}=1} ((\Sigma-\lambda_0\Gamma)w,v)_{H_2} =$$

$$= \frac{C}{\lambda_0} \sup_{\|v\|_{H_2}=1} |B(w,v)-\lambda G(w,v)| \quad ,$$

and (10.3.12) is proved.

Now we turn our attention to the adjoint problem, i.e. to the problem $I - \overline{\lambda}T^*$ . As before we write

(10.3.14) $$(\overline{\mu}I-T^*)^{-1} = \sum_{j=-1}^{\infty} (\overline{\mu}-\overline{\lambda}_0{}^{-1})Q_{*j}$$

$Q_{*-1} = Q_*$ maps $H_2$ onto the eigenvalue space $N^*(\lambda_0)$ . Further it can be shown that

(10.3.15) $$B(Qu,v) = B(u,Q_*v) \quad .$$

Now we show that

(10.3.16) $$\inf_{\substack{\|u\|_{H_1}\leq 1 \\ u\in N(\lambda_0)}} \sup_{\substack{\|v\|_{H_2}=1 \\ v\in N^*(\lambda_0)}} |B(u,v)| > C > 0 \quad .$$

Since $N(\lambda_0)$ and $N^*(\lambda_0)$ are finite dimensional, if there is no $C > 0$ in (10.3.16) then there exists $u_0 \in N(\lambda_0)$ such that

$$B(u_0, v) = 0 \quad \text{for all} \quad v \in N^*(\lambda_0) \quad .$$

But $Q_*$ maps $H_2$ onto $N^*(\lambda_0)$, consequently

$$B(u_0, Q_* v) = B(Qu_0, v) = B(u_0, v) \quad \text{for all} \quad v \in H_2 \quad ,$$

which contradicts the assumption on the form $B$. Thus the inequality (10.3.16) must hold.

The expression (10.3.16) immediately yields

$$(10.3.17) \qquad \inf_{\substack{\|u\|_{H_1} \leq 1 \\ u \in N(\lambda_0)}} \sup_{\substack{\|v\|_{H_2} = 1 \\ v \in N^*(\lambda_0)}} |G(u,v)| > C > 0 \quad .$$

In fact for any $u \in N(\lambda_0)$, $v \in N^*(\lambda_0)$, we have

$$B(u,v) = \lambda_0 G(u,v) \quad ,$$

and since $\lambda_0 \neq 0$ (10.3.17) follows immediately.

## 10.4. Associated Eigenvalue Problems.

Now we return to the general eigenvalue problem discussed in section 10.2. For the sake of simplicity we shall study only the case when in the neighborhood of the eigenvalue $\lambda_0$ we have

$$(10.4.1) \qquad \omega(\lambda) \geq C|\lambda - \lambda_0| \quad .$$

By theorem 10.2.6 we have

(10.4.2)
$$R(\lambda) = \sum_{n=-1}^{\infty} (\lambda-\lambda_0)^n R_n \quad ,$$

where $R_n$ is a bounded mapping from $H_2'$ into $H_1$. We can write

(10.4.3)
$$B(u,v) - D(\lambda,u,v) = B(u,v) - D(\lambda_0,u,v) - (\lambda-\lambda_0)D_1(\lambda_0,u,v)$$
$$- (\lambda-\lambda_0)^2 D_2(\lambda,\lambda_0,u,v) \quad ,$$

where

$$|D_2(\lambda,\lambda_0,u,v)| \leq K\|u\|_{K_1} \|v\|_{K_2} \quad ,$$

and $K$ is independent of $\lambda$ for $|\lambda-\lambda_0| < \rho$.

We define a new bilinear form

(10.4.4)
$$B_1(\lambda,u,v) = B(u,v) - D(\lambda_0,u,v) - (\lambda-\lambda_0)D_1(\lambda_0,u,v) \quad .$$

Now we show that there exists $r_1 > r_2 > 0$ such that for and $r_0 < |\lambda-\lambda_0| < r_1$, the form $B_1(\lambda,u,v)$ satisfies all conditions of theorem 5.2.1 with constants independent of $\lambda$. It is obvious that the condition (5.2.1) is satisfied. Now we show that the conditions (5.2.2) and (5.2.3) are satisfied. To this event we note that the forms $B(u,v)$ and $D(\lambda,u,v)$ may be identified with mappings of $H_1$ into $H_2'$.

Let

(10.4.5)
$$S(\lambda) = B - D(\lambda) \quad .$$

Obviously $S(\lambda)$ is a mapping of $H_1$ into $H_2'$ and $R(\lambda)$

introduced in section 10.2 is the inverse of $S$ i.e. $R = R = S^{-1}$. Considering the form $B_1(\lambda,u,v)$ as a mapping of $H_1$ into $H_2'$, we may write

(10.4.6) $$B_1(\lambda) = S(\lambda) + (\lambda-\lambda_0)^2 D_2(\lambda,\lambda_0) \quad ,$$

and therefore

(10.4.7) $$B_1(\lambda) = \left[I + (\lambda-\lambda_0)^2 D_2(\lambda,\lambda_0)R(\lambda)\right] S(\lambda) \quad ,$$

where

(10.4.8) $$I + (\lambda-\lambda_0)^2 D_2(\lambda,\lambda_0)R(\lambda) \quad ,$$

is a mapping of $H_2'$ into itself. We shall estimate the norm of the operator $\Lambda = (\lambda-\lambda_0)^2 D(\lambda,\lambda_0)R(\lambda)$. From (10.4.2) we have

(10.4.9)
$$\| |\lambda-\lambda_0|^2 D_2(\lambda,\lambda_0)R(\lambda)\|_{H_2'\to H_2'} \leq |\lambda-\lambda_0| \sum_{n=-1}^{\infty} |\lambda-\lambda_0|^{n+1}\|R_n\|$$

and therefore $\rho > 0$ exists such that for all $\lambda$ with $0 < \frac{\rho}{2} < \lambda < \rho$, we have

(10.4.10) $$\|| (\lambda-\lambda_0)|^2 D_2(\lambda-\lambda_0)R(\lambda)\| \leq \frac{1}{2} \quad .$$

Therefore there exists the mapping

(10.4.11)
$$(B_1(\lambda))^{-1} = R(\lambda)\left[I + (\lambda-\lambda_0)D_2(\lambda)R(\lambda)\right]^{-1} \quad .$$

and hence $(B_1(\lambda))^{-1}$ is uniformly bounded in norm for all $\frac{\rho}{4} < \lambda < \frac{\rho}{2}$. Now it follows immediately that $B_1$ satisfies inequalities (5.2.2) and (5.2.3).

Now we select $\lambda_1$ so that $|\lambda-\lambda_1| = \frac{3}{8}\rho$ and with $\lambda = \lambda_1 - \eta$ . Then the new eigenvalue problem may be written as

(10.4.12) $$B(u,v) - D(\lambda_1 - \eta, u, v)$$

which is identical to the original problem. The value $\eta_0 = \lambda_1 - \lambda_0$ is the eigenvalue. Now we formulate a new eigenvalue problem which we shall call the associated linear eigenvalue problem

(10.4.13) $$B_1(\lambda_1, u, v) + \eta D_1(\lambda_0, u, v) =$$

$$= B(u,v) - D(\lambda_0, u, v) - \eta_0 D_1(\lambda_0, u, v) + \eta D_1(\lambda_0, u, v)$$

Operators $B_1(\lambda_1, u, v)$ and $D_1(\lambda_0, u, v)$ satisfy all our assumptions as proved above. Further $\eta = \eta_0$ is the eigenvalue and we have a linear eigenvalue problem. The eigenfunction spaces $N(\lambda_0)$ and $N^*(\lambda_0)$ are the same for both the original and the associated linear problem.

Applying the results of the previous section we obtain e.g.

(10.4.14) $$\inf_{\substack{u \in N(\lambda_0) \\ \|u\|_{H_1} = 1}} \sup_{\substack{v \in N^*(\lambda_0) \\ \|v\|_{H_2} = 1}} D_1(\lambda_0, u, v) \geq C > 0 .$$

10.5. <u>The approximation of the eigenvalue problem.</u>

We consider the eigenvalue problem

(10.5.1) $$B(u,v) - D(\lambda, u, v) \quad ,$$

and let $\lambda_0$ be its eigenvalue, i.e. $\omega(\lambda_0) = 0$ . We assume as before that $\lambda_0$ is an isolated zero of $\omega$ .

Moreover let a one parameter family of closed subspaces $M_1(h) \subset H_1$ and $M_2(h) \subset H_2$, $0 < h < 1$ be given. We suppose that (6.2.1) and (6.2.2) hold for the form $B(u,v)$ and $d_2(M_1(h),M_2(h)) \geq C > 0$ with $C$ independent of $h$. We further assume that there exists an $\alpha > 0$ such that for any $z \in H_1$ satisfying

(10.5.2) $$B(z,v) = D(\lambda,u,v) \ , \forall v \in H_2 \ ,$$

there exists $\omega_1 \in M_1(h)$ such that

(10.5.3) $$\|z-\omega_1\|_{H_1} \leq Ch^{\alpha}\|u\|_{H_1} \ ,$$

where $C$ depends in general on $\lambda$ (it is bounded uniformly on every bounded set of $\lambda$) but it does not depend on $u$. We also assume that for every $w \in H_2$ such that

(10.5.4) $$B(u,w) = D(\lambda,u,v), \ u \quad H_1$$

there exists $\omega_2 \in M_2(h)$ such that

(10.5.5) $$\|w-\omega_2\|_{H_2} \leq C(\lambda)h^{\alpha}\|v\|_{H_2} \ .$$

with $C(\lambda)$ bounded for $\lambda$ in every compact set.

Now we introduce the functions $\omega_h(\lambda)$ and $\omega_h^*(\lambda)$ in the same way as the function $\omega(\lambda)$ and $\omega^*(\lambda)$ were introduced earlier

(10.5.6) $$\omega_h(\lambda) = \inf_{\substack{\|u\|_{H_1}=1 \\ u\in M_1(h)}} \ \sup_{\substack{\|v\|_{H_2}=1 \\ v\in M_2(h)}} |B(u,v)-D(\lambda,u,v)| \ ,$$

and

$$(10.5.7)\quad \omega_h^*(\lambda) = \inf_{\substack{\|v\|_{H_2}=1 \\ v\in M_2(h)}} \ \sup_{\substack{\|u\|_{H_1}=1 \\ u\in M_1(h)}} |B(u,v)-D(\lambda,u,v)| \ .$$

From lemma 10.2.2 it follows that $\omega_h(\lambda)$ and $\omega_h^*(\lambda)$ are continuous.

If we denote by $\Pi_h$ and $\Pi_h^*$ the set of zeros of the functions $\omega_h(\lambda)$ and $\omega_h^*(\lambda)$ respectively, then from theorem 10.2.2 we have $\Pi_h^* = \Pi_h$ . Similarly, we can apply theorem 10.2.1 and construct the eigenfunction space $N_h(\lambda_0^h)$ and $N_h^*(\lambda_0^h)$ , where $\lambda_0^h$ denotes the zero of the function $\omega_h(\lambda)$ . Furthermore on $K - \Pi_h$ the function $(\omega_h(\lambda))^{-1}$ is subharmonic.

Theorem 10.5.1. If

I) $\lambda_0$ is the eigenvalue of the problem $B(u,v) - D(\lambda,u,v) = 0$ , i.e. $\omega(\lambda_0) = 0$ .

II) $\lambda_0$ is the only zero of $\omega(\lambda)$ in the circle $|\lambda-\lambda_0| < \rho$ , $\rho > 0$ , then

III) for every $\varepsilon > 0$ , $\varepsilon < \rho$ there exists a $h_0(\varepsilon) > 0$ such that

i) For $|\lambda-\lambda_0| = \varepsilon$ and any $0 < h < h_0(\varepsilon)$ , we have

$$(10.5.8)\qquad \omega_h(\lambda) > C > 0 \ , \quad C \text{ independent on } h$$

ii) for every $0 < h < h_0(\varepsilon)$ there exists a $\lambda_0^h$ , $|\lambda_0-\lambda_0^h| < \varepsilon$ such that $\omega_h(\lambda_0^h) = 0$ .

Proof. 1. By hypothesis there is no zero of $\omega(\lambda)$ on the circumference $|\lambda-\lambda_0| = \varepsilon$ . Therefore

$$(10.5.9)\qquad \omega(\lambda) > C > 0 \ .$$

Let $u \in M_1(h)$ . Denote $z_h \in M_1(h)$ and $z \in H_1$ such that

$$B(z_h,v) = D(\lambda,u,v) , \quad \forall v \in M_2 \tag{10.5.10}$$

$$B(z,v) = D(\lambda,u,v) , \quad \forall v \in H_2 \tag{10.5.11}$$

We have

$$B(u,v) - D(\lambda,u,v) = B(u-z,v);$$

for all $\lambda$ with $|\lambda-\lambda_0| = \varepsilon$ we obtain

$$\sup_{\|v\|_{H_2}=1} |B(u-z,v)| \geq C\|u\|_{H_1} \tag{10.5.12}$$

Applying (5.2.1) to B we get from (10.5.12)

$$\|u-z\|_{H_1} \geq C\|u\|_{H_1} \tag{10.5.13}$$

From theorem 6.2.1 and (10.5.3) it follows that

$$\|z-z_h\| \leq Ch^{\alpha}\|u\|_{H_1} \tag{10.5.14}$$

Now we have by using (6.2.2)

(10.5.15)
$$\sup_{\substack{v\in M_2\\ \|v\|_{H_2}=1}} |B(u-z_h,v)| \geq C\|u-z_h\|_{H_1} \geq C\left[\|u-z\|_{H_1} - \|z_h-z\|_{H_1}\right] \geq$$

$$\geq C\left[\|u\|_{H_1} - Ch^{\alpha}\|u\|_{H_1}\right] \geq \frac{1}{2}C$$

and the first part of the theorem is proved.

2. Assume that our assection is false, then for all $h$ there is no zero of $\omega_h(\lambda)$ in the circle $|\lambda-\lambda_0| < \varepsilon$ . We showed that for all $h < h_0$ , $\omega_h(\lambda) > C > 0$ on

the circumference $|\lambda-\lambda_0| = \varepsilon$ . Since by assumption there is no zero of $\omega_h(\lambda)$ in the circle, $(\omega_h(\lambda))^{-1}$ is subharmonic and attains its maximum on the circumference $|\lambda-\lambda_0| = \varepsilon$ . Hence $\left[\omega_h(\lambda)\right]^{-1} \leq \frac{1}{C} < \infty$ everywhere in the circle $|\lambda-\lambda_0| < \varepsilon$ for all $0 < h < h_0$ .

Now let $u_0 \in N(\lambda_0)$ . Then

(10.5.16) $$B(u_0,v) = D(\lambda_0,u_0,v) \ , \forall v \in H_2 \ ,$$

therefore there exists $\omega_0 \in \overline{M_1(h)}$ such that

$$\|\omega_0-u_0\|_{H_1} \leq Ch^{\alpha}\|u_0\|_{H_1} \ .$$

On the other hand by hypotheses

(10.5.17)

$$|B(\omega_0,v)-D(\lambda_0,\omega_0,v)| \geq C\|\omega_0\|_{H_1}\|v\|_{H_2} \ , \forall v \in M(h_2) \ .$$

Therefore

(10.5.18)

$$|B(u_0,v)-D(\lambda_0,u_0,v)| = |B(\omega_0,v)-D(\lambda_0,\omega_0,v)+B(u_0-\omega_0,v) -$$

$$-D(\lambda_0,u_0-\omega_0,v)| \geq C\|v\|_{H_2}\|\omega_0\|_{H_1}\left[C-h^{\alpha}\right] \ .$$

which contradicts 10.5.16. Thus the proof of our theorem is complete.

In a similar way it is easy to prove the next theorem.

<u>Theorem 10.5.2</u>. If $\lambda_0^h \to \lambda_0$ as $h \to 0$ , then $\lambda_0$ is an eigenvalue i.e. $\omega(\lambda_0) = 0$ .

In the same vein we have the following theorem.

Theorem 10.5.3. If $\omega_h(\lambda_0^{h_i}) \to \lambda_0$ as $h_i \to 0$ and $u_0^{h_i} \in N_h(\lambda_0^{h_i})$, $\|u_0^{h_i}\|_{H_1} = 1$. Then there exists a subsequence (denoted with the same index) $u_0^{h_i}$ such that $u_0^{h_i} \to u_0$ in $H_1$ and $u_0 \neq 0$, $u_0 \in N(\lambda_0)$.

Proof. Since $\|u_0^{h_i}\| = 1$ there exists a subsequence denoted again by $u_0^{h_i}$ which converges in $K_1$ and weakly in $H_1$. Let $u_0$ be this limit. On the other hand we know that

(10.5.19) $$B(u_0^{h_i}, v) = D({}_0^{h_i}, u_0^{h_i}, v) \;, \forall v \in M_2(h) \;.$$

Let $z_i \in H_1$ be such that

(10.5.20) $$B(z_i, v) = D(\lambda_0^{h_i}, u_0^{h_i}, v) \;, \forall v \in M_2(h) \;.$$

By our assumptions on $M_1(h)$ and $M_2(h)$ we have

$$\|u_0^{h_i} - z_i\|_{H_1} \leq Ch_i^{\alpha} \;, \; \|u_0^{h_i}\|_{H_1}$$

Furthermore by the use of theorem 5.2.1 we see that $z_i \to z_0$ in $H_1$, where

(10.5.21) $$B(z_0, v) = D(\lambda_0, u_0, v) \;.$$

therefore $u_0^{h_i} \to z_0$ in $H_1$. Since $u_0^{h_i}$ converges weakly to $u_0$ in $H_1$ we have $z_0 = u_0$. This completes the proof of our theorem.

It is also easily seen that if $u_0^{h_i} \to u_0$ in $H_1$ and $\lambda_0^{h_i} \to \lambda_0$ then $u_0 \in N(\lambda_0)$.

We proved the convergence in theorems 10.5.1, 10.5.2 and 10.5.3 under the sole assumption that $\lambda_0$ was an isolated zero of $\omega(\lambda)$.

Now we assume that in the neighborhood of $\lambda_0$ we have

$$\omega(\lambda) \geq C|\lambda-\lambda_0| .$$

We let

$$\sup_{\substack{\|u\|_{H_1} \leq 1 \\ u \in N(\lambda_0)}} \inf_{\omega \in M_1(h)} \|u-\omega\|_{H_1} = \varepsilon(h) \tag{10.5.22}$$

$$\sup_{\substack{\|v\|_{H_2} \leq 1 \\ v \in N^*(\lambda_0)}} \inf_{\omega^* \in M_2(h)} \|v-\omega^*\|_{H_2} = \varepsilon^*(h) \tag{10.5.23}$$

We first observe that for every $v \in H_2$

(10.5.24)

$$B(u_0^h,v) - D(\lambda_0,u_0^h,v) = B(u_0^h,v) - D(\lambda_0^h,u_0^h,v) -$$

$$- (\lambda_0-\lambda_0^h)D_1(\lambda_0,u_0^h,v) - (\lambda_0-\lambda_0^h)^2 D_2(\lambda_0,\lambda_0^h,u_0^h,v)$$

with $D_1$ and $D_2$ as given in (10.4.3). Let $v_0$ be any element of $N^*(\lambda_0)$, then left hand side of (10.5.24) vanishes and hence

(10.5.25)

$$(\lambda_0-\lambda_0^h)\left[D_1(\lambda_0,u_0^h,v_0) + (\lambda_0-\lambda_0^h)D_2(\lambda_0,\lambda_0^h,u_0^h,v_0)\right] =$$

$$= B(u_0^h,v_0) - D(\lambda_0^h,u_0^h,v_0) \quad .$$

New we shall show that for all sufficiently small $h$ we have

(10.5.26)

$$\frac{1}{|\lambda_0^h-\lambda_0|} \inf_{\substack{\|u_0^h\|_{H_1}=1 \\ u_0^h\in N(\lambda_0^h)}} \sup_{\substack{v_0\in N^*(\lambda_0) \\ \|v_0\|_{H_2}=1}} |B(u_0^h,v_0)-D(\lambda_0^h,u_0^h,v_0)| \geq C > 0 \, .$$

Assume that there exists a sequence $h_i \to 0$ such that the left hand side of (10.5.20) converges to zero. We have shown that we may select a subsequence $h_i$ such that $u_0^{h_i} \to u_0 \in N(\lambda_0)$ . Now by using the basic properties of $D_2$ and (10.4.14) and (10.5.25) we obtain the contradiction. Therefore (10.5.26) holds.

The inequality (10.5.26) gives the first estimate. In fact we know that for every $\omega^* \in M_2(h)$ we have

$$B(u_0^h,\omega^*) - D(\lambda_0^h,u_0^h,\omega^*) = 0 \quad .$$

Now from (10.5.26) we obtain

(10.5.27) $$|\lambda_0^h-\lambda_0| \leq C\varepsilon^*(h) \quad .$$

We can improve this estimate. First we make use of this estimate for the error estimate in the eigenfunctions.

In section (10.4) we introduced the associated linear eigenvalue problem

$$B_1(\lambda_1,u,v) + \eta D_1(\lambda_0,u,v) = 0$$

Using the results of section 10.3, we can construct a projection operator $Q$ of the space $H_1$ onto $N(\lambda_0)$ such that (see (10.3.12))

(10.5.28)

$$\sup_{\substack{v\in H_2 \\ \|v\|_{H_2}\le 1}} |B_1(\lambda_1,(I-Q)u,v)+\eta_0 D_1(\lambda_0,(I-Q)u,v)| \ge C\|(I-Q)u\|_{H_2} .$$

Using (10.4.13) we also obtain

(10.5.29)

$$\sup_{\|v\|_{H_2}\le 1} |B((I-Q)u,v) - D(\lambda_0,(I-Q)u,v)| \ge C\|(I-Q)u\|_{H_1} .$$

Since $Qu \in N(\lambda_0)$ for every $u$, therefore

(10.5.30)

$$B(Qu,v) - D(\lambda_0,Qu,v) = 0 , \forall u \in H_1 , \forall v \in H_2 .$$

Hence (10.5.24) may be written in the form

(10.5.31)

$$|B((I-Q)u_0^h,v)-D(\lambda_0,(I-Q)u_0^h,v)| = |B(u_0^h,v)-D(\lambda_0^h,u_0^h,v) -$$

$$- (\lambda_0-\lambda_0^h)D_1(\lambda_0,u_0^h,v) - (\lambda_0-\lambda_0^h)^2 D_2(\lambda_0,\lambda_0^h,u_0^h,v)| .$$

From (10.5.29) and (10.5.31) we obtain

(10.5.32)

$$\|(I-Q)u_0^h\|_{H_1} \le C\Big[|\lambda_0-\lambda_0^h| + \sup_{\|v\|_{H_2}\le 1} |B(u_0^h,v)-D(\lambda_0^h,u_0^h,v)|\Big] .$$

Now we study the term $B(u_0^h,v) - D(\lambda_0^h,u_0^h,v)$ . Let

$$u_0^h = p + q \quad ,$$

where

$$p = Qu_0^h \ , \ q = (I-Q)u_0^h \quad .$$

Consider $z_1$ and $z_2$ such that for every $v \in H_2$

$$B(z_1,v) = D(\lambda_0^h,p,v) \quad ,$$

and

$$B(z_2,v) = D(\lambda_0^h,q,v)$$

Since

$$B(u_0^h,v) - D(\lambda_0^h,u_0^h,v) = 0 \ , \ \forall v \in M_2(h)$$

from theorem 6.2.1 we obtain

$$\|u_0^h-(z_1+z_2)\|_{H_1} \leq C \inf_{\omega\in M_1} \left\| \left[ z_1+z_2-\omega \right] \right\|_{H_1} \leq \inf_{\omega\in M_1}\|z-\omega\| + \inf_{\omega\in M_2}\|z_2-\omega\| \quad .$$

By hypotheses we have

$$\inf_{\omega\in M_2} \|z_2-\omega\| \leq Ch^{\alpha}\|q\|_{H_1} \quad .$$

On the other hand if we write

$$z_1 = z_1^{(1)} + z_1^{(2)} + z_1^{(3)} \quad ,$$

where $z_1^{(i)}$, $i = 1,2,3$ satisfy the relations

$$B(z_1^{(1)},v) = D(\lambda_0,p,v) \quad ,$$

$$B(z_1^{(2)},v) = (\lambda_0^h-\lambda_0)D_1(\lambda_0,p,v) \quad ,$$

$$B(z_1^{(3)},v) = (\lambda_0^h-\lambda_0)^2 D_2(\lambda_0,\lambda_0^h,p,v) \quad ,$$

then since $p \in N(\lambda_0)$ , we have $z_1^1 = p$ . Therefore

$$\inf_{\omega \in M_2} \|z_1 - \omega\| \leq C\left[\varepsilon + |\lambda_0^h - \lambda_o| h^\alpha + |\lambda_0^h - \lambda_0|^2 h^\alpha\right] .$$

Thus for all $h$ such that $0 < h < h0$ and all $v$ with $\|v\|_{H_2} = 1$ , we have

$$|B(u_0^h, v) - D(\lambda_0^h, u_0^h, v)| \leq C\left[\varepsilon + |\lambda_0^h - \lambda_0| h^\alpha + |q| h^\alpha\right] .$$

From (10.5.26) we obtain

(10.5.33) $$|\lambda_0^h, \lambda_0| \leq C \sup_{\|v\|_{H_2} = 1} |B(u_0^h, v) - D(\lambda_0^h, u_0^h, v)|$$

Therefore

(10.5.34)

$$\sup_{\|v\|_{H_2} \leq 1} |B(u_0^h, v) - D(\lambda_0^h, v_0^h, v)| \leq C\left[\varepsilon + |q| h^\alpha\right] ,$$

Thus from (10.5.32), (10.5.33) and (10.5.34) we get

$$\|q\|_{H_1} \leq C \; \varepsilon + \|q\| h^\alpha .$$

Hence for $h \leq h_0$

(10.5.35) $$\|q\|_{H_1} \leq C\varepsilon .$$

Now from (10.5.33) and (10.5.34) it follows that

(10.5.36)

$$|B(u_0^h, v) - D(\lambda_0^h, u_0^h, v)| \leq C\varepsilon \|v\|_{H_2} .$$

Now we may use (10.5.26) and the fact

(10.5.37)
$$B(u_0^h,\omega) - D(\lambda_0^h,u_0^h,\omega) = 0 \ , \ \forall\omega \in M_2(h)$$

to obtain

(10.5.38)
$$|\lambda_0^h-\lambda_0| \leq C\varepsilon \inf_{\omega\in M_2(h)} \|v_0-\omega\|_{H_2} \leq C\varepsilon(h)\varepsilon^*(h) \quad .$$

Now we summarize our result.

Theorem 10.5.4. Let $\lambda_0$ be the eigenvalue of the problem $B(u,v) = D(\lambda,u,v)$ and let $\omega(\lambda) \geq C|\lambda-\lambda_0|$ in the neighborhood of $\lambda_0$ , then for $h < h_0$ we have

I

(10.5.39)
$$|\lambda_0-\lambda_0^h| \leq C\varepsilon(h)\varepsilon^*(h) \quad ,$$

where $\varepsilon(h)$ and $\varepsilon^*(h)$ are given respectively by (10.5.22) and (10.5.23).

II If $u_0^h \in N_h(\lambda_0^h)$ , $\|u_0^h\|_{H_1} = 1$ , then there exists $u_0 \in N(\lambda_0)$ such that

(10.5.40)
$$|u_0^h-u_0| \leq C\varepsilon(h) \quad .$$

III If $v_0^h \in N_h^*(\lambda_0^h)$ , $\|u_0^h\|_{H_1} = 1$ , then there exists $v_0 \in N(\lambda_0)$ such that

(10.5.41)
$$\|v_0^h-v_0\| \leq C\varepsilon^*(h) \quad .$$

## 10.6. The approximation of the eigenvalue problem (continuation).

In section 10.5 we proved the convergence theorem for the general case and obtained error estimate for the so called simple eigenvalue problem i.e. when $\omega(\lambda) \geq C|\lambda-\lambda_0|$ .

Now we shall discuss briefly the general linear case. To this end we shall assume that $\ell > 1$ in (10.3.8), i.e., the algebraic and geometric multiplicities are different.

The first main difference is that we need to think of the generalized eigenfunction instead of the eigenfunctions in this case. With this interpretation the inequalities (10.5.39) and (10.5.40) remain valid. The proofs are very similar to those described in §10.5. It can be shown that if we restrict ourselves to eigenfunctions (as in the previous section) only, then (10.5.39) and (10.5.40) no longer hold for the general case (see [7], [9]). In this case the term $\varepsilon(h) + (\varepsilon(h)\varepsilon^*(h))^{1/\ell}$ appear in the right-hand side of (10.5.39). The inequality (10.5.39) does not hold either in this case. Instead of $\varepsilon(h)\varepsilon^*(h)$ on the right-hand side, we will have the term $[\varepsilon(h)\varepsilon^*(h)]^{1/\ell}$.

We may save the inequality (10.5.39) if, instead of $\lambda_0^h$, we look for the value $\bar{\lambda}_0^h$ which is the harmonic average value of all the eigenvalues which are converging to $\lambda_0$, i.e., we have

(10.6.1) $$\bar{\lambda}_0^h = \left(\frac{1}{n}\sum (\lambda_0^h)^{-1}\right)^{-1},$$

where we have taken $\lambda_0^h$ as many times as the multiplicity of the single eigenvalue $\lambda_0^h$. For more details in connection with this, we refer the reader to [7] and [9].

## 10.7. Examples.

In the previous sections we discussed the eigenvalue problem in a general framework based on the basic approach used in the previous chapters, in particular, in the chapters 6 and 7. This clearly indicates how to use our framework in relation to all variational principles discussed. Nonetheless, it appears to us to be useful if some typical examples which satisfy our assumptions be presented.

Let $\Omega$ be a smooth domain and consider the eigenvalue problem

$$(10.7.1) \qquad -\Delta u + e^{\lambda}\frac{\partial u}{\partial x} + \sin \lambda\cdot u = 0$$

$$(10.7.2) \qquad \frac{\partial u}{\partial n} + \cos \lambda\cdot u = 0 \ .$$

Our task is to find the value $\lambda_0$ such that for $\lambda = \lambda_0$, there exists a $u \neq 0$ satisfying equations (10.7.1) and (10.7.2). To this end we choose the spaces $H_1 = H_2 = H^1(\Omega)$. Based on the results of chapter 5 we define the bilinear form

$$(10.7.3) \qquad B(u,v) = \int_\Omega \left(\frac{\partial u}{\partial x_1}\frac{\partial v}{\partial x_1} + \frac{\partial u}{\partial x_2}\frac{\partial v}{\partial x_2} + uv\right)dx$$

$$(10.7.4) \qquad D(\lambda,u,v) = e^{\lambda}\int_\Omega \frac{\partial u}{\partial x_1} v dx + (\sin \lambda - 1)\int_\Omega uvdx$$

$$+ \cos \lambda \oint_{\dot{\Omega}} uvds \ .$$

Obviously (10.7.3) satisfies all the assumptions of theorem 5.2.1. Now we show that the bilinear form $D(\lambda,u,v)$ may be continued from $H^1(\Omega) \times H^1(\Omega)$ to $H^{1-\theta}(\Omega) \times H^{1-\theta}(\Omega)$. This continuation is obvious for the integral $\int_\Omega uvdx$. Here $\theta$ may be taken equal to $1$. The integral $\int_{\dot{\Omega}} uvds$ may be continued according to theorem 2.4.1 for any $\theta < \frac{1}{2}$.

Now we investigate the continuation of $\int \frac{\partial u}{\partial x_1} v\,dx$. The integral may be continued also on $H^{1-\theta}(\Omega) \times H^{1-\theta}(\Omega)$ for any $\theta$ with $0 < \theta < \frac{1}{2}$. However, we shall prove a weaker result which suits our purpose, i.e., we need not find a maximal $\theta$; any $\theta > 0$ will serve our purpose. Obviously the integral $\int \frac{\partial u}{\partial x_1} v\,dx$ defines a mapping $A$ of $H^1(\Omega)$ into $(H^0(\Omega))'$. From the basic interpolation theorem (see chapter 2) it follows that $A$ maps $H^{3/4+\varepsilon/2}(\Omega)$ into $(H^{\frac{1}{2}}(\Omega))'$. Therefore, the bilinear form $\int \frac{\partial u}{\partial x_1} v\,dx$ may be continued, say, to $H^{7/8}(\Omega) \times H^{7/8}(\Omega)$. The compact identity imbedding of $H^1(\Omega)$ in $H^{1-\theta}(\Omega)$ yields our desired result.

The analyticity conditions are obviously satisfied since $e^\lambda$, $\sin \lambda$ and $\cos \lambda$ are entire functions. Thus all the necessary requirements are fulfilled provided only $\omega$ has isolated zeros.

It is clear how to use the techniques developed in chapters 5 and 6 as well as the ideas of one parameter family of variational principles discussed in chapter 7.

## 10.8. Additional comments.

For a smooth domain $\Omega$, under the assumptions formulated in chapter 3, it is obvious that all the eigenfunctions and the generalized eigenfunctions are smooth. The use of the spaces $S_h^{t,k}(\Omega)$, which allows the use of the variational principle, leads to $\varepsilon(h) = O(h^{t-1})$, where $s$ is determined by the variational principle, i.e., in the example considered in §10.7 we need to have $k \geq 1$ and $s = 1$ .

In chapter 8 the problems connected with non-smooth domains and interfaces were discussed. In these cases the eigenfunctions are non-smooth also, but we may deal with this problem by using the approaches such as mesh refinement or the special element approach discussed in chapter 8. We should mention here that from what we said in the previous section it is necessary to enrich the space of trial functions by adding these special elements which lead to a good approximation of the eigenfunctions, but it is not sufficient. We must also enrich the space of test functions by the "elements" which lead to a good approximation of the eigenfunction of the adjoint problem, otherwise we may lose the accuracy.

If the problem is not self-adjoint, these "elements" are different. Therefore there is the possibility to add to both spaces, i.e., the trial functions and the test functions, the "elements" suitable to the original and the adjoint problems. We may save a few unknowns by adding suitably chosen different "elements" to the trial and test functions. We note that for a nonself-adjoint problem with non-smooth coefficients, the smoothness of the eigenfunctions may be different. We would like to emphasize that the error estimates obtained in §10.5 are sharp.

In chapter 9 we discussed the influence of the perturbations in the quadrature formulae etc. on the accuracy of the solution. This analysis can be carried over to the case of the eigenvalue problems as well. (see [12])

REFERENCES.

[1]. G. M. Vajnikko, "Asymptotic evaluation of the error of projection methods for the eigenvalue problem," Vyčisl. Mat. i Mat. Fiz., 4(1969), pp. 405-425. Engl. Transl., U.S.S.R. Comp. Math. and Math. Phys. 4 (1964), pp. 9-36.

[2]. G. M. Vajnikko, "On the rate of convergence of certain approximation methods of Galerkin type in an eigenvalue problem," Izv. Vyss. Ucebn. Zavel. Matematika 2 (1966), pp. 37-45, English Transl. Amer. Math. Soc. Transl. 2 (1970), pp. 249-259.

[3]. G. M. Vajnikko, "On the speed of convergence of approximate methods in eigenvalue problem". Ž Vycisl. Mat. i, Mat. Fiz., 7 (1967), pp. 977-987, Engl. Transl. U.S.S.R. Comp. Math. and Math. Phys. 7 (1967), pp. 18-32.

[4]. M. A. Krasnosel'skii, Approximation solution of operator equations (in Russian), Nauka, Moscow (1969).

[5]. G. Birkhoff, C. de Boor, B. Schwarz, B. Wendroff, "Rayleigh-Ritz approximation by piecewise cubic polynomials," SIAM J. Numer. Anal. 3 (1966), pp. 188-203.

[6]. G. Strang and G. Fix, An analysis of the finite Element Method, Prentice Hall, New York (to appear 1973).

[7]. J. H. Bramble, J. E. Osborn, "Rate of convergence estimates for non-selfadjoint eigenvalue approximations," MRC Tech. Sum. Rep. #1232, June 1972.

[8]. J. H. Bramble, J. E. Osborn, Approximation of Steklov Eigenvalues of non-selfadjoint Second order Elliptic Operators, this Proceedings.

[9]. G. Fix, "On the approximation of eigenvalues arising from non-selfadjoint problems," I.F.D.A.M. report to appear.

[10]. S. Agmon, L. Nirenberg, "Properties of solutions of ordinary differential equations in Banach spaces," Comm. Pure Appl. Math. 16 (1963), pp. 121-239.

[11]. K. Yosida, Functional analysis, Springer, Berlin 1965.

[12]. G. Fix, Effects of Quadrature Errors in the Finite Element Approximation of Steady State Eigenvalue and Parabolic Problems, thie Proceedings.

# CHAPTER 11.[1] THE FINITE ELEMENT METHOD FOR TIME DEPENDENT PROBLEMS

## 11.1. Introduction.

Previous chapters have dealt with elliptic differential equations such as

$$(11.1.1) \qquad -\Delta u + u = f \quad \text{in} \quad \Omega$$

$$(11.1.2) \qquad \frac{\partial u}{\partial n} = 0 \quad \text{on} \quad \dot{\Omega}$$

which are in fact Euler equations for appropriate variational principles (Theorem 5.2.1). The time dependent problem

$$(11.1.3) \qquad \Delta u - u = \frac{\partial u}{\partial t} \quad \text{in} \quad \Omega \quad , \quad t > 0 \quad ,$$

$$(11.1.4) \qquad \frac{\partial u}{\partial n} = 0 \quad \text{on} \quad \dot{\Omega} \quad , \quad t > 0 \quad ,$$

$$(11.1.5) \qquad u(0) = g \quad \text{in} \quad \Omega \quad ,$$

on the other hand, cannot be characterized in this way -- it is not the Euler equation for any variational principle (see [1])[2] -- and thus we shall approach (11.1.3)-(11.1.4) somewhat differently than the steady state problem (11.1.1)-

---

1) This chapter was written jointly with George Fix.

2) Let us mention that some other generalized principles for this problem have been proposed. See, e.g., [2].

(11.1.2). Many ideas have been suggested, but the one that appears most fruitful for finite element practice is the so-called semidiscrete method, often called the method of lines or the method of weighted residuals.

To describe this method we first recall that the solution $u \in H^1(\Omega)$ of (11.1.1) and (11.1.2) satisfies

$$(11.1.6) \qquad B(u,v) = b(f,v) \qquad \forall\ v \in H^1(\Omega)$$

where

$$(11.1.7) \qquad B(u,v) = \sum_{i=1}^{2} \int_\Omega \left[\frac{\partial u}{\partial x_i}\frac{\partial v}{\partial x_i} + uv\right] dx$$

$$(11.1.8) \qquad b(u,v) = \int_\Omega uv\,dx$$

Thus formally replacing $f$ with $-\frac{\partial u}{\partial t}$ we obtain

$$(11.1.9) \qquad B(u,v) + b\left(\frac{\partial u}{\partial t}, v\right) = 0 \quad \forall \quad v \in H^1(\Omega)$$

subject, of course, to the initial condition (11.1.5). As usual, the replacement of $H^1(\Omega)$ with finite dimensional $S_h^{t,k}(\Omega)$ defines an approximate problem which in this case is equivalent to a system of linear ordinary differential equations.

In practice it is necessary also to solve the latter approximately. While this is an interesting and challenging problem, space does not permit its treatment here, and we restrict attention entirely to the semi-discrete case which is defined continuously in time. Our goal will be to establish error estimates, and we shall do this by relating the latter to errors in associated steady state problems. This technique plus of course the approximation results (Theorem 6.2.1) forms the core of our theory.

While it is impossible to give an exhaustive list of references, we refer the reader to [3] and [4] for a discussion of semidiscretion and [5]-[10] for finite element approximations.

Throughout this chapter in referring to Hilbert spaces we shall mean complex spaces; similarly for Sobolev spaces. The parameter $t \geq 0$ will be real, and $K$ will be the space of complex numbers.

## 11.2. The variational principle.

We start with a result which in essence is a counterpart to Theorem 5.2.1. The context is the same, namely we suppose the complex Hilbert spaces $H_i$, $i = 1,2$, are given, and that $B(u,v)$, $b(u,v)$ are bilinear forms on $H_1 \times H_2$.

Theorem 11.2.1. Let $B(u,v)$ (resp. $b(u,v)$) satisfy (5.2.1)-(5.2.3) (resp. (5.2.1)). Let

$$B(z,u,v) = B(u,v) + zb(u,v) \quad . \tag{11.2.1}$$

We assume there is a sector

$$S^1_{\lambda_0,\theta_0} = \{z \in K \mid z = \lambda_0 - re^{i\theta} , \ r \geq 0 , \ |\theta| \leq \theta_0\}$$

in the complex plane -- the real number $\lambda_0$ and $|\theta_0| < \pi/2$ being fixed -- such that for each $z \notin S_{\lambda_0,\theta_0}$, $B(z,u,v)$ satisfies (5.2.1)-(5.2.3) with

$$C_2 \geq C(|z| + 1) \quad . \tag{11.2.2}$$

Then given $g \in H_1$, there exists a unique $u_0(t) \in H_1$ defined for $t \geq 0$ and $u_0'(t) \in H_1$ defined for $t > 0$ satisfying

$$\lim_{\substack{t \to 0 \\ t > 0}} \|u_0(t) - g\|_{H_1} = 0 \quad . \tag{11.2.3}$$

For each $t > 0$

$$\lim_{\tau \to 0} \left\| \frac{u_0(t+\tau) - u_0(t)}{\tau} - u_0'(t) \right\|_{H_1} = 0 \tag{11.2.4}$$

$$B(u_0, v) + b(u_0', v) = 0 \qquad \forall\ v \in H_2 \quad . \tag{11.2.5}$$

Proof. By Theorem 5.2.1 for each $z \notin S_{\lambda_0, \theta_0}$ there is a $w(z) \in H_1$ satisfying

$$B(z, w(z), v) = b(g, v) \qquad \forall\ v \in H_2 \quad . \tag{11.2.6}$$

Moreover, using (11.2.2) in Theorem 5.2.1 we obtain

$$\|w(z)\|_{H_1} \leq \left(\frac{C}{1 + |z|}\right) \|g\|_{H_1} \quad . \tag{11.2.7}$$

Let $\xi_0 > \max(\lambda_0, 0)$ and $\theta_0 < \bar{\theta}_0 < \pi/2$ . Moreover, let $\Gamma$ be the oriented curve

$$\Gamma = \{z \in K \mid z = \phi(\xi) \ , \ -\infty < \xi < \infty\}$$

where

$$\phi(\xi) = \begin{cases} \xi_0 + \xi e^{+i\bar{\theta}_0} & \text{if} \quad -\infty < \xi < 0 \\ \xi_0 - \xi e^{-i\bar{\theta}_0} & \text{if} \quad 0 \leq \xi < \infty \end{cases}$$

We define $u_0(t)$ by

$$u_0(t) = \frac{1}{2\pi i}\int_\Gamma e^{tz}w(z)dz \quad . \tag{11.2.8}$$

Our goal will be to show that this integral defines a $u_0(t) \in H_1$, $t \geq 0$, for which (11.2.2) and (11.2.4) are true, where

$$u_0'(t) = \frac{1}{2\pi i}\int_\Gamma e^{tz}w(z)zdz \quad . \tag{11.2.9}$$

We first note that by our choice of contour $\Gamma$, the integrals exist and define functions $u_0(t)$, $u_0'(t) \in H_1$ for positive values of $t$. Moreover, $u_0'(t)$ is the time derivative of $u_0(t)$ in the sense of (11.2.4) for $t > 0$.

To verify (11.2.5) for $t > 0$ we observe

$$B(u_0,v) = \frac{1}{2\pi i}\int_\Gamma e^{tz}B(w(z),v)dz \tag{11.2.10}$$

and

$$b(u_0',v) = \frac{1}{2\pi i}\int_\Gamma ze^{tz}b(w(z),v)dz \tag{11.2.11}$$

Therefore,

$$B(u_0,v) + b(u_0',v) = \frac{b(g,v)}{2\pi i}\int_\Gamma e^{tz}dz \quad . \tag{11.2.12}$$

Applying Cauchy's theorem to the last integral in (11.2.12) we see that

$$\int_\Gamma e^{tz}dz = 0 \tag{11.2.13}$$

and therefore $u_0$ defined by (11.2.8) satisfies (11.2.5). To verify (11.2.3) we use the change of variables $\xi = zt$

which gives

(11.2.14) $$u_0(t) = \frac{1}{2\pi i} \int_{\tilde{\Gamma}_t} e^{\xi} w(\tfrac{\xi}{t}) \frac{d\xi}{t}$$

where the curve $\Gamma$ is characterized by function

$$\tilde{\phi}(\xi) = t\phi(\xi);$$

then

(11.2.15) $$u_0(t) - g = \frac{1}{2\pi i} \int_{\tilde{\Gamma}_t} e^{\xi} \left\{ \frac{w(\frac{\xi}{t})}{t} - \frac{g}{\xi} \right\} d\xi \ .$$

Let us define

(11.2.16) $$\hat{w}(z) = g - zw(z)$$

then by simple computation we may verify that (see (11.2.6))

(11.2.17) $$B(z,\hat{w},v) = B(g,v)$$

$$+ zb(g,v) - zB(z,w,v) = B(g,v)$$

Furthermore it follows from (11.2.16)

(11.2.18) $$\frac{w(\xi/t)}{t} - \frac{g}{\xi} = -\frac{1}{\xi}\hat{w}(\xi/t)$$

and hence

(11.2.19) $$u_0 - g = -\frac{1}{2\pi i} \int_{\tilde{\Gamma}_t} e^{\xi} \frac{1}{\xi} \hat{w}(\xi/t)\, d\xi \quad .$$

Because $\hat{w}$ satisfies (11.2.17) we have

(11.2.20) $$\|\hat{w}\|_{H_1} \leq \frac{C}{|z|+1} \|g\|_{H_1} \quad .$$

Applying Cauchy's theorem we see that

(11.2.21) $$-\frac{1}{2\pi i} \int_{\tilde{\Gamma}_t} e^{\xi} \frac{1}{\xi} \hat{w}(\xi/t)\, d\xi = $$

$$= -\frac{1}{2\pi i} \int_{\Gamma} e^{\xi} \frac{1}{\xi} \hat{w}(\xi/t)\, d\xi$$

and therefore

(11.2.22) $$\|u_0 - g\|_{H_1} \leq C\|g\|_{H_1} \int_{\Gamma} |e^{\xi}| \frac{|d\xi|}{|\xi|\ |\frac{\xi}{t} + 1|}$$

$$\leq C\|g\|_{H_1}\ t \int_{\Gamma} |e^{\xi}| \frac{|d\xi|}{|\xi|^2}$$

$$\leq C\|g\|_{H_1}\ t \quad .$$

The condition (11.2.3) is an immediate consequence of this.

We now prove uniqueness, i.e., that $u_0(t)$ defined by (11.2.8) is the only solution. Suppose not. Then there is a $u(t)$ satisfying (11.2.3)-(11.2.5). For $0 < s < t$ define $v(s,t) \in H_1$ by

(11.2.23) $$v(t,s) = \frac{1}{2\pi i} \int_{\Gamma} e^{(t-s)z} w(s,z)\, dz$$

where $w(s,z)$ satisfies

$$(11.2.24) \quad B(z,w(s,z),v) = b(u(s),v) \qquad \forall \ v \in H_2$$

then as above

$$(11.2.25) \qquad \lim_{\substack{t\to s \\ 0<s<t}} \|v(t,s)-u(s)\|_{H_1} = 0 \ ,$$

moreover

$$\lim_{\substack{s\to 0 \\ 0<s<t}} \|v(t,s)-u(t)\|_{H_1} = 0 \ .$$

From (11.2.24) we have

$$(11.2.26) \qquad B(z, \frac{\partial w}{\partial s}, v) = b(\frac{\partial u}{\partial s}, v)$$

and therefore

$$\|\frac{\partial w}{\partial s}\|_{H_1} \leq C\|\frac{\partial u}{\partial s}\|_{H_1} \ .$$

Hence we may differentiate (11.2.23) by $s$ and get

$$(11.2.27) \qquad \frac{\partial v(t,s)}{\partial s} = \frac{1}{2\pi i} \int_\Gamma e^{(t-s)z}\left[\frac{\partial w}{\partial s} - zw\right] dz$$

Let us show that

$$(11.2.28) \qquad b(\frac{\partial v(t,s)}{\partial s}, v) = 0 \ , \forall v \in H_2$$

We start with (11.2.24) and note

$$- \int zb(w(s,z),v)e^{z(t-s)} dz =$$

$$= - \int_\Gamma b(u(s),v)e^{z(t-s)} dz + \int_\Gamma B(w(s,z),v)e^{z(t-s)} dz$$

Using Cauchy's theorem the first integral on the right side disapears and therefore

$$(11.2.29) \quad b\left(\frac{\partial v(t,s)}{\partial s}, v\right) = \frac{1}{2\pi i}\int e^{(t-s)z}\left[b\left(\frac{\partial u}{\partial s}, v\right) + \right.$$

$$\left. + B(w(s,z),v)\right] dz \quad .$$

We now define $\omega \in H_1$ by

$$(11.2.30) \qquad B(\omega,v) = -\, b\left(\frac{\partial w}{\partial s}, v\right)$$

This gives (see (11.2.26))

$$(11.2.31) \qquad B\left(z,\frac{\partial w}{\partial s},v\right) = b\left(\frac{\partial u}{\partial s},v\right) = -\, B(u,v)$$

In a manner analogous to the way we obtained (11.2.16) and (11.2.17) it is easy to show that

$$(11.2.32) \qquad \omega = \frac{1}{z}\left(u + \frac{\partial w}{\partial s}\right) = w$$

Therefore

$$(11.2.33) \qquad B(w,v) = -\, b\left(\frac{\partial w}{\partial s},v\right)$$

and hence by (11.2.29)

$$(11.2.34) \quad b\left(\frac{\partial v(t,s)}{\partial s},v\right) = 0 \quad \text{for every} \quad 0 < s < t$$

Therefore

$$(11.2.35) \qquad b(v(t,s),v) = b(v(t,t),v) =$$

$$= b(u(t),v) = b(v(t,0),v) =$$

$$= b(u_0(t),v)$$

and hence

$$(11.2.36) \qquad 0 = b\left(\frac{\partial(u-u_0)}{\partial t}, v\right) = B(u-u_0, v)$$

Using basic property (5.3.2) for the form $B(u,v)$ we get $u = u_o$. Our theorem is completely proved.

As a simple corrolary we have the following result, which is an immediate consequence of (11.2.8).

Theorem 11.2.2. Let $u_i$ satisfy (11.2.3)-(11.2.5) with $g_i$ in (11.2.3). Assume that the hypotheses of the theorem 11.2.1 are satisfied. Then for every $t > 0$ we have for $t > 0$

$$\|u_1(t)-u_2(t)\|_{H_1} \leq C(t)\|g_1-g_2\|_{H_2}$$

where $C$ generally depends on $t$.

11.3. The finite element method.

As in previous chapters we introduce $M_1 \subset H_1$, $M_2 \subset H_2$ and assume that (6.2.1) and (6.2.2) are satisfied for the form $B(u,v)$. Moreover we retain the notion of the previous section, i.e., $u_0(1)$ satisfies (11.2.3)-(11.2.5) with $w(z) \in H_1$ satisfying (11.2.6).

Theorem 11.3.1. Let all of the assumptions of theorem 11.2.1 be satisfied. Further assume that $B(z,u,v)$ satisfies (6.2.1)-(6.2.2) with $d_2(M_1,M_1) \geq C(M_1,M_1)(1+|z|)$, where $C(M_1,M_1) > 0$. Furthermore let $g_h \in \overline{M}_1$ such that

$$\|g-g_h\|_{H_1} \leq \theta$$

and denote

$$w_h(z) \in \overline{M}_1$$

(11.3.1) $$B(z,w_h(z),v) = b(g_h,v) \ , \ \forall v \in M_2$$

and assume that there exists $\omega(z)$ in $\bar{M}_1$ such that

(11.3.2) $$\|w(z)-\omega(z)\|_{H_1} \leq \theta(1+|z|)^k$$

for some $0 < k < \infty$ . Then

1) There is a unique $u_h(t) \in \bar{M}_1$ , $0 \leq t$ such that

(11.3.3) $$b(u_h(t),v) + B(u_h(t),v) = 0 \ , \ \forall v \in M_2$$

for $t > 0$ and

(11.3.4) $$\|u_h(t)-g_h\|_{H_1} \to 0 \quad \text{as} \quad t \to 0$$

2) For each $t > 0$

(11.3.5) $$u_h(t) = \frac{1}{2\pi i}\int_\Gamma e^{tz}w_h(z)\,dz$$

3) For $t > 0$

$$\|u_h(t)-u_0(t)\|_{H_1} \leq \frac{C\theta}{C(M_1,M_2)}$$

where $C$ does not depend on $M_1$ and $M_2$ but depends on $t$.

Proof. The first two statements are proved in exactly the same way as in theorem 11.2.1. In fact the calculation is identical once $H_i$ has been replaced with $M_i$ and $g$ with $g_h$ .

To prove (3) we note from theorem 6.2.1 and (11.3.2) that

$$\|w(z)-w_h(z)\|_{H_1} \leq \frac{C(1+|z|)^k}{C(M_1,M_2)}\ \theta$$

where $C$ does not depend on $M_1, M_2$ now

$$u(t) - u_h(t) = \frac{1}{2\pi i} \int_\Gamma e^{tz}(w(z)-w_h(z))dz$$

and therefore

$$\|u(t)-u_h(t)\|_{H_1} \leq \frac{C\theta}{C(M_1,M_2)} \int_\Gamma |e^{tz}|(1+|z|)^k|dz|$$

$$\leq \frac{C(t)}{C(M_1,M_2)} \theta$$

the proof is complete.

Assuming that $M_1$ and $M_2$ are finite dimensional, the calculation of $u_h(t)$ involves the solution of a system

$$M\underline{q}^{\bullet} + K\underline{q} = 0$$

of ordinary differential equations in the weights of $u_h(t)$. Interestingly, we note that if the form $b(u,v)$ is singular, i.e., $b(u_0,v) = 0$ all $v \in M_2$ for some $0 \neq v_0 \in M_1$, then the matrix $M$ will be singular. This is in fact the case when Lagrange multipliers are used (see section 6.3). Nevertheless, as we have proved, this system has a unique solution once the initial values are given.

Let us return to the problems (11.1.1) and (11.1.2) introduced in section 11.1.1. Here we take $H_1 = H_2 = H^1(\Omega)$ and bilinear form $B(u,v)$, (resp. $B(u,v)$) is given Obviously (5.2.1)-(5.2.3) is satisfied for the form $B$ and (5.2.1) for the form $b$. In this case (11.2.1) is satisfied for $\theta > 0$ arbitrarily small. For the proof of a more general case see e.g., [11].

Now let $g \in H^{\ell}(\Omega)$. Then the question is about the smoothness of $w(z)$. It may be shown see [11] that

$$\|w(z)\|_{H^{\ell+2}(\Omega)} \leq C(|z|+1)^k \|g\|_{H^{\ell}(\Omega)}$$

where $0 \leq k < \infty$ and depends generally on $\ell$. Therefore using a (t,k)-system for subspaces $M_1$ and $M_2$ we have $\theta \leq Ch^{\mu}$ with $\mu = \min(t-1,\ell+1)$ in (11.3.1) and

$$\|u_h(t)-u_0(t)\|_{H_1} \leq$$

$$\leq C(t)h^{\mu}\|g\|_{H^{\ell}(\Omega)}$$

$$\mu = \min(t-1,\ell+1)$$

Let us underline that $C(t)$ depends generaly on $t$ and it can happen that $C(t) \to \infty$ for $t \to 0$, i.e., the approximate solution is accurate if we are not interested in the case of small $t$. To solve accurately the problem for $t$ small is related in general to the boundary layer problem.

REFERENCES.

[1]. Adler G., Sulla caratterizzabilita dell' equazione del calore dal punto di vista del calcolo delle variazioni Magyar Tud. Akad. Mat. Kutató Int Koz 2 (1957), pp. 153-157.

[2]. I. Hlaváček, Variational Principles for Parabolic Equations, Apl. Mat. 14 (1969), pp. 278-297.

[3]. O. A. Liskovec, "The method of straight lines," (in Russian), Differencal'nye Uravnenija 1 (1965), pp. 1665-1678.

[4]. V. N. Abrašin, "Convergence and estimates of error of the method of straight lines in the solution of certain problems of mathematical physics," (in Russian), Differencial'nye uravnenija 7 (1968), pp. 1317-1332.

[5]. B. Schwartz, B. Wendroff, "General finite difference schemes," Math. Comp. 23 (1969), pp. 37-50.

[6]. J. Douglas, T. Dupont, "Galerkin methods for parabolic problems," SIAM J. Numer. Anal. 4 (1970), pp. 575-626.

[7]. J. Bramble, V. Thomee, "Semidescrete least squares method for a parabolic boundary value problem," Math. Comp., to appear.

[8]. H. S. Price, R. S. Varge, Error Bounds for Semi-Discrete Galerkin Approximations of Parabolic Problems with Applications to Petroleum Reservoir Mechanics, Num. Solution of Field Problems in Comp. Phys. AMS Providence, R. I., 1970, pp. 74-94.

[9]. G. Fix, N. Nasif, "On finite element approximations on time-dependent problems," Numer. Math. 19 (1972),

pp. 127-135.

[10]. J. L. Lions, Équations Différentilelles Opérationnelles, Springer, Berlin (1961).

[11]. M. S. Agranovič, M. I. Višik, "Elliptic problems with a parameter and parabolic problems of general type," (in Russian), Uspehi Mat. Nank 19 (1964), No. 3 (117), pp. 53-161.

# PART II

# INVITED HOUR LECTURES

# PIECEWISE ANALYTIC INTERPOLATION AND APPROXIMATION IN TRIANGULATED POLYGONS*

Garrett Birkhoff**

## 1. Introduction.

Many physical and engineering problems can be mapped smoothly onto a rectangular parameter domain. In such cases, piecewise bicubic Hermite and spline approximations on subdivisions into rectangular finite elements permit the achievement of high orders of accuracy in a simple and systematic way. In the next paper, Drs. Gordon and Hall will describe a very effective new procedure for rectangularly parametrizing general simply connected regions with piecewise smooth boundaries, thereby making such Hermite and spline approximations feasible.

Nevertheless, for most two-dimensional problems, triangulation is much more flexible geometrically than subdivision into rectangles. As Zienkiewicz [20, p. 115] has said: "The advantage of an arbitrary triangular shape in approximating to any boundary shape

---

* Work supported by the Office of Naval Research under Contract No. N00014 67 A 0298 0015.

** Harvard University.

has been amply demonstrated ... Its apparent superiority here over the rectangular shapes needs no further discussion." For example, one may wish to subdivide some or all of the rectangles of a rectangular mesh into triangles to provide a better approximation to the boundary of a rectangularly subdivided plane region; see Figs. 1a-1c.

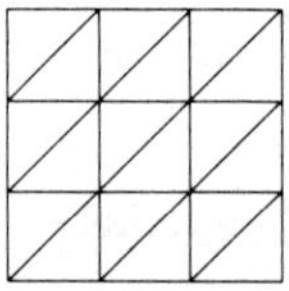

Fig. 1a

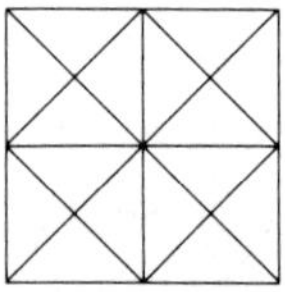

Fig. 1b

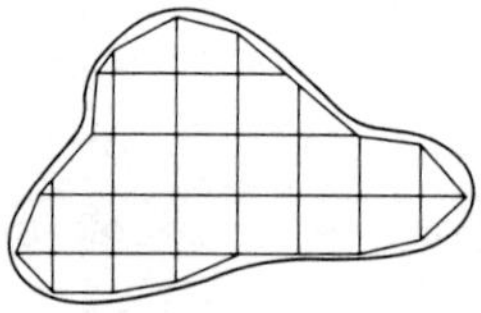

Fig. 1c

Actually, for simple variational <u>source</u> problems, even for <u>non-isotropic</u> problems minimizing an integral of the general form (cf. [13, p. 217, (4.1)]):

$$J[u] = \iint_R [(\nabla u, P\nabla u) + qu^2 - 2fu]\, dxdy \tag{1}$$

($P$ a positive definite symmetric matrix), implementation of the Rayleigh-Ritz method by simple <u>piecewise linear</u> interpolation to "nodal values" gives an adequate approximation for many engineering purposes. For this reason, it can be considered as a humble but invaluable <u>work-horse</u>, a clear understanding of whose capabilities and limitations is essential. In §2, I shall review a few basic facts about it, and add some remarks which I have not seen in the literature.

However, more sophisticated "finite elements"

than those provided by simple piecewise linear interpolants in triangles are of course more interesting mathematically.

I shall therefore devote the major portion of my talk to them. Many such more sophisticated elements have been designed by mechanical engineers over the past decade; I shall try to analyze some of them from a mathematical standpoint in §§3-4, with special reference to how well they approximate the values of the exact solutions of variational problems.

Then, in §§5-6, I shall turn to my main theme: the design, analysis, and implementation of smooth higher-order finite elements which approximate derivatives as well as values of the functions of interest. The answer to this question is intimately connected with the problem of best approximation in Sobolev spaces,† although other Hilbert and Banach spaces with topologically equivalent norms given by physically significant quantities (e.g., energy norms) are often more relevant.

Throughout, my primary concern will be with the problem of designing accurate and flexible algorithms which can be easily coded as subroutines for efficient "production codes", to be implemented on large-scale, high-speed computers. One code package, already in widespread use, is the NASTRAN code developed for NASA by R. H. MacNeal and others [11]. Other ambitious code

† See [4, 5, 14, 18, and 19]. Also F. di Guglielmo, Estratto da Calcole 6 (1969), 279-331.

packages for solving problems in structural mechanics, now comercially available, have been constructed by Argyris and Marçal.† No doubt numerous other good proprietary "finite element" codes and code packages also exist.

## 2. Piecewise linear interpolation.

Given a smooth function $u \in C^2(R)$, where $R$ is a polygonal region subdivided into triangles $T_\ell$ by a triangulation $\tau$, it is well-known that interpolation to the "nodal values" $u_k = u(P_k)$ at the nodes $P_k = (x_{i(k)}, y_{j(k)})$ approximates $u(x,y)$ to $O(h_\ell^{\,2})$ in each $T_\ell$, where $h_\ell$ is the diameter (longest side) of $T_\ell$, and the gradient $\nabla u$ to $O(h_\ell)$. Indeed, a uniform bound to the interpolation errors is easily given in terms of the maximum second derivative (i.e., in the Sobolev space $W^2_\infty$), $h = \max h_\ell = |\tau|$ (the "norm" of the triangulation), and the minimum angle $\varepsilon > 0$.

It follows that simple interpolation to the nodal values of the exact solution of (1), which exists by classic principles, will give an approximate solution $v$ with

$$J[u] \leqq J[v] \leqq J[u] + Mh^2 \ . \tag{2}$$

This follows because the minimizing property of $u$ implies that

$$J[v] = J[u] + \iint_R [(\nabla e, P\nabla e) + qe^2]\, dxdy, \tag{3}$$

where $e = v - u$ is the error bounded above.

† For Marçal's code package, see [17], pp. 433-67 and also D. Bushnell et al., Computers and Structures I (1971), 361-387.

Uniform bounds on the interpolation error $e(x,y)$ do not, of course, directly imply uniform bounds on the error on the approximate solution $w(x,y)$ found by the Rayleigh-Ritz method; we can only conclude from (2) directly that $J[w - u] \leqq Mh^2$ . For uniformly acute triangles (with maximum angle $\theta \leqq (\pi/2) - \varepsilon, \varepsilon > 0$ , bounds in the uniform norm have however been obtained by Zenicek [19] and by Ciarlet and Raviart [5], using a discrete maximum principle.

For convenient reference, I present here also a <u>formula</u> expressing the <u>Dirichlet integral</u>

$$(4) \quad D(u,u) = \iint_R (\nabla u, \nabla u)\, dxdy = \iint_R (u_x{}^2 + u_y{}^2)\, dxdy$$

taken over the triangles having a common interior vertex $P_0$ in $R$ . For a triangle in standard position, having vertices at $P_0 = (0,0)$ , $P_1 = (h,0)$ , and $P_2 = (a,b)$ , the integral is easily computed as

$$\frac{hb}{2}\left[\left(\frac{u_1 - u_0}{h}\right)^2 \left(1 + \left(\frac{a}{b}\right)^2\right) + \frac{2a}{hb}(u_1 - u_0)(u_2 - u_0) + \left(\frac{u_2 - u_0}{b}\right)\right]$$

Now setting $a = k \cos \phi$, $b = k \sin \phi$, where $\phi$ is the vertex angle at $P_0$ , we get

$$\frac{1}{2 \sin \phi}\left\{\frac{k}{h}(u_1 - u_0)^2 + 2 \cos \phi\, (u_1 - u_0)(u_2 - u_0) + \frac{h}{k}(u_2 - u_0)^2\right\}$$

Finally, changing notation so that $r_j$ represents the length of the j-th edge , and $\phi_j$ the angle between the j-th and $(j + 1)$st radius, summation expresses the Dirichlet integral (4) as

$$\sum_{j=1} \left[\left(\frac{r_j}{r_{j+1}\sin\phi_j} + \frac{r_j}{r_{j-1}\sin\phi_{j-1}}\right)(u_j - u_0)^2\right]$$
$$(4') \qquad + 2\sum_{j=1}^{n} [\cot\phi_j \,(u_j - u_0)(u_{j+1} - u_0)\,].$$

This is a minimum when $u_0$ is a suitable weighted mean of the adjacent $u_j$.

As a special case, we obtain the following generalization of a statement of Courant.[†]

Theorem 1. The nodal values of the piecewise linear function which minimizes the Dirichlet integral (4) for any triangulation by diagonals of a rectangular subdivision, one in each rectangle, satisfy the usual 5-point divided difference equation.

Unfortunately, the above statement appears not to generalize to the source problem when $q \neq 0$ in (1); neither does it give the optimal 5-point approximation when $f \neq 0$. However, for general $q \geqq 0$ and $f \neq 0$, it does yield a (diagonally dominant) Stieltjes coefficient-matrix for which the usual methods of elimination and iterative overrelaxation (e.g., cyclic Chebyshev) are applicable.

The graph of this coefficient-matrix is, moreover, typically the dual of the planar graph defined by the original mesh-points and mesh-lines of a rectangular subdivision, and the discretization defines a linear

† [8]. For the claims made here, see [3].

network flow problem on this graph.† It is only by coincidence that some coefficients cancel, giving also a coefficient-matrix with rectangular graph, when the original mesh-lines form a rectangular network triangulated by one diagonal in each rectangle.

## 3. Higher-order continuous "Lagrange" interpolants.

By interpolating to $u$ at more points, one can find continuous, piecewise polynomial, local interpolants to any smooth function $u$ in any triangulated polygon, whose order of accuracy is limited only by the smoothness of $u$. A very attractive scheme for computing such local interpolants of degree $n$ is obtained by taking as mesh-lines $n - 1$ equally spaced parallels to each side of each triangle, giving $(n+1)(n+2)/2$ mesh-points in all. See Figs. 2a - 2c for the cases $n = 2,3,4$.

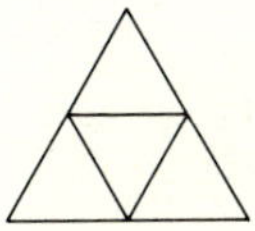

Fig. 2a

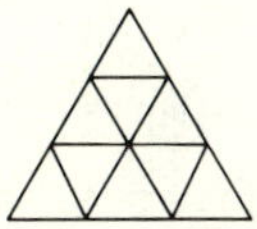

Fig. 2b

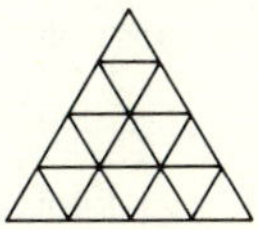

Fig. 2c

Using the barycentric coordinates of Moebius,††

† See G. Birkhoff, "Numerical solution of elliptic equations", SIAM Publications, 1971, §3.2. It follows that the coefficient-matrix is a Stieltjes matrix.

†† Birkhoff-MacLane, "Survey of Modern Algebra", 3d ed., pp. 271-3; these are called "areal" coordinates in [20].

whereby $P = w_0P_0 + w_1P_1 + w_2P_2$ ($w_j \geq 0$, $w_0 + w_1 + w_2 = 1$) expresses the coordinates of each point P as a <u>weighted average</u> of the coordinates of the vertices, the mesh-lines become the lines $w_j = k/n$ ; the j-th edge is the line $w_j = 0$ ; and the j-th vertex becomes the point $w_j = 1$ (hence all $w_k = 0$, $k \neq j$). This barycentric coordinate representation thus maps each triangle onto that spanned by the unit vectors (1,0,0), (0,1,0), and (0,0,1) in space; it has a straightforward extension to p-dimensional simplexes.

There is precisely one "Lagrange" polynomial interpolant v(x,y) of degree n which interpolates to u at all these mesh-points. Moreover Bramble and Zlámal† have shown that for $u \in W_2^{n+1}(R)$, the resulting <u>mean square error</u> $e^{(j)} = v^{(j)} - u^{(j)}$ in the j-th derivative is bounded by $Ch^{n+1-j}\|u\|_2^{n+1}$, just as for the familiar Taylor series interpolant. Furthermore, local interpolants to the same function in adjacent triangles assume the same values on their common side. Finally, similar results hold in p-dimensional simplicial complexes.

Because of its simplicity and generality, the preceding family of triangular elements is the one commonly presented first in discussions of higher-order triangular elements,†† with emphasis on the <u>quadratic</u> and

† [4]; for a good summary and further refs., see [5] and [17], pp. 22-3.

†† Zienkiewicz, [20], Sec. 7.7; Felippa and Clough, [13], pp. 221-4. Zienkiewicz (p. 120) attributes the quadratic element to Veubeke and Argyris, 1965.

cubic cases $n = 2,3$ which are those most commonly used. Of these, the quadratic case is mathematically uninteresting (though very practical!).

## 4. Cubic "Hermite" approximation.

The cubic case offers more scope for mathematical ingenuity. In particular, there is an improved (piecewise) cubic "Hermite" variant of the cubic "Lagrange" scheme of Fig. 2b, which gives a central role to the corner values and slopes $u, u_x, u_y$ of the function to be approximated.

### Unisolvence.

These 9 numbers, together with any interior value, necessarily define a well-set or "unisolvent" interpolation problem. Indeed, given 10 points $(x_j, y_j)$ at random in a triangle $T$ (or in the plane!), the probability is zero that they will lie on a single cubic curve -- i.e., that there exists a polynomial $p(x,y)$ of degree 3 or less such that $p(x_j, y_j) = 0$ at all 10 points. And unless this happens, there is precisely one

cubic polynomial $c(x,y) = \sum_{k+\ell \leqq 3} c_{k\ell} x^k y^\ell$ which assumes

given values $c(x_j, y_j) = u_j$ at these points: we have unisolvence.

In particular, interpolation is always unisolvent in a triangle $T$ if the 10 points include a vertex and three other points on the two adjacent sides, together with any three other non-collinear points, as in Fig. 3a. This follows from the observation that if $c(x,y)$ vanishes at four points on a side, then it must vanish identically on this side (extended). Hence, if $\alpha_i x + \beta_i y + \gamma_i = L_i(x,y)$ is the linear polynomial vanishing on this side, $L_i \mid c(x,y)$. The hypotheses thus imply that if $c(x,y)$ vanishes at all 10 points, then $L_1 L_2 \mid c(x,y)$, whence $c(x,y) = L_1 L_2 M$, where $M$ is a third linear polynomial vanishing on a third straight line.

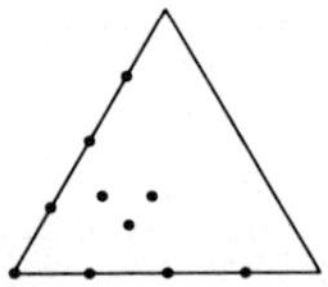

Fig. 3a

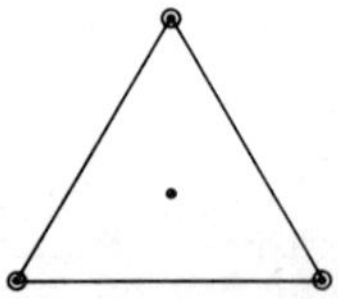

Fig. 3b

As a limiting case of the preceding, we can interpolate to the values of $u, u_x, u_y$ at the three corners, _and_ the value of $u$ at the centroid of the triangle, as indicated in Fig. 3b. This is a much better choice of unknowns than that of Fig. 2b, the reason being that triangulated polygons with many triangles and N vertices have about $2N$ triangles and $3N$ edges.[†]

---

[†] For, let $N_0 = N, N_1, N_2$ denote as usual the numbers of nodes, edges, and triangles given by the triangulation $\tau$. If $\nu$ is the number of incidence relations between edges and triangles, then $\nu = 3N_2$ and $\nu = N_1' + 2N_1'' = 2N_1 - N_1'$ where $N_1'$ is the number of boundary and $N_1''$ the number of interior edges. As $h \downarrow 0$, $N_1'/N_2 \to 0$, so that asymp-

Therefore, for the same triangulation, the scheme of Fig. 3b requires only 5N unknowns, as compared with 9N for the interpolating scheme of Fig. 2b.

Moreover, for the source problem, the value $u_{10}$ of u at the interior node in each triangle T (cf. Fig. 3b) can be eliminated by "static condensation": for any set of 9 values of $u, u_x, u_y$ at the vertices, $u_{10}$ must minimize (1), and its minimizing value and hence the integral (1) taken over T can be expressed as a quadratic function of $u_1, \cdots, u_9$ alone. The sum expresses J[u] approximately as a quadratic function of only about 3N unknowns, or about one-third as many as are required for the scheme of Fig. 2b. Since the resulting approximation has order of $\mathcal{O}(h^4)$ in u, $\mathcal{O}(h^3)$ in the gradient, and $\mathcal{O}(h)$ in the second partial derivatives in any (acute?) triangle,† the preceding discretization scheme may be presumed to be nearly optimal for source problems in triangulated polygons, to the order of accuracy indicated.††

5. Continuously differentiable interpolants.

For plate bending problems, "shape functions" u(x,y) representing normal deflections must have contin-

---

totically $2N_1 \sim 3N_2$, whence $N_1 \sim 3N$ and $N_2 \sim 2N$ since $N + N_2 = N_1 + 1$ (Euler-Poincaré formula).

† See [4, 5, 18, 19].

†† [10] shows that it is much more efficient than piecewise linear interpolation when accurate results are wanted.

uous slope (be of class $C^1(R)$) to conform to the physical condition of finite bending energy, $\iint (\nabla^2 u)^2 \, dxdy$. Whereas this property is easy to achieve for rectangular complexes, using bicubic Hermite or spline interpolation to values and derivatives, it is much harder to construct "conforming" (i.e., continuously differentiable) interpolants to given $u(P_\alpha)$ at mesh-points for triangulated polygons. For example, although prescribing $u, u_x, u_y$ at each vertex guarantees the continuity of cubic interpolating polynomials across edges, one cannot require the normal derivative $u_n$ to be interpolated linearly along each edge between its end-values without making cross-derivatives discontinuous.†

Several schemes for constructing "conforming" approximants are described in [20], pp. 198-203 and 207-9.

One such scheme, due to Clough and Tocher†† and used in NASTRAN, has 12 degrees of freedom. These correspond to the values of $u, u_x, u_y$ at the three vertices and the normal derivative $u_n$ at the midpoints of the three sides, giving about $3N + 3N = 6N$ unknowns for a fine triangulation.

The interpolating function is restricted to being cubic in each of three subtriangles (see Fig. 4a),

† See Felippa and Clough in [13], p. 226; also [20], p. 176. The same difficulty was discussed in G. Birkhoff and H.L. Garabedian, J. Math. and Phys. 39 (1960), 258-68 §5, where a rectangular element was constructed reminiscent of the Clough-Tocher finite element described below.

†† R. W. Clough and J. L. Tocher, [7]; cf. Felippa and Clough in [13], p. 226 and Fig. 8a. For its use in NASTRAN,

and continuously differentiable across the interfaces separating them. One can achieve this by taking as additional free variables the three values of $u, u_x, u_y$ at the central vertex (perhaps the centroid). Any choice of these gives a unique cubic interpolant in each subtriangle; moreover the resulting three (partial) functions have the same cubic variation along each interior common edge, and their (quadratic) normal derivatives agree at the ends of these edges. The requirement that they agree also at one other point (hence all other points) on each edge, gives three additional conditions which uniquely determine the (otherwise free) values of $u, u_x, u_y$ at the central vertex.

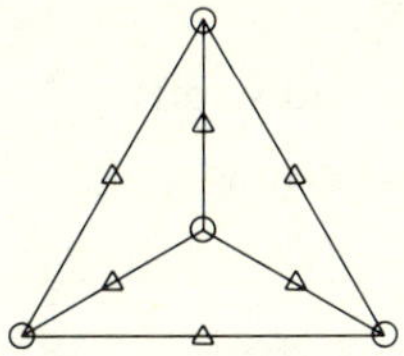

Fig. 4a

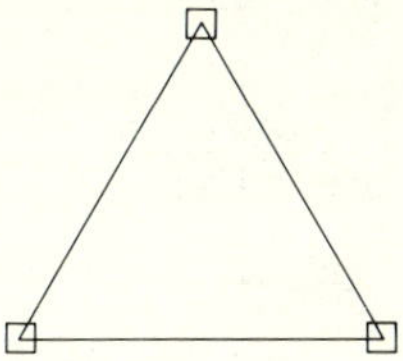

Fig. 4b

Quintic polynomials.

For high precision plate bending problems, "conforming" (class $C^1$) quintic polynomial elements have been used. Thus, using all 21 degrees of freedom of a "complete quintic polynomial", one can uniquely interpolate to $u, u_x, u_y, u_{xx}, u_{xy}, u_{yy}$ at the three corners of a triangle (match the value, the slope, and the curvature tensor), and to $u_n$ at the midpoints of the sides.

---

see [11], Sec. 5.8.3.3.

Since the normal derivative $v_n = \partial v/\partial n$ of the interpolant is a quartic polynomial, it and $v$ are determined on any edge by the above data. But $v$ determines $\partial v/\partial s$; hence the above construction gives a piecewise quintic polynomial interpolant of class $C^1$ to any function $u \in C^2(R)$. Since the (local) interpolation scheme is of "Hermite" type in the sense of Ciarlet, moreover, the error $e^{(j)}(x,y)$ is of order $O(h^{6-j})$ for $j = 1,2,\cdots,5$ if $u \in C^6(R)$. In a triangulated polygon with $h \ll$ smallest diameter, moreover, the number of unknowns is asymptotically about $6N + 3N = 9N$. The above finite element scheme is due to Argyris, Bell, and others.†

Reduced quintic polynomials.

One can constrain the preceding interpolation scheme by requiring $v_n$ to be a cubic (not merely a quartic) polynomial on each side, while still matching $u, u_x, u_y, u_{xx}, u_{xy}, u_{yy}$ at the corners. This gives an 18 degrees-of-freedom family which has been studied by Argyris and Bell (op. cit.), and by Cowper et al.†† This can match at most a general quartic, from which it follows that the error will be $O(h^5)$. For a plane domain with $N$ nodes, the dimension of the approximating subspace is only about $6N$ (as compared with $9N$ for the complete quintic discussed above). In [20], p. 208, these reduced quintic

† J. H. Argyris et al. Aer. J. Roy. Aer. Soc. 72 (1968), 701-9; K. Bell, Int. J. Num. Methods Eng. 1 (1969), 101-22; for a synopsis, see [20], Sec. 10.15 (pp. 207-9).

†† G.R. Cowper et al., AIAA J. 7 (1969), 1957-65, and [6].

polynomials are said to define a "more useful element in practice" than the full quintic polynomials discussed above. Note that the "standard" interpolation scheme of Fig. 2c for (complete) quintic polynomials requires about 25N unknowns, and is thus less efficient than either of the schemes described here without even being conforming!

Rational quartic elements.

Irons† has constructed still another ingenious "conforming" triangular element with 18 degrees of freedom. This interpolates to $u, u_x, u_y$ at the vertices, and to $u, u_n$, and $\partial^2 u/\partial s\, \partial n$ at the midpoints of the sides. The interpolating functions are taken to be linear combinations of $1, x, y, \cdots, x^2y^2, xy^3, y^4$ (the 15 monomials of degree four or less), the rational function of degree 4 given by

$$L_1 L_2^{\,2} L_3^{\,2}(L_3 - L_2)/(L_1 + L_2)(L_1 + L_3), \tag{5}$$

and its two transforms under cyclic permutations of the subscripts. Note that the denominator of (5) vanishes to order one at vertices 3 and 2, while the numerator vanishes to order 3 there. Hence the gradient is continuous, while the second derivatives are bounded but not continuous at the vertices. Irons' scheme, which involves about 12N unknowns for a fine triangulation ($h \ll 1$) with N nodes, presumably has $O(h^2)$ accuracy in $u$ and $O(h)$ in $\nabla u = (u_x, u_y)$.

† Int. J. Num. Methods Eng. 1 (1969), 29-46. For a short synopsis, see [20], pp. 202-3.

## 6. Tricubic polynomial interpolation.

Other interesting triangular elements are provided by the following scheme of "tricubic polynomial interpolation".†

Definition. A function $u(x,y)$ is tricubic on a convex domain $\Gamma$ when it is cubic on all parallels in $\Gamma$ to each of three non-parallel lines.

Lemma 1. A function $u$ is tricubic in $\Gamma$ if and only if it satisfies

$$\left(\alpha_i \frac{\partial}{\partial x} + \beta_i \frac{\partial}{\partial y}\right)^4 u = 0 \quad \text{for} \quad i = 1,2,3, \tag{6}$$

where $(\alpha_i, \beta_i)$ are three linearly independent vectors.

Proof. Since $\alpha_i(\partial/\partial x) + \beta_i(\partial/\partial y)$ is the directional derivative in the direction of the line with parametric equation $x = a + \alpha_i t, y = b + \beta_i t$, this simply restates a characteristic property of cubic polynomials.

Lemma 2. The class of tricubic functions is preserved under nonsingular affine transformations.

For example, if we take as our lines the sides of the standard triangle with vertices at $(0,0)$, $(1,0)$, $(0,1)$ and barycentric coordinates $x,y$, and $1 - x - y$, the operators of (6) are $\partial/\partial x$, $\partial/\partial y$, and $(-\partial/\partial x - \partial/\partial y)$. In this case a basis of tricubic functions is provided by the twelve monomials $1,x,y,x^2,xy,y^2,x^3,x^2y,xy^2,y^3$, $x^3y + x^2y^2, x^2y^2 + xy^3$.

From this result and Lemma 2, we obtain

Theorem 2. The class of tricubic functions forms a 12-

---

† For a more detailed description, which however contains some errors corrected here, see [2].

parameter family, consisting of all sums of a cubic polynomial and a quartic polynomial of the special form

$$(7) \qquad Q(x,y) = (Ax + By) \prod_{i=1}^{3} (\beta_i x - \alpha_i y) .$$

Theorem 3. Let $\Delta$ be a triangle with barycentric coordinates $r,s,t$ . Then there is a unique tricubic polynomial, cubic along parallels to the sides of $\Delta$ , which assumes given $u, u_r, u_s$, and $u_{rs}$ at the vertex $t = 1$ , and corresponding data at the other two vertices of $\Delta$ .

Theorem 4. Piecewise tricubic interpolation in triangulations $\pi$ of a given closed polygon $\Gamma$ approximates any function $u \in C^4(\Gamma)$, such that if $|\pi|$ is the maximum diameter of a triangle,

$$(8) \qquad |u - u_\pi| \leqq K_0 \, |\pi|^4 \quad \text{and} \quad |\nabla u - \nabla u_\pi| \leqq K_1 \, |\pi|^3 .$$

It is here assumed that the angles of all triangles involved are bounded away from $0^\circ$ , and hence from $180^\circ$. In practice, one would probably want to restrict interior angles $\theta_i$ to the range $20^\circ < \theta_i \leqq 90^\circ$ .

Proofs of Theorems 2-4 may be found in [2], where however it was incorrectly stated (in Theorem 4) that the interpolant was in $C^1(\Gamma)$ (i.e., continuously differentiable). This error will be rectified below.

We now define the piecewise tricubic polynomials with domain in the triangles having a common vertex to be vertex $C^2$-compatible there when, for some $A = u_{xx}$, $B = u_{yy}$, and $C = u_{xy}$ we have

$$(9) \qquad \frac{\partial^2 u}{\partial \varepsilon_i \partial \varepsilon_{i+1}} = A\alpha_i\alpha_{i+1} + B\beta_i\beta_{i+1} + C(\alpha_i\beta_{i+1} + \beta_i\alpha_{i+1}).$$

The class of continuous, $C^1$-compatible and vertex $C^2$-compatible piecewise tricubic polynomials has 6 free parameters at each vertex. Hence, in a triangulated polygon $(R,\tau)$ , it yields a 6N-dimensional approximating subspace (modulo the constraints imposed by the boundary conditions).

This is about twice as many as are involved in the approximating subspace of (constrained) piecewise cubic polynomials. However, it should be more efficient when extremely high precision is desired and the geometry permits the use of "large" elements.

Smoothed tricubic approximation.

We now show how to smooth the continuous, piecewise tricubic polynomial functions constructed above, so as to obtain continuously differentiable functions having the same order of accuracy. We first note the following corollary of the proof of Theorem 4.

Theorem 5. Let $v$ and $\tilde{v}$ be vertex $C^2$-compatible (piecewise) tricubic interpolants to $u$ , in adjacent triangles having as common side the segment of the x-axis from $(0,0)$ to $(h,0)$ . Then $v_y - \tilde{v}_y = x(h-x)\,\phi(x)$ along this side, where $\phi(x) = O(h)$.

We also note the following incidental lemma.

Lemma. At any vertex where two adjacent triangles have equal angles, the average $\frac{1}{2}[v_{ns} + \tilde{v}_{ns}]$ of the cross-derivatives of the tricubic interpolants is $u_{ns}$ :

Proof. We consider the case of Fig. 5a, and recall that $v_{xx} \equiv \tilde{v}_{xx}$ since $v \equiv \tilde{v}$ . Hence (assuming equilateral triangles):

$$\begin{aligned}\sqrt{3}\, v_{xy} &= u_{xx} - v_{xx} + \sqrt{3}\, u_{xy} \\ &= u_{xx} - \tilde{v}_{xx} + \sqrt{3}\, u_{xy} \\ &= -\sqrt{3}\, \tilde{v}_{xy} + \sqrt{3}\, u_{xy} + \sqrt{3}\, u_{xy}\ .\end{aligned}$$

Transposing and dividing by $2\sqrt{3}$, we get the desired result.

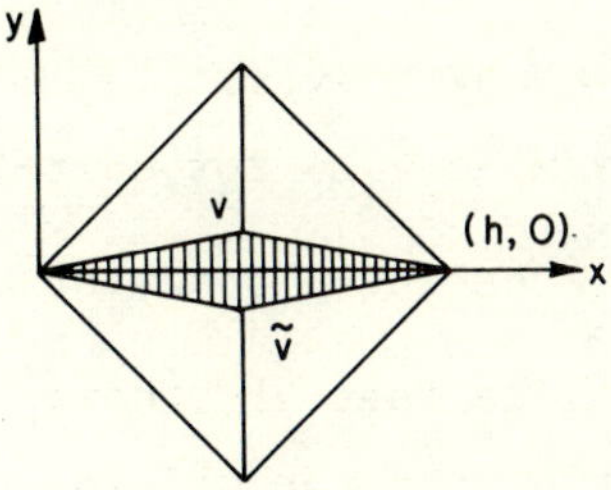


Fig. 5a

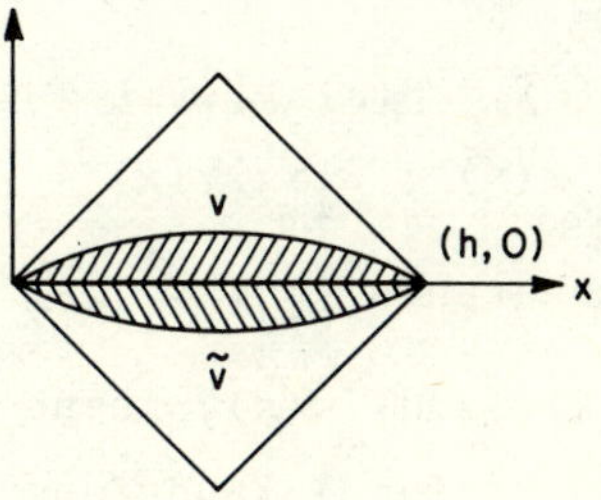


Fig. 5b

Now let $v, \tilde{v}$ be the restrictions to adjacent triangles of any continuous, vertex $C^2$-compatible, piecewise tricubic function $V(x,y)$ on a triangulated polygon. We shall use Theorem 5 to smooth out the discontinuities in $\partial V/\partial n$ across edges by smoothing fillets, one constructed across each edge independently of all others. To do this, we first trisect each triangle (as in constructing the Clough-Tocher element) by three radii (segments of medians) from the centroid, one to each corner, as in Fig. 5a. This will surround each edge by a disjoint diamond-shaped area into which we will interpolate one smoothing fillet, as follows.

First interpolate cubic curves $Y = \gamma(x)$ and $Y = -\tilde{\gamma}(x)$ tangent to the medians at each corner, where without loss of generality we have assumed the common edge to extend from (0,0) to (h,0) ; these will enclose a lens-shaped area as in Fig. 5b. Finally, displace v(x,y) by

$$w(x) = \beta(x)y\,[\gamma(x) - y]^2/2\gamma(x), \quad 0 < y < \gamma(x) ,$$

in the upper half of this area, where $\beta(x)$ is the jump in normal slope across the common edge. Likewise, displace $\tilde{v}(x,y)$ by

$$\tilde{w}(x) = \beta(x)\; y[\tilde{\gamma}(x) + y]^2/2\tilde{\gamma}(x), \quad 0 > y > -\tilde{\gamma}(x) .$$

Since $\beta(x) \sim \beta_1 x$, $\gamma(x) \sim \gamma_1 x$ , and $\tilde{\gamma}(x) \sim \tilde{\gamma}_1 x$ for suitable constants $\beta_1$, $\gamma_1$, $\tilde{\gamma}_2$ as $x \downarrow 0$ , the local behavior of $w(x)$ and $\tilde{w}(x)$, respectively, is like that of homogeneous <u>rational functions</u> of degree two near the origin, e.g., of

$$\beta y(\gamma_1 x - y)^2/2\gamma_1 x .$$

Hence $w_x, w_y = O(1)$ are certainly continous there, and $w_{xx}, w_{xy}, w_{yy}$ uniformly <u>bounded</u>.

It follows that the displaced or <u>smoothed</u> function is of class $C^1$ , with error in slope at most $O(h^2)$ . We have thus constructed a family of about 6N linearly independent "conforming" elements of class $C^1$ , in any triangulated polygon $(R,\tau)$ with N vertices.

## 7. <u>Right triangles</u>.

I shall close by analyzing briefly an important special case: that of triangulations into <u>right triangles</u> having a vertical side, a horizontal side, and a hypotenuse. This arises in two important situations: (i) when a polygon has been decomposed into rectangles and border triangles (Fig. 1c), and (ii) when each square of a network of squares has been subdivided into triangles by its two diagonals (Fig. 1b), and the resulting configuration rotated through $45^{\circ}$.

Situation (i) has been studied by C. A. Hall,[†] who has shown that in it one can achieve a continuously differentiable extension of Hermite interpolation in the rectangular elements in either of two ways (his Schemes A and B). A simplified proof of his result follows; without loss of generality, we can suppose the vertices to be at $(0,0)$, $(h,0)$, and $(0,k)$.

With a cubic interpolant, one can match $u, u_x, u_y$ at all corners; the remainder will be of the form $xy(hk - kx - hy)\phi = xyz\phi$. For any linear $\psi = (a+bx+cy)$, the product $xyz\psi = v$ will then be tricubic, i.e., cubic on parallels to the sides of the triangle. Moreover, after eliminating terms with a factor zero, we have by Leibniz' rule for differentiating a product:

$$v_{xy}(0,0) = z\psi(0,0) = ahk$$

$$v_{xy}(h,0) = -kx\psi(h,0) = -hk(a+bh) = -ahk - bh^2k$$

$$v_{xy}(0,k) = -hy\psi(0,k) = -hk(a+ck) = -ahk - chk^2 .$$

We can obviously match $u_{xy}$ at all corners by properly choosing $a,b,c$, obtaining thereby a function of class $C^{1,1}$ over the whole polygon. By a variant of this procedure, B. L. Hulme has shown how to match normal deriva-

† J. Math. Mech. 19 (1969), 1-11.

tives on the boundary.

Situation (i) has various other interesting aspects and properties, which I have discussed elsewhere;[†] I shall not repeat this discussion here.

Situation (ii) arises with Synge's subdivision [16]; it leads to a triangulation in which the vertices are alternately of degrees 4 and 8 in a checkerboard pattern. To achieve edge $C^1$-compatibility with piecewise tricubic interpolation in this configuration, it suffices to have consistent values of $u, u_x, u_y$, and $u_{xy}$ at all vertices, and consistent $u_{xx} - u_{yy}$ at all vertices of degree 8. Intuition suggests that one can achieve this by a substantial improvement on piecewise cubic Hermite interpolation in the coarse subdivision into squares, whose bisection (by one diagonal in each square) gives Synge's configuration. However, I have not seen a good way to do this other than by the rather awkward smoothed tricubic scheme described in §6, in the hour or so which I have so far been able to devote to the question.

## REFERENCES

1. Barnhill, R. E., G. Birkhoff, and W. J. Gordon, "Smooth interpolation in triangles," Report GMR-1064, General Motors Research Laboratory, Feb., 1971. (To appear in the J. Approximation Theory).
2. Birkhoff, G., "Tricubic polynomial interpolation," Proc. Nat. Acad. Sci. USA 68 (1971), 1162-64.
3. Birkhoff, G. and S. Gulati, "Optimal low-order discretizations of linear source problems," to be sub-

† See pp. 202-4 in I. J. Schoenberg (ed.), "Approximation with Special Emphasis on Spline Functions," Academic Press, 1969.

mitted to SIAM J. Numer. Anal.

4. Bramble, J. and M. Zlámal, "Triangular elements in the finite element method," Math. Comp. 24 (1970), 809-20.
5. Ciarlet, P. G. and P. A. Raviart, "Maximum Principle and Uniform Convergence for the Finite Element Method," to appear in Comp. Meth. Appl. Mech. Eng.
6. Cowper, G. R., G. M. Lindberg, and M. D. Olson, "A shallow shell element of triangular shape," Int. J. Solids Structures 6 (1970), 1133-56.
7. Proc. Conf. Matrix Methods in Structural Mechanics, Wright-Patterson Air Force Base, Dayton, Ohio, Oct. 1965.
8. Courant, Richard, "Variational methods for the solution of problems of equilibrium and vibrations," Bull. Am. Math. Soc. 49 (1943), 1-23.
9. Hall, C. A., "Bicubic interpolation over triangles," J. Math. Mech. 19 (1969), 1-11.
10. Kaper, H. G., G. K. Leaf, A. J. Lindeman, "Applications of finite element methods ...," Report ANL-7925, Argonne National Laboratory, 1972.
11. The NASTRAN Theoretical Manual, R. H. MacNeal, editor. Report NASA SP-221, Nat. Aer. Space Administration, Washington, D. C. 1970.
12. Oden, J. T., Recent Advances in Matrix Methods of Structural Analysis and Design, Univ. of Alabama Press, 1972.
13. SIAM-AMS Symposium II, 1968. Numerical Solution of Field Problems in Continuum Physics, G. Birkhoff and R. S. Varga, eds., American Mathematical Society (1970).
14. Schultz, Martin H., Finite Element Analysis, Prentice-Hall, 1973.
15. Strang, G. and G. Fix, Analysis of the Finite Element Method. (Book to be published by Prentice-Hall.)
16. Synge, J. L., The Hypercircle in Mathematical Physics, Cambridge University Press, 1957.
17. SYNSPADE, 1970. (Bert Hubbard, editor), Academic Press, 1971.
18. Varga, R. S., Functional Analysis and Approximation Theory in Numerical Analysis, SIAM Publications, Philadelphia, 1971.
19. Zenicek, A., "Interpolation polynomials on the triangle," Numer. Math. 15 (1970), 283-96.
20. Zienkiewicz, O. C., The Finite Element Method in Structural and Continuum Mechanics, 2d ed., McGraw-Hill, 1971.

# APPROXIMATION OF STEKLOV EIGENVALUES OF NON-SELFADJOINT SECOND ORDER ELLIPTIC OPERATORS*

J. H. Bramble** and J. E. Osborn***

## 1. Introduction.

In a recent paper [5] the authors develop a general approximation theory for the eigenvalues and corresponding subspaces of generalized eigenvectors for a certain class of compact operators. Because of the applications treated there the general theory was only presented for the case in which the operators were considered on Sobolev spaces $H^s(\Omega)$ with $\Omega$ a bounded domain in $R^N$ with boundary $\partial\Omega$.

In this paper we study the Galerkin method for the approximation of Steklov eigenvalues of a strongly elliptic, non-selfadjoint, second order differential operator $L$. The Steklov eigenvalues are those complex numbers $\lambda$ such that for some non-zero $u$,

$$Lu = 0 \quad \text{in } \Omega$$

and

$$\frac{\partial u}{\partial \nu} = \lambda u \quad \text{on} \quad \partial\Omega,$$

*Sponsored by the United States Army under Contract No. DA-31-124-ARO-D-462.
**Cornell University
***University of Maryland

where $\frac{\partial}{\partial\nu}$ is the conormal derivative (cf. [2]).

For this purpose we restate the main theorems of [5] in a more general setting. Since the proofs may be generalized without effort they will be omitted.

Let $\lambda$ be a Steklov eigenvalue with algebraic multiplicity $m$. There will be $m$ Galerkin eigenvalues, $\lambda_1(h),\ldots,\lambda_m(h)$, which converge to $\lambda$. Our main result, Theorem 3.2, gives a rate of convergence estimate for the convergence of

$$\left(\frac{1}{m}\sum_{j=1}^{m} 1/\lambda_j(h)\right)^{-1}$$

to $\lambda$. This estimate depends on the approximability properties of the space of trial functions and gives the same rate as that which can be established in the self-adjoint case by the methods of Birkhoff, de Boor, Swartz, and Wendroff [3].

The applications studied in [5] all involved a compact operator $T$ which was approximated by a family of operators $\{T_h\}$, each $T_h$ being of the form $P_hT$, where $P_h$ is a projection onto a finite dimensional space. Again, in the Galerkin approximation for the Steklov eigenvalue problem, there arises in a natural way a compact operator $T$ and a family of approximating operators $\{T_h\}$ but, as will be seen in the last section, $T_h$ is not in general of the form $P_hT$, where $P_h$ is a projection. This is of interest in that much work on the approximation of eigenvalues makes strong use of the assumption that $T_h = P_hT$ (at least on the range of $T_h$) for some $P_h$. In particular,

Vaĭnikko [11] obtains results similar to our Theorems 3.3 and 3.4 under such an assumption.

## 2. Preliminaries.

Let $\{H^s\}_{0\le s<+\infty}$ be a family of Hilbert spaces, with norms $\|\cdot\|_{H^s}$, with the properties

a) $H^s$ is a dense subset of $H^0$ for all $s>0$,

b) if $0 \le s_1 < s_2$ then $H^{s_2} \subset H^{s_1}$ and there is a constant $C_{s_1,s_2}$ such that

$$\|\phi\|_{H^{s_1}} \le C_{s_1,s_2} \|\phi\|_{H^{s_2}}$$

for all $\phi \in H^{s_2}$

and

c) if $0 \le s_1 < s_2$ then $I_{s_1,s_2} : H^{s_2} \to H^{s_1}$, the injection of $H^{s_2}$ into $H^{s_1}$, is compact.

Let the inner product in $H^0$ be denoted by $\langle\cdot,\cdot\rangle$. For $\psi \in H^0$ and $s < 0$ we define

$$\|\psi\|_{H^s} = \sup_{\phi \in H^{-s}} \frac{|\langle\psi,\phi\rangle|}{\|\phi\|_{H^{-s}}} .$$

Then we define $H^s$, for $s < 0$, as the completion of $H^0$ with respect to this norm (cf. [1]). Each $H^s$ is a Hilbert space. Now we have a family $\{H^s\}_{-\infty<s<+\infty}$ of

Hilbert spaces with the properties

d) if $-\infty<s_1<s_2<+\infty$ then $H^{s_2} \subset H^{s_1}$ and

$$\|\phi\|_{H^{s_1}} \leq C_{s_1,s_2} \|\phi\|_{H^{s_2}}$$

for all $\phi \in H^{s_2}$

and

e) if $s_1<s_2$ then the injection $I_{s_1,s_2}:H^{s_2} \to H^{s_1}$ is compact.

For any $A:H^s \to H^s$ we define, for $s_1 < s < s_2$,

$$\|A\|_{s_1,s_2} = \sup_{\psi \in H^{s_2}} \frac{\|A\psi\|_{H^{s_1}}}{\|\psi\|_{H^{s_2}}} . \tag{2.1}$$

If $A$ is compact then its spectrum consists of a denumerable set of non-zero complex numbers together with zero. Each non-zero $\mu$ in the spectrum of $A$ is an eigenvalue; zero may or may not be an eigenvalue.

Let $\mu$ be a non-zero eigenvalue of $A$. The least integer $\alpha$ such that $N((\mu-A)^{\alpha}) = N((\mu-A)^{\alpha+1})$, where $N$ denotes the null space, is called the ascent of $\mu-A$. If $A$ is compact $\alpha$ is finite. The integer $m = \dim N((\mu-A)^{\alpha})$ is called the algebraic multiplicity of $\mu$ and is also finite. The vectors in $N((\mu-A)^{\alpha})$ are called generalized eigenvectors of $A$ corresponding to $\mu$. The geometric multiplicity is equal to $\dim N(\mu-A)$ and is less than or equal to the algebraic multiplicity. The two multiplicities are equal if $A$ is self-adjoint.

Let $T^0: H^0 \to H^0$ be compact and $\{T_h\}_{0<h\leq 1}$ be a family of compact operators $T_h: H^0 \to H^0$ such that

$$\lim_{h\to 0} \|T^0 - T_h\|_{0,0} = 0 .$$

We will need the following well-known eigenvalue convergence result (c.f. [6]). If $\mu^1, \mu^2, \ldots$ are the non-zero eigenvalues of $T^0$ ordered by decreasing magnitude, taking account of algebraic multiplicities, then for each $h$ there is an ordering (again counting according to algebraic multiplicites) of the eigenvalues of $T_h, \mu^1(h), \mu^2(h), \ldots$ , such that $\lim_{h\to 0} \mu^j(h) = \mu^j$ for each $j$. For our purpose this can be restated as follows. Let $\mu$ be a non-zero eigenvalue of $T^0$ with algebraic multiplicity $m$ . Then exactly $m$ eigenvalues (counting according to algebraic multiplicities) of $T_h$ converge to $\mu$ . We denote these eigenvalues by $\mu_1(h), \ldots, \mu_m(h)$ . We will denote the space of generalized eigenvectors of $T^0$ corresponding to $\mu$ by $M(\mu)$ and the direct sum of the spaces of generalized eigenvectors of $T_h$ corresponding to $\mu_1(h), \ldots, \mu_m(h)$ by $M_h(\mu)$ . For $h$ sufficiently small, $\dim M_h(\mu) = \dim M(\mu)$ .

## 3. Convergence estimates.

In this section we consider a particular type of compact operator $T^0$ and a family of compact operators $\{T_h\}_{0<h\leq 1}$ which approximates it. We obtain estimates for the rate of convergence of the generalized eigenvectors and

eigenvalues of $T_h$ to the generalized eigenvectors and eigenvalues of $T^0$, respectively.

Let $s_0$ be a fixed non-negative real number and let $T$ be an operator from $H^{-s_0}$ to $H^{-s_0}$. We suppose that there is an $\varepsilon > 0$ such that

$$T: H^s \to H^{s+\varepsilon}$$

as a bounded operator for all $s \geq -s_0$. For $s \geq -s_0$ denote by $T^s$ the restriction of $T$ to $H^s$ considered as a mapping from $H^s$ to $H^s$. Then, since $I_{s+\varepsilon,s}$ is compact, we see that $T^s$ is compact. Now we shall be interested in the eigenvalues and corresponding spaces of generalized eigenvectors of $T^0$. Let $\{T_h\}_{0<h\leq 1}$ be a family of compact operators from $H^0$ to $H^0$ satisfying

$$\lim_{h\to 0} \|T^0 - T_h\|_{0,0} = 0 .$$

Let $\mu$ be a fixed non-zero eigenvalue of $T^0$ with algebraic multiplicity $m$. From Section 2 we see that exactly $m$ eigenvalues of $T_h$, $\mu_1(h),\ldots,\mu_m(h)$, converge to $\mu$ as $h \to 0$.

The following results, Theorem 3.1 and its corollary, and Theorems 3.2, 3.3 and 3.4, are direct generalizations of those given by the authors in [5]. There these results were stated and proved for the case $H^0 = L_2(\Omega)$, where $\Omega$ is a bounded domain in $R^N$, and $H^s$ was taken to be a family of Sobolev spaces. The proofs of [5] remain unchanged in this more general setting and, as mentioned above, will not be repeated here.

We can now state our first result. Throughout this section orthonormal will mean orthonormal in $H^0$.

**Theorem 3.1.** Let $\{u_j\}_{j=1}^m$ be an orthonormal basis for $M(\mu)$ and let $0 \leq s \leq s_0$, $0 \leq s_1$. Then there exist constants $C_s$ and $h_1$ and an orthonormal basis $\{w_j\}_{j=1}^m$ for $M_h(\mu)$ such that

$$\max_{1\leq j\leq m} \|u_j - w_j\|_{H^{-s}} \leq C_s \Big\{ \|T^0 - T_h\|_{-s,s_1} + \|T^0 - T_h\|_{-s,0} \|T^0 - T_h\|_{0,s_1} + \|T^0 - T_h\|^2_{0,s_1} \Big\} \tag{3.1}$$

for all $0 \leq h \leq h_1$.

We may also start with an orthonormal basis in $M_h(\mu)$ and construct a close orthonormal basis for $M(\mu)$. This is made precise in the following:

**Corollary.** For each $h$ with $0 \leq h \leq h_1$ let $\{w_j\}_{j=1}^m$ be an orthonormal basis for $M_h(\mu)$. Then there is an orthonormal basis $\{u_j\}_{j=1}^m$ for $M(\mu)$ such that

$$\max_{1\leq j\leq m} \|u_j - w_j\|_{H^{-s}} \leq C_s \Big\{ \|T^0 - T_h\|_{-s,s_1} + \|T^0 - T_h\|_{-s,0} \|T^0 - T_h\|_{0,s_1} + \|T^0 - T_h\|^2_{0,s_1} \Big\}$$

holds for all $0 \leq s \leq s_0$, where $C_s$ is a constant which does not depend on $h$.

Theorem 1 and its corollary show that the $H^s$-gap between $M(\mu)$ and $M_h(\mu)$, $\hat{\delta}(M(\mu), M_h(\mu))$, is estimated by the right-hand side of (3.1) (cf. [7]).

As before let $\mu_1(h),\ldots,\mu_m(h)$ be the eigenvalues of $T_h$ which converge to $\mu$ (counted according to algebraic multiplicity). Although each of these eigenvalues is close to $\mu$ for small $h$, their arithmetic mean is generally a closer approximation. Thus we define

$$\hat{\mu}(h) = \frac{1}{m} \sum_{j=1}^{m} \mu_j(h) .$$

In the terminology of [7] this is the weighted mean of the $\mu$-group. Our next theorem gives an estimate for $\mu - \hat{\mu}(h)$.

<u>Theorem 3.2</u>. Let $s_1 \geq 0$. Then there exist constants $C$ and $h_1$ such that for $0 < h \leq h_1$,

$$|\mu-\hat{\mu}(h)| \leq C\Big\{\|T^0 - T_h\|_{-s_0,s_1} + \|T^0 - T_h\|_{-s_0,0}\|T^0 - T_h\|_{0,s_1} + \|T^0 - T_h\|_{0,s_1}^2\Big\} .$$

Next we have an estimate for $\mu - \mu_k(h)$ for each $k$.

<u>Theorem 3.3</u>. Let $\alpha$ be the ascent of $\mu - T^0$ and let $s_1 \geq 0$. Then there are constants $C$ and $h_1$ such that for $0 < h \leq h_1$,

$$\max_{1\leq k\leq m} |\mu-\mu_k(h)|^{\alpha} \leq C\Big\{\|T^0 - T_h\|_{-s_0,s_1} + \|T^0 - T_h\|_{-s_0,0}\|T^0 - T_h\|_{0,s_1} + \|T^0 - T_h\|_{0,s_1}^2\Big\} .$$

Our final result in this section gives estimates for the rate of convergence of certain elements of $M_h(\mu)$ to certain elements of $M(\mu)$.

Theorem 3.4. Let $\mu(h)$ be an eigenvalue of $T_h$ such that $\mu(h)\to\mu$ as $h\to 0$ and let $s_1 \geq 0$. Suppose, for each $h$, $w$ is a unit vector satisfying $\left(\mu(h)-T_h\right)^k w = 0$ for some positive integer $k\leq\alpha$. Then for any integer $\ell$ with $k \leq \ell \leq \alpha$ there is a unit vector $u \in M(\mu)$ such that $(\mu-T^0)^\ell u = 0$ and

$$(3.2) \qquad \|u-w\|_{H^0} \leq C\|T^0 - T_h\|_{0,s_1}^{(\ell-k+1)/\alpha} .$$

Note that in the special case $k = \ell = 1$, (3.2) shows that, for small $h$, eigenvectors of $T_h$ will be close to eigenvectors of $T^0$.

## 4. The Steklov eigenvalue problem.

Let $\Omega$ be a bounded domain in N-dimensional Euclidean space with boundary $\partial\Omega$ which we will assume (for convenience) to be of class $C^\infty$. Let $C^\infty(\bar{\Omega})$ be the class of infinitely differentiable complex-valued functions defined on $\bar{\Omega}$, the closure of $\Omega$. For $\phi,\psi \in C^\infty(\bar{\Omega})$ we define the $L_2$-inner product by

$$(\phi,\psi) = \int_\Omega \phi\,\bar{\psi}\,dx$$

and the corresponding norm by

$$\|\phi\|_0 = \left(\int_\Omega |\phi|^2\,dx\right)^{1/2} .$$

For $\phi \in C^\infty(\bar{\Omega})$ and $j$ a non-negative integer we define the norm

$$\|\phi\|_j = \Big( \sum_{|\alpha| \leq j} \|D^\alpha \phi\|_0^2 \Big)^{1/2}$$

where $\alpha = (\alpha_1, \ldots, \alpha_N)$ is a multi-index,

$$D^\alpha = \frac{\partial^{\alpha_1}}{\partial x_1^{\alpha_1}} \cdots \frac{\partial^{\alpha_N}}{\partial x_N^{\alpha_N}} ,$$

and $|\alpha| = \alpha_1 + \ldots + \alpha_N$ . The completion of $C^\infty(\bar{\Omega})$ with respect to $\|\cdot\|_j$ will be denoted by $H^j(\Omega)$ . Note that $H^0(\Omega) = L_2(\Omega)$ . This is the Sobolev space of order $j$ and is a Hilbert space. For any positive real $s$ we define $H^s(\Omega)$ by interpolation between successive integers following Definition 2.1 of [8]. For $\phi \in L_2(\Omega)$ and $s < 0$ define the norm

$$\|\phi\|_s = \sup_{\psi \in H^{-s}(\Omega)} \frac{|(\phi, \psi)|}{\|\psi\|_{-s}} .$$

$H^s(\Omega)$ , for $s < 0$ , is defined as the completion of $L_2(\Omega)$ with respect to $\|\cdot\|_s$ (cf. [1]). Now $H^s(\Omega)$ is a Hilbert space for each real $s$ .

We shall also need the spaces $H^s(\partial\Omega)$ , the Sobolev spaces of order $s$ on the boundary. They can be defined as follows. Let $\Delta_{\partial\Omega}$ denote the Laplace-Beltrami operator on $\partial\Omega$ . This operator has an increasing sequence of positive eigenvalues $\{\lambda_j\}$ which tend to infinity and corresponding eigenfunctions $\{w_j\}$ with $w_j \in C^\infty(\partial\Omega)$ . The $w_j$ are orthonormal with respect to the

$L_2$ -inner product on $\partial\Omega$ ,

$$<\phi,\psi> = \int_{\partial\Omega} \phi\bar{\psi}d\sigma .$$

Here $\sigma$ is the surface measure. We use this notation since $H^0$ will be taken in the sequel to be $L_2(\partial\Omega)$ . For any real $s$ , $H^s(\partial\Omega)$ is defined as the completion of $C^\infty(\partial\Omega)$ , the space of infinitely differentiable complex-valued functions on $\partial\Omega$ , with respect to the norm

$$|u|_s = \left( \sum_{j=1}^{\infty} \lambda_j^s |<u,w_j>|^2 \right)^{1/2} .$$

$H^s(\partial\Omega)$ is a Hilbert space for each real $s$ . Note that $H^0(\partial\Omega) = L_2(\partial\Omega)$ . Also for $s < 0$ we have

$$(4.1) \qquad |u|_s = \sup_{\phi\in H^{-s}(\partial\Omega)} \frac{|<u,\phi>|}{|\phi|_{-s}} .$$

It is clear that the spaces $H^s(\partial\Omega)$ , $s \geq 0$ , satisfy conditions a), b) and c) in Section 2 and that $H^s(\partial\Omega)$ for $s < 0$ is constructed from $H^0$ and $H^{-s}$ as in Section 2.

Let $L$ be a second order partial differential operator given by

$$L\phi = -\sum_{i,j=1}^{N} \frac{\partial}{\partial x_i}(a_{ij}\frac{\partial\phi}{\partial x_j}) + \sum_{i=1}^{N} b_i \frac{\partial\phi}{\partial x_1} + c\phi$$

where $a_{ij}$, $b_i$ and $c$ are in $C^\infty(\bar{\Omega})$ and $a_{ij} = a_{ji}$ . We assume that $L$ is uniformly strongly elliptic; i.e.

there is a positive constant $a_0$ such that

$$\operatorname{Re} \sum_{i,j=1}^{N} a_{ij}(x)\xi_i\xi_j \geq a_0 \sum_{i=1}^{N} \xi_i^2$$

for all real $\xi_1,\ldots\xi_N$ and all $x \in \Omega$ . The sesquilinear form on $H^1(\Omega) \times H^1(\Omega)$ associated with L is given by

$$B(\phi,\psi) = \sum_{i,j=1}^{N} \int_\Omega a_{ij} \frac{\partial\phi}{\partial x_i} \overline{\frac{\partial\psi}{\partial x_j}} \, dx + \sum_{i=1}^{N} \int_\Omega b_i \frac{\partial\phi}{\partial x_i} \bar{\psi} \, dx +$$

$$+ \int_\Omega c \, \phi \, \bar{\psi} \, dx \; .$$

We shall assume that B is coercive over $H^1(\Omega)$; i.e. there is a positive constant C such that

$$\operatorname{Re} B(\phi,\phi) \geq C\|\phi\|_1^2 \tag{4.2}$$

for all $\phi \in H^1(\Omega)$ . Since the coefficients in L are bounded it is clear that there is another constant C such that

$$|B(\phi,\psi)| \leq C\|\phi\|_1\|\psi\|_1 \tag{4.3}$$

for all $\phi, \psi \in H^1(\Omega)$ .

We consider now the Steklov eigenvalue problem for L . The complex number $\lambda$ is called a Steklov eigenvalue of L if there is a function $u \in H^2(\Omega)$ which is not identically zero such that

$$L\,u = 0 \quad \text{in} \quad \Omega$$

and

$$\frac{\partial u}{\partial \nu} = \lambda u \quad \text{on} \quad \Omega \quad ,$$

where $\frac{\partial}{\partial \nu} = \sum_{i,j=1}^{N} a_{ij}\, n_j \frac{\partial}{\partial x_i}$ and $n_j$ is the jth component of the outward unit normal on $\partial\Omega$ . This problem has the following equivalent weak formulation:
$\lambda$ is a Steklov eigenvalue if

(4.4) $$B(u,\phi) = \lambda\langle u,\phi\rangle$$

for some non-zero $u \in H^1(\Omega)$ and all $\phi \in H^1(\Omega)$ .

Next we wish to identify the Steklov eigenvalues of $L$ with the eigenvalues of a certain compact operator. To this end we consider the Neumann problem associated with $L$ ; i.e. the problem of finding $u \in H^1(\Omega)$ such that

(4.5) $$B(u,\phi) = \langle g,\phi\rangle$$

for all $\phi \in H^1(\Omega)$ where $g$ is a given function in $L_2(\partial\Omega)$. Since $B$ is coercive this prolem is uniquely solvable. For $g \in L_2(\partial\Omega)$ the solution $u$ will in fact be in $H^{3/2}(\Omega)$ . The solution operator $A: L_2(\partial\Omega) \to H^{3/2}(\Omega)$ is defined by $Ag = u$ . Now we define $T$ by $Tg = (Ag)'$ , where the prime denotes the restriction to $\partial\Omega$ . It follows from the theory of trace (cf.[8]) that $(Ag)' \in H^1(\partial\Omega)$ for each $g \in L_2(\partial\Omega)$ . Hence $T: L_2(\partial\Omega) \to H^1(\partial\Omega)$ .

The following estimate was proved in [10] (cf. also [9]). For any real $s$ there is a constant $C_s$ such that

$$|Tg|_{s+1} \leq C_s |g|_s \tag{4.6}$$

for all $g \in H^s(\partial\Omega) \cap L_2(\partial\Omega)$. Inequality (4.6) allows us to extend $T$ from $L_2(\partial\Omega)$ to $H^s(\partial\Omega)$ for any $s < 0$. Now let $s_0 \geq 0$ and consider $T$ as being defined on $H^{-s_0}(\partial\Omega)$. Hence $T$ is an operator of the general type discussed in Section 3 with $\varepsilon = 1$. We write $T^s$ for the restriction of $T$ to $H^s(\partial\Omega)$ for $s \geq -s_0$.

Now suppose $\mu$ is an eigenvalue of $T^0$; i.e. there is a nonzero $w$ in $L_2(\partial\Omega)$ such that $T^0 w = \mu w$. Then $\mu \neq 0$ and

$$\begin{aligned} B(Aw,\phi) &= \langle w,\phi\rangle \\ &= \frac{1}{\mu}\langle T^0 w,\phi\rangle \\ &= \frac{1}{\mu}\langle Aw,\phi\rangle \end{aligned}$$

for all $\phi \in H^1(\Omega)$. Thus $1/\mu$ is an eigenvalue of (4.4) with $Aw$ the corresponding eigenvector. Conversely suppose $B(u,\phi) = \lambda\langle u,\phi\rangle$ with $u \neq 0$ for all $\phi \in H^1(\partial\Omega)$. Then $\lambda \neq 0$ and

$$T^0 u' = (Au')' = \frac{1}{\lambda} u' \; ;$$

i.e. $1/\lambda$ is an eigenvalue of $T^0$ with $u'$ the corresponding eigenvector. Hence the Steklov eigenvalues are the reciprocals of the eigenvalues of $T^0$.

In order to define approximate eigenvalues we must discuss certain classes of finite dimensional spaces of trial functions. Let $\{S_h\}_{0 < h \leq 1}$ be a one parameter family of finite dimensional vector spaces. For given integers $k$ and $r$ with $0 \leq k \leq r$ we say that $\{S_h\}_{0 < h \leq 1}$ is of class $\mathcal{S}_{k,r}$ if $S_h \subset H^k(\Omega)$ for each $h$ and there is a constant $C$ independent of $h$ such that for any $v \in H^t(\Omega)$ with $k \leq t \leq r$ ,

$$\text{(4.7)} \qquad \inf_{\chi \in S_h} \sum_{j=0}^{k} h^j \|v-\chi\|_j \leq C\, h^t \|v\|_t .$$

It was proved in [4] that if (4.7) holds for $t = r$ it then holds for $k \leq t \leq r$ .

Let $\{S_h\}_{0 < h \leq 1}$ belong to $\mathcal{S}_{1,r}$ . For $0 < h \leq 1$ we define the approximate solution of the Neumann problem as the unique $u_h \in S_h$ such that

$$\text{(4.8)} \qquad B(u_h, \chi) = \langle g, \chi \rangle$$

for all $\chi \in S_h$ . Again, since $B$ is coercive, this problem has a unique solution for each $h$ . We define the approximate solution operator $A_h : L_2(\partial\Omega) \to S_h$ by $A_h g = u_h$. Letting $\delta S_h$ denote the functions defined on $\partial\Omega$ which are restrictions to $\partial\Omega$ of functions in $S_h$ we define $T_h : L_2(\partial\Omega) \to \delta S_h \subset L_2(\partial\Omega)$ by $T_h g = (A_h g)'$ .

We want now to consider Galerkin eigenvalues of the Steklov problem; i.e. complex numbers $\lambda(h)$ which satisfy

(4.9) $$B(w,\chi) = \lambda(h) \langle w,\chi\rangle$$

for some non-zero $w \in S_h$ and all $\chi \in S_h$. These eigenvalues will be considered as approximations to the Steklov eigenvalues. If (4.9) holds then $\lambda(h) \neq 0$ and

$$T_h w' = (A_h w')' = \frac{1}{\lambda(h)} w' ;$$

i.e. $1/\lambda(h)$ is an eigenvalue of $T_h$ and $w' \in \delta S_h$ is a corresponding eigenvector. Conversely, if $T_h w = \mu(h) w$ with $\mu(h) \neq 0$ then

$$B(A_h w,\chi) = \langle w,\chi\rangle = \frac{1}{\mu(h)} \langle T_h w,\chi\rangle = \frac{1}{\mu(h)} \langle A_h w,\chi\rangle$$

for all $\chi \in S_h$; i.e. $1/\mu(h)$ is a Galerkin eigenvalue and $A_h w$ is a corresponding eigenvector. Thus we may compare the eigenvalues of (4.4) with the Galerkin approximations by comparing the eigenvalues of $T^0$ with those of $T_h$. In order to apply the theorems of Section 3 to obtain eigenvalue estimates we need two further estimates from [10].

For any real $s$ there is a constant $C_s$ such that

(4.10) $$\|Ag\|_{s+3/2} \leq C_s |g|_s$$

for all $g \in H^s(\partial\Omega) \cap L_2(\partial\Omega)$. Corresponding to the adjoint problem we have the solution operator $A^*: L_2(\partial\Omega) \to H^{3/2}(\partial\Omega)$ which is characterized by

$$B(\phi, A^*g) = \langle \phi, g\rangle$$

for all $\phi \in H^1(\Omega)$ . For $A^*$ we have the same type of estimate, namely

$$\|A^*g\|_{s+3/2} \leq C_s^* \, |g|_s \tag{4.11}$$

for all $g \in H^s(\partial\Omega) \cap L_2(\partial\Omega)$ .

We are now ready to derive the required estimates for $T^0 - T_h$ . Let $g \in H^{t-1/2}(\partial\Omega)$ where $1/2 \leq t \leq r-1$ . Using (4.2), (4.5), (4.8) and (4.3) we have

$$\begin{aligned}\|(A-A_h)g\|_1^2 &\leq C\left|B\Big((A-A_h)g,\ (A-A_h)g\Big)\right| \\ &= C\left|B\Big((A-A_h)g, Ag-\chi_1\Big)\right| \\ &\leq C\|(A-A_h)g\|_1\|Ag-\chi_1\|_1\end{aligned}$$

for any $\chi_1 \in S_h$ . Hence

$$\|(A-A_h)g\|_1 \leq C \inf_{\chi_1 \in S_h} \|Ag-\chi_1\|_1 \ . \tag{4.12}$$

Now let $\psi \in H^{s-1/2}(\partial\Omega)$ be arbitrary where $1/2 \leq s \leq r-1$ . Then

$$\begin{aligned}\left|\Big\langle (T^0 - T_h)g, \psi \Big\rangle\right| &= \left|B\Big((A-A_h)g, A^*\psi\Big)\right| \\ &= \left|B\Big((A-A_h)g, A^*\psi-\chi_2\Big)\right| \\ &\leq C\|(A-A_h)g\|_1\|A^*\psi-\chi_2\|_1\end{aligned} \tag{4.13}$$

for any $\chi_2 \in S_h$ . From (4.12), (4.13) and the approximability assumption (4.7) we thus have

$$\left|\Big\langle (T^0 - T_h)g, \psi \Big\rangle\right| \leq Ch^{s+t}\|Ag\|_{t+1}\|A^*\psi\|_{s+1} \ .$$

Combining this result with (4.10) and (4.11) we obtain

$$|\langle (T^0 - T_h)g, \psi \rangle| \leq Ch^{s+t} |g|_{t-1/2} |\psi|_{s-1/2} .$$

Taking $H^s = H^s(\partial\Omega)$ is follows immediately from this inequality, (4.1) and (2.1) that

$$\|T^0 - T_h\|_{1/2-s,t-1/2} \leq Ch^{s+t}$$

for $1/2 \leq s$, $t \leq r-1$ .

It remains only to apply the results of Section 3. Notice first, letting $s = t = 1/2$ , that $\|T^0 - T_h\|_{0,0} \leq Ch$ . Thus all of the results of Section 3 hold for this problem. Now taking $s = t = r-1$ we have

$$\|T^0 - T_h\|_{3/2-r,r-3/2} \leq Ch^{2r-2}$$

taking $s = 1/2$ , $t = r-1$ we have

$$\|T^0 - T_h\|_{0,r-3/2} \leq Ch^{r-1/2}$$

and taking $s = r-1$ , $t = 1/2$ we have

$$\|T^0 - T_h\|_{3/2-r,0} \leq Ch^{r-1/2} .$$

Let $\lambda$ be an eigenvalue with algebraic multiplicity $m$ of (4.4) and let $\mu = 1/\lambda$ . The eigenvalues $\mu_1(h),\ldots,\mu_m(h)$ which converge to $\mu$ are the reciprocals of the eigenvalues $\lambda_1(h),\ldots,\lambda_m(h)$ of (4.9). From Theorem 2 we thus have the estimate

$$\left|\lambda - \left(\frac{1}{m}\sum_{j=1}^{m} 1/\lambda_j(h)\right)^{-1}\right| \leq Ch^{2r-2} .$$

Using Theorem 3.1 and its corollary we see that the $L_2(\partial\Omega)$-gap between $M(1/\lambda)$ and $M_h(1/\lambda)$ , $\hat{\delta}\left(M(1/\lambda), M_h(1/\lambda)\right)$ , satisfies

$$\hat{\delta}\left(M(1/\lambda), M_h(1/\lambda)\right) \leq Ch^{r-1/2} .$$

Further, from Theorem 3.3, we obtain the estimate

$$\max_{1\leq k\leq m} |\lambda - \lambda_k| \leq Ch^{(2r-2)/\alpha} ,$$

where $\alpha$ is the ascent of $1/\lambda - T^0$ . Finally for $w$ an eigenvector of $T_h$ in $M_h(1/\lambda)$ we see, using Theorem 3.4, that there is an eigenvector $u$ of $T^0$ in $M(1/\lambda)$ such that

$$|u-w|_0 \leq Ch^{(r-1/2)/\alpha} .$$

## 5. On the structure of $T_h$.

By the term projection onto $\delta S_h$ we shall mean an operator $P_h$ (not necessarily bounded) defined on a linear manifold in $L_2(\partial\Omega)$ , with range equal to $\delta S_h$ and such that $P_h^2 = P_h$ .

We first prove a lemma which gives a necessary condition, when the form $B$ is Hermitian, that $T_h = P_h T^0$ for some projection $P_h$ .

Lemma. Let $B$ be Hermitian and $\{S_h\}_{0<h\leq 1} \in S_{2,r}$ . Suppose there is a projection $P_h$ onto $\delta S_h$ such that $T_h = P_h T^0$ . Let $\tilde{u}_h \in S_h$ and $u$ satisfy

$$L\,u = 0 \quad \text{in } \Omega$$

and

$$u = \tilde{u}_h \quad \text{on } \partial\Omega.$$

Then $u \in S_h$ .

Proof: Set $u_h = A_h \frac{\partial u}{\partial \nu}$ . Then

(5.1) $$u_h' = T_h \frac{\partial u}{\partial \nu} = P_h T^0 \frac{\partial u}{\partial \nu} = P_h u' = P_h \tilde{u}_h' = \tilde{u}_h' = u' .$$

From (4.8) we have

$$B(u_h, \chi) = \langle \frac{\partial u}{\partial \nu}, \chi \rangle$$

for all $\chi \in S_h$ and from (4.5),

$$B(u, \phi) = \langle \frac{\partial u}{\partial \nu}, \phi \rangle$$

for all $\phi \in H^1(\Omega)$ . Thus $B(u_h - u, \chi) = 0$ for all $\chi \in S_h$ and hence

(5.2) $$B(u_h, u_h - u) = \overline{B(u_h - u, u_h)} = 0$$

Now $L u = 0$ and from (5.1) we have $u_h - u = 0$ on $\partial\Omega$ . By integration by parts,

(5.3) $$B(u, u_h - u) = 0 .$$

Thus from (4.2), (5.2) and (5.3) we see that

$$\|u_h - u\|_1^2 \leq C\, B(u_h - u, u_h - u) = 0$$

which implies that $u = u_h \in S_h$ .

Using this result it is easy to see, for many families of spaces $\{S_h\}_{0<h\leq 1}$ which are used in the finite element method, that in general there is no projection $P_h$ onto $\delta S_h$ such that $T_h = P_h T^0$ . More

precisely we have:

Theorem 5.1. Let $B$ be Hermitian, $\{S_h\}_{0<h\leq 1} \in \mathcal{S}_{2,r}$ and suppose that $S_h$ consists of piecewise polynomials and that there is at least one $\chi_0 \in S_h$ such that $\chi_0'$ is not the restriction to $\partial\Omega$ of a polynomial. Then there does not exist a projection onto $\delta S_h$ such that $T_h = P_h T^0$.

Proof: Let $\chi_0 \in S_h$ be such that $\chi_0' \neq p'$ for all polynomials $p$ and let $u$ satisfy

$$Lu = 0 \quad \text{in } \Omega$$

and

$$u = \chi_0 \quad \text{on } \partial\Omega.$$

Suppose now that there exists a projection $P_h$ with $T_h = P_h T^0$. Then by the lemma, $u \in S_h$. Since the coefficients of $L$ are of class $C^\infty$ we see by elliptic regularity theory that $u \in C^\infty(\Omega)$. Thus $u$ must in fact be a polynomial. But this implies that $\chi_0'$ is the restriction of a polynomial to $\partial\Omega$ which is a contradiction.

REFERENCES


1. Berezanskii, Ju. M., Expansion in Eigenfunctions of Self Adjoint Operators, Naukova Dumka, Kiev, 1965; English transl., Transl. Math. Monographs 17, Amer. Math. Soc., Providence, R.I., 1968.
2. Bergman, S., and M. Schiffer, Kernel Functions and Elliptic Differential Equations in Mathematical Physics, Academic Press, New York. 1953.
3. Birkhoff, G., C. de Boor, B. Swartz, and B. Wendroff, "Rayleigh-Ritz approximation by piecewise cubic polynomials," SIAM J. Numer. Anal. 3 (1966), 188-203.
4. Bramble, J.H., T. Dupont and V. Thomée, "Projection methods for Dirichlet's problem in approximating polygonal domains with boundary corrections," Math.

Comp., to appear.

5. Bramble, J. H. and J. E. Osborn, "Rate of convergence estimates for non-selfadjoint eigenvalue approximations," MRC Tech. Summary Report #1232, University of Wisconsin, 1972.
6. Dunford, N. and J.T. Schwartz, Linear Operators, Part II, Spectral Theory, Interscience, New York, 1963.
7. Kato, T., Perturbation Theory for Linear Operators, Springer-Verlag, New York, 1966.
8. Lions, J. L. and E. Magenes, Problèmes aux Limites non Homogènes et Applications, Vol. I, Travaux et Recherches Mathématiques, No. 17. Dunod, Paris, 1968.
9. Schechter, M., "On $L^p$ estimates and regularity I," Amer. J. Math. 85 (1963), 1-13.
10. Schechter, M., "On $L^p$ estimates and regularity II," Math. Scand. 13(1963), 47-69.
11. Vaĭnikko, G. M., "On the speed of convergence of approximate methods in the eigenvalue problem," Zh. Vych. Mat. 7(1967), 977-987; English transl., U.S.S.R. Comput. Math. and Math. Phys. 7 (1967), 18-32.

# THE COMBINED EFFECT OF CURVED BOUNDARIES AND NUMERICAL INTEGRATION IN ISOPARAMETRIC FINITE ELEMENT METHODS

by

P. G. Ciarlet* and P.-A. Raviart**

*Abstract*

Consider the model problem

$$-\sum_{i,j=1}^{n} \frac{\partial}{\partial x_j}\left(a_{ij}\frac{\partial u}{\partial x_i}\right) = f \quad \text{in } \Omega \,, \qquad u = 0 \quad \text{on } \Gamma \,,$$

where $\Omega$ is a bounded open subset of $R^n$ with a *curved* boundary $\Gamma$. The finite element method with *isoparametric* finite elements is applied to this problem, with *curved* finite elements along the boundary, in connection with a *numerical quadrature scheme* which is used to compute the coefficients of the resulting linear system. However, except in some very special cases, the interior $\Omega_h$ of the union of the finite elements is not equal to $\Omega$ (although the boundary of $\Omega_h$ is very close to $\Gamma$), and thus,

*Service de Mathématiques, Laboratoire Central des Ponts et Chaussées, 58, Boulevard Lefebvre, 75732 PARIS CEDEX 15, France.

**Analyse Numérique, Tour 55, Université de Paris VI, 11, Quai Saint-Bernard, 75230 PARIS CEDEX 05, France.

taking also numerical integration into account, we obtain an approximate solution $u_h$ in the space $H_0^1(\Omega_h)$.

For several types of isoparametric finite elements (of simplicial and quadrilateral type) commonly used by the Engineers, we prove asymptotic error estimates of the form

$$\|\tilde{u}-u_h\|_{H^1(\Omega_h)} = O(h^{r+1}) \quad ,$$

$$\|\tilde{u}-u_h\|_{L^2(\Omega_h)} = O(h^{r+2}) \quad ,$$

where $r = 0, 1$ or $2$ depending upon the type of finite elements which are used, $h$ is the greatest diameter of the finite elements, and either $\tilde{u}$ is an extension of the exact solution $u$ to the set $\Omega_h$ when $\Omega_h \not\subset \Omega$, or $\tilde{u} = u$ if $\Omega_h \subset \Omega$.

## 1. Introduction and preliminaries

Let there be given a bounded open subset $\Omega$ of $R^n$, with boundary $\Gamma$. We consider the *model problem*

$$(1.1) \qquad Au \equiv -\sum_{i,j=1}^{n} \frac{\partial}{\partial x_j}\left(a_{ij}\frac{\partial u}{\partial x_i}\right) = f \quad \text{in } \Omega,$$

$$(1.2) \qquad u = 0 \quad \text{on } \Gamma,$$

where the functions $a_{ij}$ and $f$ are continuous over $\bar{\Omega}$. At this stage, we shall not make further assumptions about the operator $A$ itself (additional smoothness and ellipticity assumptions will be made in §§ 4 and 5). Let us remark at this point that we could add a term of the form $+au$ in the definition (1.1) of the operator $A$, with $a \in C^0(\bar{\Omega})$ and $a \geq 0$ over $\bar{\Omega}$ (for instance). All that follows applies equally well to this case, with a straightforward supplementary analysis.

If we approximate the solution $u$ of (1.1)-(1.2) by a finite element method, we first construct a finite number of finite elements (how these finite elements are actually constructed and pieced together will be explained in §4), and we denote by $\Omega_h$ the interior of their union.

*Except in some special cases* (such as when $\bar{\Omega}$ is a polyhedral domain), *the boundary* $\Gamma_h$ *of the set* $\Omega_h$ *does not coincide in general with the boundary* $\Gamma$ *of* $\Omega$ (cf. Fig. 3 and 4 in § 4).

Then a finite-dimensional space $V_h$ is constructed which satisfies

$$V_h \subset H_0^1(\Omega_h) \quad . \tag{1.3}$$

Since the approximate solution $u_h$ will be sought in the space $V_h$, the boundary condition (1.2) is thus "approximated" by the boundary condition $u_h = 0$ on $\Gamma_h$.

Next, the natural way of defining the discrete problem associated with the space $V_h$ consists in finding a function $\tilde{u}_h \in V_h$ such that

$$\int_{\Omega_h} \sum_{i,j=1}^{n} \tilde{a}_{ij} \frac{\partial \tilde{u}_h}{\partial x_i} \frac{\partial v_h}{\partial x_j} \, dx = \int_{\Omega_h} \tilde{f} v_h \, dx \tag{1.4}$$

$$\text{for all } v_h \in V_h \quad ,$$

where the functions $\tilde{a}_{ij}$ and $\tilde{f}$ are extensions of the functions $a_{ij}$ and $f$ to the set $\Omega_h$, when $\Omega_h \not\subset \Omega$. In other words, what we are really approximating is a solution of

$$(1.5) \qquad \tilde{A}_h u \equiv - \sum_{i,j=1}^{n} \frac{\partial}{\partial x_j}\left(\tilde{a}_{ij} \frac{\partial u}{\partial x_i}\right) = \tilde{f} \quad \text{in} \quad \Omega_h,$$

$$(1.6) \qquad u = 0 \quad \text{on} \quad \Gamma_h .$$

Since we will of course consider a *family* of discrete problems, with the defining paramater $h$ approaching zero in the limit, we will make the natural assumption that there exists a bounded open subset $\tilde{\Omega}$ of $R^n$ such that

$$(1.7) \qquad \Omega \subset \tilde{\Omega} \quad \text{and} \quad \Omega_h \subset \tilde{\Omega} \quad \text{for all } h.$$

By the Tietze-Urysohn theorem, the functions $a_{ij}$ and $f$ have (non unique) continuous extensions to $R^n$ and we may use any one of these in (1.4). That these extensions are not unique is irrelevant since as will be shown later on, these extensions play only an "intermediary" role: the results which we shall eventually obtain *do not* depend upon them. So let us assume that one choice for these extensions has been made once and for all. Then the bilinear form $\tilde{a}_h(.,.)$ defined over $H^1(\Omega_h)\times H^1(\Omega_h)$ by

$$(1.8) \qquad \tilde{a}_h(u,v) = \int_{\Omega_h} \sum_{i,j=1}^{n} \tilde{a}_{ij} \frac{\partial u}{\partial x_i} \frac{\partial v}{\partial x_j}\, dx$$

is continuous and moreover by (1.7), there exists a constant $\tilde{M}$ independent of $h$ such that

(1.9) $$|\tilde{a}_h(u,v)| \leq \tilde{M}\|u\|_{H^1(\Omega_h)} \|v\|_{H^1(\Omega_h)}$$

for all $u,v \in H^1(\Omega_h)$.

Since it is either too costly or simply impossible to exactly evaluate the integrals which appear in (1.4) (whether $\Omega = \Omega_h$ or not !), we must now take into account the fact that *approximate integration* is used for their computation. As will be explained in details in § 4, this amounts to replacing the discrete problem (1.4) by the following: *Find* $u_h \in V_h$ *such that*

(1.10) $$a_h(u_h,v_h) = (f_h,v_h)_h \quad \text{for all} \quad v_h \in V_h \quad ,$$

where $a_h(.,.)$ is a bilinear form which "approximates" the bilinear form $\tilde{a}_h(.,.)$ of (1.8) over $V_h \times V_h$ (cf. (4.3)), and similarly, if we denote by $(.,.)_h$ the pairing between the space $V_h$ and its dual $V'_h$ (with the convention that if $g \in H^{-1}(\Omega_h)$, we shall still denote by $g$ the corresponding element in $V'_h$), the element $f_h \in V'_h$ "approximates" the linear form $v \in V_h \rightarrow \int_{\Omega_h} \tilde{f}v\, dx$ (cf. 4.4)).

We now evaluate the error. Our result generalizes a result of STRANG [22, Lemma IV.1 and Theorem IV], where the effect of numerical integration was analyzed (however, the possibility that $\Omega_h$ may differ from $\Omega$ was not considered), and, likewise, it generalizes a result of STRANG & BERGER [23] who have analyzed the change in domain in a special case, however without taking numerical integration into account (see also [2]).

**Theorem 1.** *Let there be given a family of discrete problems* (1.10) *such that the assumption of* (1.7) *holds and such that there exists a constant* $\beta > 0$ *independent of* $h$ *with*

$$a_h(v_h, v_h) \geq \beta \|v_h\|^2_{H^1(\Omega_h)} \quad \text{for all } v_h \in V_h \text{ and all } h. \tag{1.11}$$

*Then the unique solution* $u_h$ *of the discrete problem* (1.10) *satisfies*

$$\|\tilde{u}-u_h\|_{H^1(\Omega_h)} \leq C \left( \inf_{v_h \in V_h} \{ \|\tilde{u}-v_h\|_{H^1(\Omega_h)} + \sup_{w_h \in V_h} \frac{|\tilde{a}_h(v_h,w_h)-a_h(v_h,w_h)|}{\|w_h\|_{H^1(\Omega_h)}} \} + \sup_{w_h \in V_h} \frac{|(\tilde{A}_h\tilde{u}-f_h, w_h)_h|}{\|w_h\|_{H^1(\Omega_h)}} \right), \tag{1.12}$$

*where*

$\tilde{u}$ = *any function in* $H^1(\tilde{\Omega})$,

$C$ = *a constant independent of* $\tilde{u}$ *and* $h$.

*Proof.* First, the assumption of (1.11) insures that the discrete problem (1.10) has a unique solution $u_h$ for all $h$. Next, for a fixed $\tilde{u} \in H^1(\Omega)$, the linear form $w \in H^1_0(\Omega_h) \rightarrow \tilde{a}_h(\tilde{u},w)$ is continuous by (1.9), and the element in $H^{-1}(\Omega_h)$ which it defines is precisely $\tilde{A}_h\tilde{u}$, so that we have in particular

$$(1.13) \qquad \tilde{a}_h(\tilde{u},w_h) = (\tilde{A}_h\tilde{u},w_h)_h \quad \text{for all} \quad w_h \in V_h \;.$$

Given any function $w_h \in V_h$, let $w_h = u_h - v_h$. Then, using (1.9), (1.10), (1.11) and (1.13), we obtain

$$\begin{aligned}
\beta\|w_h\|^2_{H^1(\Omega_h)} &\le a_h(w_h,w_h) \\
&= \tilde{a}_h(\tilde{u}-v_h,w_h) + \tilde{a}_h(v_h,w_h) - a_h(v_h,w_h) - (\tilde{A}_h\tilde{u}-f_h,w_h)_h \\
&\le \tilde{M}\|\tilde{u}-v_h\|_{H^1(\Omega_h)}\|w_h\|_{H^1(\Omega_h)} + |\tilde{a}_h(v_h,w_h) - a_h(v_h,w_h)| \\
&\qquad + |(\tilde{A}_h\tilde{u}-f_h,w_h)_h| \;,
\end{aligned}$$

and thus

$$\begin{aligned}
\|\tilde{u}-u_h\|_{H^1(\Omega_h)} &\le \left(\frac{\tilde{M}}{\beta}+1\right)\|\tilde{u}-v_h\|_{H^1(\Omega_h)} \\
&\quad + \frac{1}{\beta}\sup_{w_h\in V_h}\frac{|\tilde{a}_h(v_h,w_h)-a_h(v_h,w_h)|}{\|w_h\|_{H^1(\Omega_h)}} \\
&\quad + \frac{1}{\beta}\sup_{w_h\in V_h}\frac{|(\tilde{A}_h\tilde{u}-f_h,w_h)_h|}{\|w_h\|_{H^1(\Omega_h)}} \;,
\end{aligned}$$

from which the inequality of (1.12) follows, since $v_h$ is an arbitrary function in the space $V_h$. ∎

Several comments about this result are now in order. At the same time, we briefly indicate the content of the subsequent §§.

As we pointed at the beginning of this §, no ellipticity assumption has yet been made upon the "original" operator $A$, but such an assumption is indeed implicitly present here since (1.11) is exactly a "uniform" ellipticity assumption (i.e., independent of h) made upon the discrete problems, and that such an assumption is valid will follow from the ellipticity of the operator $A$ (see proof of Theorem 9).

It was clear *a priori* that to estimate the error one had to compare the solution $u_h$ of (1.10) with a function $\tilde{u}$ defined over $\Omega_h$ for all $h$. Since this is not true of the solution $u$ of (1.1)-(1.2) when $\Omega_h \not\subset \Omega$, a natural candidate for the function $\tilde{u}$ of Theorem 1 will be of course any *extension* $\tilde{u} \in H^1(\tilde{\Omega})$ *of the function* $u$. Notice that in this case, $\tilde{A}_h \tilde{u} = f$ in $\Omega$ so that $\tilde{A}_h \tilde{u}$ is an extension of $f$ to $\Omega_h$.

The term $\inf_{v_h \in V_h} \|\tilde{u}-v_h\|_{H^1(\Omega_h)}$ appearing in (1.12) is the generalization of the usual *"approximation theory" term:* indeed, if $\Omega = \Omega_h$, $\tilde{u} = u$ and no numerical integration is used, we obtain the well-known upper bound for the error. We will evaluate this term in §2 (Theorem 2) for subspaces $V_h$ corresponding to (possibly curved) *isoparametric finite elements* of either *simplicial* (*triangular* if $n = 2$) or *quadrilateral* type, commonly used by the Engineers.

Each one of the last two terms in the right-hand side of (1.12) represent the contribution to the error of the *approximate integration* used respectively on the left-hand sides and right-hand sides of (1.10). The basic techniques for estimating this type of error (in particular for curved finite elements) are developed in §3 (Theorems 4, 5, 6, 7 and 8) where we have followed, and extended, crucial ideas due to G. STRANG.[1]

The beginning of §4 is devoted to the complete description of the discrete problem (1.10) associated with the model problem (1.1)-(1.2), and then the various estimates of §§2 and 3 are "pieced together" so as to obtain (cf. Theorem 9) an asymptotic estimate of the form

$$\|\tilde{u}-u_h\|_{H^1(\Omega_h)} = O(h^{r+1}) \quad , \tag{1.14}$$

for some integer $r \geq 0$. For analogous results for one-dimensional problems or multidimensional problems on rectangular regions, see HERBOLD [14] and HERBOLD and VARGA [15].

In §5, we then obtain (cf. Theorems 10 and 11) an asymptotic estimate of the form

$$\|\tilde{u}-u_h\|_{L^2(\Omega_h)} = O(h^{r+2}) \quad , \tag{1.15}$$

using the now well-known *duality approach*, which yields an increase of one in the order of accuracy.

Let us now describe some of the notations used in the text. Given an integer $m \geq 0$ and an open subset $\Omega$ of $R^n$,

$$W^{m,p}(\Omega) = \{u \in L^p(\Omega); \quad \partial^\alpha u \in L^p(\Omega) \quad \text{for all} \quad |\alpha| \leq m\}$$

$$H^m(\Omega) = W^{m,2}(\Omega) \quad ,$$

[1] In addition to the references quoted in the sequel, see also the papers presented by G. FIX and by G. STRANG at this Symposium.

denote, for $1 \leq p \leq +\infty$, the usual Sobolev spaces, where we use the multi-index notation for the derivatives, i.e., if $\alpha = (\alpha_1,\ldots,\alpha_n) \in \mathbb{N}^n$, then

$$\partial^\alpha = \left(\frac{\partial}{\partial x_1}\right)^{\alpha_1} \cdots \left(\frac{\partial}{\partial x_n}\right)^{\alpha_n} \quad \text{and} \quad |\alpha| = \sum_{i=1}^{n} \alpha_i \ .$$

We shall use the norms and semi-norms defined by

$$\|u\|_{W^{m,p}(\Omega)} = \left( \sum_{|\alpha| \leq m} \|\partial^\alpha u\|^p_{L^p(\Omega)} \right)^{1/p} ,$$

$$|u|_{m,p,\Omega} = \left( \sum_{|\alpha| = m} \|\partial^\alpha u\|^p_{L^p(\Omega)} \right)^{1/p} ,$$

$$[u]_{m,p,\Omega} = \left( \sum_{i=1}^{n} \left\| \frac{\partial^m u}{\partial x_i^m} \right\|^p_{L^p(\Omega)} \right)^{1/p} ,$$

for $1 \leq p < +\infty$, with the standard modification for $p = +\infty$.

Given a subset $A \subset R^n$, a mapping $G : A \to R^n$ is *affine* if there exists an $n \times n$ matrix $B$ and a vector $b$ such that $G(x) = Bx+b$ for all $x \in A$.

For ease of exposition, we shall often not go into technical details related to the smoothness of the various data of the model problem (1.1)-(1.2).

## 2. Isoparametric finite elements

This § depends heavily upon two previous papers by the same authors [8,9] (see also [7] for a "survey"), and no proofs will be given. Also, for the sake of brevity, we will confine ourselves to finite elements corresponding to *Lagrange* interpolation, but the present analysis applies as well to finite elements corresponding to *Hermite* interpolation.

An *isoparametric finite element* $K$, in the terminology of the Engineers [26, §8], can be described as follows.

First, we are given:

(a) A set $\hat{\Sigma} = \bigcup_{i=1}^{N} \{\hat{a}_i\}$ of $N$ distinct points of $R^n$.

(b) A finite dimensional space $\hat{P}$ of functions defined over the closed convex hull $\hat{K}$ of $\hat{\Sigma}$ with $\dim \hat{P} = N$, and such that the set $\hat{\Sigma}$ is $\hat{P}$-*unisolvent*, i.e., the Lagrange interpolation problem: "Find $\hat{p} \in \hat{P}$ such that $\hat{p}(\hat{a}_i) = \alpha_i$, $1 \leq i \leq N$," has one and only one solution for any real numbers $\alpha_i$.

(c) A set $\Sigma = \bigcup_{i=1}^{N} \{a_i\}$ of $N$ distinct points of $R^n$.

Then *the finite element* $K$ *is the image*

$$K = F(\hat{K}) \tag{2.1}$$

*of the set* $\hat{K}$ *through the unique mapping* $F : \hat{K} \to R^n$ *which satisfies*

$$F \in \hat{P}^n \text{ and } F(\hat{a}_i) = a_i,\ 1 \leq i \leq N \ , \tag{2.2}$$

where the notation $F \in \hat{P}^n$ means that

$$F : \hat{x} \in \hat{K} \to F(\hat{x}) = (F_1(\hat{x}), \ldots, F_n(\hat{x})) \in R^n \tag{2.3}$$

$$\text{with } F_i \in \hat{P},\ 1 \leq i \leq n.$$

Notice that if we denote by $\hat{p}_i$, $1 \leq i \leq N$, the *basis functions* over the set $\hat{K}$ (i.e., $\hat{p}_i \in \hat{P}$ and $\hat{p}_i(\hat{a}_j) = \delta_{ij}$, $1 \leq j \leq N$), then it is easily seen that the mapping $F$ is given by $F = \sum_{i=1}^{N} \hat{p}_i a_i$.

Denoting by $\cdot$ the usual composition of functions, we associate with the finite element $K$ the space (of dimension $N$)

$$(2.4) \qquad P = \{p : K \to R;\ p = \hat{p} \cdot F^{-1}, \forall\, \hat{p} \in \hat{P}\} \ ,$$

assuming of course that the mapping $F^{-1} : K \to \hat{K}$ exists. Then, by definition, the restrictions to the set $K$ of the "trial functions" (associated with a partition $\mathcal{C}_h$ in which the finite element $K$ is included; cf. §4) belong to the space $P$. In particular, the $K$-*interpolate* $\Pi u$ of a given function $u : K \to R$ is the unique function which satisfies

$$(2.5) \qquad \Pi u \in P \text{ and } \Pi u(a_i) = u(a_i),\ 1 \leq i \leq N \ .$$

The mappings $F$ and $F^{-1}$ determine one-to-one correspondences between the points $\hat{x}$ of $\hat{K}$ and the points $x$ of $K$, by

$$(2.6) \qquad x = F(\hat{x}) \text{ and } \hat{x} = F^{-1}(x) \ ,$$

and, likewise, between the functions $\hat{p} \in \hat{P}$ and the functions $p \in P$, by

$$(2.7) \qquad p = \hat{p} \cdot F^{-1} \text{ and } \hat{p} = p \cdot F \ ,$$

in such a way that

$$(2.8) \qquad \hat{p}(\hat{x}) = p(x),\ \forall\, \hat{x} \in \hat{K},\ x \in K,\ \hat{p} \in \hat{P},\ p \in P \ .$$

Also notice that $\widehat{pp'} = \hat{p}\hat{p}', \forall p,p' \in P$.

*We shall constantly use the above correspondences.*

There are two types of isoparametric finite elements which are commonly used, which we shall refer to as *simplicial* and *quadrilateral* finite elements.

Example 1: *Simplicial finite elements.* A finite element $K = F(\hat{K})$ is *simplicial* (*triangular* if $n = 2$, *tetrahedral* if $n = 3$) if the set $\hat{K}$ is a non degenerate n-simplex of $R^n$, and $\hat{P}(1) \subset \hat{P}$, where, for any integer $k \geq 0$, $\hat{P}(k)$ denotes the space of the restrictions over the set $\hat{K}$ of all polynomials of degree $\leq k$ in the $n$ variables $\hat{x}_1, \hat{x}_2, \ldots, \hat{x}_n$. Since the vertices of the n-simplex $\hat{K}$ necessarily belong to the set $\hat{\Sigma} = \bigcup_{i=1}^{N} \{\hat{a}_i\}$, we may assume that they are the first $(n+1)$ points $\hat{a}_i$, $1 \leq i \leq n+1$. Next, we let

$$\hat{\Sigma}(k) = \{\hat{x} \in \hat{K};\ \hat{x} = \sum_{i=1}^{n+1} \lambda_i \hat{a}_i,\ \sum_{i=1}^{n+1} \lambda_i = 1,$$

$$\lambda_i \in \{0, \frac{1}{k}, \cdots, \frac{k-1}{k}, 1\} \quad \text{for} \quad 1 \leq i \leq n+1\}$$

denote the *principal lattice of order* $k$ of the n-simplex $\hat{K}$. Then *the set* $\hat{\Sigma}(k)$ *is* $\hat{P}(k)$-*unisolvent* for any integer $k \geq 1$ [18], and we shall say that a finite element $K$ is *of type* $(k,S)$ if it is the finite element associated with (a) the set $\hat{\Sigma}(k)$, (b) the space $\hat{P}(k)$, and (c) any set $\Sigma$ of $N = \binom{n+k}{k}(=\dim\hat{P}(k))$ points of $R^n$ (cf. Fig. 1 and 2 for $n = 2$ and $k = 2,3$). ∎

Example 2: *Quadrilateral finite elements.* A finite element $K = F(\hat{K})$ is *quadrilateral* if the set $\hat{K}$ is the unit hypercube $[0,1]^n$ of $R^n$, K is convex, and $\hat{Q}(1) \subset \hat{P}$, where, for any integer $k \geq 1$, we let

$$\hat{Q}(k) = \{\hat{q} : \hat{K} \to R;\ \hat{q}(\hat{x}) = \sum_{i_1,\ldots,i_n \leq k} b(i_1,\ldots,i_n)\hat{x}_1^{i_1}\ldots\hat{x}_n^{i_n} \text{ for } \hat{x} \in \hat{K}\},$$

so that

$$\hat{P}(k) \subset \hat{Q}(k) \subset \hat{P}(nk) \tag{2.9}$$

with strict inclusions for $n \geq 2$. Then the set

$$\hat{E}(k) = \{\hat{x} \in \hat{K};\ \hat{x} = (\hat{x}_1,\ldots,\hat{x}_n) \text{ with } \hat{x}_i \in \{0,\tfrac{1}{k},\ldots,\tfrac{k-1}{k},k\} \text{ for } 1 \leq i \leq n\}$$

*is* $\hat{Q}(k)$*-unisolvent,* and we shall say that a finite element if *of type* (k,Q) it is the finite element associated with (a) the set $\hat{E}(k)$, (b) the space $\hat{Q}(k)$ and (c) any set of $N = (k+1)^n$ (=dim $\hat{Q}(k)$) points of $R^n$ (cf. Fig. 1 and 2 for $n = 2$ and $k = 1,2$). ∎

In both cases, we could have also considered finite elements where the "internal" parameters are eliminated, such as the value at the point $a_{10}$ for type (3,S) or the value at the point $a_9$ for type (2,Q) (cf. Fig. 1 and 2) which correspond to a well-known device used by the Engineers. For details, see [8] and [9]. For brevity, those cases will not be studied here, although they are covered by the subsequent results.

In view of establishing asymptotic error estimates (Theorem 2), we consider a *family* $(K_h)$ of finite elements, that is, we are given once and for all a *reference set* $\hat{K}$, *a space* $\hat{P}$, and isoparametric finite elements

$$(2.10) \qquad K_h = F_h(\hat{K}), \quad \text{with} \quad F_h \in \hat{P}^n \quad \text{for all} \quad h \quad ,$$

where the defining parameter $h$ is a geometrical quantity attached to the finite element $K_h$. As will be seen later (cf. (2.27) and (2.30)), this parameter is either equal to, or asymptotically equal to, the diameter of $K_h$. Also, it is understood that $h$ *approaches zero in the limit.*

At first glance it might appear that, apart from the geometrical parameter $h$, a family of isoparametric finite elements is solely dependent upon the space $\hat{P}$, but this is too crude a consideration. To make this point clear, consider for instance a family $(K_h)$ of finite elements of type $(k,S)$ (resp. $(k,Q)$) (cf. Examples 1 and 2). Without further specification, all that can be said is that

$$(2.11) \qquad F_h \in (\hat{P}(k))^n \text{ (resp. } F_h \in (\hat{Q}(k))^n) \quad .$$

However, since $\hat{P}(1) \subset \hat{P}$ (resp. $\hat{P}(1) \subset \hat{Q}(1) \subset \hat{P}$), it might happen that

$$(2.12) \qquad F_h \in (\hat{P}(1))^n \quad \text{for all} \quad h \quad ,$$

and this corresponds to the important special case of what we shall henceforth call *straight finite elements*, in which case *the mappings* $F_h : \hat{K} \to R^n$ *are affine*, while the "general" isoparametric case corresponds to what we shall call *curved* finite elements. These considerations are illustrated in Fig.1 and 2 (one word of caution: for n = 2, the general isoparametric finite element of type (1,Q) is a quadrilateral and is therefore not curved in the usual sense!).

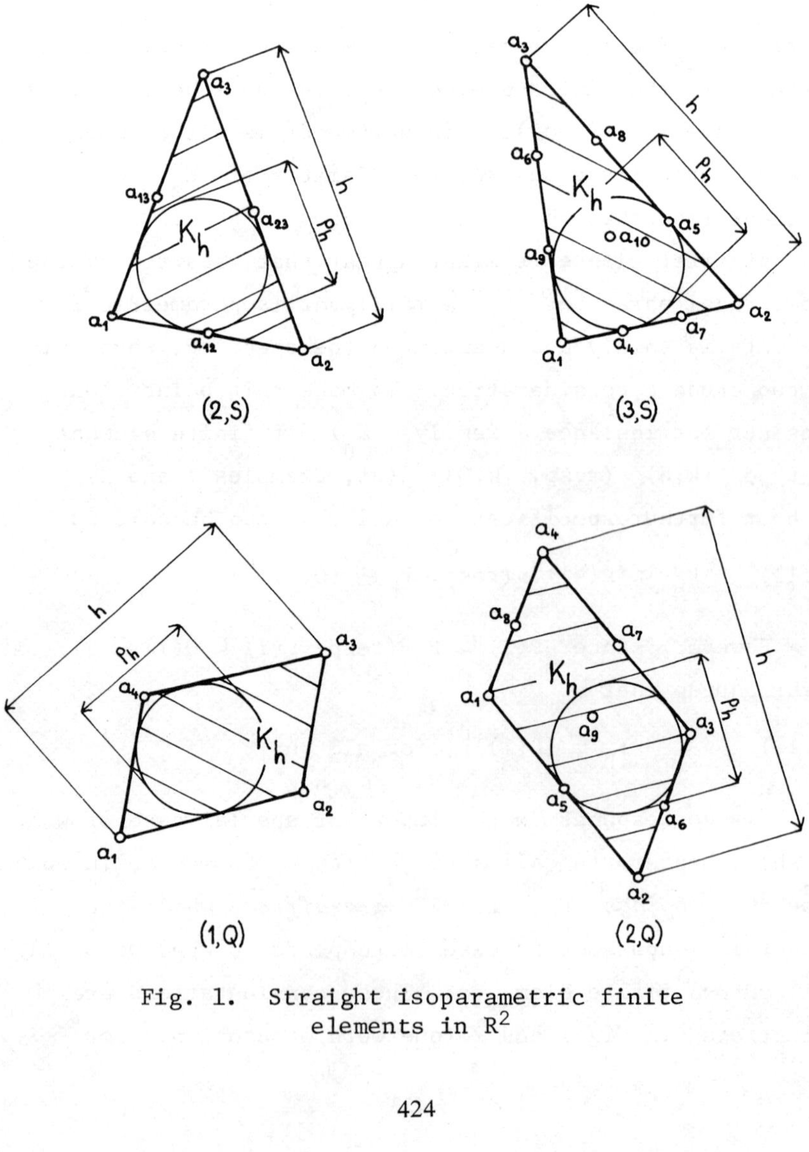


Fig. 1. Straight isoparametric finite elements in $R^2$

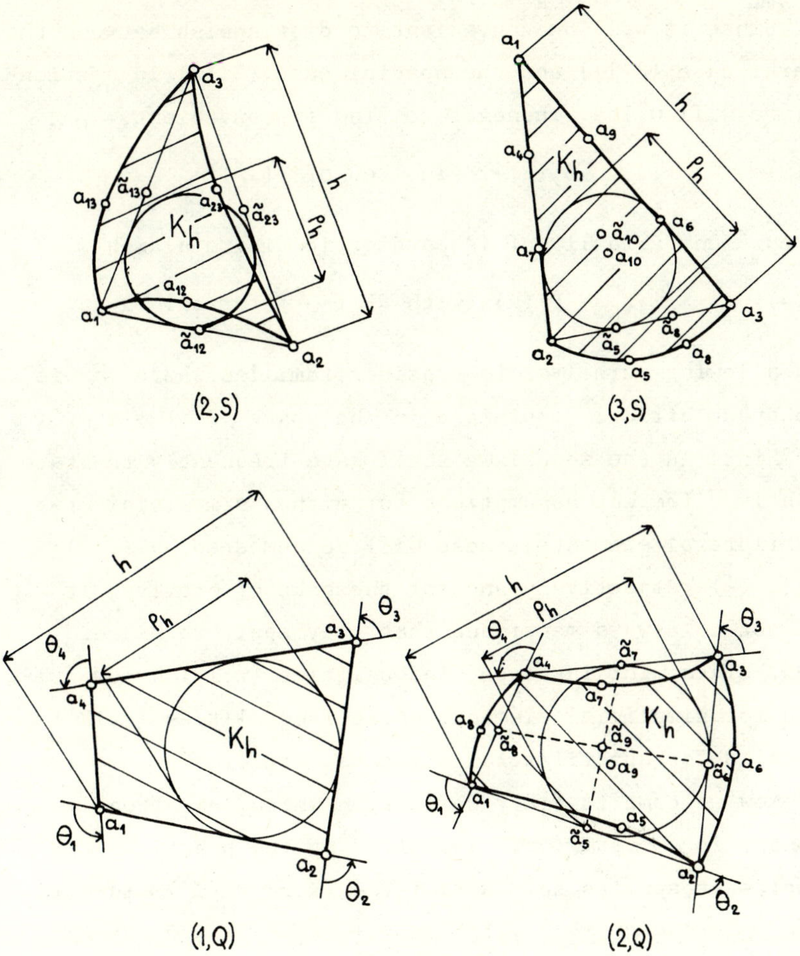


Fig. 2. Curved isoparametric finite elements in $R^2$

Since it will be convenient to distinguish between the general case (2.11) and the special case (2.12) in particular, we will write, whenever we find it convenient,

(2.13) $$\mathcal{K}(P;\mathcal{F}) \quad \text{in lieu of} \quad (K_h)$$

to represent a family of isoparametric finite elements

(2.14) $$K_h = F_h(\hat{K}) \quad \text{with} \quad F_h \in \mathcal{F} \quad \text{for all } h \quad ,$$

thus allowing ourselves to consider families where $\mathcal{F}$ is a subspace strictly contained in the space $\hat{P}^n$.

Since in the sequel we shall have frequently to state slightly different assumptions for either simplicial or quadrilateral elements, these will be numbered by (...) and (...') respectively and for the sake of brevity, it will not be always mentioned that they apply to either cases. Thus, for instance, inequalities (2.17) below are valid for simplicial elements while inequalities (2.17') are valid for quadrilateral elements, etc...

Now we come to a key definition which, as often happens, is *a posteriori* justified by the practical examples it applies to. A family $\mathcal{K}(P;\mathcal{F}) = (K_h)$ is said to be a k-*regular family* for some integer $k \geq 0$ if the following conditions are simultaneously satisfied:

(d) The space $\hat{P}$ satisfies

(2.15) $$\hat{P} \subset C^{k+1}(K) \quad .$$

(e) The finite elements $K_h$ are either simplicial or quadrilateral.

(f) For all $h$, the mapping $F_h \in \mathcal{F}$ is a $C^{k+1}$-*diffeomorphism* (in particular, $F_h$ is invertible) and there exist constants $c_\ell$, $-1 \leq \ell \leq k+1$, *independent* of $h$, such that *for all* $h$:

$$(2.16) \qquad |||DF_h^{-1}|||_{\hat{K}} \leq c_{-1} h^{-1} \quad ,$$

$$(2.17) \qquad |||D^\ell F_h|||_{\hat{K}} \leq c_\ell h^\ell, \quad 1 \leq \ell \leq k+1 \quad ,$$

$$(2.17') \qquad |[D^\ell F_h]|_{\hat{K}} \leq c_\ell h^\ell, \quad 1 \leq \ell \leq k+1 \quad ,$$

where

$$(2.18) \qquad |||D^\ell F_h|||_{\hat{K}} = \sup_{\hat{x}\in\hat{K}} \max_{|\alpha|=\ell} |\partial^\alpha F_h(\hat{x})|, \text{ for any } \ell \geq 1,$$

$$(2.19) \qquad |[D^\ell F_h]|_{\hat{K}} = \sup_{\hat{x}\in\hat{K}} \max_{1\leq i\leq n} \left|\frac{\partial^\ell F_h}{\partial x_i^\ell}(\hat{x})\right|, \text{ for any } \ell \geq 1.$$

(notice that $|||DF_h|||_{\hat{K}} = |[DF_h]|_{\hat{K}}$), and finally *for all* $h$:

$$(2.20) \qquad 0 < \frac{1}{c_0} \leq \frac{J_h(\hat{x})}{J_h(\hat{y})} \leq c_0 \quad \text{for all } \hat{x},\hat{y} \in \hat{K} \quad ,$$

where

(2.21) $J_h(\hat{x})$ = Jacobian of the mapping $F_h$ at the point $\hat{x} \in \hat{K}$.

Notice that a $k$-regular family is a $k'$-regular family for any $k' \leq k$. Also, we will often use the following technical result:

Lemma 1. *Let there be given a function* $\hat{u} \in W^{k+1,q}(\hat{K})$ *and the functions* $u = \hat{u} \cdot F_h^{-1}$ *associated with a* $k$-*regular family of finite elements* $K_h = F_h(\hat{K})$. *Then*

$u \in W^{k+1,q}(K_h)$ *and there exists a constant* $C$ *independent of* $h$ *such that*

$$(2.22)\quad |\hat{u}|_{k+1,q,\hat{K}} \leq C(\inf_{\hat{x}\in\hat{K}} J_{F_h}(\hat{x}))^{-1/q} h^{k+1} \|u\|_{W^{k+1,q}(K_h)}$$

$$\text{for all } \hat{u} \in W^{k+1,q}(\hat{K}) \quad ,$$

$$(2.22')\quad [\hat{u}]_{k+1,q,\hat{K}} \leq C(\inf_{\hat{x}\in\hat{K}} J_{F_h}(\hat{x}))^{-1/q} h^{k+1} \|u\|_{W^{k+1,q}(K_h)}$$

$$\text{for all } \hat{u} \in W^{k+1,q}(\hat{K}) \quad .$$

*Proof*. Since $F_h : \hat{K} \to K_h$ is a $C^{k+1}$-diffeomorphism by definition of a k-regular family, we may apply Lemma 3 of [9]: there exists a constant $C = C(K)$ such that for any function $\hat{u} \in W^{k+1,q}(\hat{K})$,

$$(2.23)\quad |\hat{u}|_{k+1,q,\hat{K}} \leq C(\inf_{\hat{x}\in\hat{K}} J_{F_h}(\hat{x}))^{-1/q} \sum_{\ell=1}^{k+1} |u|_{\ell,q,K_h}$$

$$\sum_{j\in I(k+1,\ell)} |||DF_h|||_{\hat{K}}^{j_1} \cdots |||D^{k+1}F_h|||_{\hat{K}}^{j_{k+1}} \quad ,$$

$$(2.23')\quad [\hat{u}]_{k+1,q,\hat{K}} \leq C(\inf_{\hat{x}\in\hat{K}} J_{F_h}(\hat{x}))^{-1/q} \sum_{\ell=1}^{k+1} |u|_{\ell,q,K_h}$$

$$\sum_{j\in I(k+1,\ell)} |[DF_h]|_{\hat{K}}^{j_1} \cdots |[D^{k+1}F_h]|_{\hat{K}}^{j_{k+1}} \quad ,$$

respectively, where for all $1 \leq \ell \leq k+1$,

(2.24) $$I(k+1,\ell) = \{j = (j_1,\ldots,j_{k+1});\ j_1+\ldots+j_{k+1} = \ell,\ j_1+2j_2+\ldots+(k+1)j_{k+1} = k+1\},$$

so that the inequality of (2.22) (resp. (2.22')) is a direct consequence of the inequality of (2.23) (resp. (2.23')), the definition (2.24) of the sets $I(k+1,\ell)$ of indices, and the inequalities (2.17) (resp. (2.17')) satisfied by a k-regular family. ∎

Now we state a basic interpolation theory result for k-regular families. The key assumption is that some space of polynomials (either $\hat{P}(k)$ or $\hat{Q}(k)$) is preserved by interpolation *over the reference set* $\hat{K}$, (and thus *not* on $K_h$, except for straight finite elements !), and it is this space that determines the order of convergence.

**Theorem 2.** *Let there be given a* k-*regular family* $\mathcal{K}(P;\mathcal{F}) = (K_h)$ *of finite elements such that*

(2.25) *either* $\hat{P}(k) \subset \hat{P}$ *for simplicial elements,*

(2.25') *or* $\hat{Q}(k) \subset \hat{P}$ *for quadrilateral elements.*

*Then for any integer* $m$ *with* $0 \leqslant m \leqslant k+1$, *and any* $p$ *with* $1 \leqslant p \leqslant +\infty$, *there exists a constant* $C$ *independent of* $h$ *such that for any function* $u \in W^{k+1,p}(K_h)$ *with* $0 < k+1 - \frac{n}{p}$ *if* $p < +\infty$, *we have*

(2.26) $$|u-\Pi_h u|_{m,p,K} \leqslant C\, h^{k+1-m} \|u\|_{W^{k+1,p}(K_h)} ,$$

*where* $\Pi_h u$ *is the unique* $K_h$*-interpolate of the function* $u$ *(defined as in* (2.5)).

*Proof*. This is a straightforward consequence of the results of [9, §§3 and 4] and of the hypotheses (d), (e) and (f) made for a k-regular family. ∎

Let us now examine some examples of k-regular families. In so doing, we give a unified treatment of various results found in the literature (cf. [3, 5, 6, 10, 21, 24, 27, 28] for straight elements; very few people seem to have considered curved finite elements, with the exception of FRIED [12] and ZLÁMAL [29]).

Example 3: k-*regular families of straight finite elements*. Given a family $(K_h) = \mathcal{F}(\hat{P};(\hat{P}(1))^n)$ of straight finite elements, we let

$$h = \text{diam } K_h \tag{2.27}$$

$$\rho_h = \sup\{\text{diameters of the spheres contained in } K_h\}. \tag{2.28}$$

Then, *if there exists a constant* $\alpha$ *such that*

$$0 < \alpha \leq \frac{\rho_h}{h} \quad \textit{for all} \quad h\ , \tag{2.29}$$

*the family* $(K_h)$ *is a* k-*regular family for any* $k \geq 0$.

To see this, it suffices to observe that the mappings $F_h$ are affine and thus, all quantities (2.18) or (2.19) are 0 for $\ell \geq 2$. That inequalities (2.16) and (2.17) with $\ell = 1$ are valid is a consequence of Lemma 2 of [8].

In fact, the above result applies to *any* type of straight isoparametric elements and they need not be of either simplicial or quadrilateral type. All that is required is that $\hat{P}(1) \subset \hat{P}$. ∎

When we turn to the general isoparametric case, the situation is more complicated and although general results can be obtained [9, Theorems 5 and 6], we will only mention here some examples of practical interest. For the proofs, see [9, §§5 and 6].

Example 4: *Some* k*-regular families of curved finite elements.* First consider finite elements of type (2,S) (cf. Example 1). A finite element of this type is determined by the data of the set

$$\hat{\Sigma}(2) = \left(\bigcup_{i=1}^{n+1} \{\hat{a}_i\}\right) \cup \left(\bigcup_{i<j} \{\hat{a}_{ij}\}\right) \text{ with } \hat{a}_{ij} = (\hat{a}_i + \hat{a}_j)/2$$

where the points $\hat{a}_i$ are the vertices of a nondegenerate n-simplex of $R^n$. Then a finite element $K_h$ of type (2,S) is determined by a set

$$\Sigma_h = \left(\bigcup_{i=1}^{n+1} \{a_i\}\right) \cup \left(\bigcup_{i<j} \{a_{ij}\}\right)$$

of $\binom{n+2}{2}$ points of $R^n$. We introduce the n-simplex $\tilde{K}_h$ whose vertices are the points $a_i$, $1 \leq i \leq n+1$, and we let (cf. Fig. 2 for $n = 2$)

$$h = \text{diam } \tilde{K}_h \quad , \tag{2.30}$$

$$\rho_h = \text{diam (inscribed sphere of } \tilde{K}_h) \tag{2.31}$$

$$\tilde{a}_{ij} = (a_i + a_j)/2 \quad \text{for} \quad i < j \quad .$$

Then we have: *a family* $(K_h)$ *of curved finite elements of type* (2,S) *is a* 2*-regular family if there exists a constant* $\alpha$ *such that*

$$0 < \alpha \leq \frac{\rho_h}{h} \text{ for all } h \tag{2.32}$$

*and*

(2.33) $$\|a_{ij}-\tilde{a}_{ij}\| = 0(h^2),\ 1 \leqslant i < j \leqslant n+1 \ .$$

Next, consider a finite element $K_h$ of type (3,S) for $n = 2$ (cf. Example 1). It is determined by the data of 10 points $a_i$ whose first three will play the role of "vertices". Again we introduce the triangle $\tilde{K}_h$ whose vertices are the points $a_i$, $1 \leqslant i \leqslant 3$, and we define $h$ and $\rho_h$ as in (2.30)-(2.31). We let (cf. Fig. 2)

$$\tilde{a}_5 = (2a_2+a_3)/3,\ \tilde{a}_8 = (2a_3+a_2)/3 \text{ and } \tilde{a}_{10} = (a_1+a_2+a_3)/3.$$

Then we have: *a family* $(K_h)$ *of finite elements of type* (3,S) *for* $n = 2$ *is a 3-regular family if there exists a constant* $\alpha$ *such that*

(2.34) $$0 < \alpha \leqslant \frac{\rho_h}{h} \text{ for all } h$$

*and*

(2.35) $$\|a_5-\tilde{a}_5\| = \|a_8-\tilde{a}_8\| = 0(h^2) \quad ,$$

(2.36) $$\|(a_5-\tilde{a}_5)-(a_8-\tilde{a}_8)\| = 0(h^3) \quad ,$$

(2.37) $$a_{10} = \tilde{a}_{10} + \frac{1}{4}\{(a_5-\tilde{a}_5)+(a_8-\tilde{a}_8)\} \quad ,$$

(2.38) $$\begin{cases} a_4 = (2a_1+a_2)/3,\ a_7 = (2a_2+a_1)/3 \quad , \\ a_6 = (2a_3+a_1)/3,\ a_9 = (2a_1+a_3)/3 \quad . \end{cases}$$

In other words, *only one side is allowed to be curved* but as we will see in §4, this is not a restriction for 2-dimensional problems.

Consider now finite elements of type (k,Q) for $n = 2$, $k = 1,2$ (cf. Example 2). We begin with a finite element $K_h$ of type (1,Q) which is determined by the data of 4 points $a_i$ : $K_h$ is the quadrilateral with vertices $a_i$, $1 \leq i \leq 4$. We define (cf. Fig. 2)

$$h = \text{diam } K_h \quad , \tag{2.39}$$

$$\rho_h = \sup\{\text{diameter of the spheres contained in } K_h\}, \tag{2.40}$$

and in addition, we let (cf. Fig. 2)

$$\gamma_h = \max \{|\cos \theta_i|, \ 1 \leq i \leq 4\} \quad . \tag{2.41}$$

Then we have: *a family* $(K_h)$ *of finite elements of type* (1,Q) *for* $n = 2$ *is a* 1-*regular family if there exist constants* $\alpha$ *and* $\gamma$ *such that*

$$0 < \alpha \leq \frac{\rho_h}{h} \textit{ and } \gamma_h \leq \gamma < 1 \textit{ for all } h. \tag{2.42}$$

Finally, consider a finite element $K_h$ of type (2,Q) for $n = 2$ which is determined by the data of 9 points $a_i$, whose first four will play the role of "vertices". We let $\tilde{K}_h$ denote the quadrilateral whose vertices are the points $a_i$, $1 \leq i \leq 4$, and we define $h$, $\rho_h$, $\gamma_h$ as in (2.39), (2.40) and (2.41) *with* $K_h$ *replaced by* $\tilde{K}_h$ (cf. Fig. 2). In addition, we let

$$\tilde{a}_5 = (a_1+a_2)/2, \ \tilde{a}_6 = (a_2+a_3)/2, \ \tilde{a}_7 = (a_3+a_4)/2,$$

$$\tilde{a}_8 = (a_1+a_4)/2, \ \tilde{a}_9 = (a_1+a_2+a_3+a_4)/4 \ .$$

Then we have: *a family* $(K_h)$ *of finite elements of type* (2,Q) *for* $n = 2$ *is a* 2-*regular family if there exist constants* $\alpha$ *and* $\gamma$ *such that*

(2.43) $$0 < \alpha \leq \frac{\rho_h}{h} \text{ and } \gamma_h \leq \gamma < 1 \text{ for all } h \quad ,$$

*and*

(2.44) $$\|a_i - \tilde{a}_i\| = 0(h^2),\ 5 \leq i \leq 9 \quad .$$

## 3. "Isoparametric" numerical integration

In this §, we give a description of the numerical integration technique used over the isoparametric finite elements described in the previous §. The key idea is that *it is sufficient to have a numerical quadrature scheme over the reference set* $\hat{K}$, which leaves invariant a specified space of polynomials. Thus, for handling practical examples, we need to approximately integrate functions either over an n-simplex or over the unit hypercube. In this respect, see for instance [16], [26, §8] and the survey of HABER [13].

The notations used in this § are similar to those of §2 although, for notational convenience, we shall freely drop the index h and replace it by the index F in various places.

We assume that we have at our disposal a *quadrature formula* over the *reference set* $\hat{K}$. In other words,

(3.1) $$\int_{\hat{K}} \hat{\phi}(\hat{x})\,d\hat{x} \text{ is approximated by } \sum_{\ell=1}^{L} \hat{\omega}_\ell \hat{\phi}(\hat{b}_\ell) \quad ,$$

for some specified points $\hat{b}_\ell \in \hat{K}$ and weights $\hat{\omega}_\ell$ which will be assumed *once and for all* to satisfy

(3.2) $$\hat{\omega}_\ell > 0,\ 1 \leq \ell \leq L \quad .$$

This assumption is by no means necessary but it yields simpler proofs. With the quadrature scheme (3.1), we associate the *error*

$$(3.3)\qquad \hat{E}(\hat{\phi}) = \int_{\hat{K}} \hat{\phi}(\hat{x})\,d\hat{x} - \sum_{\ell=1}^{L} \hat{\omega}_\ell \hat{\phi}(\hat{b}_\ell) \quad .$$

Given an isoparametric finite element $K = F(\hat{K})$ corresponding to an invertible mapping $F : \hat{K} \to K$, we may, and will, assume that

$$(3.4)\qquad J_F(\hat{x}) = \det(DF(\hat{x})) > 0 \quad \text{for all} \quad \hat{x} \in \hat{K} \quad .$$

Using the standard formula for the change of variables in multiple integrals, we find that

$$(3.5)\qquad \int_K \phi(x)\,dx \quad \textit{is approximated by} \quad \sum_{\ell=1}^{L} \omega_\ell \phi(b_\ell) \quad ,$$

with

$$(3.6)\qquad \omega_\ell = \hat{\omega}_\ell\, J_F(\hat{b}_\ell) \quad \text{and} \quad b_\ell = F(\hat{b}_\ell),\ 1 \leq \ell \leq L \quad .$$

In other words, the quadrature scheme (3.1) over the reference set $\hat{K}$ automatically induces the quadrature scheme (3.5) over the finite element $K$, a circumstance which we call *"isoparametric" numerical integration.* With the scheme (3.5), we associate the *error*

$$(3.7)\qquad E(\phi) = \int_K \phi(x)\,dx - \sum_{\ell=1}^{L} \omega_\ell \phi(b_\ell) \quad ,$$

so that we have

$$(3.8)\qquad E(\phi) = \hat{E}(\hat{\phi} J_F) \quad \text{and} \quad \hat{E}(\hat{\phi}) = E(\phi J_F^{-1}) \quad ,$$

using again the correspondences of (2.7).

We first examine conditions which will eventually be used in Theorem 9 of §4 for proving the uniform ellipticity assumption of (1.11). Theorem 3 is based upon, and is a generalization of, an idea of G. STRANG [22].

**Theorem 3.** *Let there be given a quadrature scheme over the reference set* $\hat{K}$ *which satisfies*

$$\sum_{\ell=1}^{L} \hat{\omega}_\ell \sum_{i=1}^{n} \left(\frac{\partial \hat{p}}{\partial \hat{x}_i}(\hat{b}_\ell)\right)^2 \geq \hat{\gamma}|\hat{p}|^2_{1,2,\hat{K}} \text{ for all } \hat{p} \in \hat{P}, \tag{3.9}$$

*for some constant* $\hat{\gamma} > 0$. *Given a 0-regular family of isoparametric finite elements* $K_h$, *there exists a constant* $\gamma$ *independent of* h *such that*

$$\sum_{\ell=1}^{L} \omega_\ell \left(\sum_{i=1}^{n} \frac{\partial p}{\partial x_i}(b_\ell)\right)^2 \geq \gamma|p|^2_{1,2,K_h} \text{ for all } p \in P \tag{3.10}$$

*and for all* h.

*Proof.* We will use the fact that if A is a non-singular matrix, the matrix $AA^T$ is positive definite and therefore for any row vector $\xi$, we have $\xi AA^T\xi \geq \frac{1}{\|A^{-1}\|^2}\xi\xi^T$. Since $Dp(x) = D\hat{p}(\hat{x})\cdot DF^{-1}(x)$, where $Dp(x)$ is the row vector $\left(\frac{\partial p}{\partial x_i}(x)\right)_{i=1}^{n}$, and $DF^{-1}(x)$ is the Jacobian matrix of the mapping $F^{-1}$ at the point $x \in K$, we have for all $x \in K$,

$$\sum_{i=1}^{n} \left(\frac{\partial p}{\partial x_i}(x)\right)^2 = Dp(x)\cdot Dp(x)^T \geq \frac{1}{\|DF(\hat{x})\|^2} \sum_{i=1}^{n} \left(\frac{\partial \hat{p}}{\partial \hat{x}_i}(\hat{x})\right)^2, \tag{3.11}$$

and likewise, for all $\hat{x} \in \hat{K}$,

$$(3.12)\qquad \sum_{i=1}^{n}\left(\frac{\partial\hat{p}}{\partial\hat{x}_i}(\hat{x})\right)^2 \geq \frac{1}{\|DF^{-1}(\hat{x})\|^2}\sum_{i=1}^{n}\left(\frac{\partial p}{\partial x_i}(x)\right)^2 .$$

By (3.6) and (3.9),

$$\sum_{\ell=1}^{L}\omega_\ell\sum_{i=1}^{n}\left(\frac{\partial p}{\partial x_i}(b_\ell)\right)^2 = \sum_{\ell=1}^{L}\hat{\omega}_\ell J_F(\hat{b}_\ell)\sum_{i=1}^{n}\left(\frac{\partial p}{\partial x_i}(b_\ell)\right)^2$$

$$\geq \hat{\gamma}\inf_{\hat{x}\in\hat{K}} J_F(\hat{x})\frac{1}{|||DF|||_{\hat{K}}^2}\int_{\hat{K}}\sum_{i=1}^{n}\left(\frac{\partial\hat{p}}{\partial\hat{x}_i}(\hat{x})\right)^2 d\hat{x}\quad\text{(by 3.11)},$$

where $|||\cdot|||_K$ (and $|||\cdot|||_{\hat{K}}$ in what follows) are defined as in (2.18). By letting $\hat{x} = F^{-1}(x)$ in the last integral and using (3.12), we find

$$\int_{\hat{K}}\sum_{i=1}^{n}\left(\frac{\partial\hat{p}}{\partial\hat{x}_i}(\hat{x})\right)^2 d\hat{x} \geq \frac{1}{\sup\limits_{x\in K} J_F(x)}\,\frac{1}{|||DF^{-1}|||_K^2}\int_{K_h}\sum_{i=1}^{n}\left(\frac{\partial p}{\partial x_i}(x)\right)^2 dx,$$

so that we obtain

$$\sum_{\ell=1}^{L}\omega_\ell\left(\sum_{i=1}^{n}\frac{\partial p}{\partial x_i}(b_\ell)\right)^2$$

$$\geq \hat{\gamma}\left(\frac{\inf\limits_{\hat{x}\in\hat{K}} J_F(\hat{x})}{\sup\limits_{\hat{x}\in\hat{K}} J_F(\hat{x})}\right)\frac{1}{(|||DF|||_{\hat{K}}|||DF^{-1}|||_K)^2}|p|_{1,2,K_h}^2$$

for all $p\in P$,

and the inequality of (3.10) follows by using inequalities (2.16), (2.17) with $\ell = 1$, and (2.20), so that we may choose $\gamma = \hat{\gamma}(c_{-1}c_0c_1)^{-1}$. ∎

Assumption (3.9) can be interpreted in various ways:

First, it is clear that the left hand side of (3.9) represents an approximation of the integral $|\hat{p}|^2_{1,2,K}$. Thus *if* $\hat{P} \subset \hat{P}(\ell)$ *for some integer* $\ell \geq 1$ *and if* $\hat{E}(\hat{p}) = 0$ *whenever* $\hat{p} \in \hat{P}(2\ell-2)$, *the inequality* (3.9) *becomes an equality with* $\hat{\gamma} = 1$.

Next, this assumption could have been equivalently stated as follows: the quantity $\left\{\sum_{\ell=1}^{L} \hat{\omega}_\ell \left(\sum_{i=1}^{n} \frac{\partial \hat{p}}{\partial \hat{x}_i}(\hat{b}_\ell)\right)^2\right\}^{1/2}$ is a norm over the (finite-dimensional) quotient space $\hat{P}/\hat{P}(0)$. This is the case, as is easily seen, *if* $\hat{P} \subset \hat{P}(\ell)$ (resp. $\hat{P} \subset \hat{Q}(\ell)$) *for some integer* $\ell \geq 1$ *and if the set* $\bigcup_{\ell=1}^{L} \{\hat{b}_\ell\}$ *contains a* $\hat{P}(\ell-1)$-*unisolvent subset* (*resp. a* $\hat{Q}(\ell) \cap \hat{P}(2\ell-1)$-*unisolvent subset*). This second condition is a generalization of a similar condition of STRANG [22].

In view of §§4 and 5, we now obtain (Theorems 4 to 8) upper bounds for the error $E(\phi)$ for various forms of the argument $\phi$. For the sake of brevity, the proofs will be given for families of simplicial finite elements only, the modifications for quadrilateral finite elements being minor. Likewise, some arguments will be systematically omitted: for instance, the inequality $r+1-\frac{n}{q} > 0$ insures by the Sobolev's imbedding theorem that linear forms such as $\hat{v} \in W^{r+1,q}(\hat{K}) \to \hat{E}(\hat{v}\ldots)$ are continuous. Also, various constants $\gamma_i$ appear in the proof: they can be easily seen to be solely dependent upon a subset of $\{n,r,q,\hat{K},\hat{P},\hat{\omega}_1,\ldots,\hat{\omega}_L\}$ and thus they are independent of $h$.

For a given number $q$ with $1 \leq q \leq +\infty$, we denote by $q'$ its conjugate exponent $(1/q+1/q' = 1)$. For a family

$\mathcal{K}(P;\mathcal{F})$ of finite elements, we denote by $J_F$ the Jacobians of the mappings $F \in \mathcal{F}$.

Finally, frequent use will be made of the *Bramble-Hilbert lemma* [4,5] in the form given in [8, Lemma 6].

We recall that hypotheses numbered as (...) and (...)' refer to simplicial and quadrilateral finite elements, respectively.

**Theorem 4.** *For some integer* $r \geq 0$, *let there be given an* r-*regular family* $(K_h) = \mathcal{K}(P;\mathcal{F})$ *of finite elements and a quadrature scheme over the reference set* $\hat{K}$ *which satisfies*

(3.13) $\hat{E}(\hat{v}J_F) = 0$ *for all* $\hat{v} \in \hat{P}(r)$ *and* $F \in \mathcal{F}$ ,

*and, in addition if* $r \geq 1$,

(3.14) $\hat{E}(\hat{v}\hat{p}J_F) = 0$ *for all* $\hat{v} \in \hat{P}(r-1)$, $\hat{p} \in \hat{P}$ *and* $F \in \mathcal{F}$, *if* $r \geq 1$,

(3.13') $\hat{E}(\hat{v}J_F) = 0$ *for all* $\hat{v} \in \hat{Q}(r)$ *and* $F \in \mathcal{F}$ ,

*and, in addition if* $r \geq 1$,

(3.14') $\hat{E}(\hat{v}\hat{p}J_F) = 0$ *for all* $\hat{v} \in \hat{Q}(r-1)$, $\hat{p} \in \hat{P}$ *and* $F \in \mathcal{F}$, *if* $r \geq 1$ .

*Then for any number* $q$ *with* $1 \leq q \leq +\infty$ *and* $r+1 - \frac{n}{q} > 0$, *there exists a constant* $C$ *independent of* $h$ *such that*

(3.15) $$|E(vp)| \leq Ch^{r+1} \|v\|_{W^{r+1,q}(K_h)} \|p\|_{W^{1,q'}(K_h)}$$

*for all* $v \in W^{r+1,q}(K_h)$ *and* $p \in P$.

*Proof.* Let us first assume that $r \geq 1$. Since $r+1 - \frac{n}{q} > 0$, we may choose a number $s$ with $1 \leq s \leq +\infty$ and $r - \frac{n}{s} > 0$, such that

$$W^{r+1,q}(\hat{K}) \subset W^{r,s}(\hat{K}) \subset C^{o}(\hat{K}) \quad ,$$

with continuous injections (if $1 \leq q < n$, let $s = \frac{nq}{n-q}$, if $n \leq q \leq +\infty$, let $s$ be so that $r - \frac{n}{s} > 0$; if $q = +\infty$, let $s = +\infty$) and moreover

$$(3.16) \qquad |\hat{v}|_{r,s,\hat{K}} \leq \gamma_1 \left( |\hat{v}|_{r,q,\hat{K}} + |\hat{v}|_{r+1,q,\hat{K}} \right)$$

$$\text{for all } \hat{v} \in W^{r+1,q}(\hat{K})$$

(we use the fact that $\hat{v} \in W^{r+1,q}(\hat{K})$ implies $D^{\alpha}u \in W^{1,q}(\hat{K})$ for $|\alpha| = r$ and the inclusion $W^{1,q}(\hat{K}) \subset L^{s}(\hat{K})$).

We let the operator $\hat{\Pi}_0 : L^1(\hat{K}) \to \hat{P}(0)$ be defined by

$$(3.17) \qquad \hat{\Pi}_0 \hat{u} = \frac{1}{\text{meas } \hat{K}} \int_{\hat{K}} \hat{u}(\hat{x}) \, d\hat{x} \quad .$$

Consider next, for fixed $\hat{p} \in \hat{P}$ and $F \in \mathcal{F}$, the linear form

$$\hat{v} \in W^{r,s}(\hat{K}) \to \hat{E}(\hat{v}(\hat{p}-\hat{\Pi}_0\hat{p})J_F) \quad .$$

It is continuous with norm $\leq \gamma_2 \| (\hat{p}-\hat{\Pi}_0\hat{p})J_F \|_{L^{q'}(\hat{K})}$ and moreover, by (3.14), it vanishes over $\hat{P}(r-1)$. Thus, by the Bramble-Hilbert lemma,

$$(3.18) \qquad |\hat{E}(\hat{v}(\hat{p}-\hat{\Pi}_0\hat{p})J_F)| \leq \gamma_3 |\hat{v}|_{r,s,\hat{K}} \| (\hat{p}-\hat{\Pi}_0\hat{p})J_F \|_{L^{q'}(\hat{K})}$$

for all $\hat{v} \in W^{r,s}(\hat{K})$, $\hat{p} \in \hat{P}$ and $F \in \mathcal{F}$. Using again the Bramble-Hilbert lemma, we obtain

$$(3.19) \qquad \|(\hat{p}-\hat{\Pi}_0\hat{p})J_F\|_{L^{q'}(\hat{K})} \leq \gamma_4 \sup_{\hat{x}\in\hat{K}} J_F(\hat{x})|\hat{p}|_{1,q',\hat{K}}$$

for all $\hat{p}\in \hat{P}$,

since the operator $\hat{\Pi}_0$ of (3.17) leaves the space $\hat{P}(0)$ invariant. Combining (3.16), (3.18) and (3.19), we have

$$(3.20) \qquad |\hat{E}(\hat{v}(\hat{p}-\hat{\Pi}_0\hat{p})J_F)| \leq \gamma_5 \sup_{\hat{x}\in\hat{K}} J_F(\hat{x}) (|\hat{v}|_{r,q,\hat{K}} + |\hat{v}|_{r+1,q,\hat{K}})|\hat{p}|_{1,q',\hat{K}}$$

for all $\hat{v} \in W^{r+1,q}(\hat{K})$, $\hat{p} \in \hat{P}$ and $F \in \mathcal{F}$.

Likewise, for fixed $\hat{p} \in \hat{P}$ and $F\in\mathcal{F}$, the continuous linear form

$$\hat{v} \in W^{r+1,q}(\hat{K}) \to E(\hat{v}(\hat{\Pi}_0\hat{p})J_F)$$

vanishes over $\hat{P}(r)$ by (3.13), and thus,

$$(3.21) \qquad |\hat{E}(\hat{v}(\hat{\Pi}_0\hat{p})J_F)| \leq \gamma_6 \sup_{\hat{x}\in\hat{K}} J_F(\hat{x})|\hat{v}|_{r+1,q,\hat{K}}\|p\|_{L^{q'}(\hat{K})}$$

for all $\hat{v} \in W^{r+1,q}(\hat{K})$. Combining (3.20) and (3.21), we obtain next

$$(3.22) \quad |\hat{E}(\hat{v}\hat{p}J_F)| \leq \gamma_7 \sup_{\hat{x}\in\hat{K}} J_F(\hat{x})\{(|\hat{v}|_{r,q,\hat{K}} + |\hat{v}|_{r+1,q,\hat{K}}) |\hat{p}|_{1,q',\hat{K}} + |\hat{v}|_{r+1,q,\hat{K}}\|\hat{p}\|_{L^{q'}(\hat{K})}\},$$

and the inequality of (3.15) follows by observing that $\hat{E}(\hat{v}\hat{p}J_F) = E(vp)$ and by using inequality (2.22) of Lemma 1 in the right-hand side of (3.22).

Finally, consider the case $r = 0$. Since $\hat{E}(\hat{v}J_F) = 0$ for all $\hat{v} \in \hat{P}(0)$, we have

$$|\hat{E}(\hat{v}J_F)| \leqslant \gamma_8 \sup_{\hat{x}\in\hat{K}} J_F(\hat{x})|\hat{v}|_{1,q,\hat{K}}$$

for all $\hat{v} \in W^{1,q}(\hat{K})$, and thus

$$|\hat{E}(\hat{v}\hat{p}J_F)| \leqslant \gamma_9 \sup_{\hat{x}\in\hat{K}} J_F(\hat{x})(|\hat{v}|_{1,q,\hat{K}}\|\hat{p}\|_{L^{q'}(\hat{K})}$$

$$+ \|\hat{v}\|_{L^{q'}(\hat{K})}|\hat{p}|_{1,q',\hat{K}})$$

for all $\hat{v} \in W^{1,q}(\hat{K})$ and $\hat{p} \in \hat{P}$. The end of the proof consists in applying again inequality (2.22) of Lemma 1. ∎

Theorem 5. *For some integer* $r \geqslant 0$, *let there be given an* $(r+1)$*-regular family* $(K_h) = \mathcal{K}(P;\mathcal{F})$ *of finite elements and a quadrature scheme over the reference set* $\hat{K}$ *which satisfies*

(3.23) $\hat{E}(\hat{v}J_F) = 0$ *for all* $\hat{v} \in \hat{P}(r+1)$ *and* $F \in \mathcal{F}$ ,

*and, in addition if* $r \geqslant 1$,

(3.24) $\hat{E}(\hat{v}\hat{p}J_F) = 0$ *for all* $\hat{v} \in \hat{P}(r-1)$, $\hat{p} \in \hat{P}$ *and* $F \in \mathcal{F}$ *if* $r \geqslant 1$ ,

(3.23') $\hat{E}(\hat{v}J_F) = 0$ *for all* $\hat{v} \in \hat{Q}(r+1)$ *and* $F \in \mathcal{F}$ ,

*and, in addition if* $r \geqslant 1$,

(3.24') $\hat{E}(\hat{v}\hat{p}J_F) = 0$ *for all* $\hat{v} \in \hat{Q}(r-1)$, $\hat{p} \in \hat{P}$ *and* $F \in \mathcal{F}$ *if* $r \geqslant 1$ .

*Then for any number* $q$ *with* $1 \leqslant q \leqslant +\infty$ *and* $r+2 - \frac{n}{q} > 0$, *there exists a constant* $C$ *independent of* $h$ *such that*

(3.25) $$|E(vp)| \leq Ch^{r+2} \|v\|_{W^{r+2,q}(K_h)} \|p\|_{W^{2,q'}(K_h)}$$

*for all* $v \in W^{r+2,q}(K_h)$ *and* $p \in P$.

*Proof.* The proof will be similar to that of Theorem 4, and for this reason, it will be only sketched, in the case $r \geq 1$. Let the operator $\hat{\Pi}_1 : \hat{u} \in L^1(\hat{K}) \to \hat{\Pi}_1\hat{u} \in \hat{P}(1)$ be defined by

$$\int_{\hat{K}} (\hat{\Pi}_1\hat{u})v \, dx = \int_{\hat{K}} uv \, dx \quad \text{for all} \quad v \in \hat{P}(1) \quad ,$$

so that $\hat{P}(1)$ is left invariant by $\hat{\Pi}_1$. Choosing $s$ so that $W^{r+2,q}(\hat{K}) \subset W^{r,s}(\hat{K}) \subset C^0(\hat{K})$ with continuous injections, we find that

$$|\hat{E}(\hat{v}(\hat{p}-\hat{\Pi}_1\hat{p})J_F)| \leq \gamma_1 \sup_{\hat{x}\in\hat{K}} J_F(\hat{x}) |\hat{v}|_{r,s,\hat{K}} |\hat{p}|_{2,q',\hat{K}}$$

for all $\hat{v} \in W^{r,s}(\hat{K})$, $\hat{p} \in \hat{P}$, and $F \in \mathcal{F}$, taking into account the assumption of (3.24), and thus,

(3.26) $$|\hat{E}(\hat{v}(\hat{p}-\hat{\Pi}_1\hat{p})J_F)| \leq \gamma_2 \sup_{\hat{x}\in\hat{K}} J_F(\hat{x}) \left( \sum_{\nu=0}^{2} |\hat{v}|_{r+\nu,q,\hat{K}} \right) |\hat{p}|_{2,q',\hat{K}}$$

for all $\hat{v} \in W^{r+2,q}(\hat{K})$, $\hat{p} \in \hat{P}$ and $F \in \mathcal{F}$. Likewise, using the assumption of (3.23), we have

(3.27) $$|E(\hat{v}(\hat{\Pi}_1\hat{p})J_F)| \leq \gamma_3 \sup_{\hat{x}\in\hat{K}} J_F(\hat{x}) (|\hat{v}|_{r+2,q,\hat{K}} \|\hat{p}\|_{L^{q'}(\hat{K})} + |\hat{v}|_{r+1,q,\hat{K}} |\hat{p}|_{1,q',\hat{K}})$$

for all $\hat{v} \in W^{r+2,q}(\hat{K})$, $\hat{p} \in \hat{P}$ and $F \in \mathcal{F}$, and the proof is achieved by combining inequalities (3.26) and (3.27). ∎

Given a function $p : K \to R$ which is everywhere differentiable over the set $K$, we may consider the functions $\widehat{\frac{\partial p}{\partial x_i}} : \hat{K} \to R$, $1 \leqslant i \leqslant n$, via the correspondence (2.7); actually, it is easily seen that

$$(3.28) \quad \widehat{\frac{\partial p}{\partial x_i}}(\hat{x}) = \frac{1}{J_F(\hat{x})} \sum_{j=1}^{n} \frac{\partial \hat{p}}{\partial \hat{x}_j}(\hat{x}) \det\left(\frac{\partial F}{\partial \hat{x}_1}(\hat{x}), \ldots, \frac{\partial F}{\partial \hat{x}_{j-1}}(\hat{x}),\right.$$

$$\left. e_i, \frac{\partial F}{\partial \hat{x}_{j+1}}(\hat{x}), \ldots, \frac{\partial F}{\partial \hat{x}_n}(\hat{x})\right) ,$$

where $e_i$ is the i-th vector of the canonical basis of $R^n$. For a given family $(K_h) = \mathcal{K}(\hat{P};\mathcal{F})$ of finite elements, we introduce the finite-dimensional space

$$(3.29) \quad [\hat{P};\mathcal{F}] = \{\widehat{\frac{\partial p}{\partial x_i}} J_F : \hat{K} \to R ; \forall \hat{p} \in \hat{P}, \forall F \in \mathcal{F}, 1 \leqslant i \leqslant n\}$$

so that, in view of (3.28), the space $[\hat{P};\mathcal{F}]$ does *not* depend upon $h$. Moreover in spite of its apparently complicated definition, the space $[\hat{P};\mathcal{F}]$ consists of polynomials as long as the space $\hat{P}$ also consists of polynomials. For instance,

$$(3.30) \quad [\hat{P}(k);(\hat{P}(1))^n] = \hat{P}(k-1) \quad ,$$

$$(3.31) \quad [\hat{P}(k);(\hat{P}(k))^n] \subset \hat{P}(n(k-1)) \quad ,$$

$$(3.32) \quad [\hat{Q}(k);(\hat{Q}(1))^2] = \hat{Q}(k) \cap \hat{P}(2k-1) \text{ for } n = 2 \quad ,$$

$$(3.33) \quad [\hat{Q}(k);(\hat{Q}(k))^2] \subset \hat{Q}(2k-1) \text{ for } n = 2, \text{ etc...}$$

Theorem 6. *For some integer* $r \geqslant 0$, *let there be given an* r*-regular family* $(K_h) = \mathcal{K}(\hat{P};\mathcal{F})$ *of finite elements and a quadrature scheme over the reference set* $\hat{K}$ *which satisfies*

(3.34) $\hat{E}(\hat{v}\hat{\omega}) = 0$ *for all* $\hat{v} \in \hat{P}(r)$ *and* $\hat{\omega} \in [\hat{P};\mathcal{F}]$ ,

(3.34') $\hat{E}(\hat{v}\hat{\omega}) = 0$ *for all* $\hat{v} \in \hat{Q}(r)$ *and* $\hat{\omega} \in [\hat{P};\mathcal{F}]$ .

*Then for any number* $q$ *with* $1 \leq q \leq +\infty$ *and* $r+1 - \frac{n}{q} > 0$, *there exists a constant* $C$ *independent of* $h$ *such that*

(3.35) $$\left|E\left(v\frac{\partial p}{\partial x_i}\right)\right| \leq Ch^{r+1}\|v\|_{W^{r+1,q}(K_h)} \left\|\frac{\partial p}{\partial x_i}\right\|_{L^{q'}(K_h)}$$

*for all* $v \in W^{r+1,q}(K_h)$ *and all* $p \in P$.

*Proof*. For fixed $\hat{p} \in \hat{P}$ and $F \in \mathcal{F}$, the linear form

$$\hat{v} \in W^{r+1,q}(\hat{K}) \to \hat{E}\left(\hat{v}\widehat{\frac{\partial p}{\partial x_i}} J_F\right)$$

is continuous with norm $\leq \gamma_1 \left\|\widehat{\frac{\partial p}{\partial x_i}} J_F\right\|_{L^{q'}(K)}$ and it vanishes when $\hat{v} \in \hat{P}(r)$ by assumption (3.34) since $\widehat{\frac{\partial p}{\partial x_i}} J_F \in [\hat{P};\mathcal{F}]$, by definition. Thus, by the Bramble-Hilbert lemma,

$$\left|\hat{E}\left(\hat{v}\widehat{\frac{\partial p}{\partial x_i}} J_F\right)\right| \leq \gamma_2 |\hat{v}|_{r+1,q,\hat{K}} \left\|\widehat{\frac{\partial p}{\partial x_i}} J_F\right\|_{L^{q'}(\hat{K})},$$

$$\forall \hat{v} \in W^{r+1,q}(\hat{K}), \forall \hat{p} \in \hat{P} .$$

Since $E\left(v\frac{\partial p}{\partial x_i}\right) = E\left(\hat{v}\widehat{\frac{\partial p}{\partial x_i}} J_F\right)$ by (3.8), the conclusion follows by observing that

$$\left\|\widehat{\frac{\partial p}{\partial x_i}} J_F\right\|_{L^{q'}(\hat{K})} \leq \left(\sup_{\hat{x}\in\hat{K}} J_F(\hat{x})\right)^{1/q} \left\|\frac{\partial p}{\partial x_i}\right\|_{L^{q'}(K)} ,$$

and applying inequality (2.22) of Lemma 1 combined with the inequality of (2.20). ∎

Theorem 7. *Let there be given a* 0*-regular family* $(K_h) = \mathcal{K}(\hat{P};\mathcal{F})$ *of finite elements and a quadrature scheme over the reference set* $\hat{K}$ *which satisfies*

$$(3.36) \qquad \hat{E}(\hat{\omega}) = 0 \text{ for all } \hat{\omega} \in [\hat{P};\mathcal{F}] \ .$$

*Then there exists a constant* C *independent of* h *such that*

$$(3.37) \qquad \left| E\left(a_{ij} \frac{\partial p}{\partial x_i} \frac{\partial p'}{\partial x_j}\right) \right|$$

$$\leq Ch \|a_{ij}\|_{W^{1,\infty}(K_h)} \left\|\frac{\partial p}{\partial x_i}\right\|_{L^2(K_h)} \left\|\frac{\partial p'}{\partial x_j}\right\|_{H^1(K_h)}$$

*for all* $a_{ij} \in W^{1,\infty}(K_h)$, $p \in P$ *and* $p' \in P$.

*Proof.* Consider first the expression $\hat{E}\left(\hat{a}_{ij} \widehat{\frac{\partial p}{\partial x_i}} \left(\widehat{\frac{\partial p'}{\partial x_j}} - \hat{\Pi}_0 \widehat{\frac{\partial p'}{\partial x_j}}\right) J_F\right)$, $\hat{p}$, $\hat{p}' \in P$, $F \in \mathcal{F}$, where the operator $\hat{\Pi}_0$ is defined as in (3.17). We obtain

$$\left| \hat{E}\left(\hat{a}_{ij} \widehat{\frac{\partial p}{\partial x_i}} \left(\widehat{\frac{\partial p'}{\partial x_j}} - \hat{\Pi}_0 \widehat{\frac{\partial p'}{\partial x_j}}\right) J_F\right) \right|$$

$$\leq \gamma_1 \left(\inf_{\hat{x}\in\hat{K}} J_F(\hat{x})\right)^{-1} \|\hat{a}_{ij}\|_{L^\infty(\hat{K})} \left\|\widehat{\frac{\partial p}{\partial x_i}} J_F\right\|_{L^2(\hat{K})} \left\|\left(\widehat{\frac{\partial p'}{\partial x_j}} - \hat{\Pi}_0 \widehat{\frac{\partial p'}{\partial x_j}}\right) J_F\right\|_{L^2(\hat{K})} .$$

Using the Bramble-Hilbert lemma,

$$\left\|\left(\widehat{\frac{\partial p'}{\partial x_j}} - \hat{\Pi}_0 \widehat{\frac{\partial p'}{\partial x_j}}\right) J_F\right\|_{L^2(\hat{K})} \leq \gamma_2 \left(\sup_{\hat{x}\in\hat{K}} J_F(\hat{x})\right) \left|\widehat{\frac{\partial p'}{\partial x_j}}\right|_{1,2,\hat{K}} ,$$

so that

$$(3.38)\quad \hat{E}\left(\hat{a}_{ij}\,\widehat{\frac{\partial p}{\partial x_i}}\,\left(\widehat{\frac{\partial p'}{\partial x_j}} - \hat{\Pi}_0\,\widehat{\frac{\partial p'}{\partial x_j}}\right)J_F\right)$$

$$\leq \gamma_3\left(\sup_{\hat{x}\in\hat{K}} J_F(\hat{x})/\inf_{\hat{x}\in\hat{K}} J_F(\hat{x})\right)\|\hat{a}_{ij}\|_{L^\infty(\hat{K})}\left\|\widehat{\frac{\partial p}{\partial x_i}}J_F\right\|_{L^2(\hat{K})}\left|\widehat{\frac{\partial p'}{\partial x_j}}\right|_{1,2,\hat{K}}\quad .$$

Now, for fixed $\hat{p}, \hat{p}' \in \hat{P}$ and $F \in \mathcal{F}$, the linear form

$$\hat{a}_{ij} \in W^{1,\infty}(\hat{K}) \to \hat{E}\left(\hat{a}_{ij}\,\widehat{\frac{\partial p}{\partial x_i}}\,\left(\hat{\Pi}_0\,\widehat{\frac{\partial p'}{\partial x_j}}\right)J_F\right)$$

is continuous with norm $\leq \gamma_4\left\|\widehat{\frac{\partial p}{\partial x_i}}J_F\right\|_{L^2(\hat{K})}\left\|\widehat{\frac{\partial p'}{\partial x_j}}\right\|_{L^2(\hat{K})}$ and, by assumption (3.36), it vanishes on $\hat{P}(0)$. Thus, by the Bramble-Hilbert lemma,

$$(3.39)\quad \left|\hat{E}\left(\hat{a}_{ij}\,\widehat{\frac{\partial p}{\partial x_i}}\left(\Pi_0\,\widehat{\frac{\partial p'}{\partial x_j}}\right)J_F\right)\right|$$

$$\leq \gamma_5\,|\hat{a}_{ij}|_{1,\infty,\hat{K}}\left\|\widehat{\frac{\partial p}{\partial x_i}}J_F\right\|_{L^2(\hat{K})}\left\|\widehat{\frac{\partial p'}{\partial x_j}}\right\|_{L^2(\hat{K})}\quad .$$

By combining inequalities (3.38)-(3.39) and inequalities (2.20) and (2.22), we obtain the desired estimate (3.37). ▮

Theorem 8. *For some integer* $r \geq 0$, *let there be given an* $(r+1)$-*regular family* $(K_h) = \mathcal{K}(P;\mathcal{F})$ *of finite elements and a quadrature scheme over the reference set* $\hat{K}$ *which satisfies*

$$(3.40)\quad \begin{cases} \hat{E}(\hat{v}\hat{w}) = 0 \text{ for all } \hat{v} \in \hat{P}(r) \text{ and } \hat{w} \in [\hat{P},\mathcal{F}]\quad , \\ \hat{E}(\hat{v}J_F) = 0 \text{ for all } \hat{v} \in \hat{P}(r+1) \text{ and } F \in \mathcal{F}\quad , \end{cases}$$

$$(3.40')\quad \begin{cases} \hat{E}(\hat{v}\hat{w}) = 0 \text{ for all } \hat{v} \in \hat{Q}(r) \text{ and } \hat{w} \in [\hat{P},\mathcal{F}]\quad , \\ \hat{E}(\hat{v}J_F) = 0 \text{ for all } \hat{v} \in \hat{Q}(r+1) \text{ and } F \in \mathcal{F}\, . \end{cases}$$

*Then for any number* $q$ *with* $1 \leq q \leq +\infty$ *and* $r+1 - \frac{n}{q} > 0$, *there exists a constant* $C$ *independent of* $h$ *such that*

$$(3.41)\quad \left|E\left(a_{ij}\frac{\partial p}{\partial x_i}\frac{\partial p'}{\partial x_j}\right)\right|$$

$$\leq C h^{r+2}\|a_{ij}\|_{W^{r+2,\infty}(K_h)}\left\|\frac{\partial p}{\partial x_i}\right\|_{W^{r+1,q}(K_h)}\left\|\frac{\partial p'}{\partial x_j}\right\|_{W^{1,q'}(K_h)}$$

*for all* $a_{ij} \in W^{r+2,\infty}(K_h)$, $p \in P$ *and* $p' \in P$.

*Proof*. Consider first, for fixed $\hat{p}' \in \hat{P}$, $F \in \mathcal{F}$, the linear form

$$\hat{v} \in W^{r+1,q}(\hat{K}) \to \hat{E}\left(\hat{v}\left(\frac{\widehat{\partial p'}}{\partial x_j} - \hat{\Pi}_0\frac{\widehat{\partial p'}}{\partial x_j}\right)J_F\right)$$

which is continuous with norm $\leq \gamma_1\left(\sup_{\hat{x}\in\hat{K}} J_F(\hat{x})\right)\left|\frac{\widehat{\partial p'}}{\partial x_j}\right|_{1,q',\hat{K}}$ and which vanishes over $\hat{P}(r)$ by (3.40). Thus,

$$\left|\hat{E}\left(\hat{v}\left(\frac{\widehat{\partial p'}}{\partial x_j} - \hat{\Pi}_0\frac{\widehat{\partial p'}}{\partial x_j}\right)J_F\right)\right| \leq \gamma_2\left(\sup_{\hat{x}\in\hat{K}} J_F(\hat{x})\right)|\hat{v}|_{r+1,q,\hat{K}}\left|\frac{\widehat{\partial p'}}{\partial x_j}\right|_{1,q',\hat{K}}$$

so that

$$(3.42)\quad \left|\hat{E}\left(\hat{a}_{ij}\frac{\widehat{\partial p}}{\partial x_i}\left(\frac{\widehat{\partial p'}}{\partial x_j} - \hat{\Pi}_0\frac{\widehat{\partial p'}}{\partial x_j}\right)J_F\right)\right|$$

$$\leq \gamma_2\left(\sup_{\hat{x}\in\hat{K}} J_F(\hat{x})\right)\left|\hat{a}_{ij}\frac{\widehat{\partial p}}{\partial x_i}\right|_{r+1,q,\hat{K}}\left|\frac{\widehat{\partial p'}}{\partial x_j}\right|_{1,q',\hat{K}}$$

for all $\hat{a}_{ij} \in W^{r+1,\infty}(\hat{K})$, $\hat{p}, \hat{p}' \in \hat{P}$, $F \in \mathcal{F}$.

Likewise, for fixed $\hat{p}, \hat{p}' \in \hat{P}$, $F \in \mathcal{F}$, the linear form

$$\hat{a}_{ij} \in W^{r+1,\infty}(\hat{K}) \to \hat{E}\left(\hat{a}_{ij}\left(\widehat{\frac{\partial p}{\partial x_i}} - \hat{\Pi}_0 \widehat{\frac{\partial p}{\partial x_i}}\right)\left(\hat{\Pi}_0 \widehat{\frac{\partial p'}{\partial x_j}}\right) J_F\right)$$

is continuous with norm $\leq \gamma_3\left(\sup_{\hat{x}\in\hat{K}} J_F(\hat{x})\right) \left|\widehat{\frac{\partial p}{\partial x_i}}\right|_{1,q,\hat{K}} \left\|\widehat{\frac{\partial p'}{\partial x_j}}\right\|_{L^{q'}(\hat{K})}$

and, by assumption (3.40), it vanishes over $\hat{P}(r)$. Thus,

$$\left|\hat{E}\left(\hat{a}_{ij}\left(\widehat{\frac{\partial p}{\partial x_i}} - \hat{\Pi}_0 \widehat{\frac{\partial p}{\partial x_i}}\right)\left(\hat{\Pi}_0 \widehat{\frac{\partial p'}{\partial x_j}}\right) J_F\right)\right| \leq \gamma_4\left(\sup_{\hat{x}\in\hat{K}} J_F(\hat{x})\right) |\hat{a}_{ij}|_{r+1,\infty,\hat{K}} \left|\widehat{\frac{\partial p}{\partial x_i}}\right|_{1,q,\hat{K}} \left\|\widehat{\frac{\partial p'}{\partial x_j}}\right\|_{L^{q'}(\hat{K})} . \tag{3.43}$$

Finally, for fixed $\hat{p}, \hat{p}' \in \hat{P}$, $F \in \mathcal{F}$, the linear form

$$\hat{a}_{ij} \in W^{r+2,\infty}(\hat{K}) \to \hat{E}\left(\hat{a}_{ij}\left(\hat{\Pi}_0 \widehat{\frac{\partial p}{\partial x_i}}\right)\left(\hat{\Pi}_0 \widehat{\frac{\partial p'}{\partial x_j}}\right) J_F\right)$$

is continuous with norm $\leq \gamma_5\left(\sup_{\hat{x}\in\hat{K}} J_F(\hat{x})\right) \left\|\widehat{\frac{\partial p}{\partial x_i}}\right\|_{L^q(\hat{K})} \left\|\widehat{\frac{\partial p'}{\partial x_j}}\right\|_{L^{q'}(\hat{K})}$

and, by assumption (3.40), it vanishes over $\hat{P}(r+1)$. Thus,

$$\left|\hat{E}\left(\hat{a}_{ij}\left(\hat{\Pi}_0 \widehat{\frac{\partial p}{\partial x_i}}\right)\left(\hat{\Pi}_0 \widehat{\frac{\partial p'}{\partial x_j}}\right) J_F\right)\right| \leq \gamma_6\left(\sup_{\hat{x}\in\hat{K}} J_F(\hat{x})\right) |\hat{a}_{ij}|_{r+2,\infty,\hat{K}} \left\|\widehat{\frac{\partial p}{\partial x_i}}\right\|_{L^q(\hat{K})} \left\|\widehat{\frac{\partial p'}{\partial x_j}}\right\|_{L^{q'}(\hat{K})} . \tag{3.44}$$

Combining (3.42), (3.43) and (3.44), we obtain

$$\left|\hat{E}\left(\hat{a}_{ij} \widehat{\frac{\partial p}{\partial x_i}} \widehat{\frac{\partial p'}{\partial x_j}} J_F\right)\right| \leq \gamma_7\left(\sup_{\hat{x}\in\hat{K}} J_F(\hat{x})\right) \left\{\left|\hat{a}_{ij} \widehat{\frac{\partial p}{\partial x_i}}\right|_{r+1,q,\hat{K}} \left|\widehat{\frac{\partial p'}{\partial x_j}}\right|_{1,q',\hat{K}} + \left(|\hat{a}_{ij}|_{r+1,\infty,\hat{K}} \left|\widehat{\frac{\partial p}{\partial x_i}}\right|_{1,q,\hat{K}} + |\hat{a}_{ij}|_{r+2,\infty,\hat{K}} \left\|\widehat{\frac{\partial p}{\partial x_i}}\right\|_{L^q(\hat{K})}\right) \left\|\widehat{\frac{\partial p'}{\partial x_j}}\right\|_{L^{q'}(\hat{K})}\right\}.$$

The conclusion (3.41) follows by using the same arguments as in the previous theorems. ∎

## 4. The discrete problem: error bound in $H^1(\Omega_h)$.

We first give a complete description of the discrete problem (1.10). We begin by constructing a set $\bar{\Omega}_h$ as a finite union of isoparametric finite elements $K = F_K(\hat{K})$ where, following §2, $\hat{K}$ is a fixed reference set and $F_K \in \hat{P}^n$, where $\hat{P}$ is a fixed finite-dimensional space of functions defined over the set $\hat{K}$. Notice that we adopt here a new notation for the mapping $F$, namely $F_K$, which is more appropriate now. Also notice that we implicitly allow only one type of isoparametric finite elements by fixing $\hat{K}$ and $\hat{P}$ but this is not a restriction: it just makes matters simpler!

Each element $K$ is determined by the data of $N$ points $a_{i,K}$ *which will always be assumed to belong to the set* $\bar{\Omega}$. Then we distinguish two cases:

(i) The corresponding mapping $F_K$ is either *affine*, i.e., $F_K \in (\hat{P}(1))^n$, for simplicial elements, or *bilinear*, i.e., $F_K \in (\hat{Q}(1))^2$ for *quadrilateral* elements (in $R^2$). We will denote by $\mathcal{E}_h^1$ the union of the corresponding finite elements.

(ii) We have the *general isoparametric case*, i.e., $F_K \in \hat{P}^n$. We will denote by $\mathcal{E}_h^2$ the union of the corresponding finite elements.

In both cases, some points $a_{i,K}$ may belong to the boundary $\Gamma$. However in case (ii), at least one "face" of the set $K$ will be curved. These considerations are perhaps better illustrated by Fig. 3 and 4 where finite elements of type (2,S) and (2,Q) have been chosen for definiteness.

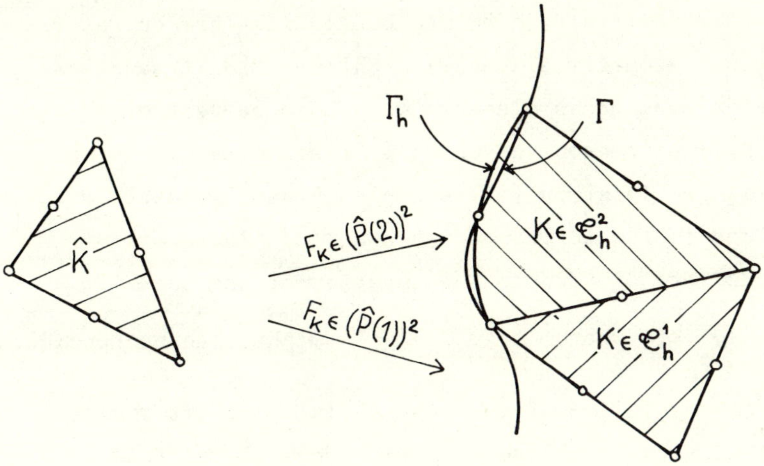


Fig. 3. Boundary and interior finite elements of type (2,S)

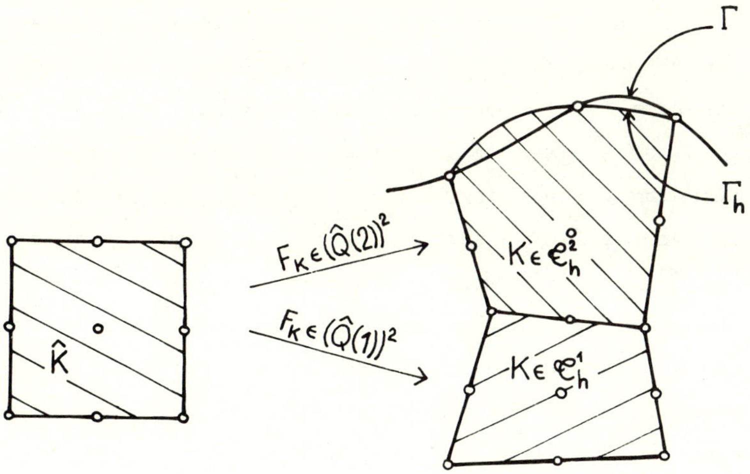


Fig. 4. Boundary and interior finite elements of type (2,Q)

In practice, the set $\mathcal{E}_h^2$ will often consist of *boundary* finite elements whereas the set $\mathcal{E}_h^1$ will rather consist of *interior* finite elements. Let us add, however, that interior finite elements whose sides are curved (in the usual geometrical sense) might be used if we please, in which case those elements will also be part of the set $\mathcal{E}_h^2$. Serious reasons to do so must exist though, since as we will see later on (cf. Examples 6 and 7), extra work is required in the corresponding numerical integration.

We let $\Omega_h$ denote the interior of the set $\bar{\Omega}_h = \bigcup_{K \in \mathcal{E}_h} K$, where $\mathcal{E}_h = \mathcal{E}_h^1 \cup \mathcal{E}_h^2$ might then be considered as a *triangulation* of the set $\bar{\Omega}$ (although it really is a triangulation of the set $\bar{\Omega}_h$ !). We will use this terminology for problems in $R^n$ with $n > 2$ as well as for "triangulations" with finite elements of quadrilateral type!

If $K_1$ and $K_2$ are two "adjacent" finite elements, their intersection must be an $(n-1)$-dimensional "face" of each other or, equivalently, the geometrical shape of any "face" $\Delta$ of a given finite element $K$ must be completely determined by those points $a_{i,K}$ which belong to $\Delta$. This is true for the finite elements we are considering in the examples for the following reason: if we call $\hat{\Delta}$ the corresponding face of the reference set $\hat{K}$, the set $\hat{\Sigma} \cap \hat{\Delta}$ contains a $\hat{P}_{\hat{\Delta}}$-unisolvent set, where $\hat{P}_{\hat{\Delta}}$ is the space of the restrictions of the functions of $\hat{P}$ over the set $\hat{\Delta}$ (cf. also Theorem 1, p. 137 of [26]).

We say that a *family* $(\mathcal{C}_h)$ of triangulations of a given domain $\Omega$ is a k-*regular family* if the family $(\bigcup_{K\in\mathcal{C}_h} K)$ is itself a k-regular family, in the sense of §2.

Example 5: *Some* k-*regular families of triangulations*. For finite elements of type (2,S) in $R^2$ (cf. Fig. 5), it suffices to pick the point $a_{23}$ at the intersection of the boundary $\Gamma$ and the normal to the segment $[a_2,a_3]$ passing through the point $\tilde{a}_{23} = (a_2+a_3)/2$. If the boundary $\Gamma$ is smooth enough, conditions (2.33) are thus automatically satisfied. This construction is readily extensible to finite elements of type (2,S) in $R^n$, and also to finite elements of type (2,Q) for plane problems (cf. Fig. 6), thus yielding in both cases 2-*regular families*. The (1,Q) type corresponds trivially to a 1-*regular family* (cf. Fig. 6).

For finite elements of type (3,S) in $R^2$, it suffices to choose the points $a_5$ and $a_8$ at the intersections of $\Gamma$ and the normals to the segment $[a_2,a_3]$ passing through the points $\tilde{a}_5 = (2a_2+a_3)/3$ and $\tilde{a}_8 = (2a_3+a_2)/3$, respectively (cf. Fig. 5). Then conditions (2.35) and (2.36) are automatically satisfied if the boundary $\Gamma$ is smooth enough, and then the point $a_{10}$ is chosen so as to satisfy (2.37). In this fashion we obtain a 3-*regular family* (apparently, this construction is not extensible to higher-dimensional problems). ∎

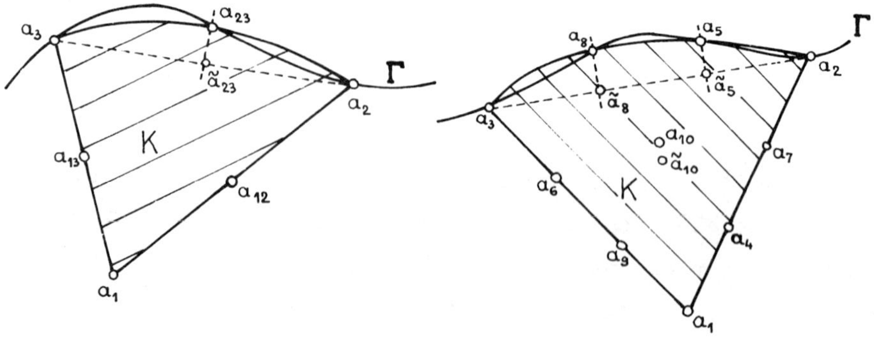


Fig. 5. Boundary finite elements of type (2,S) and (3,S)

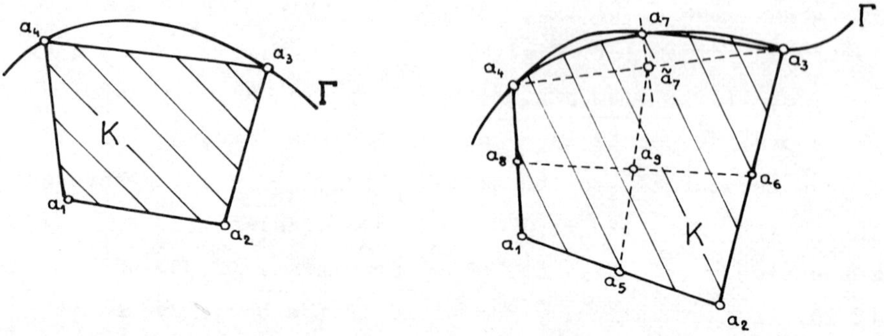


Fig. 6. Boundary finite elements of type (1,Q) and (2,Q)

With a given triangulation $\mathcal{C}_h$, we associate the space $V_h$ of the *trial functions*, defined by

$$V_h = \{v \in C^0(\bar{\Omega}_h);\ v_K \in P_K \text{ for all } K \in \mathcal{C}_h,\ u = 0 \text{ on } \Gamma_h\} ,$$

where $v_K$ is the restriction of the function $v$ over the set $K$, the space $P_K$ is defined for each finite element $K$ as in (2.4), and $\Gamma_h$ denotes the boundary of the set $\Omega_h$. A function $v \in V_h$ is thus uniquely determined by its values at the points $a_{i,K}$, $1 \leq i \leq N$, $K \in \mathcal{E}_h$.

We notice at this point that the assumption of (1.7) is surely satisfied by our examples. Also because of this assumption, there exists a constant $\delta$ *independent of* $h$ such that

$$(4.1) \qquad \|u\|_{H^1(\Omega_h)} \leq \delta |u|_{1,2,\Omega_h} \quad \text{for all} \quad u \in V_h$$

$$\text{and for all} \quad h \quad .$$

Now we introduce *numerical integration*, and we follow the approach and notations of §3. We make the basic assumption that *for all* $h$ *and all* $K \in \mathcal{E}_h$, *all the integration points* $b_{\ell,K}$ *lie in the set* $\bar{\Omega}$. In our examples, this will be the case for $h$ small enough, if for any integration point $b_\ell$, $1 \leq \ell \leq L$, over the reference set $\hat{K}$,

- either $\hat{b}_\ell = \hat{a}_i$ for some $1 \leq i \leq N$,
- or if $\hat{b}_\ell \neq \hat{a}_i$ for $1 \leq i \leq N$, then $\hat{b}_\ell$ is in the *interior* of $\hat{K}$.

That there exist schemes satisfying this assumption along with $> 0$ quadrature weights $\hat{\omega}_\ell$ follows from a result of TCHAKALOFF [25] (see also DAVIS [11] for a constructive proof).

Following §1, the discrete problem associated with the triangulation $\mathcal{E}_h$ consists in approximating the problem (1.4) using numerical integration, that is, the integrals $\int_{\Omega_h} \cdots \, dx$ appearing in (1.4) will be written as $\sum_{K \in \mathcal{E}_h} \int_K \cdots \, dx$ and each integral $\int_K \cdots \, dx$ will be computed using the integration procedure described in §3. As it will become apparent later, we might have to use *several* integration schemes *over the reference set* $\hat{K}$; this possibility is implicitly contained in the dependence upon the sets $K$ appearing in (4.3) and (4.4). Then *the discrete problem consists in finding* $u_h \in V_h$ *such that*

$$(4.2) \qquad a_h(u_h, v_h) = (f_h, v_h)_h \quad \textit{for all} \quad v_h \in V_h \quad ,$$

*where for each* $u_h$, $v_h \in V_h$ *we have*

$$(4.3) \qquad a_h(u_h, v_h) = \sum_{K \in \mathcal{E}_h} \sum_{\ell=1}^{L_K} \omega_{\ell,K} \left( \sum_{i,j=1}^{n} a_{ij} \frac{\partial u_h}{\partial x_i} \frac{\partial v_h}{\partial x_j} \right) (b_{\ell,K}),$$

$$(4.4) \qquad (f_h, v_h)_h = \sum_{K \in \mathcal{E}_h} \sum_{\ell=1}^{L_K} \omega_{\ell,K} (f v_h)(b_{\ell,K}) \quad .$$

Since all the integration points $b_{\ell,K}$ have been assumed to belong to the set $\bar{\Omega}$, it is indeed true that *the discrete problem is independent of possible extensions* $\tilde{a}_{ij}$ *and* $\tilde{f}$ *of the functions* $a_{ij}$ *and* $f$, as was announced in §1.

We now come to a key result. For the sake of brevity, smoothness assumptions will not be explicitly given in the statement of Theorem 9, but rather in its proof.

Theorem 9. *We assume that the operator* $A$ *of* (1.1) *is strongly elliptic: for some constant* $\alpha > 0$,

(4.5) $$\sum_{i,j=1}^{n} a_{ij}(x)\xi_i\xi_j \geq \alpha \sum_{i=1}^{n} \xi_i^2 \text{ for all } x \in \bar{\Omega} \text{ and } \xi \in R^n,$$

*and that the functions* $a_{ij}$ *and* f, *as well as the boundary* $\Gamma$ *are sufficiently smooth.*

*For a fixed integer* $r \geq 0$, *we consider an* (r+1)-*regular family* $(\mathcal{C}_h)$ *of triangulations such that*

(4.6) *either* $\hat{P}(r+1) \subset \hat{P}$ *for simplicial elements,*

(4.6') *or* $\hat{Q}(r+1) \subset \hat{P}$ *for quadrilateral elements,*

*and we assume that a finite number of quadrature schemes are used over the reference set* $\hat{K}$ *which satisfy the assumptions* (3.9) *of Theorem* 3, (3.13) *and* (3.14) *of Theorem* 4 *(resp.* (3.13') *and* (3.14')), *and* (3.34) *of Theorem* 6 *(resp.* (3.34')) *for some number* q *satisfying* $2 \leq q \leq +\infty$ *and* $r+1 - \frac{n}{q} > 0$.

*Then there exists a constant* C *independent of* h *such that*

(4.7) $$\|\tilde{u}-u_h\|_{H^1(\Omega_h)} \leq Ch^{r+1}\|\tilde{u}\|_{W^{r+3,q}(\Omega_h)}$$

*where* $\tilde{u}$ *is a sufficiently smooth extension over* $\tilde{\Omega}$ *of the solution* u *of* (1.1)-(1.2), *and* $u_h$ *is the solution of the discrete problem* (1.10).

*Proof.* Since we will apply Theorem 1, we need first to check the "uniform ellipticity" condition of (1.11). Using (4.3), (4.5), inequality (3.10) of Theorem 3 and inequality (4.1), we obtain, for any function $v_h \in V_h$,

$$a_h(v_h,v_h) \geq \alpha \sum_{K\in\mathcal{C}_h} \sum_{\ell=1}^{L_K} \omega_{\ell,K} \sum_{i=1}^{n} \left(\frac{\partial v_h}{\partial x_i}(b_{\ell,K})\right)^2 \geq \frac{\alpha\gamma}{\delta}\|v_h\|^2_{H^1(\Omega_h)},$$

and thus inequality (1.11) is satisfied.

Next, with any function $v \in C^o(\bar{\Omega})$ which vanishes over $\Gamma_h$, we may associate its unique $V_h$-interpolate $r_h v$, which satisfies

$$r_h v \in V_h \text{ and } r_h v(a_{i,K}) = v(a_{i,K}),\ 1 \leq i \leq N,\ K \in \mathcal{E}_h \ .$$

If $\tilde{u} \in W^{r+2,q}(\tilde{\Omega})$, inequality (2.26) of Theorem 2 used in conjunction with assumption (4.6) (resp. (4.6')) yields

$$(4.8) \qquad \|\tilde{u}-r_h\tilde{u}\|_{H^1(\Omega_h)} \leq \gamma_1 h^{r+1} \|\tilde{u}\|_{W^{r+2,q}(\Omega_h)} \ ,$$

for some constant $\gamma_1$ independent of $h$ and $\tilde{u}$, since $2 \leq q$.

Let us now examine the second term appearing in the right-hand side of inequality (1.12). For any $v_h, w_h \in V_h$, we have by (4.3),

$$(4.9) \quad \tilde{a}_h(v_h,w_h) - a_h(v_h,w_h) = \sum_{K\in\mathcal{E}_h} E_K\left(\sum_{i,j=1}^{n} \tilde{a}_{ij} \frac{\partial v_h}{\partial x_i} \frac{\partial w_h}{\partial x_j}\right) .$$

By Theorem 6, there exist some constants $\gamma_2$, $\gamma_3$ independent of $h$ and $K$ such that, if we assume for instance that $\tilde{a}_{ij} \in W^{r+1,\infty}(\tilde{\Omega})$,

$$\left|E_K\left(\sum_{i,j=1}^{n} \tilde{a}_{ij} \frac{\partial v_h}{\partial x_i} \frac{\partial w_h}{\partial x_j}\right)\right| \leq \gamma_2 h^{r+1} \left\|\tilde{a}_{ij} \frac{\partial v_h}{\partial x_i}\right\|_{W^{r+1,q}(K)} \left\|\frac{\partial w_h}{\partial x_j}\right\|_{L^{q'}(K)}$$

$$\leq \gamma_3 h^{r+1} \max_{i,j} \|\tilde{a}_{ij}\|_{W^{r+1,\infty}(\tilde{\Omega})} \|v_h\|_{W^{r+2,q}(K)} |\omega_h|_{1,q',K} \ ,$$

and thus, using $2 \leq q$,

(4.10) $$\Big|\sum_{K\in\mathcal{E}_h} E_K\Big(\sum_{i,j=1} \tilde{a}_{ij}\,\frac{\partial v_h}{\partial x_i}\,\frac{\partial w_h}{\partial x_j}\Big)\Big|$$

$$\leq \gamma_4 h^{r+1}\Big(\sum_{K\in\mathcal{E}_h} \|v_h\|^q_{W^{r+2,q}(K)}\Big)^{1/q} |\omega_h|_{1,2,\Omega_h} \quad ,$$

for some constant $\gamma_4$ independent of $h$.

*Since* $r_h\tilde{u} = r_h u$, it follows that $r_h\tilde{u} = 0$ on $\Gamma_h$, and therefore $r_h\tilde{u} \in V_h$. Thus, we deduce from (4.8), (4.9) and (4.10) that

(4.11) $$\inf_{v_h\in V_h} \{\|\tilde{u}-v_h\|_{H^1(\Omega_h)} + \sup_{w_h\in V_h} \frac{|\tilde{a}_h(v_h,w_h)-a_h(v_h,w_h)|}{\|w_h\|_{H^1(\Omega_h)}}\}$$

$$\leq \gamma_5 h^{r+1}\Big(\|\tilde{u}\|_{W^{r+2,q}(\Omega_h)} + \Big(\sum_{K\in\mathcal{E}_h} \|r_h\tilde{u}\|^q_{W^{r+2,q}(K)}\Big)^{1/q}\Big)$$

$$\leq \gamma_6 h^{r+1}\|\tilde{u}\|_{W^{r+2,q}(\Omega_h)} \quad ,$$

using inequality (2.26) with $m = r+2$, for some constants $\gamma_5$ and $\gamma_6$ independent of $h$.

Finally, we estimate the last term in the right-hand side of inequality (1.12). *Since* $\tilde{u} = u$ *over* $\Omega$ *and all the points* $b_{\ell,K}$ *belong to* $\bar{\Omega}$, *we have*

$$A_h\tilde{u}(b_{\ell,K}) = A_h u(b_{\ell,K}) = f(b_{\ell,K})$$

for all the points $b_{\ell,K}$, and therefore, using (1.13), (1.18) and (4.4), we may write, for any function $w_h \in V_h$,

(4.12) $$(\tilde{A}_h\tilde{u}-f_h,\, w_h)_h = \sum_{K\in\mathcal{E}_h} E_K\Big((\tilde{A}_h\tilde{u})w_h\Big) \quad .$$

Thus, applying inequality (3.15) of Theorem 4 combined with $2 \leq q$, we obtain

$$(4.13) \quad \left| \sum_{K \in \mathcal{C}_h} E_K\left((\tilde{A}_h \tilde{u}) w_h\right) \right|$$

$$\leq \gamma_7 h^{r+1} \left( \sum_{K \in \mathcal{C}_h} \|\tilde{A}_h \tilde{u}\|^q_{W^{r+1,q}(K_h)} \right)^{1/q} \|w_h\|_{H^1(\Omega_h)}$$

for some constant $\gamma_7$ independent of $h$. If $\tilde{u} \in W^{r+3,q}(\tilde{\Omega})$, we have, by (4.12) and (4.13),

$$(4.14) \qquad \sup_{w_h \in V_h} \frac{|(\tilde{A}_h \tilde{u} - f_h, w_h)|}{\|\omega_h\|_{H^1(\Omega_h)}} \leq \gamma_8 h^{r+1} \|\tilde{u}\|_{W^{r+3,q}(\Omega_h)} ,$$

for some constant $\gamma_8$ independent of $h$.

The proof is now achieved by combining inequalities (4.11) and (4.14). ∎

Remark 1. If the set $\Omega$ has the $(r+3,q)$-*extension property*, i.e., there exists $\mathcal{P} \in \mathcal{L}(W^{r+3,q}(\Omega); W^{r+3,q}(\tilde{\Omega}))$ such that $\mathcal{P}u = u$ over $\Omega$, then we obtain as an immediate consequence of (4.7):

$$\|\tilde{u} - u_h\|_{H^1(\Omega_h)} \leq Ch^{r+1} \|u\|_{W^{r+3,q}(\Omega)} . \qquad ∎$$

Remark 2. The estimates related to approximate integration require the family $(\mathcal{C}_h)$ to be only r-regular. The (r+1)-regularity will be fully used in Theorem 11. ∎

Example 6. Let us assume we are using *finite elements of type* (2,S) for solving problem (1.1)-(1.2) in $R^n$, for arbitrary $n$. It is well known that if $\Omega = \Omega_h$ for all $h$ and numerical integration is not used, then one has

$\|u-u_h\|_{H^1(\Omega)} = O(h^2)$. To obtain the same asymptotic order of convergence in the present case, we thus have to apply Theorem 9 with $r = 1$. We leave to the reader to verify, using (3.13), (3.14), (3.30), (3.31) and (3.34) that

$$\|\tilde{u}-u_h\|_{H^1(\Omega_h)} = O(h^2)$$

*if the quadrature scheme over the reference set* $\hat{K}$

(i) *is exact for* $\hat{P}(2)$ for the *straight* finite elements $K \in \mathcal{C}_h^1$,

(ii) *is exact for* $\hat{P}(n+2)$ for the *curved* finite elements $K \in \mathcal{C}_h^2$.

Likewise, if we are using *finite elements* of type (3,S) for solving problem (1.1)-(1.2) in $R^2$, we obtain

$$\|\tilde{u}-u_h\|_{H^1(\Omega_h)} = O(h^3)$$

*if the quadrature scheme over the reference set* $\hat{K}$

(i) *is exact for* $\hat{P}(4)$ for the *straight* finite elements $K \in \mathcal{C}_h^1$.

(ii) *is exact for* $\hat{P}(8)$ for the *curved* finite elements $K \in \mathcal{C}_h^2$.

In both cases, the assumption (3.9) of Theorem 3 is automatically satisfied since it is a consequence of condition (i).

Also, a careful analysis of the proof of Theorem 6 shows that for finite elements of type (2,S) (resp. (3,S)), exact integration for $\hat{P}(n+1)$ (resp. $\hat{P}(7)$) is sufficient for computing the coefficients of the *matrix* of the resulting linear system, the conditions (ii) stated above being necessary for computing the components of the *vector* of the resulting linear system, in view of Theorem 4.

In other words, it might be of interest to use *different* quadrature schemes, not only for straight vs. curved elements, but also depending upon which coefficients of the linear system, i.e., those of the matrix vs. those of the vector, are computed.

Needless to say, the above analysis (including numerical quadrature) can be also applied to the "simplest" finite elements, i.e., those of type (1,S) (which incidentally are always straight !).

Consider now *finite elements of type* (1,Q) for solving problem (1.1)-(1.2) in $R^2$. By using (3.9), (3.13'), (3.14') and (3.34') we obtain

$$\|\tilde{u}-u_h\|_{H^1(\Omega_h)} = O(h)$$

*if the quadrature scheme over the reference set* $\hat{K}$

(i) is such that $\{\hat{b}_\ell\}_{\ell=1}^{L}$ *contains a* $\hat{P}(1)$*-unisolvent subset,*

(ii) *is exact for* $\hat{Q}(1)$ *for any quadrilateral* $K \in \mathcal{E}_h$ $(= \mathcal{E}_h^1 = \mathcal{E}_h^2)$.

Finally, if we are using *finite elements of type* (2,Q) for solving problem (1.1)-(1.2) in $R^2$, we get

$$\|\tilde{u}-u_h\|_{H^1(\Omega_h)} = O(h^2)$$

*if the quadrature scheme over the reference set* $\hat{K}$

(i) is such that $\{\hat{b}_\ell\}_{\ell=1}^{L}$ *contains a* $\hat{P}(3) \cap \hat{Q}(2)$ *unisolvent subset,*

(ii) *is exact for* $\hat{Q}(3)$ for the quadrilaterals $K \in \mathcal{C}_h^1$,

(iii) *is exact for* $\hat{Q}(5)$ for the *curved* quadrilaterals $K \in \mathcal{C}_h^2$. ∎

As the above examples show, there is a similarity between the "classical" finite-difference method and the present one: when the boundary is curved, it is well-known that special care (in the choice of mesh points and discrete formulae) must be taken in a finite difference method. That this is also the case here is reflected by the extra work which is required in the numerical quadrature scheme associated with boundary finite elements.

We would go even further about this relationship between the two methods. A finite element method, whether of Lagrange type or of Hermite type (in which case we may always, at least theoretically, eliminate the unknown which behave like derivatives) can be trivially viewed as a finite-difference method. Conversely, *it is conjectured here* (and it has been verified on several examples) *that most finite-difference methods can be viewed as appropriate finite element methods associated with proper numerical quadrature schemes*. If this is the case, the approach developed here would thus be a *unifying approach* for many discrete schemes used for approximating solutions of partial differential equations.

## 5. Error bound in $L^2(\Omega_h)$

The notations are the same as those of §1. In addition,

$$A^* \equiv - \sum_{i,j=1}^{n} \frac{\partial}{\partial x_j}\left(a_{ji} \frac{\partial}{\partial x_i}\right) \text{ and } \tilde{A}_h^* \equiv - \sum_{i,j=1}^{n} \frac{\partial}{\partial x_j}\left(\tilde{a}_{ji} \frac{\partial}{\partial x_i}\right)$$

denote the adjoint operators of the operators $A$ of (1.1) and $\tilde{A}_h$ of (1.5), respectively. Also, we denote by $\frac{\partial}{\partial \nu_h^*}$ the conormal derivative associated with the operator $\tilde{A}_h^*$ along the boundary $\Gamma_h$ of $\Omega_h$.

The following abstract result generalizes the approach of AUBIN [1] and NITSCHE [19, 20].

Theorem 10. *We assume that the ellipticity condition of* (4.5) *is satisfied and that the data:* $a_{ij}$, $f$ *and* $\Gamma$, *are sufficiently smooth. Then the unique solution* $u_h$ *of the discrete problem* (1.10) *satisfies*

$$(5.1)\quad \|\tilde{u}-u_h\|_{L^2(\Omega_h)} \leq C\Big\{ \|\tilde{u}-u_h\|_{L^2(\Omega_h \cap \complement\Omega)} + \|\tilde{u}\|_{L^2(\Gamma_h)} \sup_{\phi \in H_0^1(\Omega)\cap H^2(\Omega)} (\|\phi\|_{H^2(\Omega)})^{-1} \left\|\frac{\partial \mathcal{P}\phi}{\partial \nu_h^*}\right\|_{L^2(\Gamma_h)}$$

$$+ \sup_{\phi\in H_0^1(\Omega)\cap H^2(\Omega)} (\|\phi\|_{H^2(\Omega)})^{-1}\Big\{ \inf_{\phi_h \in V_h} \Big[\|\tilde{u}-u_h\|_{H^1(\Omega_h)} \|\mathcal{P}\phi-\phi_h\|_{H^1(\Omega_h)}$$

$$+ |\tilde{a}_h(u_h,\phi_h)-a_h(u_h,\phi_h)| + |(\tilde{A}_h\tilde{u}-f_h,\phi_h)_h|\Big]\Big\}\Big\},$$

*where*

$\tilde{u}$ = *any function in* $H^1(\tilde{\Omega})$,

$C$ = *a constant independent of* $\tilde{u}$ *and* $h$.

$\mathcal{P}$ = *a 2-extension operator, i.e.,* $\mathcal{P} \in \mathcal{L}(H^2(\Omega);H^2(\tilde{\Omega}))$ *and* $\mathcal{P}u = u$ *in* $\Omega$ *for all* $u \in H^2(\Omega)$.

*Proof*. In what follows the constants $\gamma_i$ are solely dependent upon $\Omega$ and $\tilde{\Omega}$. We have

(5.2) $$\|\tilde{u}-u_h\|_{L^2(\Omega_h)} = \sup_{g\in L^2(\Omega_h)} \frac{|\int_{\Omega_h}(\tilde{u}-u_h)g\,dx|}{\|g\|_{L^2(\Omega_h)}} .$$

Given $g \in L^2(\Omega_h)$, we let $\tilde{g} \in L^2(\tilde{\Omega})$ be defined by $\tilde{g} = g$ over $\Omega_h$ and $\tilde{g} = 0$ over $\tilde{\Omega}-\Omega_h$. If the data are smooth enough, the ellipticity condition of (4.5) implies that the Dirichlet problem $A^*\phi = \tilde{g}$ in $\Omega$, $\phi = 0$ on $\Gamma$, has a unique solution $\phi \in H_0^1(\Omega) \cap H^2(\Omega)$; cf. NEČAS [17, Theorem 2.1, p. 201]. Moreover,

(5.3) $$\|\phi\|_{H^2(\Omega)} \leq \gamma_1 \|g\|_{L^2(\Omega_h)} .$$

Likewise, enough smoothness will imply the existence of a 2-*extension operator* $\mathcal{P} \in \mathcal{L}(H^2(\Omega);\, H^2(\tilde{\Omega}))$ such that $\mathcal{P}\phi = \phi$ in $\Omega$ for all $\phi \in H^2(\Omega)$; cf. NEČAS [17, Theorem 3.9, p. 75]. Since $\tilde{A}_h^*\mathcal{P}\phi = g$ over $\Omega_h \cap \Omega$, we may write

(5.4) $$\int_{\Omega_h}(\tilde{u}-u_h)g\,dx = \int_{\Omega_h}(\tilde{u}-u_h)\tilde{A}_h^*\mathcal{P}\phi\,dx + \int_{\Omega_h\cap\complement\Omega}(\tilde{u}-u_h)(g-\tilde{A}_h^*\mathcal{P}\phi)\,dx.$$

Using Green's formula, we obtain

(5.5) $$\int_{\Omega_h}(\tilde{u}-u_h)\tilde{A}_h^*\mathcal{P}\phi\,dx = \tilde{a}_h(\tilde{u}-u_h,\mathcal{P}\phi) + \int_{\Gamma_h}\tilde{u}\,\frac{\partial\mathcal{P}\phi}{\partial\nu_h^*}\,d\sigma ,$$

since $u_h = 0$ on $\Gamma_h$. On the one hand,

(5.6) $$\left|\int_{\Gamma_h}\tilde{u}\,\frac{\partial\mathcal{P}\phi}{\partial\nu_h^*}\,d\sigma\right| \leq \|\tilde{u}\|_{L^2(\Gamma_h)}\left\|\frac{\partial\mathcal{P}\phi}{\partial\nu_h^*}\right\|_{L^2(\Gamma_h)} ,$$

and on the other hand,

$$(5.7)\qquad \left|\int_{\Omega_h \cap \complement\Omega} (\tilde{u}-u_h)(g-\tilde{A}_h^{*}\mathcal{P}\phi)\,dx\right|$$

$$\leq \|\tilde{u}-u_h\|_{L^2(\Omega_h \cap \complement\Omega)} \|g-\tilde{A}_h^{*}\phi\|_{L^2(\Omega_h)}$$

$$\leq \gamma_2 \|\tilde{u}-u_h\|_{L^2(\Omega_h \cap \complement\Omega)} \|g\|_{L^2(\Omega_h)} \quad ,$$

since $\|\tilde{A}_h^{*}\mathcal{P}\phi\|_{L^2(\Omega_h)} \leq \gamma_3 \|\mathcal{P}\phi\|_{H^2(\Omega_h)} \leq \gamma_1\gamma_3 \|\mathcal{P}\| \|g\|_{L^2(\Omega_h)}$ .

Using (5.2), (5.3), ..., (5.7), and the identity

$$\tilde{a}_h(\tilde{u}-u_h,\mathcal{P}\phi) = \tilde{a}_h(\tilde{u}-u_h,\mathcal{P}\phi-\phi_h) - \tilde{a}_h(u_h,\phi_h) + a_h(u_h,\phi_h)$$
$$+ (\tilde{A}_h\tilde{u}-f_h,\phi_h)_h$$

valid for any function $\phi_h \in V_h$, we obtain (5.1). ∎

In order to evaluate (in Theorem 11) the first two terms appearing in the right-hand side of inequality (5.1), we need two technical results.

Lemma 2. *Let there be given a family of triangulations* $(\mathcal{C}_h)$ *made up either of finite elements of type* (k,S), *with* k = 2,3 (n=2 *if* k=3) *or of finite elements of type* (k,Q) *with* k=1,2 *for* n=2. *Then there exists a constant* C *independent of* h *such that*

$$(5.8)\qquad \|v\|_{L^2(\Omega_h \cap \complement\Omega)} \leq Ch^{k+1} |v|_{1,2,\Omega_h \cap \complement\Omega}$$

*either for all functions* $v \in H_0^1(\Omega_h)$ *or for all functions* $v \in H^1(\tilde{\Omega})$ *such that* v = 0 *on* Γ.

*Proof*. The notations will be those of Fig. 7, where we have considered

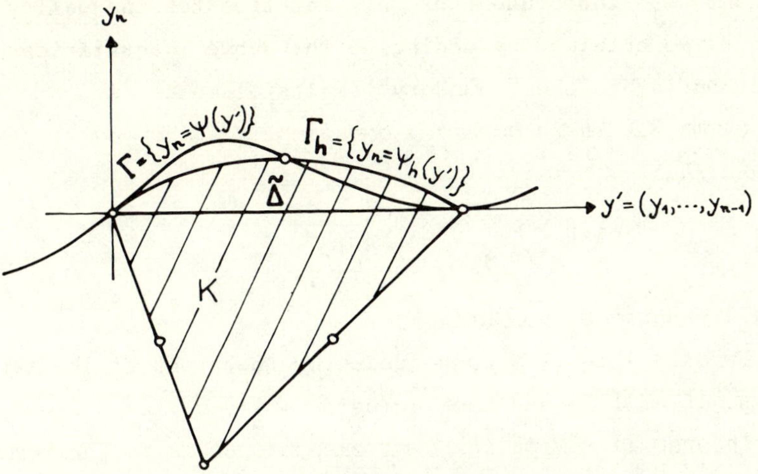


Fig. 7. Notations of Lemma 2

a finite element $K$ of type (2,S). As usual, it is sufficient to prove inequality (5.7) for smooth functions $v$ which vanish over $\Gamma_h$ (the proof is analogous for functions $v$ which vanish on $\Gamma$).

Whenever $\psi(y') \leqslant y_n \leqslant \psi_h(y')$, we may write

$$|v(y',y_n)|^2 = \left|\int_{\psi(y')}^{y_n} \frac{\partial v}{\partial y_n}(y',\xi_n)\,d\xi_n\right|^2$$

$$\leqslant (y_n-\psi_h(y'))\int_{\psi(y')}^{y_n} \left|\frac{\partial v}{\partial y_n}(y',\xi_n)\right|^2 d\xi_n \quad ,$$

and thus,

$$\|v\|^2_{L^2(\Omega_h\cap\complement\Omega\cap K)} \leqslant \frac{1}{2}\max_{y'\in\tilde{\Delta}}|\psi(y')-\psi_h(y')|^2|v|^2_{1,2,\Omega_h\cap\complement\Omega\cap K}.$$

For the families considered in Lemma 2, it is easily established that $\max_{y'\in\tilde{\Delta}}|\psi(y')-\psi_h(y')| \leqslant C_1h^{k+1}$ for some constant $C_1$ independent of $h$, and thus the inequality of (5.8) is obtained by adding up the above inequalities corresponding to the "boundary" finite elements $K$. ∎

Lemma 3. *There exists a constant* $C$ *independent of* $h$ *such that*

$$(5.9) \qquad \|v\|_{L^2(\Gamma_h)} \leq C \|v\|_{H^1(\Omega_h)}$$

*for all functions* $v \in H^1(\Omega_h)$.

*Proof.* This is a consequence of the proof of Theorem 1.2, p. 15 of [17] and the inclusion $\Omega_h \subset \tilde{\Omega}$ for all $h$. ∎

Theorem 11. *With the same assumptions as in Theorem 9, we assume that, in addition, the quadrature schemes over the reference set* $\hat{K}$ *satisfy the assumptions* (3.23)-(3.24) (*resp.* (3.23')-(3.24')) *of Theorem 5, the assumption* (3.36) *of Theorem 7, and the assumptions* (3.40) (*resp.* (3.40')) *of Theorem 8, and finally that the family of triangulations satisfy the assumption* (5.8) *of Lemma 2 with* $k = r+1$.

*Then there exists a constant* $C$ *independent of* $h$ *such that*

$$(5.10) \qquad \|\tilde{u}-u_h\|_{L^2(\Omega_h)} \leq Ch^{r+2} \|\tilde{u}\|_{W^{r+4,q}(\tilde{\Omega})}$$

*where* $\tilde{u}$ *is a sufficiently smooth extension over* $\tilde{\Omega}$ *of the solution* $u$ *of* (1.1)-(1.2) *and* $u_h$ *is the solution of the discrete problem* (1.10).

*Proof.* We give the proof for $n \leq 3$ so that $H^2(\cdot) \subset C^0(\cdot)$; the proof for $n > 3$ can be achieved by using a smoothing procedure, following an idea of STRANG [21]. As usual, the constants $\gamma_i$ are independent of $h$. From Lemma 2, we obtain

$$(5.11)\quad \|\tilde{u}-u_h\|_{L^2(\Omega_h \cap \complement\Omega)} \leq \gamma_0 h^{r+2}\left(\|\tilde{u}\|_{H^1(\Omega_h)} + \|u_h\|_{H^1(\Omega_h)}\right)$$

$$\leq \gamma_1 h^{r+2}\|\tilde{u}\|_{H^1(\Omega_h)} \quad ,$$

and from Lemma 3,

$$(5.12)\quad \left\|\frac{\partial \mathcal{P}\phi}{\partial \nu_h^*}\right\|_{L^2(\Gamma_h)} \leq \gamma_2 \|\mathcal{P}\phi\|_{H^2(\Omega_h)} \leq \gamma_3 \|\phi\|_{H^2(\Omega)} \quad .$$

Also it easily follows from the proof of Lemma 2 that since $\tilde{u} = 0$ on $\Gamma$ ($\tilde{u}$ is an extension of $u$), then $|\tilde{u}(x)| \leq \gamma_4 h^{r+2} |\tilde{u}|_{1,\infty,\tilde{\Omega}}$ for $x \in \Gamma_h$, and thus

$$(5.13)\quad \|\tilde{u}\|_{L^2(\Gamma_h)} \leq \gamma_4 h^{r+2} \|\tilde{u}\|_{W^{1,\infty}(\tilde{\Omega})} \quad .$$

Combining (5.12) and (5.13), we obtain

$$(5.14)\quad \|\tilde{u}\|_{L^2(\Gamma_h)} \sup_{\phi \in H_0^1(\Omega) \cap H^2(\Omega)} \left(\|\phi\|_{H^2(\Omega)}\right)^{-1} \left\|\frac{\partial \mathcal{P}\phi}{\partial \nu_h^*}\right\|_{L^2(\Gamma_h)}$$

$$\leq \gamma_5 h^{r+2} \|\tilde{u}\|_{W^{1,\infty}(\tilde{\Omega})} \quad .$$

From now on, we let $\phi_h = r_h \mathcal{P}\phi$, so that $\phi_h \in V_h$, because $\phi \in H_0^1(\Omega) \cap H^2(\Omega)$. Then we obtain

$$(5.15)\quad \|\mathcal{P}\phi - \phi_h\|_{H^1(\Omega_h)} \leq \gamma_6 h \|\phi\|_{H^2(\Omega)} \quad ,$$

$$(5.16)\quad \left(\sum_{K \in \mathcal{C}_h} \|\phi_h\|^2_{H^2(K)}\right)^{1/2} \leq \gamma_7 \|\phi\|_{H^2(\Omega)} \quad ,$$

as easy consequences of Theorem 2 and of the inequality $\|\mathcal{P}\phi\|_{H^2(\Omega_h)} \leq \|\mathcal{P}\| \|\phi\|_{H^2(\Omega)}$. Thus, from Theorem 9 and (5.15), we get

$$(5.17)\quad \left\{\|\phi\|_{H^2(\Omega)}\right\}^{-1}\|\tilde{u}-u_h\|_{H^1(\Omega_h)}\|\phi-\phi_h\|_{H^1(\Omega_h)} \leq \gamma_8 h^{r+2}\|\tilde{u}\|_{W^{r+2,q}(\Omega_h)} .$$

As in (4.8), we have

$$\tilde{a}_h(r_h u,\phi_h) - a_h(r_h u,\phi_h) = \sum_{K\in\mathcal{C}_h} E_K\left(\sum_{i,j=1}^{n} \tilde{a}_{ij}\frac{\partial r_h u}{\partial x_i}\frac{\partial \phi_h}{\partial x_j}\right) .$$

By Theorem 8,

$$\left|E_K\left(\sum_{i,j=1}^{n} \tilde{a}_{ij}\frac{\partial r_h u}{\partial x_i}\frac{\partial \phi_h}{\partial x_j}\right)\right| \leq \gamma_9 h^{r+2}\max_{i,j}\|\tilde{a}_{ij}\|_{W^{r+2,\infty}(K)}\|r_h u\|_{W^{r+2,q}(K)}\|\phi_h\|_{W^{2,q'}(K)} ,$$

and therefore, using (5.16),

$$(5.18)\quad |\tilde{a}_h(r_h u,\phi_h)-a_h(r_h u,\phi_h)| \leq \gamma_{10} h^{r+2}\left(\sum_{K\in\mathcal{C}_h}\|r_h u\|^q_{W^{r+2,q}(K)}\right)^{1/q}\|\phi\|_{H^2(\Omega)} .$$

Likewise, by Theorem 7,

$$\left|E_K\left(\sum_{i,j=1}^{n} \tilde{a}_{ij}\frac{\partial(u-r_h u)}{\partial x_i}\frac{\partial \phi_h}{\partial x_j}\right)\right| \leq \gamma_{11} h\|\tilde{a}_{ij}\|_{W^{1,\infty}(K)}\|u-r_h u\|_{H^1(K)}\|\phi_h\|_{H^2(K)} ,$$

and therefore

$$(5.19)\quad |\tilde{a}_h(u_h - r_h u, \phi_h) - a_h(u_h - r_h u, \phi_h)|$$

$$\leq \gamma_{12} h \|u_h - r_h u\|_{H^1(\Omega_h)} \left\{ \sum_{K \in \mathcal{E}_h} \|\phi_h\|^2_{H^2(K)} \right\}^2$$

$$\leq \gamma_{13} h \|u_h - r_h u\|_{H^1(\Omega_h)} \|\phi\|_{H^2(\Omega)}$$

Combining (5.18) and (5.19), we obtain

$$(5.20)\quad \frac{|\tilde{a}_h(u_h, \phi_h) - a_h(u_h, \phi_h)|}{\|\phi\|_{H^2(\Omega)}}$$

$$\leq \gamma_{14}(h\|u_h - r_h u\|_{H^1(\Omega_h)} + h^{r+2}\|\tilde{u}\|_{W^{r+2,q}(\Omega_h)}) \quad .$$

Since $r_h u = r_h \tilde{u}$,

$$(5.21)\quad \|u_h - r_h u\|_{H^1(\Omega_h)} \leq \|\tilde{u} - r_h \tilde{u}\|_{H^1(\Omega_h)} + \|\tilde{u} - u_h\|_{H^1(\Omega_h)}$$

$$\leq \gamma_{15} h^{r+1} \|\tilde{u}\|_{W^{r+3,q}(\Omega_h)} \quad ,$$

by Theorems 2 and 9, and thus

$$(5.22)\quad \frac{|\tilde{a}_h(u_h, \phi_h) - a_h(u_h, \phi_h)|}{\|\phi\|_{H^2(\Omega)}} \leq \gamma_{16} h^{r+2} \|\tilde{u}\|_{W^{r+3,q}(\Omega_h)} \quad .$$

Finally, as in (4.11),

$$\left(\tilde{A}_h \tilde{u} - f_h, \phi_h\right)_h = \sum_{K \in \mathcal{E}_h} E_K\left((\tilde{A}_h \tilde{u}) \phi_h\right) \quad .$$

By Theorem 5,

$$\left| E_K\left((\tilde{A}_h \tilde{u}) \phi_h\right) \right| \leq \gamma_{17} h^{r+2} \|\tilde{A}_h \tilde{u}\|_{W^{r+2,q}(K)} \|\phi_h\|_{W^{2,q'}(K)} \quad ,$$

and therefore, combining with (5.16),

$$\frac{|(\mathcal{A}_h\tilde{u}-f_h,\phi_h)_h|}{\|\phi\|_{H^2(\Omega)}} \leq \gamma_{18}h^{r+2}\|\tilde{u}\|_{W^{r+4,q}(\Omega_h)} . \tag{5.23}$$

The inequality of (5.10) is then obtained by combining inequalities (5.11), (5.14), (5.17), (5.22) and (5.23). ∎

Example 7. *With the same assumptions as in Example 6*, we obtain

$$\|\tilde{u}-u_h\|_{L^2(\Omega_h)} = 0(h^2)$$

for finite elements of type (1,Q) with $\hat{Q}(2)$ instead of $\hat{Q}(1)$ in (ii),

$$\|\tilde{u}-u_h\|_{L^2(\Omega_h)} = 0(h^3)$$

for finite elements of type (2,S) and (2,Q), and

$$\|u-u_h\|_{L^2(\Omega_h)} = 0(h^4)$$

for finite elements of type (3,S). ∎

## REFERENCES

1. Aubin, J.P., "Behavior of the error of the approximate solutions of boundary value problems for linear elliptic operators by Galerkin's and finite difference methods," Ann. Scuola Norm. Sup. Pisa 21 (1967), 599-637.
2. Berger, A., R. Scott, and G. Strang, "Approximate boundary conditions in the finite element method," Proc. Symp. on Numerical Analysis, Istituto Nazionale di Alta Matematica, Rome, Jan. 1972, to appear.
3. Birkhoff, G., M. H. Schultz, and R. S. Varga, "Piecewise Hermite interpolation in one and two variables with applications to partial differential equations," Numer. Math. 11 (1968), 232-256.
4. Bramble, J. H., and S. R. Hilbert, "Estimation of linear

functions on Sobolev spaces with applications to Fourier transforms and spline interpolation," SIAM J. Numer. Anal. 7 (1970), 112-124.
5. Bramble, J.H., and S. R. Hilbert, "Bounds for a class of linear functionals with applications to Hermite interpolation," Numer. Math. 16 (1971), 362-369.
6. Bramble, J.H., and M. Zlámal, "Triangular elements in the finite element method, Math. Comp. 24 (1970), 809-820.
7. Ciarlet, P.G., "Orders of convergence in finite element methods," Proc. of Conference on the Mathematics of Finite Elements and Applications, Brunel University, April 1972, Academic Press, to appear.
8. Ciarlet, P.G., and P.-A. Raviart, "General Lagrange and Hermite interpolation in $R^n$ with applications to finite element methods," Arch. Rational Mech. Anal. 46 (1972), 177-199.
9. Ciarlet, P.G., and P.-A. Raviart, "Interpolation theory over curved elements, with applications to finite element methods," Computer Methods in Appl. Mech. and Engineering 1 (1972), 217-249.
10. Ciarlet, P.G., and C. Wagschal, "Multipoint Taylor formulas and applications to the finite element method," Numer. Math. 17 (1971), 84-100.
11. Davis, P.J., "A construction of nonnegative approximate quadratures," Math. Comp. 21 (1967), 578-582.
12. Fried, I., Discretization and Round-Off Errors in the Finite Element Analysis of Elliptic Boundary Value Problems and Eigenvalue Problems, Ph.D. Dissertation, Massachusetts Institute of Technology, 1971.
13. Haber, S., "Numerical evaluation of multiple integrals," SIAM Rev. 12 (1970), 481-526.
14. Herbold, R.J., Consistent Quadrature Schemes for the Numerical Solution of Boundary Value Problems by Variational Techniques, Ph.D. Dissertation, Case Western Reserve University, 1968.
15. Herbold, R.J., and R. S. Varga, "The effect of quadrature errors in the numerical solution of two-dimensional boundary value problems by variational techniques," Aequationes Math. 7 (1972), 36-58.
16. Irons, B.M., "Quadrature rules for brick based finite elements," Int. J. Numer. Mech. Eng. 3 (1971), 293-294.
17. Nečas, J., Les Méthodes Directes en Théorie des Equations Elliptiques, Masson, Paris, 1967.

18. Nicolaides, R. A., "On a class of finite elements generated by Lagrange interpolation," to appear.
19. Nitsche, J., "Ein Kriterium für die quasi-optimalitat des Ritzchen Verfahrens," Numer. Math. 11 (1968), 346-348.
20. Nitsche, J., "Verfahren von Ritz und Spline Interpolation bei Sturm-Liouville Randwertproblemen," Numer. Math. 13 (1969), 260-265.
21. Strang, G., "Approximation in the finite element method," Numer. Math. 19 (1972), 81-98.
22. Strang, G., "Numerical integration," to appear in G. Strang and G. Fix, An Analysis of the Finite Element Method, Prentice Hall.
23. Strang, G., and A.E. Berger, "The change in solution due to change in domain," to appear.
24. Strang, G., and G. Fix, An Analysis of the Finite Element Method, Prentice Hall, to appear.
25. Tchakaloff, V., "Formules de cubatures mécaniques à coefficients non négatifs," Bull. Sci. Math. (2) 81 (1957), 123-134.
26. Zienkiewicz, O.C., The Finite Element Method in Engineering Science, McGraw-Hill, London, 1971.
27. Zlámal, M., "On the finite element method," Numer. Math. 12 (1968), 384-409.
28. Zlámal, M., "A finite element procedure of the second order of accuracy," Numer. Math. 14 (1970), 394-402.
29. Zlámal, M., "Curved elements in the finite element method," to appear.

## A SUPERCONVERGENCE RESULT FOR THE APPROXIMATE SOLUTION OF THE HEAT EQUATION BY A COLLOCATION METHOD

Jim Douglas, Jr.*

### 1. Introduction.

This lecture will present a special case of some research of Todd Dupont and the author on the numerical solution of partial differential equations by collocation methods using piecewise-polynomial spaces [6]. For the sake of brevity the differential problem to be approximated will be the simplest parabolic problem:

$$
\begin{aligned}
&u_t - u_{xx} = 0 , \quad && x \in I = [0,1] , \; t \in J = [0,T] , \\
(1.1)\quad &u(0,t) = u(1,t) = 0 , && t \in J , \\
&u(x,0) = f(x) , && x \in I .
\end{aligned}
$$

Let $h = M^{-1}$ , $\Delta t = TN^{-1}$ , $x_i = ih$ , $t_j = j\Delta t$ , $I_i = [x_{i-1}, x_i]$ , $J_k = [t_{k-1}, t_k]$ . Let $P_r(E)$ denote the class of polynomials of degree not greater than $r$ on the set $E$ . Let

* University of Chicago.

$$M_1(r,h) = \{v \in C^1(I) \mid v \in P_r(I_i), i=1,\dots,M\}, \quad r \geq 3,$$

$$(1.2) \quad M_0(s,\Delta t) = \{w \in C^0(J) \mid w \in P_s(J_k), k=1,\dots,N\}, \quad s \geq 1,$$

$$M = M_1(r,h) \otimes M_0(s,\Delta t) \quad .$$

Note that $M$ is the set of linear combinations of products $v(x)w(t)$, $v \in M_1(r,h)$, $w \in M_0(s,\Delta t)$. Let $0 < \xi_1 < \xi_2 < \dots < \xi_{r-1} < 1$ and $w_j > 0$, $j = 1,\dots,r-1$, be the Gaussian points and weights, respectively, so that

$$\int_0^1 p(x)\,dx = \sum_{j=1}^{r-1} p(\xi_j) w_j \quad , \quad p \in P_{2r-3}(I) \quad ;$$

similarly, let $0 < \tau_1 < \dots < \tau_s < 1$ and $w_\ell^* > 0$, $\ell = 1,\dots,s$, be such that

$$\int_0^1 p(x)\,dx = \sum_{\ell=1}^{s} p(\tau_\ell) w_\ell^* \quad , \quad p \in P_{2s-1}(I) \; .$$

Set

$$(1.3) \quad \begin{aligned} \xi_{ij} &= x_{i-1} + h\xi_j \; , \quad i = 1,\dots,M \; , \; j = 1,\dots,r-1 \; , \\ \tau_{k\ell} &= t_{k-1} + \Delta t \tau_\ell \; , \quad k = 1,\dots,N \; , \; \ell = 1,\dots,s \; . \end{aligned}$$

The collocation method to be studied is defined as follows. A function $U \in M$ is to be determined so that

$$(1.4) \quad \begin{aligned} &(U_t - U_{xx})(\xi_{ij},\tau_{k\ell}) = 0 \; , \; i = 1,\dots,M \; , \; j = 1,\dots,r-1, \\ &\qquad\qquad k = 1,\dots,N \; , \; \ell = 1,\dots,s \; , \\ &U(0,t) = U(1,t) = 0 \quad , \quad t \in J \; , \\ &U(x,0) = \hat{f}(x) \in M_1(r,h) \quad , \end{aligned}$$

where $\hat{f} \in M_1(r,h)$ will be determined from $f$ by a quasi-interpolation process to be specified below. These equations permit the evaluation of $U$ to be done locally in time; i.e., one time step at a time.

Elsewhere Dupont and the author [6] will treat the generalization of (1.4) to nonlinear parabolic equations, along with a number of collocation methods employing the space $M_1(r,h)$ in space and either no discretization in time or various means of discretization in time. In particular, it has been shown [6] that

$$\|u - U\|_{L^\infty(I\times J)} = O(h^{r+1} + (\Delta t)^{s+1}) \tag{1.5}$$

for sufficiently smooth $u$, provided that there exist $c_1, c_2 > 0$ such that

$$c_1 h^3 \leq \Delta t \leq c_2 h \tag{1.6}$$

as $h$ and $\Delta t$ tend to zero. Estimates of the same form as (1.5) hold on the grid lines without (1.6). Previously, Douglas-Dupont [4,5] had obtained similar results for the Hermite cubic case $r = 3$ for the continuous-time collocation method based on $M_1(3,h)$.

Recently, several authors [1,6-10] have observed that there can be identifiable points at which the solution of a finite element approximation can converge at rates that exceed the rate that can exist globally. This superconvergence phenomenon occurs for (1.4) if $U(x,0)$ is chosen properly. Under much more general circumstances than (1.4), it has been shown [6] that

(1.7) $$\max_{\substack{i=0,\ldots,M \\ k=0,\ldots,N}} |(u - U)(x_i, t_k)| = O(h^{2r-2} + (\Delta t)^{2s})$$

for sufficiently smooth $u$. This result is a direct generalization of those of de Boor-Swartz [1], who treated both the initial value problem and the two-point boundary problem for ordinary differential equations (though not limited to first or second order). The method of proof is totally different and can be used to give an independent, but noticeably longer, proof of their results.

The object of this paper is to give a proof of (1.7) and at the same time a method of constructing $U(x,0)$. The superconvergence phenomenon first occurs for $r = 4$ and $s = 2$. Since the general argument of [6] requires a somewhat complicated machinery of notation and lemmas, the analysis will be given for the case $r = 4$, $s = 2$.

## 2. Error estimates.

Let

(2.1) $$B(x) = \frac{1}{6!}\frac{d}{dx}(x^3(x-1)^3), \quad 0 \leq x \leq 1,$$

and set

(2.2) $$C_q(x) = \frac{5!}{(4+q)!}\left(x - \frac{1}{2}\right)^{q-1} B(x), \quad q = 1,2,\ldots, \quad 0\leq x\leq 1 .$$

Note that $B(x)$ has double roots at $x = 0$ and $x = 1$ and a simple root at $x = \frac{1}{2}$. Define an interpolation operator $T : C^1(I) \to P_4(I)$ by the relations

$$(2.3)\qquad \begin{aligned} (Tv)(x) &= v(x) \quad , \quad x = 0\ ,\ \tfrac{1}{2}\ ,\ \text{or}\ \ 1\ , \\ (Tv)'(x) &= v'(x)\ ,\quad x = 0\ \ \text{or}\ \ 1\ . \end{aligned}$$

The Peano kernel theorem [3] has the simple consequence that

$$(2.4)\qquad (v - Tv)(x) = \sum_{q=1}^{p} v^{(4+q)}\left(\frac{1}{2}\right)C_q(x) + S_p(x)\ ,$$

and

$$(2.5)\qquad S_p(x) = \int_0^1 k_p(x,y)v^{(5+p)}(y)\,dy\ ,$$

for every $v \in H^{5+p}(I)$, the Sobolev space of functions having $L^2$-derivatives of order $5 + p$. The remainder $S_p$ can be differentiated formally $4 + p$ times; only two derivatives will be used.

Let $x_{i-1/2} = (i - 1/2)h$, and let $T_h : C^1(I) \to \mathcal{M}_1(4,h)$ be the interpolation operator resulting from the affine transformation of $T$ to the intervals $I_i$ :

$$(2.6)\qquad \begin{aligned} (T_h v)(x_i) &= v(x_i)\ , & i &= 0,\ldots,M\ , \\ (T_h v)(x_{i-1/2}) &= v(x_{i-1/2})\ , & i &= 1,\ldots,M\ , \\ (T_h v)'(x_i) &= v'(x_i)\ , & i &= 0,\ldots,M\ . \end{aligned}$$

Then, it follows from (2.3)-(2.5) that, for $x \in I_i$,

$$(2.7)\qquad \begin{aligned} (v - T_h v)^{(\alpha)}(x) = &\sum_{q=1}^{p} v^{(4+q)}(x_{i-1/2})C_q^{(\alpha)}\left(\frac{x - x_{i-1}}{h}\right)h^{4+q-\alpha} \\ &+ O(\|v^{(5+p)}\|_{L^\infty(I)}\,h^{5+p-\alpha})\ ,\quad \alpha = 0,1,2\ , \end{aligned}$$

if $v \in C^{5+p}(I)$.

It is easy to see that $B''$ is a multiple of the Legendre polynomial of degree three for the interval $I$ ; it then follows [3] that

$$C_1''(\xi_j) = 0 \quad , \qquad j = 1,2,3 \quad . \tag{2.8}$$

Let

$$\begin{aligned} {}_h\langle v, z\rangle &= \sum_{j=1}^{3} v(h\xi_j) z(h\xi_j) w_j h \ , \quad 0 < h \leqslant 1 \ , \\ {}_h|v|^2 &= {}_h\langle v, v\rangle \ . \end{aligned} \tag{2.9}$$

It is easy to see [6] that

$$ {}_1\langle B^{(\alpha)}, x^{\beta}\rangle = 0 \ , \quad \alpha = 0,1,2 \ , \quad \beta \leqslant \alpha \ . \tag{2.10}$$

Set

$$\|v\|_{p,q} = \sup \left\{ \frac{\partial^{\alpha+\beta} v}{\partial x^{\alpha} \partial t^{\beta}} (x,t) \ \Big| \ x \in I, \ t \in J, \ \alpha \leqslant p \ , \ \beta \leqslant q \right\} . \tag{2.11}$$

It follows from (2.7) and (2.8) that, if

$$L = \frac{\partial}{\partial t} - \frac{\partial^2}{\partial x^2} \ ,$$

then

$$L(u - T_h u)(\xi_{ij},t)$$

(2.12)
$$\begin{aligned}
&= -\frac{\partial^6 u}{\partial x^6}(x_{i-1/2},t)C_2''(\xi_j)h^4 - \frac{\partial^7 u}{\partial x^7}(x_{i-1/2},t)C_3''(\xi_j)h^5 \\
&\quad + \frac{\partial^6 u}{\partial x^5 \partial t}(x_{i-1/2},t)C_1(\xi_j)h^5 + \mathcal{O}(\|u\|_{8,0} + \|u\|_{6,1})h^6 \\
&= -\frac{\partial^6 u}{\partial x^6}(x_{i-1/2},t)C_2''(\xi_j)h^4 \\
&\quad + \frac{\partial^7 u}{\partial x^7}(x_{i-1/2},t)(C_1(\xi_j) - C_3''(\xi_j))h^5 \\
&\quad + \mathcal{O}(\|u\|_{8,0})h^6 \quad .
\end{aligned}$$

The method of analysis begins with a construction of a modification $\tilde{u}$ of $T_h u$ that reduces $L(u - \tilde{u})(\xi_{ij},t)$ to $\mathcal{O}(h^6)$ . The $\mathcal{O}(h^4)$ term can be eliminated by using a locally defined correction $D_0(x,t)$ . For $t \in J$ let $D_0(\cdot,t) \in M_1(4,h)$ be determined by the relations

(2.13)
$$\begin{aligned}
h^2 \frac{\partial^2}{\partial x^2} D_0(\xi_{ij},t) &= \frac{\partial^6 u}{\partial x^6}(x_{i-1/2},t)C_2''(\xi_j) \quad , \\
& \qquad i = 1,\ldots,M \ , \ j = 1,2,3 \quad , \\
D_0(x_i,t) = \frac{\partial}{\partial x} D_0(x_i,t) &= 0 \quad , \quad i = 0,\ldots,M \quad .
\end{aligned}$$

The conditions at the knots imply that $D_0(x,t) = c_i(t)(x - x_{i-1})^2(x - x_i)^2$ on $I_i$ ; a simple calculation shows that

(2.14)
$$\begin{aligned}
D_0(x,t) &= (2^3 6! h^4)^{-1} \frac{\partial^6 u}{\partial x^6}(x_{i-1/2},t) \\
&\quad \times (x-x_{i-1})^2(x-x_i)^2 \quad , \qquad x \in I_i \quad .
\end{aligned}$$

Thus,

$$D_0 = O(\|u\|_{6,0}) \ , \qquad \frac{\partial}{\partial t} D_0 = O(\|u\|_{8,0}) \ , \tag{2.15}$$

and, if $\tilde{u}^{(1)} = T_h u + h^6 D_0$ ,

$$\begin{aligned} L(u-\tilde{u}^{(1)})(\xi_{ij},t) &= \frac{\partial^7 u}{\partial x^7}(x_{i-1/2},t)(C_1(\xi_j) - C_3''(\xi_j))h^5 \\ &\quad + O(\|u\|_{8,0})h^6 \ . \end{aligned} \tag{2.16}$$

The second correction $D_1(x,t) \in M_1(4,h)$ cannot be defined locally. Let

$$V(y) = \begin{cases} 0 & , \quad y \leq 0 \ , \\ 3y^2 - 2y^3 & , \quad 0 \leq y \leq 1 \ , \\ 1 & , \quad 1 \leq y \ , \end{cases} \tag{2.17}$$

and set

$$E_i(x,t) = Q\frac{\partial^7 u}{\partial x^7}(x_{i-1/2},t)V\left(\frac{x-x_{i-1}}{h}\right) \ , \qquad x \in I \ , \tag{2.18}$$

where

$$Q = \frac{C_3''(\xi_j) - C_1(\xi_j)}{V''(\xi_j)} \ , \qquad j = 1,3 \ . \tag{2.19}$$

Note that the oddness of $C_1$ , $C_3$ and $V - \frac{1}{2}$ about $\frac{1}{2}$ implies that $Q$ is unambiguously defined. Let

$$\begin{aligned} C_1(x,t) &= \sum_{i=1}^{M} E_i(x,t) - x \sum_{i=1}^{M} E_i(1,t) \\ &= Q \sum_{i=1}^{M} \frac{\partial^7 u}{\partial x^7}(x_{i-1/2},t)\left[V\left(\frac{x-x_{i-1}}{h}\right) - x\right] \ . \end{aligned} \tag{2.20}$$

It is clear that $D_1 \in M_1(4,h)$ and that

$$(2.21)\qquad D_1 = O(h^{-1}\|u\|_{7,0}) \quad , \quad \frac{\partial D_1}{\partial t} = O(h^{-1}\|u\|_{9,0}) \quad .$$

Moreover,

$$(2.22)\quad h^2 \frac{\partial^2}{\partial x^2}D_1(\xi_{ij},t) = -\frac{\partial^7 u}{\partial x^7}(x_{i-1/2},t)(C_1(\xi_j) - C_3''(\xi_j)) \ .$$

Let

$$(2.23)\qquad \tilde{u} = T_h u + h^6 D_0 + h^7 D_1 \quad ;$$

then

$$(2.24)\qquad L(u - \tilde{u})(\xi_{ij},t) = O(\|u\|_{9,0})h^6 \quad .$$

Note that

$$(2.25)\qquad \begin{aligned} &\|\tilde{u}\|_{0,q} \leq C[\|u\|_{0,q} + h\|u\|_{1,q} + h^6\|u\|_{7,q}] \quad , \\ &\left|\frac{\partial^{\alpha+\beta}\tilde{u}}{\partial x^{\alpha}\partial t^{\beta}}(\xi_{ij},t)\right| \leq \|u\|_{2,\beta} + C\|u\|_{7,\beta}h^4 \quad . \end{aligned}$$

The next step is to consider a Lagrange piecewise-quadratic interpolation of $\tilde{u}$ in time. Let $S_{\Delta t}$ be the interpolation operator determined by the following:

$$(2.26)\qquad \begin{aligned} &\text{i)}\quad (S_{\Delta t}w)(t) = w(t) \ , \quad t = t_{k-1}, t_{k-1/2}, t_k \ , \ k=1,\ldots,N \ , \\ &\text{ii)}\quad S_{\Delta t}w \in P_2(J_k) \quad , \quad k = 1,\ldots,N \quad . \end{aligned}$$

If

$$(2.27)\qquad C_q^*(s) = \frac{1}{(q+2)!}\, s\left(s - \frac{1}{2}\right)^q (s-1) \quad , \quad q = 1,2 \quad ,$$

then

$$(\tilde{u} - S_{\Delta t}\tilde{u})(x,t) = \frac{\partial^3 \tilde{u}}{\partial t^3}(x, t_{k-1/2}) C_1^*\left(\frac{t-t_{k-1}}{\Delta t}\right)(\Delta t)^3$$

$$(2.28) \qquad + \frac{\partial^4 \tilde{u}}{\partial t^4}(x, t_{k-1/2}) C_2^*\left(\frac{t-t_{k-1}}{\Delta t}\right)(\Delta t)^4$$

$$+ O(\|u\|_{0,5})(\Delta t)^5 \quad , \qquad t \in J_k \quad .$$

Since $C_1^{*\prime}(\tau_\ell) = 0$ , $\ell = 1,2$ , it follows from (2.25) and (2.28) that

$$L(\tilde{u} - S_{\Delta t}\tilde{u})(\xi_{ij}, \tau_{k\ell}) = -(I - S_{\Delta t})\frac{\partial^2 \tilde{u}}{\partial x^2}(\xi_{ij}, \tau_{k\ell})$$

$$+ \frac{\partial^4 \tilde{u}}{\partial t^4}(\xi_{ij}, t_{k-1/2}) C_2^{*\prime}(\tau_\ell)(\Delta t)^3$$

$$+ O(\|\tilde{u}\|_{0,5})(\Delta t)^4$$

$$(2.29) \qquad = \left\{ -\frac{\partial^5 \tilde{u}}{\partial x^2 \partial t^3}(\xi_{ij}, t_{k-1/2}) C_1^*(\tau_\ell) + \frac{\partial^4 \tilde{u}}{\partial t^4}(\xi_{ij}, t_{k-1/2}) C_2^{*\prime}(\tau_\ell)\right\}(\Delta t)^3$$

$$+ O(\|u\|_{0,5} + h\|u\|_{1,5} + h^4\|u\|_{7,4} + h^6\|u\|_{7,5})(\Delta t)^4 \quad .$$

The function $S_{\Delta t}\tilde{u} \in M$ can be altered so that the $O(\Delta t)^3$ term can be removed. Let

$$(2.30) \qquad F(s) = s(s-1) \quad , \qquad 0 \leq s \leq 1 \quad ,$$

and let

$$(2.31)\qquad G(x,t) = \left[Q_1^* \frac{\partial^4 \tilde{u}}{\partial t^4}(x,t_{k-1/2}) + Q_2^* \frac{\partial^4 \tilde{u}}{\partial t^4}(x,t_{k-1/2})\right] F\left(\frac{t-t_{k-1}}{\Delta t}\right), \quad t \in J_k ,$$

where

$$(2.32)\qquad Q_1^* = -\frac{C_1^*(\tau_\ell)}{F'(\tau_\ell)}, \quad Q_2^* = \frac{C_2^{*\prime}(\tau_\ell)}{F'(\tau_\ell)}, \quad \ell = 1 \text{ and } 2 .$$

It is easy to see that $Q_\alpha^*$ is well-defined and that

$$(2.33)\qquad \begin{aligned} \Delta t \frac{\partial}{\partial t} G(\xi_{ij},\tau_{k\ell}) &= -\frac{\partial^5 \tilde{u}}{\partial x^2 \partial t^3}(\xi_{ij},t_{k-1/2}) C_1^*(\tau_\ell) \\ &\quad - \frac{\partial^4 \tilde{u}}{\partial t^4}(\xi_{ij},t_{k-1/2}) C_2^{*\prime}(\tau_\ell) + O(h^6 \|u\|_{9,4}) , \\ G(x,t_k) &= 0 . \end{aligned}$$

Moreover, by (2.25), (2.31), and (1.1),

$$(2.34)\qquad \left| \frac{\partial^2}{\partial x^2} G(\xi_{ij},t) \right| \le C(\|u\|_{0,5} + h^4 \|u\|_{7,4}) .$$

Set

$$(2.35)\qquad \hat{u} = S_{\Delta t}\tilde{u} + (\Delta t)^4 G .$$

Then

$$(2.36)\qquad L(u - \hat{u})(\xi_{ij},\tau_{k\ell}) = \rho(\xi_{ij},\tau_{k\ell}) ,$$

where

$$|\rho(\xi_{ij},\tau_{k\ell})| \leq C[h^6\|u\|_{9,0} + (\|u\|_{0,5} \tag{2.37}$$
$$+ h\|u\|_{1,5} + h^4\|u\|_{7,4} + h^6\|u\|_{7,5})(\Delta t)^4] \quad .$$

Let

$$\vartheta = U - \hat{u} \quad . \tag{2.38}$$

Since $L(u - U)(\xi_{ij},\tau_{k\ell}) = 0$ ,

$$L\vartheta(\xi_{ij},\tau_{k\ell}) = \rho(\xi_{ij},\tau_{k\ell}) \quad . \tag{2.39}$$

Define the discrete inner product (real functions only are needed)

$$\langle v,z\rangle = \sum_{i=1}^{M}\sum_{j=1}^{3} v(\xi_{ij})z(\xi_{ij})w_j h \ , \quad |v|^2 = \langle v,v\rangle \quad . \tag{2.40}$$

The relation between this inner product and the standard $L^2$ inner product on $M_1(4,h)$ has been studied [6]; in particular, for $v \in M_1(4,h)$ such that $v(0) = v(1) = 0$ ,

$$(v',v') \leq -\langle v'',v\rangle \leq 2(v',v') \quad ,$$
$$\frac{1}{4}\|v\|^2_{H^1(I)} \leq (v',v') + |v|^2 \leq C\|v\|^2_{H^1(I)} \quad , \tag{2.41}$$
$$|v| \leq C\|v\|_{L^2(I)} \quad .$$

Also, if $v(0,t) = v(1,t) = 0$ and $v(\cdot,t) \in M_1(4,h)$ ,

$$\frac{1}{2}\frac{d}{dt}\langle v_{xx},v\rangle = \langle v_{xx},v_t\rangle \quad . \tag{2.42}$$

Since $\vartheta \in M$ and $\vartheta(0,t) = \vartheta(1,t) = 0$ ,

$$\begin{aligned}\langle\rho,\vartheta_t\rangle(\tau_{k\ell}) &= \langle\vartheta_t - \vartheta_{xx},\vartheta_t\rangle(\tau_{k\ell})\\ &= |\vartheta_t|^2(\tau_{k\ell}) - \frac{1}{2}\frac{d}{dt}\langle\vartheta_{xx},\vartheta\rangle(\tau_{k\ell}) \quad .\end{aligned} \tag{2.43}$$

Hence,

$$|\vartheta_t|^2(\tau_{k\ell}) - \frac{d}{dt}\langle\vartheta_{xx},\vartheta\rangle(\tau_{k\ell}) \leq |\rho|^2(\tau_{k\ell}) \quad . \tag{2.44}$$

Note that $|\vartheta_t|^2 \in P_2(J_k)$ and $\frac{d}{dt}\langle\vartheta_{xx},\vartheta\rangle \in P_3(J_k)$ , $k = 1, \ldots, N$. Consequently, the Gauss quadrature in time is exact for both terms:

$$\begin{aligned}&\sum_{\ell=1}^{2}\{|\vartheta_t|^2(\tau_{k\ell}) - \frac{d}{dt}\langle\vartheta_{xx},\vartheta\rangle(\tau_{k\ell})\}w_\ell^*\Delta t\\ &\qquad = \int_{J_k}|\vartheta_t|^2(\tau)d\tau - \langle\vartheta_{xx},\vartheta\rangle\Big|_{t_{k-1}}^{t_k} \quad .\end{aligned} \tag{2.45}$$

It is apparent that

$$\begin{aligned}&\int_0^T|\vartheta_t|^2(\tau)d\tau + \max_{k=1,\ldots,N}\{-\langle\vartheta_{xx},\vartheta\rangle(t_k)\}\\ &\quad\leq \sum_{k=1}^{N}\sum_{\ell=1}^{2}|\rho|^2(\tau_{k\ell})w_\ell^*\Delta t - \langle\vartheta_{xx},\vartheta\rangle(0)\\ &\quad\leq C[\|u\|_{9,0}h^6 + (\|u\|_{0,5} + h\|u\|_{1,5} + h^4\|u\|_{7,4}\\ &\qquad + h^6\|u\|_{7,5})(\Delta t)^4]^2 - \langle\vartheta_{xx},\vartheta\rangle(0) \quad .\end{aligned} \tag{2.46}$$

It follows from (2.41), (2.46) and the inequality

$$|\vartheta|^2(t) \leq C(|\vartheta|^2(0) + \int_0^t|\vartheta_t|^2(\tau)d\tau) \ , \ 0 \leq t \leq T \ , \tag{2.47}$$

that

$$\max_{k=1,\dots,N} \|\vartheta(t_k)\|_{H^1(I)} \leq C[\|\vartheta(0)\|_{H^1(I)}$$

$$(2.48)\qquad + \|u\|_{9,0}h^6 + (\|u\|_{0,5} + h\|u\|_{1,5} + h^4\|u\|_{7,4}$$

$$+ h^6\|u\|_{7,5})(\Delta t)^4] \quad .$$

In order to obtain maximal accuracy it is necessary to choose $U(x,0)$ so that $\vartheta(0) = \mathcal{O}(h^6)$ in $H^1(I)$ . Since $\hat{u}(0) = \tilde{u}(0)$ , $\vartheta(0) = U(0) - \tilde{u}(0)$ . Note that

$$(2.49)\qquad \frac{\partial}{\partial x} D_1(x,t) = \mathcal{O}(h^{-1}\|u\|_{7,0}) \quad .$$

Thus, it is sufficient to take

$$(2.50)\qquad U(x,0) = (T_h f)(x) + h^6 D_0(x,0) \quad ;$$

note that this function can be constructed from $f$ alone. With this choice of $U(0)$ , it follows that

$$\max_{k=1,\dots,N} \|(U - \tilde{u})(t_k)\|_{L^\infty(I)} \leq C[\|u\|_{9,0}h^6$$

$$(2.51)\qquad + (\|u\|_{0,5} + h\|u\|_{1,5} + h^4\|u\|_{7,4}$$

$$+ h^6\|u\|_{7,5})(\Delta t)^4] \quad .$$

It is easy to see that

$$(2.52)\qquad |(u - \tilde{u})(x_i,t_k)| = h^7|D_1(x_i,t_k)| \leq C\|u\|_{7,0}h^6 \quad .$$

Consequently,

$$\max_{\substack{i=0,\ldots,M \\ k=0,\ldots,N}} |(u - U)(x_i, t_k)| \leq C[\|u\|_{9,0} h^6$$

$$+ (\|u\|_{0,5} + h\|u\|_{1,5} + h^4\|u\|_{7,4} \tag{2.53}$$

$$+ h^6\|u\|_{7,5})(\Delta t)^4] \quad .$$

This establishes the validity of (1.7) for the special case $r = 4$ and $s = 2$ .

It is possible to simplify the smoothness requirements in the estimates above using an approximation process of Bramble, Dupont, and Thomée [2]; the multiplier of $(\Delta t)^4$ can be replaced by $\|u\|_{1,5} = \|u\|_{11,0}$ . This will be discussed in [6]. The proof in the case of arbitrary $r \geq 4$ and $s \geq 2$ and a much more general differential equation follows the same general line as the one given here, although the construction of $\hat{u}$ is much more complicated and it is necessary to use a different test function than $\vartheta_t$ in the analogue of (2.43).

REFERENCES

1. deBoor, C., and B. Swartz, "Collocation at Gaussian points," to appear.
2. Bramble, J. H., T. Dupont, and V. Thomée, "Projection methods for Dirichlet's problem in approximating polygonal domains with boundary value corrections," to appear in Mathematics of Computation.
3. Davis, P.J., Interpolation and Approximation, Blaisdell, New York, 1963.
4. Douglas, J., Jr., and T. Dupont, "A finite element collocation method for the heat equation," to appear in Proceedings of the Convegno su Problemi di Analisi Numerica, Rome, 1972.

5. Douglas, J., Jr., and T. Dupont, "A finite element collocation method for quasilinear parabolic equations," to appear in Mathematics of Computation.
6. _____, _____, "The approximate solution of nonlinear parabolic equations by collocation methods using $C^1$-piecewise-polynomial spaces," Parts I-III, to appear.
7. _____, _____, "Some superconvergence results for Galerkin methods for the approximate solution of the two-point boundary problem," to appear in Proceedings of the Conference on Numerical Analysis, Royal Irish Academy, Dublin, 1972.
8. _____, _____, "Superconvergence for Galerkin methods for the two point boundary problem via local projections, to appear.
9. Hulme, B. L., "One-step piecewise polynomial Galerkin methods for initial value problems," Math. Comp., April 1972.
10. Thomée, V., article in these Proceedings.

# SOME $L^2$ ERROR ESTIMATES FOR PARABOLIC GALERKIN METHODS

by

Todd Dupont*

## 1. Introduction

The purpose of this note is to present some estimates of the $L^2$ norms of the errors that result from using Galerkin procedures to approximate solutions of a linear parabolic problem. In this linear case these estimates refine those (for nonlinear problems) of Wheeler [6] and Douglas-Dupont [3,4] in that they require less stringent regularity hypotheses on the solution of the differential problem. Some regularity theory for parabolic equations will be presented and utilized to provide error estimates in terms of the data of the problem. The use of a negative Sobolev norm in the parabolic estimate and the associated elliptic estimates gives bounds which mesh very naturally with the regularity theory. Certain special cases of these results are very similar to specific cases of the results of Bramble-Thomée [2], but the techniques are quite different; the results of Bramble-Thomée provided important insight and motivation.

---

*University of Chicago, Chicago, Ill. 60637

Let $\Omega$ be a bounded domain in $\mathbb{R}^p$ (p-dimensional Euclidean space) with a smooth boundary $\partial\Omega$. Suppose that $A(x) = (a_{ij}(x))$ is a $p \times p$ symmetric positive-definite matrix for each $x = (x_1,\cdots,x_p) \in \bar{\Omega}$ and that $A \in C^\infty(\bar{\Omega})$. (A function is in $C^\infty(\bar{\Omega})$ if it can be extended to be infinitely differentiable on $\mathbb{R}^p$.) Let $b(x) = (b_1(x),\cdots,b_p(x))$ and $c(x)$ be $C^\infty(\bar{\Omega})$ functions with values in $\mathbb{R}^p$ and $\mathbb{R}$, respectively. We shall consider the approximate solution of

$$\frac{\partial u}{\partial t} - Lu = f(x,t) \quad \text{on} \quad \Omega \times (0,T] ,$$

(1.1)

$$\frac{\partial u}{\partial \nu_A} = 0 \quad \text{on} \quad \partial\Omega \times (0,T], \quad u(\cdot,0) = u_0 \quad \text{on} \quad \Omega,$$

where $L$ is the uniformly elliptic operator

$$\text{(1.2)} \qquad Lw = \nabla\cdot(A\nabla w) - b\cdot\nabla w - cw$$

$$= \sum_{\ell,k=1}^{p} \frac{\partial}{\partial x_k}\left(a_{k\ell}\frac{\partial w}{\partial x_\ell}\right) - \sum_{\ell=1}^{p} b_\ell \frac{\partial w}{\partial x_\ell} - cw ,$$

and

$$\text{(1.3)} \qquad \frac{\partial w}{\partial \nu_A} = \sum_{\ell,k=1}^{p} a_{\ell k}\frac{\partial w}{\partial x_\ell}\nu_k ,$$

where $\nu = (\nu_1,\cdots,\nu_p)$ is the outward normal to $\partial\Omega$.

For nonnegative integers $s$, the Sobolev space $H^s(\Omega)$ is the set of all functions in $L^2(\Omega)$ whose distribution derivatives of order less than $s+1$ are also in $L^2(\Omega)$. These spaces will be normed by

(1.4) $$\|\phi\|_s^2 = \sum_{|\alpha|\leq s} \|D^\alpha\phi\|^2 ,$$

where $\alpha = (\alpha_1,\cdots,\alpha_p)$ is a p-tuple of nonnegative integers,

$$D^\alpha\phi = \left(\frac{\partial}{\partial x_1}\right)^{\alpha_1} \cdots \left(\frac{\partial}{x_p}\right)^{\alpha_p} \phi , \quad |\alpha| = \alpha_1 + \cdots + \alpha_p ,$$

and the $L^2$ norm is

(1.5) $$||\phi||^2 = (\phi,\phi) ,$$

where

(1.6) $$(\phi,\psi) = \int_\Omega \phi(x)\psi(x)dx .$$

We shall also need the norm on $H^{-1} = (H^1)'$; let

(1.7)
$$\|\phi\|_{-1} = \sup\{(\phi,\psi)/\|\psi\|_1 ; \|\psi\|_1 \neq 0 , \psi \in H^1\} ,$$

where $( , )$ has been extended to give the duality between $H^1$ and $H^{-1}$. For $X$ a normed space with norm $\| \|_X$ and $\phi:[0,T] \to X$, adopt the notations

$$\|\phi\|^2_{L^2(X)} = \int_0^T \|\phi(t)\|_X^2 dt ,$$

(1.8)

$$\|\phi\|_{L^\infty(X)} = \operatorname{ess\,sup}_{0\leq t\leq T} \|\phi(t)\|_X .$$

We shall bound the error for both continuous-time and Crank-Nicolson Galerkin approximations to (1.1). Let $M$

be a finite-dimensional subspace of $H^1(\Omega)$. A differentiable map $U:[0,T] \to \mathcal{M}$ is a <u>continuous-time Galerkin approximation</u> if

$$(1.9)\qquad \begin{aligned} &\left(\frac{\partial U}{\partial t}, V\right) + B(U,V) = (f,V) \quad \text{on} \quad (0,T]\ , \qquad V \in \mathcal{M}\ , \\ &U(0) - u_0 \quad \text{small}\ , \end{aligned}$$

where

(1.10)

$$B(\psi,\phi) = \sum_{\ell,k=1}^{p} \left(a_{\ell k} \frac{\partial \psi}{\partial x_k}, \frac{\partial \phi}{\partial x_\ell}\right) + \sum_{\ell=1}^{p} \left(b_\ell \frac{\partial \psi}{\partial x_\ell}, \phi\right) + c(\psi,\phi)\ .$$

It is easily checked that for any $U(0) \in \mathcal{M}$ there is a unique solution $U$ of (1.9). The <u>Crank-Nicolson Galerkin approximation</u> is a sequence $\{U_n\}_{n=0}^{N}$ in $\mathcal{M}$ such that

$$(1.11)\qquad \begin{aligned} &(\partial_t U_{n+\frac12}, V) + B(U_{n+\frac12}, V) = (f_{n+\frac12}, V)\ , \qquad V \in \mathcal{M}\ , \\ &U_0 - u_0 \quad \text{small}\ , \end{aligned}$$

where

$$(1.12)\qquad \begin{aligned} &\Delta t = T/N\ , \qquad \phi_n = \phi(n\Delta t) \\ &\phi_{n+\frac12} = \frac{1}{2}(\phi_n + \phi_{n+1})\ , \qquad \partial_t \phi_{n+\frac12} = (\Delta t)^{-1}(\phi_{n+1} - \phi_n)\ . \end{aligned}$$

The $U(0)$ and $U_0$ can be chosen many ways; for our purposes $L^2$ projection of $u_0$ into $\mathcal{M}$ is fine.

We shall assume throughout that the space $\mathcal{M}$ is taken from a family $\{\mathcal{M}_h\}_{0<h\leq 1}$ of subspaces of $H^1(\Omega)$ which

satisfy the following approximation assumption: There is a constant $C_0$ and an integer $r \geq 1$, both independent of $h$, such that for $1 \leq s \leq 2r$ and $v \in H^s(\Omega)$

$$\inf_{\chi \in M_h} \{ \|v-\chi\| + h\|v-\chi\|_1 \} \leq C_0 h^s \|v\|_s . \tag{1.13}$$

Define the following norms and semi-norms for functions in $C^\infty(\overline{\Omega} \times [0,T])$:

$$\|\phi\|_{W^s} = \sum_{j=0}^{s} \left\| \left(\frac{\partial}{\partial t}\right)^j \phi \right\|_{L^\infty(H^{2s-2j})} + \sum_{j=0}^{s+1} \left\| \left(\frac{\partial}{\partial t}\right)^j \phi \right\|_{L^2(H^{2s+1-2j})} , \quad s \geq 0 ,$$

$$\|\phi\|_{W^{-1}} = \|\phi\|_{L^2(H^{-1})} ,$$

$$\|\phi\|_{G^s} = \|\phi(\cdot,0)\|_{2s} + \left\|\frac{\partial\phi}{\partial t} - L\phi\right\|_{W^{s-1}} , \quad s \geq 0 .$$

Let $G^s$ be the completion of $\{\phi \in C^\infty(\overline{\Omega} \times [0,T]): \frac{\partial\phi}{\partial\nu_A} \equiv 0$ on $\partial\Omega \times [0,T]\}$ with respect to the norm $\| \ \|_{G^s}$. (Note that on this class it is a norm and not just a semi-norm.) We shall see that on $G^s$ the norm $\| \ \|_{G^s}$ is stronger than $\| \ \|_{W^s}$; this is the parabolic regularity result that we shall use. Let

$$D^s = \{(u_0,f): \exists\ u \in G^s \text{ such that } u_0 = u(\cdot,0) \text{ on } \Omega \text{ and } f = \frac{\partial u}{\partial t} - Lu \text{ on } \Omega \times (0,T)\} .$$

The set $D^s$ is the set of data which give solutions in $G^s$. For a pair $(u_0,f)$ to belong to $D^s$ it is necessary that $\|u_0\|_{2s}$ and $\|f\|_{W^{s-1}}$ be finite, and if $s \geq 1$, it is also necessary that certain compatability conditions be satisfied between $f$ and $u_0$.

The principal results of this paper are the following two theorems.

Theorem 1. There are constants $C_1$ and $C_2$ such that if $U$ is the continuous-time Galerkin approximation in $M_h$, $U(0)$ is properly chosen, and the pair $(u_0,f)$ of (1.1) belongs to $D^r$, then

$$\|U-u\|_{L^\infty(L^2)} \leq C_1 h^{2r} \|u\|_{W^r}$$
$$\leq C_2 h^{2r} [\|u_0\|_{2r} + \|f\|_{W^{r-1}}] .$$

Theorem 2. There are positive constants $\tau_0, C_1$, and $C_2$ such that if $\{U_n\}_{n=0}^N$ is the Crank-Nicolson Galerkin approximation, with $U_0$ properly chosen, and the pair $(u_0,f)$ of (1.1) belongs to $D^s$, where $s = \max\{2,r\}$, then for $0 < \Delta t \leq \tau_0$,

$$\max_{0 \leq n \leq N} \|U_n - u_n\| \leq C_1 [h^{2r} \|u\|_{W^r} + (\Delta t)^2 \|u\|_{W^2}]$$
$$\leq C_2 [h^{2r} + (\Delta t)^2] [\|u_0\|_{2s} + \|f\|_{W^{s-1}}] .$$

Note that in the case $r \geq 2$ these estimates have the weakest norm on $u_0$ that will guarantee that we can choose $\|U(0)-u_0\|$ or $\|U_0-u_0\|$ to be bounded by $Ch^{2r}$.

In section 2 we shall indicate the derivation of the parabolic regularity result, and in section 3 we shall prove Theorems 1 and 2.

## 2. Parabolic Regularity

In this section we use the usual energy estimate [5] for parabolic equations to show the following

**Lemma 1.** For $s \geq 0$, there is a constant $C = C(s)$ such that

$$\|\phi\|_{W^s} \leq C\,\|\phi\|_{G^s}\,, \qquad \phi \in G^s\,. \tag{2.1}$$

**Proof.** If we can show (2.1) for $\phi \in \{\psi \in C^\infty(\overline{\Omega} \times [0,T]):$ $\frac{\partial\psi}{\partial\nu_A} \equiv 0$ on $\partial\Omega \times [0,T]\}$, it then follows that $\|u\|_{W^s}$ can be defined for all elements $G^s$ and that (2.1) holds on all of $G^s$.

The demonstration of (2.1) is by induction on $s$. Let $f = \frac{\partial\phi}{\partial t} - L\phi$. Note that since $\frac{\partial\phi}{\partial\nu_A} \equiv 0$ on $\partial\Omega \times [0,T]$ we see that

$$\frac{1}{2}\frac{d}{dt}(\|\phi\|^2) + B(\phi,\phi) = (f,\phi) \leq \|f\|_{-1}\|\phi\|_1\,.$$

It then follows from the uniform positive-definiteness of the matrix $A(x)$ that there is an $m > 0$ such that

$$\frac{d}{dt}(\|\phi\|^2) + m\|\phi\|_1^2 \leq C[\|f\|_{-1}^2 + \|\phi\|^2]\,.$$

Gronwall's lemma then implies that

$$\|\phi\|_{L^\infty(L^2)} + \|\phi\|_{L^2(H^1)} \leq C[\|f\|_{L^2(H^{-1})} + \|\phi(\cdot,0)\|]\,. \tag{2.2}$$

Next, for $\psi \in H^1$ and $0 \leq t \leq T$

$$\left(\frac{\partial \phi}{\partial t}, \psi\right) = (f,\psi) - B(\phi,\psi) \leq C\|\psi\|_1 [\|f\|_{-1} + \|\phi\|_1] .$$

Thus

$$\|\frac{\partial \phi}{\partial t}\|_{-1} \leq C[\|f\|_{-1} + \|\phi\|_1] ;$$

hence

$$\|\frac{\partial \phi}{\partial t}\|_{L^2(H^{-1})} \leq C[\|f\|_{L^2(H^{-1})} + \|\phi\|_{L^2(H^1)}] . \tag{2.3}$$

It follows from (2.2) and (2.3) that

$$\|\phi\|_{W^0} \leq C[\|\phi(\cdot,0)\| + \|f\|_{W^{-1}}] = C\|\phi\|_{G^0} .$$

Next note that

(2.4)

$$\|\phi\|_{W^{s+1}} = \|\frac{\partial \phi}{\partial t}\|_{W^s} + \|\phi\|_{L^\infty(H^{2s+2})} + \|\phi\|_{L^2(H^{2s+3})} .$$

It follows from the induction hypothesis that

$$\begin{aligned}\|\frac{\partial \phi}{\partial t}\|_{W^s} &\leq C[\|\frac{\partial \phi}{\partial t}(\cdot,0)\|_{2s} + \|\frac{\partial f}{\partial t}\|_{W^{s-1}}] \\ &\leq C[\|f(\cdot,0)\|_{2s} + \|L\phi(\cdot,0)\|_{2s} + \|\frac{\partial f}{\partial t}\|_{W^{s-1}}] \\ &\leq C[\|f\|_{L^\infty(H^{2s})} + \|\phi(\cdot,0)\|_{2s+2} + \|f\|_{W^s}] .\end{aligned}$$

Hence

$$\|\frac{\partial \phi}{\partial t}\|_{W^s} \leq C\|\phi\|_{G^{s+1}} . \tag{2.5}$$

For each $t \in [0,T]$, elliptic regularity and the homogeneous boundary conditions imply that

$$\|\phi\|_{2s+2} \leq C[\|L\phi\|_{2s} + \|\phi\|_{2s}] .$$

Hence

$$\|\phi\|_{L^\infty(H^{2s+2})} \leq C[\|f\|_{L^\infty(H^{2s})} + \|\frac{\partial\phi}{\partial t}\|_{L^\infty(H^{2s})} + \|\phi\|_{L^\infty(H^{2s})}]$$

(2.6)
$$\leq C[\|f\|_{W^s} + \|\frac{\partial\phi}{\partial t}\|_{W^s} + \|\phi\|_{W^s}]$$

$$\leq C\|\phi\|_{G^{s+1}} ,$$

where we used (2.5) in the last step. Similarly

$$\|\phi\|_{L^2(H^{2s+3})} \leq C[\|f\|_{L^2(H^{2s+1})} + \|\frac{\partial\phi}{\partial t}\|_{L^2(H^{2s+1})} + \|\phi\|_{L^2(H^{2s+1})}]$$

(2.7)
$$\leq C[\|f\|_{W^s} + \|\frac{\partial\phi}{\partial t}\|_{W^s} + \|\phi\|_{W^s}]$$

$$\leq C\|\phi\|_{G^{s+1}} ,$$

It now follows from (2.4), (2.5), (2.6) and (2.7) that

(2.8)
$$\|\phi\|_{W^{s+1}} \leq \|\phi\|_{G^{s+1}} .$$

Thus the lemma is proved.

## 3. Error Estimates

In order to prove Theorems 1 and 2 we shall first need to study an elliptic projection of the solution $u$ of (1.1) into the space $M = M_h$. For each $t$ let $W(t) \in M$ be defined by

$$(3.1) \qquad B(u-W,V) + \lambda(u-W,V) = 0 , \qquad V \in M ,$$

where $\lambda$ is some sufficiently large constant; take $\lambda$ such that $[B(\phi,\phi) + \lambda\|\phi\|^2]^{1/2}$ is equivalent to the $H^1$ norm. The proof of the following lemma can be found, for example, in Bramble-Osborn [1].

<u>Lemma 2</u>. There is a constant C, independent of h and u, such that for each t and for $1 \leqslant s \leqslant 2r$,

$$(3.2) \qquad \|W-u\| + h^{-1}\|W-u\|_{-1} \leqslant Ch^s\|u\|_s .$$

Note that $\frac{\partial W}{\partial t}$ satisfies a corresponding estimate with u replaced by $\frac{\partial u}{\partial t}$. From Lemma 2 we can obtain

<u>Lemma 3</u>. There is a constant C such that if $u \in G^r$ and $\eta = W-u$, then

$$(3.3) \qquad \|\eta\|_{L^\infty(L^2)} + \|\frac{\partial \eta}{\partial t}\|_{L^2(H^{-1})} \leqslant Ch^{2r}\|u\|_{W^r} .$$

<u>Proof</u>. From Lemma 2 we see that

$$\|\eta\|_{L^\infty(L^2)} + \|\frac{\partial \eta}{\partial t}\|_{L^2(H^{-1})} \leqslant Ch^{2r}[\|u\|_{L^\infty(H^{2r})} + \|\frac{\partial u}{\partial t}\|_{L^2(H^{2r-1})}] ;$$

the result now follows.

<u>Proof of Theorem 1</u>. Let $\zeta = U-u$, $\eta = W-u$, $\vartheta = U-W$. Note that $\vartheta$ satisfies

$$\left(\frac{\partial \vartheta}{\partial t}, v\right) + B(\vartheta, v) = -\left[\left(\frac{\partial \eta}{\partial t}, v\right) + B(\eta, v)\right]$$

$$= -\left[\left(\frac{\partial \eta}{\partial t}, v\right) - \lambda(\eta, v)\right], \quad v \in M,$$

where we used (1.9) and (1.1) to get the first equation and (3.1) to get the second. With $v = \vartheta$ we see that for some $m > 0$

$$\frac{d}{dt}\|\vartheta\|^2 + m\|\vartheta\|_1^2 \leq C[\|\vartheta\|^2 + \|\eta\|_{-1}^2 + \|\frac{\partial \eta}{\partial t}\|_{-1}^2] .$$

Hence, by Gronwall's Lemma, we see that

$$\|\vartheta\|_{L^\infty(L^2)} \leq C[\|\vartheta(0)\| + \|\eta\|_{L^2(H^{-1})} + \|\frac{\partial \eta}{\partial t}\|_{L^2(H^{-1})}] .$$

The triangle inequality then gives

(3.4)
$$\|\zeta\|_{L^\infty(L^2)} \leq C[\|\zeta(0)\| + \|\eta\|_{L^\infty(L^2)} + \|\frac{\partial \eta}{\partial t}\|_{L^2(H^{-1})}] .$$

Lemma 3 and (3.4) imply that

$$\|\zeta\|_{L^\infty(L^2)} \leq C[\|\zeta(0)\| + h^{2r}\|u\|_{W^r}] .$$

If $U(0)$ is obtained in any fashion so that

$$\|\zeta(0)\| \leq Ch^{2r}\|u_0\|_{2r} ,$$

such as $L^2$ or $H^1$ projection, the conclusion follows.

It is interesting to note that if $U(0) = W(0)$ (i.e., $\vartheta(0) = 0$), and $u$ is sufficiently smooth, then $\|\vartheta\|_{L^\infty(L^2)} = O(h^{2r+1})$.

<u>Proof of Theorem 2</u>. If we average the differential equation at times $t_n$ and $t_{n+1}$ we can see that

(3.5)

$$(\partial_t u_{n+\frac{1}{2}}, v) + B(u_{n+\frac{1}{2}}, v) = (f_{n+\frac{1}{2}}, v) + (\rho_{n+\frac{1}{2}}, v), \qquad v \in H^1,$$

where $\rho_{n+\frac{1}{2}} = \partial_t u_{n+\frac{1}{2}} - \left(\frac{\partial u}{\partial t}\right)_{n+\frac{1}{2}}$. Using the notation of the previous proof we see that

$$\begin{aligned}(\partial_t \vartheta_{n+\frac{1}{2}}, v) + B(\vartheta_{n+\frac{1}{2}}, v) &= -[(\partial_t \eta_{n+\frac{1}{2}}, v) + B(\eta_{n+\frac{1}{2}}, v) \\ &\qquad - (\rho_{n+\frac{1}{2}}, v)] \\ &= -[(\partial_t \eta_{n+\frac{1}{2}}, v) - \lambda(\eta_{n+\frac{1}{2}}, v) - (\rho_{n+\frac{1}{2}}, v)] , \\ &\qquad v \in M .\end{aligned}$$

With $v = \vartheta_{n+\frac{1}{2}}$ we then see that for some $m > 0$

$$\begin{aligned}\frac{1}{\Delta t}[\|\vartheta_{n+1}\|^2 - \|\vartheta_n\|^2] + m\|\vartheta_{n+\frac{1}{2}}\|_1^2 &\leq C[\|\vartheta_n\|^2 + \|\vartheta_{n+1}\|^2 \\ &+ \|\eta_{n+\frac{1}{2}}\|_{-1}^2 + \|\partial_t \eta_{n+\frac{1}{2}}\|_{-1}^2 + \|\rho_{n+\frac{1}{2}}\|_{-1}^2] .\end{aligned}$$

The descrete analogue of Gronwall's inequality says there is a $\tau_0 > 0$ and a $C$ such that for $0 < \Delta t < \tau_0$

(3.6)

$$\max_{0\leq n\leq N} \|\vartheta_n\| \leq C\left\{ \|\vartheta_0\| + \left(\sum_{\ell=0}^{N-1} [\|\eta_{n+\frac12}\|_{-1}^2 + \|\partial_t\eta_{n+\frac12}\|_{-1}^2 + \|\rho_{n+\frac12}\|_{-1}^2]\Delta t\right)^{\frac12}\right\} .$$

Thus the triangle inequality gives

(3.7)

$$\max_{0\leq n\leq N} \|\zeta_n\| \leq C\left\{ \|\zeta_0\| + \|\eta\|_{L^\infty(L^2)} + \left(\sum_{\ell=0}^{N-1} [\|\partial_t\eta_{n+\frac12}\|_{-1}^2 + \|\rho_{n+\frac12}\|_{-1}^2]\Delta t\right)^{\frac12}\right\} .$$

Since

$$\partial_t\eta_{n+\frac12} = \frac{1}{\Delta t}\int_{t_n}^{t_{n+1}} \frac{\partial\eta}{\partial t}\, dt , \tag{3.8}$$

it follows that

$$\|\partial_t\eta_{n+\frac12}\|_{-1}^2 \leq \frac{1}{\Delta t}\int_{t_n}^{t_{n+1}} \left\|\frac{\partial\eta}{\partial t}\right\|_{-1}^2 dt . \tag{3.9}$$

Hence

$$\sum_{\ell=0}^{N-1} \|\partial_t\eta_{\ell+\frac12}\|_{-1}^2\Delta t \leq \left\|\frac{\partial\eta}{\partial t}\right\|_{L^2(H^{-1})}^2 . \tag{3.10}$$

It is easily checked that

$$\rho_{n+\frac12} = \frac{1}{2\Delta t}\int_{t_n}^{t_{n+1}} (t_{n+1}-\tau)(t_n-\tau)\,\frac{\partial^3 u}{\partial t^3}(\tau)\, d\tau . \tag{3.11}$$

This in turn implies that

$$(3.12) \qquad \|\rho_{n+\frac{1}{2}}\|_{-1}^{2} \leq C(\Delta t)^{3} \int_{t_n}^{t_{n+1}} \left\|\frac{\partial^3 u}{\partial t^3}\right\|_{-1}^{2} d\tau \ ;$$

hence

$$(3.13) \qquad \sum_{\ell=0}^{N-1} \|\rho_{\ell+\frac{1}{2}}\|_{-1}^{2} \Delta t \leq C(\Delta t)^{4} \left\|\frac{\partial^3 u}{\partial t^3}\right\|_{L^2(H^{-1})}^{2} .$$

Combining (3.7), (3.10) and (3.13) we see that

$$\max_{0\leq n\leq N} \|\zeta_n\| \leq C\{\|\zeta_0\| + \|\eta\|_{L^\infty(L^2)} + \left\|\frac{\partial \eta}{\partial t}\right\|_{L^2(H^{-1})} + (\Delta t)^2\|u\|_{W^2}\}$$

The result now follows from Lemma 3.

REFERENCES


1. J.H. Bramble and J.E. Osborn, Rate of convergence for nonselfadjoint eigenvalue approximations, MRC Report, Univ. Wisconsin, 1972.
2. J.H. Bramble and V. Thomée, Discrete time Galerkin methods for a parabolic boundary value problem, MRC Report 1240, Univ. Wisconsin, 1972.
3. J. Douglas, Jr., and T. Dupont, Galerkin methods for parabolic equations, SIAM Numer. Anal. 4 (1970), 575-626.
4. _____ and _____, Galerkin methods for parabolic equations with nonlinear boundary conditions, to appear, Numer. Math.
5. J.L. Lions, Equations differentieles opérationelles et problèmes aux limites, Springer-Verlag, Berlin, 1961.
6. M.F. Wheeler, A priori $L^2$ error estimates for nonlinear parabolic equations, Thesis, Rice University, 1971.

# COMPUTATIONAL ASPECTS OF THE FINITE ELEMENT METHOD*

S. C. Eisenstat** and M. H. Schultz**

## 1. Introduction.

In this paper we consider a particular implementation of the finite element procedure for approximating the solution $u(x,y)$ of the boundary value problem

$$(1.1) \quad -D_x[p(x,y)D_x u] - D_y[q(x,y)D_y u] + c(x,y)u = f(x,y)$$

for all $(x,y) \in S \equiv \{(x,y) \mid 0 < x , y < 1\}$ , and

$$(1.2) \quad u(x,y) = 0$$

for all $(x,y) \in \partial S \equiv$ the boundary of $S$ . We assume that the functions $p(x,y)$ , $q(x,y)$ , $c(x,y)$ , and $f(x,y)$ are smooth and that there exists a positive constant $\gamma$ such that

$$(1.3) \quad \gamma \leq p(x,y) , \quad \gamma \leq q(x,y) , \quad 0 \leq c(x,y)$$

for all $(x,y) \in S$ .

Our finite element procedure uses basis functions consisting of piecewise bicubic Hermite polynomials defined

---

* This research was supported in part by the Office of Naval Research, N0014-67-A-0097-0016.

** Department of Computer Science, Yale University.

on a mesh which is refined (in a well-defined manner) in a neighborhood of each corner. The coefficients and right-hand side of the resulting linear algebraic system of equations involve integrals (over two-dimensional rectangular elements) which are approximated by the local nine-point product Gaussian quadrature scheme (the tensor product of the one-dimensional three-point Gaussian quadrature schemes). Finally, the approximate linear algebraic system of equations is symmetric and positive definite and is solved by either the band Cholesky or profile Cholesky decomposition procedure.

In sections 2 and 3, we will give theoretical justifications for our procedure. We will show that asymptotically our procedure is far more efficient than the combination of the five-point central difference approximation and SOR (successive overrelaxation) cf. [2]. Moreover, for constant coefficient problems, i.e., $p(x,y)$, $q(x,y)$ , and $c(x,y)$ are constants in $S$ , it is asymptotically more efficient than the five-point central difference approximation coupled with the fast direct methods, cf. [2]. In fact, a version of our procedure with a slightly less efficient quadrature scheme has been shown to be asymptotically optimal in a well-defined sense. In section 4, we will present the results of some numerical experiments which demonstrate the effectiveness of our procedure for even modest sized problems. The results of this paper can be directly generalized to other boundary value problems and other piecewise polynomial basis functions. The complete details will be given elsewhere.

We introduce the tensor product mesh $\rho = \Delta_x \times \Delta_y$ where $\Delta_x : 0 = x_0 < x_1 < \ldots < x_n = 1$ and $\Delta_y : 0 = y_0 < y_1 < \ldots < y_n = 1$. Moreover, for each $1 \leq i \leq n-1$, we define the basis functions

$$\text{(1.4)} \qquad \phi_i(x) \equiv \begin{cases} -2(x_i-x_{i-1})^{-3}(x-x_{i-1})^3 \\ \quad + 3(x_i-x_{i-1})^{-2}(x-x_{i-1})^2 , & x \in [x_{i-1},x_i] , \\ 2(x_{i+1}-x_i)^{-3}(x-x_i)^3 \\ \quad - 3(x_{i+1}-x_i)^{-2}(x-x_i)^2 + 1 , & x \in [x_i,x_{i+1}] , \\ 0 , & x \in [0,x_{i-1}] \cup [x_{i+1},1] , \end{cases}$$

$$\text{(1.5)} \qquad \xi_i(x) \equiv \begin{cases} (x_i-x_{i-1})^{-2}(x-x_{i-1})^2(x-x_i) , & x \in [x_{i-1},x_i] , \\ (x_{i+1}-x_i)^{-2}(x-x_i)(x_{i+1}-x)^2 , & x \in [x_i,x_{i+1}] , \\ 0 , & x \in [0,x_{i-1}] \cup [x_{i+1},1] , \end{cases}$$

$$\text{(1.6)} \qquad \xi_0(x) \equiv \begin{cases} x_1^{-2}x(x_1-x)^2 , & x \in [0,x_1] , \\ 0 , & x \in [x_1,1] , \end{cases}$$

and

$$(1.7) \qquad \xi_n(x) \equiv \begin{cases} (1-x_{n-1})^{-2}(x-x_{n-1})^2(x-1) \ , & x \in [x_{n-1},1] \ , \\ 0 \ , & x \in [0,x_{n-1}] \ . \end{cases}$$

Similarly, we define $\phi_i(y)$, $1 \leq i \leq n-1$, and $\xi_i(y)$, $0 \leq i \leq n$. Our piecewise bicubic Hermite basis functions are defined to be $\{\phi_i(x)\phi_j(y)\ ,\ \phi_i(x)\xi_k(y)\ ,\ \xi_\ell(x)\phi_j(y)\ ,\ \xi_\ell(x)\xi_k(y) \mid 1 \leq i\ ,\ j \leq n-1 \text{ and } 0 \leq k\ ,\ \ell \leq n\}$.

There are $4n^2$ such basis functions. Each interior mesh point $(x_i,y_j)$, $1 \leq i$, $j \leq n-1$, has four basis functions associated with it; each corner mesh point has one basis function associated with it; and each of the other boundary mesh points has two basis functions associated with it.

If we consecutively order the basis functions associated with each mesh point of $\rho$ and consecutively order the mesh points along rows, we may rename our basis functions as $\{B_i(x,y) \mid 1 \leq i \leq 4n^2 \equiv m\}$ and seek our approximation of the form

$$w(x,y) \equiv \sum_{i=1}^{m} \beta_i B_i(x,y) \quad ,$$

We determine the vector of coefficients $\underline{\beta} \in R^m$ by means of the Rayleigh-Ritz-Galerkin procedure, cf. [12]. This leads to a characterization of $\underline{\beta}$ as the solution of the $m \times m$ linear system

$$(1.8) \qquad A\underline{\beta} = \underline{k} \quad ,$$

where

$$A \equiv [a_{ij}] \equiv \left[\int_o^1 \int_o^1 \{p(x,y) D_x B_i D_x B_j + q(x,y) D_y B_i D_y B_j + c(x,y) B_i B_j\} dxdy\right]$$

and

$$\underline{k} \equiv [k_i] \equiv \left[\int_o^1 \int_o^1 f(x,y) B_i \, dxdy\right] .$$

The matrix $A$ is symmetric, positive definite and hence the linear system has a unique solution.

Because of the local nature of the basis functions, each entry of $A$ and $\underline{k}$ is the sum of integrals over at most four contiguous elements. In place of (1.8), we consider the approximate system

$$(1.9) \qquad \tilde{A}\underline{\tilde{\beta}} = \underline{\tilde{k}} \quad ,$$

where the entries of $\tilde{A}$ and $\underline{\tilde{k}}$ are obtained from the corresponding entries of $A$ and $\underline{k}$ by using the nine-point product Gaussian quadrature scheme over each element $[x_i, x_{i+1}] \times [y_j, y_{j+1}]$ , $0 \leq i$ , $j \leq n - 1$ . To be precise, every integral of the form

$$\int_{y_j}^{y_{j+1}} \int_{x_i}^{x_{i+1}} g(x,y) dxdy$$

is approximated by

$$\frac{(x_{i+1}-x_i)(y_{j+1}-y_j)}{4} \sum_{k=1}^{3} \sum_{\ell=1}^{3} w_k w_\ell \times$$

$$\times\ g\left(\frac{1}{2}[x_{i+1}-x_i]\theta_k + \frac{1}{2}[x_{i+1}+x_i], \frac{1}{2}[y_{j+1}-y_j]\theta_\ell + \frac{1}{2}[y_{j+1}+y_j]\right)$$

where

(1.10) $$w_1 = w_3 = 5/9 \quad , \quad w_2 = 8/9 \quad ,$$

and

(1.11) $$\theta_1 = -0.774597 \ , \quad \theta_2 = 0 \ , \quad \theta_3 = 0.774597 \quad .$$

We will show in section 2 that the matrix $\tilde{A}$ is symmetric positive definite and hence that (1.9) has a unique solution.

The matrix $\tilde{A}$ is a band matrix with band width $4(n + 2)$ . We solve (1.9) by the profile Cholesky decomposition procedure.

## 2. The asymptotic rate of convergence.

In this section we study the asymptotic rate of convergence of the approximation $u_n$ generated by our finite element procedure to the solution $u$ of the boundary value problem (1.1)-(1.2).

The standard analysis of the finite element method shows that the $L^2$ error in the finite element approximation is of the same order as the $L^2$ error in the best approximation to the solution $u$ from the subspace, or equivalently in the interpolant. Thus, for piecewise bicubic Hermite polynomials:

Theorem (cf. [12]): If $u \in W^{k,p}(S)$ $(1 \leq k \leq 4,$ $1 \leq p \leq 2)$ , then

$$\|u - u_n\|_{0,2} \leq K\bar{h}^{t+1-(2/p)}$$

where $\bar{h} = \max_{i,j} (x_{i+1} - x_i, y_{j+1} - y_j)$.

In particular, if $u \in W^{4,2}(S)$ , as is generally assumed in the literature, then $\|u - u_n\|_{0,2} = O(\bar{h}^4)$ . However, even for Poisson's equation

$$\Delta u = f \text{ in } S , \quad u = 0 \text{ on } \partial S ,$$

the solution is likely to have logarithmic singularities in the second derivative at each corner (cf. [9], [13]). In such a case, $u \in W^{3,2-\varepsilon}(S)$ whence $\|u - u_n\|_2 = O(\bar{h}^{3-\varepsilon})$ for any $\varepsilon > 0$ , but $u \notin W^{3,2}(S)$ whence $\|u - u_n\|_2 \neq O(\bar{h}^3)$ ; convergence is no better than third order (see Table I).

Several ideas have been proposed to overcome this loss of accuracy. Babuska [1] has given an overlay technique to locally refine the mesh near each corner, and Fix [4] and Schultz [11] have suggested adding singular corner solutions to the piecewise polynomial basis. But the implementation of these algorithms is rather complicated, the approximate quadrature made more difficult, and the condition number and nonzero structure of the finite element matrix greatly increased. Instead, we use a non-uniform tensor product mesh appropriately graded near the corners which does not create implementation problems versus using a uniform mesh. Under mild assumptions on the rate of growth of high derivatives of the solution $u$ near the corners, fourth order accuracy is restored. We state our result for the case where there is only one singular corner:

Theorem: Let $r = \sqrt{x^2 + y^2}$ and assume that $\|r^{k+\varepsilon-2}D^k u\|_{0,2} \leq K$ for $2 \leq k \leq 4$ $(\varepsilon > 0)$. Then for the mesh $\rho = \Delta_x \times \Delta_y$ with

$$x_i = \left(\frac{i}{n}\right)^{3/2}, \quad y_j = \left(\frac{j}{n}\right)^{3/2}, \quad i,j = 0,\ldots,n$$

we have

$$\|u - u_n\|_{0,2} \leq K'\left(\frac{1}{n}\right)^{4-\varepsilon} .$$

The condition number $K(A)$ of the finite element matrix $A$ is not greatly increased:

Theorem: If the coordinate basis functions $\phi_i(x)$, $\xi_i(x)$, $\phi_i(y)$, $\xi_i(y)$ are locally orthonormalized such that

$$\int_0^1 \phi_i^2 = 1, \quad \int_0^1 \xi_i^2 = 1, \quad \int_0^1 \phi_i \xi_i = 0$$

then the condition number of the Gram matrix $G \equiv [g_{ij}]$

$$\equiv \left[\iint_S B_i(x,y)B_j(x,y)\,dxdy\right]$$

is bounded independent of the mesh. Consequently,

$$K(A) \leq K\underline{h}^{-2}$$

where $\underline{h} = \min_{i,j}(x_{i+1}-x_i, y_{j+1}-y_j)$ .

Thus for the uniform mesh, $K(A) \sim \mathcal{O}(n^2)$ ; whereas for the graded mesh, $K(A) \sim \mathcal{O}(n^3)$ .

The remaining question to be answered is how to compute the integrals appearing in the coefficients of the finite element matrix $A$ and the right-hand side $\underline{k}$ . The integrals must be done numerically but the quadrature rule must be efficient and the approximate finite element matrix $\tilde{A}$ positive definite, yet fourth order convergence must be maintained. Our procedure uses the nine-point product Gaussian quadrature scheme.

<u>Theorem</u>: The matrix $\tilde{A}$ is positive definite.

<u>Proof</u>: Let $Q$ denote the approximate quadrature operator. Given $\underline{\beta} \neq \underline{\theta}$ , let

$$w(x,y) = \sum_{i=1}^{4n^2} \beta_i B_i(x,y) \quad .$$

Then

$$\underline{\beta}^T \tilde{A} \underline{\beta} = Q(p(D_x w)^2 + q(D_y w)^2 + cw^2) \geq \gamma Q((D_x w)^2 + (D_y w)^2) \quad .$$

If $\underline{\beta}^T \tilde{A} \underline{\beta} = 0$ , then the piecewise bicubic Hermite polynomials $D_x w$ , $D_y w$ vanish at the nine quadrature points in each mesh element. From this and the fact that $w$ vanishes on $\partial S$ , we conclude that $w \equiv 0$ , a contradiction.

Q.E.D.

A nine point quadrature seems to be minimal; any lower degree scheme experimentally seems not to preserve the high order accuracy. Yet our scheme maintains fourth order convergence: following Fix [5] we introduce the approximate bilinear forms

$$\tilde{a}(u_n, v_n) = Q(pD_x u_n D_x v_n + qD_y u_n D_y v_n + cu_n v_n) \quad ,$$

$$\tilde{b}(f, v_n) = Q(fv_n) \quad ,$$

and note that $\underline{\tilde{A}}\tilde{\beta} = \underline{\tilde{k}}$ if and only if

$$\tilde{a}(\tilde{u}_n, v_n) = \tilde{b}(f, v_n)$$

for all $v_n$ in the subspace, where

$$\tilde{u}_n = \sum_{i=1}^{4n^2} \tilde{\beta}_i B_i \quad .$$

The form $\tilde{a}(u_n, u_n)$ is strongly coercive

$$\tilde{a}(v_n, v_n) \geq c\|v_n\|_{1,2}^2$$

and

$$\begin{aligned}\tilde{a}(u_n - \tilde{u}_n, u_n - \tilde{u}_n) &= \{\tilde{a}(u_n, u_n - \tilde{u}_n) - a(u_n, u_n - \tilde{u}_n)\} \\ &\quad + \{b(f, u_n - \tilde{u}_n) - \tilde{b}(f, u_n - \tilde{u}_n)\} \\ &\leq O(n^{-4})\|u_n - \tilde{u}_n\|_{1,2}\end{aligned}$$

at least for a wide class of nonuniform meshes. Thus

$$\|u_n - \tilde{u}_n\|_{0,2} \leq \|u_n - \tilde{u}_n\|_{1,2} \sim O(n^{-4}) \quad .$$

The result is not true for an arbitrary nonuniform mesh.

## 3. Operation counts.

In this section, we give operation counts for our finite element procedure. In particular, we analyze the number of arithmetic operations needed (1) to generate the finite element matrix $\tilde{A}$ and the right-hand side $\underline{\tilde{k}}$ by numerical quadrature and (2) to solve the resulting linear system $\tilde{A}\underline{\tilde{\beta}} = \underline{\tilde{k}}$ . Throughout this section we should keep in mind the comparable results for the five-point central difference scheme coupled with either SOR (for variable coefficient problems) or the fast direct methods (for constant coefficient problems). By the results of section 2 we know that the number $N$ of mesh points in each coordinate for the five-point difference scheme is comparable to $n^2$ for comparable accuracies.

We consider the quadratures first. The quadrature question is particularly important since the quality of the implementation of this part of the procedure can easily make a difference of a factor of forty or fifty in the number of arithmetic operations. By symmetry we need compute at most $80n^2$ nonzero entries in $\tilde{A}$ and each such entry is on the average the sum of quadratures over at most two elements. Hence, it suffices to bound the number of arithmetic operations needed for the quadratures over the individual elements. For each fixed element, $[x_i,x_{i+1}] \times [y_j,y_{j+1}]$ , let $\alpha_1(x) \equiv \phi_i(x)$ , $\alpha_2(x) \equiv \xi_i(x)$, $\alpha_3(x) \equiv \phi_{i+1}(x)$ , $\alpha_4(x) \equiv \xi_{i+1}(x)$ , $\alpha_5(y) \equiv \phi_j(y)$ , $\alpha_6(y) \equiv \xi_j(y)$ , $\alpha_7(y) \equiv \phi_{j+1}(y)$ , and $\alpha_8(y) \equiv \xi_{j+1}(y)$ .

Then, by symmetry, the quantities which we must compute are

$$\text{(3.1)}\quad \tilde{a}(\alpha_s\alpha_t,\alpha_r\alpha_z)\ ,\quad 1 \leq s \leq r \leq 4 \ \text{ and } \ 5 \leq t \leq z \leq 8 \ .$$

To compute the quantities (3.1), we first compute for each $1 \leq \ell \leq 3$ and $1 \leq s \leq r \leq 4$

$$\text{(3.2)}\quad \psi^p_{sr}(y_{j\ell}) \equiv \left(\frac{x_{i+1} - x_i}{2}\right) \times \sum_{k=1}^{3} w_k p(x_{ik},y_{j\ell}) D_x\alpha_s(x_{ik}) D_x\alpha_r(x_{ik}) \ ,$$

$$\text{(3.3)}\quad \psi^q_{sr}(y_{j\ell}) \equiv \left(\frac{x_{i+1} - x_i}{2}\right) \times \sum_{k=1}^{3} w_k q(x_{ik},y_{j\ell}) \alpha_s(x_{ik}) \alpha_r(x_{ik}) \ ,$$

and

$$\text{(3.4)}\quad \psi^c_{sr}(y_{j\ell}) \equiv \left(\frac{x_{i+1} - x_i}{2}\right) \times \sum_{k=1}^{3} w_k c(x_{ik},y_{j\ell}) \alpha_s(x_{ik}) \alpha_r(x_{ik}) \ ,$$

where $x_{ik} \equiv \frac{1}{2}[x_{i+1} - x_i]\theta_k + \frac{1}{2}[x_{i+1} + x_i]$ , $1 \leq k \leq 3$ , and $y_{j\ell} \equiv \frac{1}{2}[y_{j+1} - y_j]\theta_\ell + \frac{1}{2}[y_{j+1} + y_j]$ . The computations require approximately a total of $3(10\cdot3\cdot3\cdot2)n^2 = 540n^2$ arithmetic operations for all the elements, where we assume that we have already done the $\mathcal{O}(n)$ operations needed to evaluate the quantities

$$\left(\frac{x_{i+1}-x_i}{2}\right) w_k D_x\alpha_s(x_{ik}) D_x\alpha_r(x_{ik}) \ , \quad \left(\frac{x_{i+1}-x_i}{2}\right) w_k \alpha_s(x_{ik})\alpha_r(x_{ik})$$

$$1 \leqslant s \leqslant r \leqslant 4 \ , \quad 1 \leqslant k \leqslant 3 \ , \quad 1 \leqslant i \leqslant n-1 \ .$$

Finally we compute

$$\tilde{a}(\alpha_s\alpha_t, \alpha_r\alpha_z) = \frac{(y_{j+1} - y_j)}{2} \times$$

$$\times \sum_{\ell=1}^{3} \{w_\ell(\psi^p_{sr}(y_{j\ell}) + \psi^c_{sr}(y_{j\ell}))\alpha_t(y_{j\ell})\alpha_z(y_{j\ell})$$

$$+ w_\ell \psi^q_{sr}(y_{j\ell}) D_y\alpha_t(y_{j\ell}) D_y\alpha_z(y_{j\ell})\}$$

which requires $10\cdot 10\cdot 3\cdot 4 = 1200$ arithmetic operations for each element or $1200n^2$ arithmetic operations for all the elements. Likewise, the number of operations needed to compute $\underline{k}$ is bounded by $180n^2$ . Thus, the total of all arithmetic operations to do the quadratures is bounded by $2000n^2$ .

Using the results of George, cf. [7], the number of arithmetic operations for the profile Cholesky decomposition procedure is bounded by the number for the band Cholesky decomposition procedure which in turn is bounded by $\sim (4n^2)(4n)^2 = 64n^4$ . Thus we have the following table:

Finite Element Procedure

| n | operations for quadrature = $2000n^2$ | operations for solution = $64n^4$ | total |
|---|---|---|---|
| 5 | 50,000 | 38,000 | 88,000 |
| 10 | 200,000 | 640,000 | 840,000 |
| 20 | 800,000 | 10,240,000 | 11,040,000 |

We see that for coarse meshes most of the operations are used for the quadratures while for moderate or fine meshes most of the operations are used for solving the linear system. To give comparable work estimates for the five-point difference scheme and SOR, we assume perfect knowledge of the optimal relaxation factor and that we iterate until the initial error is reduced by a factor of $N^2$.

Comparable Five-Point Central Difference Schemes

| $N \approx n^2$ | SOR† $\left(\frac{2}{3} N^3 \log_2 N\right)$ | fast methods $\left(\frac{9}{2} N^2 \log_2 N\right)$ |
|---|---|---|
| 32 | 100,000+ | 23,040 |
| 128 | 9,000,000+ | 504,000 |
| 512 | 1,000,000,000+ | 16,000,000 |

cf. [2].

We can see from these figures that our finite element algorithm is much more efficient than the five-point difference scheme and SOR even for coarse meshes and that the factor of improvement becomes larger as the mesh becomes finer.

† Better results can be obtained using SSOR, cf. [14].

Moreover, for constant coefficient problems, we need do only the quadratures to form $\underline{\tilde{k}}$ and we have essentially only the arithmetic operations for solving the system. Hence, our finite element algorithm is more efficient than the fast methods for fine meshes and the factor of improvement becomes larger as the mesh becomes finer.

We should point out that the quadrature work estimates appear to be relatively sharp. However, we can substantially improve the solution work estimates by improving our direct method; e.g., by using George's nested ordering scheme, cf. [8], or by using a fast iterative method, cf. [6].

## 4. Numerical results.

In this section we present the results of some numerical experiments which illustrate the preceding theoretical results. All computations were performed on a PDP-10 in single precision arithmetic (27 bit mantissa).

Let $S$ denote the unit square $[0,1] \times [0,1]$ in the $(x,y)$-plane with boundary $\partial S$. We introduce the tensor product mesh $\rho$

$$\rho = \Delta \times \Delta \; ; \quad \Delta : 0 = x_0 < x_1 < \dots < x_n = 1$$

where

$$x_k = \frac{1}{2}\left(\frac{2k}{n}\right)^{\beta}, \quad x_{n-k} = 1 - x_k, \quad k = 0,1,\dots,\left[\frac{n+1}{2}\right] .$$

For $\beta = 1$, $\rho$ is the uniform $n \times n$ mesh; for $\beta = \frac{3}{2}$, $\rho$ is the graded mesh referred to in §2. Our basis functions are the piecewise bicubic Hermite polynomials described in §1. The integrals in the finite element matrix $\tilde{A}$ and the right-hand side $\underline{\tilde{k}}$ were computed numerically using the nine-point Gaussian quadrature scheme on each mesh element as described in §1. The system of linear equations $\tilde{A}\underline{\tilde{\beta}} = \underline{\tilde{k}}$ were solved using a band Cholesky routine with the natural ordering of unknowns. The $L^2$ error estimates were computed numerically using the 25-point product Gaussian quadrature scheme on each mesh element; the $L^\infty$ error estimate is the maximum error at the quadrature points. The estimated rate of convergence is given by

$$\alpha_n = -\frac{\ln(\varepsilon_n/\varepsilon_{n-1})}{\ln(n/(n-1))}$$

where $\varepsilon_n$ is the estimated error for an $n \times n$ mesh.

Our first example is the Poisson equation:

$$\Delta u = 1 \quad \text{in} \quad S \; ; \quad u = 0 \quad \text{on} \quad \partial S \quad .$$

There is no closed form representation for the solution† but the second derivatives are known to have logarithmic singularities at each corner. Thus we would expect $\sim \mathcal{O}(n^{-3})$ convergence using a uniform mesh (see Table I). By grading the mesh near the corners $\left(\beta = \frac{3}{2}\right)$, we improve the convergence to $\sim \mathcal{O}(n^{-4})$ (see Table II).

† A numerical solution was generated by the method of particular solutions (cf. [3], [10]). Using singular particular solutions and taking advantage of the fourfold symmetry, ten terms were sufficient for $\sim 10^{-9}$ accuracy.

By way of comparison, we also present the results of solving the familiar five-point finite difference approximation to the differential equation using the Fast Fourier Transform (FFT) method to solve the difference equations (see Table III). This example is somewhat biased in favor of the finite difference method; the $\mathcal{O}(h^2)$ convergence follows from the fact that the finite difference equations are precisely equivalent to the finite element equations for piecewise linear splines on a triangular mesh and not from the classical analysis. Yet with a graded mesh, the error in the finite element approximation on a $10 \times 10$ mesh is slightly better than the error in the finite difference approximation on a $128 \times 128$ mesh.

Our second example is a variable coefficient problem:

$$D_x(e^{xy}D_x u) + D_y(e^{x+y}D_y u) + \frac{1}{1 + x + y}\, u = f \quad \text{in} \quad S \quad ,$$

$$u = 0 \quad \text{on} \quad \partial S \quad ,$$

where the right-hand side $f$ is chosen to make the solution

$$u(x,y) = \sin \pi x \cdot \sin \pi y \quad .$$

Both the coefficients and the solution are analytic so that it is not necessary to grade the mesh. The results (see Table IV) indicate fourth order convergence, confirming the fact that the nine-point product Gaussian quadrature scheme is sufficient.

## REFERENCES

1. Babuška, I., "Finite element method for domains with corners," Computing 6 (1970), 264-273.
2. Dorr, F. W., "The direct solution of the discrete Poisson equation on a rectangle," SIAM Review 12 (1970), 248-261.
3. Eisenstat, S.C., "On the rate of convergence of the Bergman-Vekua method for the numerical solution of elliptic boundary value problems," Doctoral thesis, Stanford University, 1972.
4. Fix, G.J., "Higher-order Rayleigh-Ritz approximations," J. Math. Mech. 18 (1969), 645-658.
5. Fix, G.J., "Perturbation of matrices in the numerical solution of differential equations: I. Quadrature errors," Gatlinburg V, Los Alamos, New Mexico, 1972.
6. Fix, G.J., and K. Larsen, "On the convergence of SOR iterations for finite element approximations to elliptic boundary value problems," SIAM J. Numer. Anal. 8 (1971), 536-547.
7. George, J.A., "Computer implementation of the finite element method," Doctoral thesis, Stanford University, 1971.
8. George, J.A., "Block elimination on finite element systems of equations," Proceedings of the IBM Symposium on Sparse Matrices and their Applications, 1972.
9. Lehman, R.S., "Developments at an analytic corner of solutions of partial differential equations, J. Math. Mech. 8 (1959), 727-760.
10. Schryer, N.L., "Constructive approximation of solutions to linear elliptic boundary value problems, " SIAM J. Numer. Anal. (to appear).
11. Schultz, M.H., "The computational complexity of elliptic partial differential equations," IBM Symposium on Complexity of Computer Computations, 1972.
12. Schultz, M.H., Spline Analysis, Prentice-Hall, Englewood Cliffs, New Jersey, 1973.
13. Wigley, N., "Asymptotic expansions at a corner of solutions of mixed boundary value problems, " J. Math. Mech. 13 (1964), 549-576.
14. Young, D.M., "Second-degree iterative methods for the solution of large linear systems," J. Approximation Theory 5 (1972), 137-148.

TABLE I

| n | $\|u - u_n\|_2$ | $\alpha_n$ | $\|u - u_n\|_\infty$ | $\alpha_n$ |
|---|---|---|---|---|
| 1 | 1.88E-03 | | 4.45E-03 | |
| 2 | 2.81E-04 | 2.7 | 6.27E-04 | 2.7 |
| 3 | 8.03E-05 | 3.1 | 2.20E-04 | 3.1 |
| 4 | 3.24E-05 | 3.2 | 1.14E-04 | 3.2 |
| 5 | 1.61E-05 | 3.1 | 7.15E-05 | 3.1 |
| 6 | 9.18E-06 | 3.1 | 4.93E-05 | 3.1 |
| 7 | 5.72E-06 | 3.1 | 3.62E-05 | 3.1 |
| 8 | 3.81E-06 | 3.0 | 2.77E-05 | 3.0 |
| 9 | 2.66E-06 | 3.0 | 2.19E-05 | 3.0 |
| 10 | 1.94E-06 | 3.0 | 1.77E-05 | 3.0 |

TABLE II

| n | $\|u - u_n\|_2$ | $\alpha_n$ | $\|u - u_n\|_\infty$ | $\alpha_n$ |
|---|---|---|---|---|
| 1 | 1.88E-03 | | 4.45E-03 | |
| 2 | 2.81E-04 | 2.7 | 6.27E-04 | 2.7 |
| 3 | 9.73E-05 | 2.6 | 2.78E-04 | 2.6 |
| 4 | 3.86E-05 | 3.2 | 9.54E-05 | 3.2 |
| 5 | 1.83E-05 | 3.4 | 5.14E-05 | 3.4 |
| 6 | 9.51E-06 | 3.6 | 2.60E-05 | 3.6 |
| 7 | 5.43E-06 | 3.6 | 1.59E-05 | 3.6 |
| 8 | 3.30E-06 | 3.7 | 9.55E-06 | 3.7 |
| 9 | 2.12E-06 | 3.8 | 6.39E-06 | 3.8 |
| 10 | 1.42E-06 | 3.8 | 4.35E-06 | 3.8 |

TABLE III

| n | $\|u - u_n\|_2$ | $\|u - u_n\|_\infty$ |
|---|---|---|
| 4 | 1.95E-03 | 3.36E-03 |
| 8 | 5.27E-04 | 8.89E-04 |
| 16 | 1.35E-04 | 2.26E-04 |
| 32 | 3.39E-05 | 5.67E-05 |
| 64 | 8.87E-06 | 1.50E-05 |
| 128 | 2.34E-06 | 4.00E-06 |

TABLE IV

| n | $\|u - u_n\|_2$ | $\alpha_n$ | $\|u - u_n\|_\infty$ | $\alpha_n$ |
|---|---|---|---|---|
| 1 | 3.31E-02 | | 6.88E-02 | |
| 2 | 2.41E-03 | 3.8 | 8.27E-03 | 3.8 |
| 3 | 5.98E-04 | 3.4 | 2.15E-03 | 3.4 |
| 4 | 2.10E-04 | 3.6 | 8.41E-04 | 3.6 |
| 5 | 9.05E-05 | 3.8 | 3.54E-04 | 3.8 |
| 6 | 4.50E-05 | 3.8 | 1.84E-04 | 3.8 |
| 7 | 2.47E-05 | 3.9 | 9.93E-05 | 3.9 |
| 8 | 1.47E-05 | 3.9 | 6.01E-05 | 3.9 |
| 9 | 9.24E-06 | 3.9 | 3.74E-05 | 3.9 |
| 10 | 6.10E-06 | 3.9 | 2.50E-05 | 3.9 |

# EFFECTS OF QUADRATURE ERRORS IN FINITE ELEMENT APPROXIMATION OF STEADY STATE, EIGENVALUE AND PARABOLIC PROBLEMS*

George J. Fix**

## 1. Introduction.

A finite element method approximates steady state problems $Lu = f$, eigenvalue problems $Lu = \lambda u$, and parabolic problems $u_t + Lu = f$ by matrix problems of the form $KQ = F$, $KQ = \lambda MQ$, and $M\frac{dQ}{dt} + KQ = F$. The entries of the "stiffness" and "mass" matrices $K,M$ as well as the source vector $F$ involve integrals of basis functions and coefficients of the differential equation. Typically these integrals cannot be evaluated exactly, and approximate methods are required. In this paper we shall estimate the size of the errors in the finite element approximation that result from the use of quadrature formulas for these integrals.

The steady state case has been discussed in [6] in terms of simple models. In this paper we briefly review and extend these results. They are of central importance since the strategy for studying eigenvalue and parabolic problems will be to bound the errors in terms of errors for

* This work was supported in part by the National Science Foundation under Grant No. GP 18064.

** Institute for Fluid Dynamics and Applied Mathematics, University of Maryland.

associated steady state problems. For eigenvalues this is done by techniques recently developed by the author [5] and I. Babuška [1], which in turn is based on the work of Vainikko [8]. For parabolic problems this is done by semi-group representations.

## 2. A simple example.

Since the bulk of this paper will be involved in developing the appropriate abstract machinery to study the effects of quadrature errors, it is perhaps important to start with a concrete example. Let $\Omega$ be a polygon with boundary $\Gamma$ and consider the second order steady state problem

(2.1) $$-\nabla(p\nabla u) + qu = f \text{ in } \Omega$$

(2.2) $$u = 0 \text{ on } \Gamma .$$

Let $a(v,w)$ denote the bilinear form associated with the second order operator, i.e.,

$$a(v,w) = \int_\Omega (p(x)\nabla v\nabla w + q(x)vw)dx$$

for $v,w$ in the Sobolev space $H = H_0^1(\Omega)$, and $b(v,w)$ denote the $L_2$ inner product

$$b(v,w) = \int_\Omega vwdx .$$

For simplicity let us assume that $p$ is positive in $\overline{\Omega}$ and $q$ is nonnegative there so that $a(v,w)$ is positive definite in the sense that

(2.2') $$a(v,v) \gtrsim C\|v\|_1^2 \text{ all } v \in H ,$$

where, as elsewhere in this section, we have put

$$\|v\|_r = \left\{ \sum_{|\alpha| \leq r} \int_\Omega |D^\alpha v|^2 \right\}^{\frac{1}{2}} .$$

The weak form of (1.1)-(1.2) is to seek a $u_0 \in H$ such that

(2.3) $$a(u_0, v) = b(f, v) \quad \text{all } v \in H .$$

To approximate this problem we first subdivide $\Omega$ into triangles (or rectangles) $e_\alpha^h$, $h_\alpha > 0$ denoting the diameter of $e_\alpha^h$ and $h = \max_\alpha h_\alpha$. We then choose a subspace $S^h \subset H$ of continuous functions which are polynomials of degree $k-1$ over each element $e_\alpha^h$. An approximate solution $u_h \in S^h$

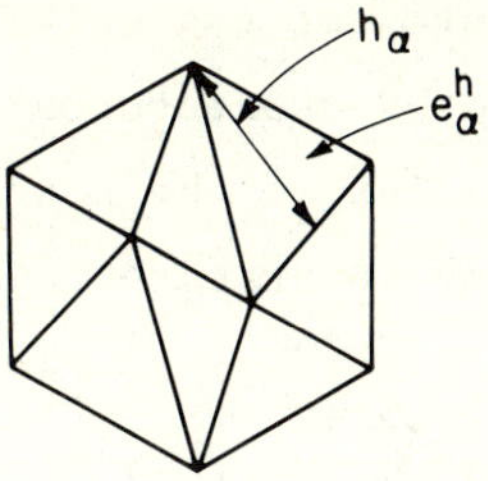


Fig. 1. The region $\Omega$

is determined from

(2.4) $$a(u_h, v^h) = b(f, v^h) \quad \text{all } v^h \in S^h .$$

Letting $\phi_1^h, \cdots, \phi_N^h$ be a basis for $S^h$, this is equivalent to the matrix problem

(2.5) $$K^h Q^h = F^h$$

in the vector of weights $Q^h$ of

(2.6) $$u_h(x) = \sum_{j=1}^{N} Q_j^h \phi_j^h(x) \ .$$

The (i,j) entry of the stiffness matrix is

(2.7) $$\int_\Omega (p\nabla\phi_i^h\nabla\phi_j^h + q\phi_i^h\phi_j^h)dx \ ,$$

and the jth entry of the source vector $F^h$ is

(2.8) $$\int_\Omega f(x)\phi_j^h(x)dx \ .$$

Under the usual regularity assumptions, the error $u-u_h$ is of order $\mathcal{O}(h^k)$ and the error in the first derivative is of order $\mathcal{O}(h^{k-1})$ [7]. The question we wish to ask is what additional errors are made when the integrals (1.7)-(1.8) are not evaluated exactly but rather are approximated by quadratures?

Specifically, let e be a unit triangle (rectangle) and consider the quadrature rule

(2.9) $$\int_e r(\xi)d\xi = \sum_{\ell=1}^{L} w_\ell r(\xi_\ell) = I(r) \ .$$

After an affine transformation this becomes

(2.10) $$\int_{e_\alpha^h} R(x)dx = \sum_{\ell=1}^{L} w_{\ell,\alpha}^h R(\xi_{\alpha,\ell}^h) = I_\alpha^h(R)$$

for a generic element $e_\alpha^h$ . With this we introduce a new stiffness matrix $\tilde{K}^h$ whose (i,j) entry is

(2.11) $$\sum_{e_\alpha^h} I_\alpha^h(p\nabla\phi_i^h\nabla\phi_j^h + q\phi_i^h\phi_j) ,$$

and a new source vector $\tilde{F}^h$ whose ith entry is

(2.12) $$\sum_{e_\alpha^h} I_\alpha^h(t\phi_i^h)^{\dagger} .$$

Thus instead of (2.6), we use

(2.13) $$\tilde{u}_h(x) = \sum_{j=1}^{N} \tilde{Q}_j^h \phi_j^h(x)$$

to approximate $u_0(x)$ , where the weights $\tilde{Q}^h = [\tilde{Q}_j^h]$ are obtained from

(2.14) $$\tilde{K}^h\tilde{Q}^h = \tilde{F}^h .$$

The problem is to estimate the error $u_h - \tilde{u}_h$ for a given quadrature rule (29) and finite element space $S^h$ .

The first step in the error analysis is to introduce a measure of the accuracy for the quadrature formula. To this end we let $P_I$ denote the set of polynomials $x^\gamma = x_1^{\gamma_1}x_2^{\gamma_2}$ such that the rule is exact; i.e.,

(2.15) $$x^\gamma \in P_I \leftrightarrow I(x^\gamma) = \int_e x^\gamma dx .$$

---

† We assume that the data p,q,f are defined at each point in Ω. If they are only square-integrable, it is therefore necessary to smooth them before applying the quadrature rule.

Let $\ell_0$ denote the length of the smallest exponent $\gamma$ of polynomials $x^\gamma$ which are not in $P_I$; i.e.,

(2.16) $$\ell_0 = \min\left\{\gamma_1 + \gamma_2 \mid x^\gamma \notin P_I\right\} .$$

For example, the midpoint rule over a triangle is exact for $1, x_1, x_2$ and hence $\ell_0 = 2$ .

The second item of interest is the type of polynomials that occur in the representation of $v^h \varepsilon S^h$ over the elements $e_\alpha^h$ . In particular, let us consider the highest derivatives appearing in the quadratic form $a(v,w)$, namely $\frac{\partial}{\partial x_i}$ $(i=1,2)$ , and let $P_p$ denote the polynomials $x^\gamma$ appearing in $\frac{\partial v^h}{\partial x_i}$ $(i=1,2)$ on $e_\alpha^h$ ; i.e.,

(2.17) $$\frac{\partial v^h}{\partial x_i} = \sum_{x^\gamma \in P_p} a_\gamma x^\gamma \ , \ x \in e_\alpha^h \ .$$

We put

(2.18) $$k_0 = \max\left\{\gamma_1 + \gamma_2 \mid x^\gamma \in P_p\right\} .$$

Therefore if $S^h$ consists of piecewise linear functions, i.e., functions of the form $a + bx_1 + cx_2$ over the elements $e_\alpha^h$ , then $k_0 = 0$ . For bilinear polynomials $a + bx_1 + cx_2 + dx_1x_2$ , $k_0$ is $1$ .

We are now in a position to estimate the errors due to quadrature. One technique for doing this is to treat it as a matrix perturbation problem; i.e., we estimate the size of the changes in $K^h$ and $F^h$ and then use standard perturbation theorems to bound $Qh - \tilde{Q}^h$ . This technique, however, invariably leads to inaccurate estimates [6] since it does not use crucial property that the solution $u_0$ of (2.3) is smooth. To exploit the latter, we work not with the perturbations in the matrices, but rather with perturbations in the forms $a(v,w)$, $b(f,w)$ on the function space $S^h$. Specifically, we introduce the new forms

$$\tilde{a}(v^h,w^h) = \sum_{e_\alpha^h} Q_\alpha^h(p\nabla v^h \cdot \nabla w^h + qv^h w^h) \tag{2.19}$$

$$\tilde{b}(f,w^h) = \sum_{e_\alpha^h} Q_\alpha^h(fw^h) \tag{2.20}$$

defined for $v^h, w^h \in S^h$, and observe that the solution $\tilde{u}_h$ of (2.14) is characterized by

$$\tilde{a}(\tilde{u}_h,v^h) = \tilde{b}(f,v^h) \quad \text{all } v^h \in S^h . \tag{2.21}$$

We now estimate the changes in $a$ and $b$, namely

$$\varepsilon_a(v^h,w^h) = \tilde{a}(v^h,w^h) - a(v^h,w^h) \tag{2.22}$$

$$\varepsilon_b(f,v^h) = \tilde{b}(f,v^h) - b(f,v^h) . \tag{2.23}$$

Recall [12] that the difference between $\int_e r(\xi)d\xi$ and the quadrature rule (2.9) for an analytic integrand $r(\xi)$ is the sum

$$\sum_{\gamma \notin I_p} C_\gamma \int_e |D^\gamma r| .$$

Hence, $\varepsilon_b(f,v^h)$ is dominated by terms of the form

$$(2.24) \quad \sum_{e_\alpha^h} h_\alpha^{\ell_0} \int_{e_\alpha^h} |D^\gamma (fv^h)| , \quad |\gamma| = \gamma_1 + \gamma_2 = \ell_0 .$$

Let us suppose that we have normalized $v^h$ by $a(v^h,v^h) = 1$. Since $v^h$ is an arbitrary function in $S^h$, every time we differentiate it past the first we lose a power of $h_\alpha$. Moreover, we can continue to do so until $v^h$ is annihilated, which by (2.18) means that $v^h$ can be differentiated up to order $k_0 + 1$. We conclude that (2.24) is of order $O(h^{\ell_0 - k_0})$, and hence

$$(2.25) \quad |\varepsilon_a(f,v^h)| \lesssim Ch^{\ell_0 - k_0} \|f\|_{\ell_0 - k_0} \|v^h\|_1 .^\dagger$$

The estimate (2.25) is the order of the change in the source vector $F^h$ -- assuming the basis $\phi_j^h$ has been normalized by $\|\phi_j^h\|_{H_1} = 1$ -- hence, if there were no changes in the stiffness matrix $K^h$, a straightforward perturbation analysis would give

$$\|u_h - \tilde{u}_h\|_1 \sim O(h^{\ell_0 - k_0}) ,$$

---

† If $f$ has only $s < \ell_0 - h_0$ derivatives, then the right side of (2.25) is replaced with $h^s \|f\|_s \|v^h\|_1$.

which is sharp [6].[†] However, such elementary techniques break down when the changes in $K^h$ are considered. Indeed, $\varepsilon_a(v^h,w^h)$ is bounded by terms of the form

$$(2.26) \qquad \sum_{e_\alpha^h} h_2^{\ell_0} \int_{e_\alpha^h} |D^\gamma[p\nabla v^h \cdot \nabla w^h + qv^h w^h]| .$$

We can now differentiate <u>both</u> $v^h$ and $w^h$, losing a power of $h_\alpha$ each time this is done. It follows that

$$(2.27) \qquad |\varepsilon_a(v^h,w^h)| \leq Ch^{\ell_0-2k_0}|v^h|_1|w^h|_1$$

where the constant $C$ depends on

$$(2.27') \qquad \max_{|\gamma|=\ell_0-2k_0} \sup_{x\in\Omega} |D^\gamma p(x)| , \qquad \max_{|\gamma|=\ell_0-2k_0} \sup_{x\in\Omega} |D^\gamma q(x)| ,$$

but is independent of $h$. It follows that $K_{ij}^h$ differs from $\tilde{K}_{ij}^h$ by order $(h^{\ell_0-2k_0})$, and hence a straightforward perturbation estimate would show $u_h - \tilde{u}_h$ is order $(h^{\ell_0-2k_0})$ instead of $(h^{\ell_0-k_0})$. This result is not sharp for the reason cited earlier -- nowhere was the smoothness of the solution $u_0$ used.

To derive sharp estimates we start with the identity

$$(2.28) \qquad \tilde{a}(u_h-\tilde{u}_h,u_h-\tilde{u}_h) = \varepsilon_a(u_h,u_h-\tilde{u}_h) - \varepsilon_b(f,u_h-\tilde{u}_h) .$$

The second term on the right-hand side has been estimated and is given by (2.25) with $v^h = u_h - \tilde{u}_h$. The estimate

---

† This particular result has also been obtained by Herbolt and Varga [9], [10].

(2.27) is not a sharp bound for the first term, however, and we improve it by using the properties of the finite element approximation $u_h$ . In particular, the latter satisfies

$$\sum_{e_\alpha^h} \sum_{|x| \leq k} \int_{e_\alpha^h} |D^\gamma u_h|^2 dx \leq C\|u\|_k^2 \tag{2.29}$$

provided the right-hand side is finite and provided the inverse inequality

$$\|v^h\|_j \leq Ch^{\ell-j}\|v^h\|_\ell \ , \ \ell<j \ , \tag{2.30}$$

is satisfied for all $v^h \in S^h$ . Thus in (2.26), where $v^h = u_h$ and $w^h = u_h - \tilde{u}_h$ , we lose powers of $h$ only by differentiating $w^h$ past the first and $u_h$ past the kth derivative. It follows that

$$\varepsilon_a(u_h, u_h - \tilde{u}_h) \leq Ch^{\ell_0 - k_0}[1 + h^{k-1-k_0}]\|u_h - \tilde{u}_h\|_1 \ . \tag{2.31}$$

Let us assume that $\ell_0 > 2k_0$ . Then

$$\tilde{a}(v_h, v_h) \geq Ca(v_h, v_h) \geq C'\|v_h\|_1^2$$

for all $v_h \in S^h$ , and hence (2.29) implies that $\|u_h - \tilde{u}_h\|_1$ is of order $h^{\ell_0+k-1-2k_0}$ . We summarize these results in

<u>Theorem 2.1</u>. Let $S^h$ consist of piecewise polynomials of degree $k-1$ , and satisfy the inverse condition (2.30). Let $k_0$ be defined by (2.18), and let the index $\ell_0$ associated with quadrature satisfy (2.16). Then if $\ell_0 > 2k_0$ ,

(2.32) $\|u_h - \tilde{u}_h\|_1 \leq Ch^{\ell_0 - k_0}[1 + h^{k-1-k_0}]\|u\|_k$ ,

where the constant $C$ depends on $\|f\|_r$ and (2.27') but is independent of $h$ .

We note that if $S^h$ consists of piecewise polynomial functions of <u>total degree</u> $k-1$ , i.e.,

$$\sum_{0 \leq \gamma_1 + \gamma_2 < k} a_\gamma x^\gamma ,$$

such as triangular elements, then $k_0 = k - 2$ , and hence (2.32) is of order $O(h^{\ell_0 - k_0}) = O(h^{\ell_0 + 2 - k})$ . For polynomials of <u>individual degree</u> $k-1$ , i.e.,

$$\sum_{i=1}^{2} \sum_{0 \leq \gamma_i < k} a_\gamma x^\gamma ,$$

$k_0 = 2k - 3$ , and (2.32) is of order $O(h^{\ell_0 + 5 - 3k})$ . Of course in one spatial dimension the distinction between total and individual degree disappears and the errors are of order $O(h^{\ell_0 + 2 - k})$ . Examples with specific quadrature rules are given in [6].

## 3. Steady state problems.

We formulate the steady state problem $Lu = f$ abstractly by selecting two Hilbert spaces $H_1, H_2$ and two bilinear forms $a(u,v)$, $b(u,v)$ on $H_1 \times H_2$ . We seek a $u_0 \in H_1$ such that

$$(3.1)\qquad a(u_0,v) = b(f,v) \quad \text{all } v \in H_2 \;.$$

To insure that this problem is well posed we assume that the forms are bounded in the sense that

$$(3.2)\quad |a(u,v)| \leq C_1 \|u\|_{H_1}\|v\|_{H_2} \;,\quad |b(u,v)| \leq C_1 \|u\|_{H_1}\|v\|_{H_2}$$

holds for all $u \in H_1$, $v \in H_2$. In addition, we require that $a$ is invertible; i.e.,

$$(3.3)\quad \begin{aligned} &\inf_{\|u\|_{H_1}=1}\ \sup_{\|v\|_{H_2}=1} |a(u,v)| \geq C_0 \;, \\ &\inf_{\|v\|_{H_2}=1}\ \sup_{\|u\|_{H_1}=1} |a(u,v)| \geq C_0 \;, \end{aligned}$$

(see [1] and [2]). Observe that by the Riesz Representation Theorem (3.2)-(3.3) implies that there is a bounded operator $T$ on $H_1$ such that

$$(3.4)\qquad a(Tf,v) = b(f,v) \quad \text{all } v \in H_2 \;.$$

Hence the necessarily unique solution of (3.1) is $u_0 = Tf$.

In the model problem of the previous section $H_1 = H_2 = H^1(\Omega)$. The inequalities (3,2) are clearly true in this case as are (3.3) since $a(v,w)$ is positive definite on $H \times H$ (see (2.2')).

To approximate this problem we introduce two finite dimensional subspaces $S_i^h \subset H_i$, i=1,2, parameterized by h>0, and solve (3.1) on $S_1^h \times S_2^h$. That is, we seek a $u_h \in S_1^h$ such that

$$(3.5)\qquad a(u_h,v^h) = b(f,v^h) \quad \text{all } v^h \in S_2^h \;.$$

For this problem to have a solution we need to assume that

a is invertible on $S_1^h \times S_2^h$ in the sense that

$$(3.6) \qquad \inf_{\|v^h\|_{H_1}=1} \; \sup_{\|w^h\|_{H_2}=1} |a(v^h,w^h)| \geq C_0'$$

where $C_0'$ is independent of $h$ and $v^h$ varies over $S_1^h$, with $w^h$ varying over $S_2^h$.† This inequality is true for the problem in §2 since in this case $a(v,w)$ is positive definite on $H \times H$ and in particular positive definite on the smaller set $S^h \times S^h$.

Observe that (3.6) implies that there is an operator $T^h$ from $H_1$ to $S_1^h$ whose norm is bounded independent of $h$ such that

$$(3.6') \qquad a(T^h f, v^h) = b(f, v^h) \quad \text{all } v^h \in S_2^h .$$

Hence the necessarily unique solution $u_h$ of (3.5) is $T^h f$.

To model the effect of quadrature we introduce the new forms

$$(3.7) \qquad \tilde{a}(v^h,w^h) = a(v^h,w^h) + \varepsilon_a(v^h,w^h)$$

$$(3.8) \qquad \tilde{b}(v^h,w^h) = b(v^h,w^h) + \varepsilon_b(v^h,w^h)$$

defined on $S_1^h \times S_2^h$ and consider the new problem

$$(3.9) \qquad \tilde{a}(\tilde{u}_h, v^h) = \tilde{b}(f_h, v^h) \quad \text{all } v^h \in S_2^h$$

where $f^h$ is some approximation to $f$ in $S_1^h$. (In §2 we took $f^h = f$ for simplicity.) To study this approximation

† Since $S_1^h$ and $S_2^h$ are finite dimensional, it follows that a similar inequality holds if $v^h$ and $w^h$ are interchanged.

to (3.1) we first introduce the maximal errors

$$(3.10)\qquad \eta_a^h = \sup_{\|v^h\|_{H_1}=1} \sup_{\|w^h\|_{H_2}=1} |\varepsilon_a(v^h,w^h)|$$

$$(3.11)\qquad \eta_b^h = \sup_{\|v^h\|_{H_1}=1} \sup_{\|w^h\|_{H_2}=1} |\varepsilon_b(v^h,w^h)|$$

and the error

$$(3.12)\qquad \eta_f^h = \|f - f^h\|_{H_1}$$

in the approximation $f^h$ to $f$. In the model problem of §2, (3.10) was of order $O(h^{\ell_0-2k_0})$ and measured the difference in the coefficient matrices in the matrix representation of (3.5) and (3.9). A trivial but very important remark is that if $\eta_a^h \to 0$ as $h \to 0$, then $\tilde{a}(v,w)$ is invertible in the sense that it satisfies an inequality like (3.6). Because of its importance we record this fact in

Lemma 1. Let $\eta_a^h \to 0$ as $h \to 0$. Therefore there is a $h_0 > 0$ and a constant $C$ depending only on $h_0$ such that

$$(3.13)\qquad \inf_{\|v^h\|_{H_1}=1} \sup_{\|w^h\|_{H_2}=1} |\tilde{a}(v^h,w^h)| \geq C \quad \text{for } 0<h\leq h_0 .$$

It follows from (3.13) that there is an operator $\tilde{T}^h$ from $S_1^h$ to $S_1^h$ whose norm is bounded independent of $h$ satisfying

$$(3.14)\qquad \tilde{a}(\tilde{T}^h f^h,v^h) = \tilde{b}(f^h,v^h) \quad \text{all } v^h \in S_2^h .$$

Thus (3.9) has the unique solution $\tilde{u}_h = \tilde{T}^h f^h$ .

To bound the error $u_h - \tilde{u}_h$ , we start with the identity

$$\tilde{a}(u_h - \tilde{u}_h, v^h) = \varepsilon_a(u_h, v^h) - \varepsilon_b(f^h, v^h) + b(f - f^h, v^h) , \tag{3.15}$$

which is true for all $v^h$ in $S_2^h$ . Put

$$\delta_a^h(u_h) = \sup_{\|v^h\|_{H_2}=1} |\varepsilon_a(u_h, v^h)| \tag{3.16}$$

$$\delta_b^h(u_h) = \sup_{\|v^h\|_{H_2}=1} |\varepsilon_b(f^h, v^h)| . \tag{3.17}$$

Then taking the sup over $v^h$ in (3.15) and using (3.13) we obtain

<u>Theorem 3.1</u>. Define $\eta_a^h$ by (3.10) and assume $\eta_a^h \to 0$ as $h \to 0$. Then with $\delta_a^h$ , $\delta_b^h$ defined by (3.16)-(3.17), we have

$$\|u_h - \tilde{u}_h\|_{H_1} \leq C_1 \delta_a^h(u_h) + C_2 \delta_b^h(f_h) + C_3 \|f - f_h\|_{H_1} \tag{3.18}$$

for constants $C_i$ , $1 \leq i \leq 3$ , independent of $h$ .

For the model problem in §2

$$\delta_a^h(u_h) = O[h^{\ell_0 - k_0}(1 + h^{k-1-k_0})] , \quad \delta_b^h(f_h) = O(h^{\ell_0 - k_0}) \tag{3.19}$$

and

$$\eta_a^h = O(h^{\ell_0 - 2k_0}) . \tag{3.19'}$$

Hence (3.18) is exactly (2.32) in this case, and the condition $\eta_a^h \to 0$ means $\ell_o > 2k_0$ as in Theorem 2.1.

Similar estimates hold for 2mth order problems

$$Lu = \sum_{|\alpha|\leq m} \sum_{|\beta|\leq m} (-)D^\alpha p_{\alpha\beta} D^\beta u = f$$

in n dimensions provided the integers $\ell_0$ and $k_0$ have been suitably defined. In particular, suppose the forms associated with $Lu = f$ are

$$\text{(3.20)} \qquad a(u,v) = \sum_{|\alpha|\leq m} \sum_{|\beta|\leq m} \int_\Omega p_{\alpha\beta} D^\alpha u D^\beta v$$

$$\text{(3.21)} \qquad b(f,v) = \int_\Omega fv$$

for $u,v \in H_0^m(\Omega)$ . Instead of (2.18) we associate $k_0$ with the polynomial representation of the highest derivatives appearing in (3.40). More precisely, we define the index set $P_p$ by

$$D^\gamma v^h(x) = \sum_{\beta \in P_p} a_\beta x^\beta$$

for x in a generic element $e_\alpha^h$ , and the integer $k_0$ by

$$k_0 = \max\{\beta_1 + \cdots + \beta_n \mid \beta = (\beta_1, \cdots, \beta_n) \in P_p\}.$$

The index $\ell_0$ associated with the quadrature rule, on the other hand, is not changed and in particular is defined by (2.16). With $k_0$ and $\ell_0$ the argument given in §2 can be used to show that

$$(3.19') \quad \delta_a^h(u_h) = O[h^{\ell_0-k_0}(1+h^{k-m-k_0})] , \quad \delta_a^h(f_h) = O(h^{\ell_0-k_0})$$

and hence $u_h - \tilde{u}_h$ is of this order in the norm $\|\cdot\|_m$, provided $\ell_0 > 2k_0$.

To spell out the dependence of the error $u_h - \tilde{u}_h$ on the order $2m$ of the operator and the number $n$ of spatial dimension, let us first suppose $S^h = S_1^h = S_2^h$ consists of piecewise polynomials of <u>total degree</u> $k-1$. Then $k_0 = k - m - 1$, and hence $u_h - \tilde{u}_h = O(h^{\ell_0+m+1-k})$ in $H^m(\Omega)$. The other extreme is when $S^h$ consists of piecewise polynomials of degree $k-1$ in each variable. In this case $k_0 = n(k - 1) - m$, and hence $u_h - \tilde{u}_h$ is of order $h^{\ell_0+m-n(k-1)} + h^{\ell_0+m+k-2n(k-1)}$.

In the sequel we shall call

$$(3.22) \qquad \eta_a^h \to 0 \quad \text{as} \quad h \to 0$$

the <u>stability condition</u> since it insures the invertibility of (3.9) on $S_1^h \times S_2^h$. Observe that it necessarily implies

$$(3.23) \qquad \delta_a^h \to 0 \quad \text{as} \quad h \to 0 ,$$

which with $\delta_b^h \to 0$ and $\|f - f_h\|_{H_1} \to 0$ are called the <u>accuracy conditions</u>. The converse is not true, and in [6] are given examples where (3.23) holds and yet (3.22) is not true. In some of these examples the matrix representation of (3.9) with respect to the natural patch basis is not invertible, and in others it is invertible but very poorly conditioned.

## 4. The Eigenvalue Problem.

We turn now to the eigenvalue problem $a(u,v) = \lambda b(u,v)$ defined on $H_1 \times H_2$. We seek those complex numbers $\lambda_0$ such that there is a $0 \neq u_0 \in H_1$ satisfying

$$a(u_0,v) = \lambda_0 b(u_0,v) \quad \text{all} \quad v \in H_2 . \tag{4.1}$$

Our goal is to show that if the stability condition (3.22) in §3 is satisfied then the errors in both the eigenvalues and eigenfunctions due to quadrature have the same order of accuracy as the steady state errors.

To do this we shall need in addition to (3.2)-(3.3) and (3.6), the compactness of the operator $T$ defined by (3.4). Observe that $\lambda_0$ satisfies (4.1) for some nontrivial $v_0 \in H_1$ if and only if $1/\lambda_0$ is an eigenvalue of $T$.† Hence it follows from the compactness of $T$ that the eigenvalues of (4.1) form a countable set with accumulation only at infinity [11]. For simplicity we shall deal only with simple eigenvalues whose geometric and algebraic multiplicities coincide.

Approximations are obtained by computing the eigenvalues of $a$ with respect to $b$ on $S_1^h \times S_2^h$; i.e., those numbers $\lambda_h$ for which there is a nontrivial $u_h \in S_1^h$ (normalize $u_h$ by $\|u_h\|_{H_1} = 1$) satisfying

$$a(u_h,v^h) = \lambda_h b(u_h,v^h) \quad \text{all} \quad v^h \in S_2^h . \tag{4.2}$$

Observe that (3.6) implies (4.2) is equivalent to

† The inequality (3.3) implies that $\lambda_0 = 0$ is not an eigenvalue of (4.1).

$$\lambda_h T^h u_h = u_h$$

where the operator $T^h: H_1 \to S_1^h$ is defined by (3.6').

It was shown in [5] (also [3]) that if

$$(4.3) \qquad \|T^h - T\|_{L(H_1)} = \sup_{\|u\|_{H_1}=1} \|(T^h - T)u\|_{H_1} \to 0$$

as $h \to 0$, then there is an eigenvalue $\lambda_h$ and eigenfunction $u_h$ of (4.2) such that

$$|\lambda_h - \lambda_0| \to 0 \ , \ \|u - u_h\|_{H_1} \to 0$$

as $h \to 0$. To describe the error bounds we introduce the eigenspace

$$E_0 = \{u_0 \in H_1 \,|\, u_0 \text{ satisfies } (4.1)\}$$

associated with $\lambda_0$, and the adjoint eigenspace

$$E_0^* = \{v_0 \in H_2 \,|\, a(u, v_0) = \lambda_0 b(u, v_0) \quad \text{all } u \in H_1\} .$$

Then the errors $\lambda_0 - \lambda_h$, $u_0 - u_h$ are related to the approximation errors

$$\varepsilon_h = \sup_{u_0 \in H_1} \inf_{x^h \in S_1^h} (\|u_0 - x^h\|_{H_1} / \|u_0\|_{H_1})$$

$$\varepsilon_L^* = \sup_{v_0 \in H_2} \inf_{x^h \in S_2^h} (\|v_0 - x^h\|_{H_2} / \|v_0\|_{H_2}) ,$$

which are typically of order $O(h^{k-m})$ for 2mth order problems and piecewise polynomial functions of degree k-1 [7]. In particular, it was proved in [5] that

$$|\lambda_0 - \lambda_h| \leq C\varepsilon_h \varepsilon_h^* \ , \ \|u_0 - u_h\|_{H_1} \leq C\varepsilon_h$$

for a constant $C$ independent of $h$.

To model the effect of quadrature errors we turn to the forms $\tilde{a}(v,w)$, $\tilde{b}(v,w)$ defined on $S_1^h \times S_2^h$, and consider the eigenvalue problem

$$(4.4) \qquad \tilde{a}(\tilde{u}_h, v^h) = \tilde{\lambda}_h \tilde{b}(\tilde{u}_h, v^h) \quad \text{all} \quad v^h \in S_2^h .$$

Assuming that the stability condition (3.22) is valid, the eigenvalues $\tilde{\lambda}_h$ of (4.4) satisfy

$$\tilde{\lambda}_h \tilde{T}^h \tilde{u}_h = \tilde{u}_h$$

where $\tilde{T}^h$ is defined by (3.14). Our goal of course is to show that (4.4) has an eigenset $\tilde{\lambda}_h, \tilde{u}_h$ such that $\lambda_h - \tilde{\lambda}_h \to 0$, $\|u_h - \tilde{u}_h\|_{H_1} \to 0$ as $h \to 0$. In addition, we want to bound the errors $\lambda_h - \tilde{\lambda}_h$, $\|u_h - \tilde{u}_h\|_{H_1}$ in terms of $\delta_a^h$, $\delta_b^h$.

To do this we first show that $T^h - \tilde{T}^h$ converges to zero in the operator norm

$$(4.5) \qquad \|T^h - \tilde{T}^h\|_{L(S_1^h)} = \sup_{\|v_a^h\|_{H_1}=1} \|(T^h - \tilde{T}^h)v^h\| ,$$

and for this we need the stability condition for both the forms $\tilde{a}$ and $\tilde{b}$.

<u>Lemma 4.1</u>. Let $\eta_a^h$ and $\eta_b^h$ approach zero as $h \to 0$. Then (4.5) approaches zero with $h$.

<u>Proof</u>. For $v^h$ in $S_1^h$ and $w^h$ in $S_2^h$ we have

$$\tilde{a}([\tilde{T}^h - T^h]v^h, w^h) = \varepsilon_b(v^h, w^h) - \varepsilon_a(T^h v^h, w^h)$$

where $\varepsilon_a$ and $\varepsilon_b$ are defined by (3.7)-(3.8). Taking the sup as $w^h$ varies over $S_2^h$ with $\|w^h\|_{H_2} = 1$ and using (3.13) we obtain

$$\|[T^h - \tilde{T}^h]v^h\|_{H_1} \leq [C\eta_a^h + C'\eta_b^h]\|v^h\|_{H_1}$$

(from (3.10)-(3.11)) . Taking the sup over a vector $v^h$ of unit length in $S_1^h$ and letting h approach, we obtain the desired conclusion.

Since $T^h$ and $\tilde{T}^h$ are compact (they have finite dimensional range) and $T^h$ has an eigenvalue $\lambda_h$ converging to $\lambda_0$ , it follows from Lemma 4.1 and the usual compactness arguments that there is an eigenvalue $\tilde{\lambda}_h$ of (4.4) such that $|\tilde{\lambda}_h - \lambda_h| \to 0$ as $h \to 0$ (see [1], [5] and also [4]). In addition, there is an eigenfunction $\tilde{u}_h$ of (4.4) such that $\|u_h - \tilde{u}_h\|_{H_1} \to 0$ . We shall prove that $|\lambda_h - \tilde{\lambda}_h|$ and $\|u_h - \tilde{u}_h\|_{H_1}$ are of order $O(\delta_a^h) + O(\delta_b^h)$ .

To do this we put

$$a_\lambda(v,w) = a(v,w) - \lambda b(v,w)$$

(also $\tilde{a}_\lambda = \tilde{a} - \lambda\tilde{b}$) , and observe that

$$a_{\lambda_h}(\tilde{u}_h,v^h) = \tilde{a}_{\tilde{\lambda}_h}(\tilde{u}_h,v^h) - \varepsilon_a(\tilde{u}_h,v^h)$$
(4.6)
$$+ \tilde{\lambda}_h\varepsilon_b(\tilde{u}_h,v^h) + (\tilde{\lambda}_h - \lambda_h)b(\tilde{u}_h,v^h)$$

holds for each $v^h$ in $S_2^h$ . We now make a special choice for this function, namely we let $v^h$ be the adjoint eigenfunction $v_h$ satisfying

$$a_{\lambda_h}(u^h,v_h) = a(u^h,v_h) - \lambda_h b(u^h,v_h) = 0 \quad \text{all} \quad u^h \in S_1^h .$$

For this choice (4.6) reduces to

$$(4.7)\quad (\tilde{\lambda}_h - \lambda_h)b(\tilde{u}_h, v_h) = \varepsilon_a(\tilde{u}_h, v^h) - \tilde{\lambda}_h \varepsilon_b(\tilde{u}_h, v^h) \ .$$

Now, $\tilde{u}_h$ converges in $H_1$ to the eigenfunction $u_0$ associated with $\lambda_0$ and $v_h$ converges in $H_2$ to an adjoint eigenfunction $v_0$ satisfying

$$a(u, v_0) = \lambda_0 b(u, v_0) \quad \text{all} \quad u \in H_1 \ .$$

Moreover, we can choose $v_0$ such that

$$b(u_0, v_0) = 1$$

(see [5]). Hence (4.7) gives

$$|\tilde{\lambda}_h - \lambda_h| \leq C\delta_a^h + C'\delta_b^h$$

as $h \to 0$ .

To derive the eigenfunction estimates we introduce the projection $\tilde{Q}^h$ of $S_1^h$ onto the space of eigenfunctions associated with $\tilde{\lambda}_h$ . Since the latter converges to $\lambda_0$, $\tilde{Q}^h$ is given by

$$(4.7')\qquad \tilde{Q}^h = \frac{1}{2\pi i} \int_\Gamma (z - \tilde{T}^h)^{-1} dz$$

where $\Gamma$ is a circle about $\frac{1}{\lambda_0}$ containing $\frac{1}{\tilde{\lambda}_h}$ and no other eigenvalues of $\tilde{T}^h$ ([4], [11]). Since $\tilde{Q}^h u_h$ is an eigenfunction of (4.4), we have

$$(4.8)\qquad \tilde{a}_{\tilde{\lambda}_h}((I - \tilde{Q}_h)u_h, v^h) = (\varepsilon_a - \tilde{\lambda}_h \varepsilon_b)(u_h, v^h) + (\lambda_h - \tilde{\lambda}_h) b(u_h, v^h)$$

Thus, since $\|u_h\|_{H_1} = 1$ (by normalization)

(4.9) $$\sup_{\|v^h\|_{H_2}=1} |\tilde{a}_{\tilde{\lambda}_h}((I - \tilde{Q}_h)u_h, v^h)| \leq C\delta_a^h + C'\delta_b^h .$$

We now show that the left-hand side of (4.9) bounds $\|(I - \tilde{Q}_h)u_h\|_{H_1}$ . To see this we introduce the operator

(4.10) $$\tilde{Q}_0^h = \frac{1}{2\pi i}\int_\Gamma (z - \tilde{T}^h)^{-1}(z - \frac{1}{\tilde{\lambda}_h})^{-1}dz .$$

Then

(4.11) $$(I - \tilde{\lambda}_h\tilde{T}^h)\tilde{Q}_0^h = I - \tilde{Q}^h$$

(see [11] p. 229). Thus

(4.12) $$\begin{aligned}\|(I - \tilde{Q}^h)u_h\|_{H_1} &= \|\tilde{Q}_0^h (I - \tilde{\lambda}_h\tilde{T}^h)u_h\|_{H_1} \\ &\leq \|Q_0^h\|_{L(\delta_1^h)}\|(I - \tilde{\lambda}_h\tilde{T}^h)u_h\|_{H_1} .\end{aligned}$$

It follows from (4.10) that $\|\tilde{Q}_0^h\|_{L(\delta_1^h)}$ is bounded independent of $h$ , and hence from (3.13)

(4.13) $$\begin{aligned}\|(I - \tilde{Q}^h)u_h\|_{H_1} &\leq C\|(I - \tilde{\lambda}_h\tilde{T}^h)u_h\|_{H_1} \\ &\leq C' \sup_{\|v^h\|_{H_2}=1} |\tilde{a}((I - \tilde{\lambda}_h\tilde{T}^h)u_h, v^h)| .\end{aligned}$$

But $\tilde{Q}^h$ is a projection whose norm is bounded independent of $h$ (this follows immediately from (4.7')), hence

$$\|(I - \tilde{Q}^h)u_h\|_{H_1} \leq C'\delta_a^h + C''\delta_b^h .$$

It follows that there is an eigenfunction $\tilde{u}_h$ associated with $\tilde{\lambda}_h$ such that $\|\tilde{u}_h\|_{H_1} = 1$ and

$$\|u_h - \tilde{u}_h\|_{H_1} \leq C'\delta_a^h + C''\delta_b^h .$$

## 5. Parabolic problems.

The last class of problems we shall treat are time dependent problems of the form

(5.1) $$b\left(\frac{\partial u_0}{\partial t}, v\right) + a(u_0, v) = 0 \quad \text{all } v \in H_2 .$$

We seek a family $\{u_0(t)\}_{t>0} \subset H_1$ which is continuously differentiable with respect to time $t$ and satisfies (5.1) for $t > 0$. Given the initial data $g \in H_1$, we also require that

(5.2) $$\|u_0(t) - g\|_{H_1} \to 0 \quad \text{as} \quad t \downarrow 0 .$$

As usual it is assumed that (3.2)-(3.3) and (3.6) are valid. In addition, we assume that the spectrum of $a(v,w)$ with respect to $b(v,w)$ is sectorial in the following sense.

Definition 5.1. The form $a(v,w)$ is said to be sectorial with respect to $b(v,w)$ on $H_1 \times H_2$ if there is a sector

(5.3) $$S = \{\xi - \lambda_0 \mid |\arg \xi| > \alpha, \ \alpha > \frac{\pi}{2},$$

in the complex plane such that if $z \notin S$, then

(5.4) $$a(u,v) + zb(u,v) = b(f,v) \quad \text{all} \quad v \in H_2$$

has a unique solution $u \in H_1$ for each $f \in H_1$. Moreover,

(5.5) $$\|u\|_{H_1} \leq K(z)\|f\|_{H_1}$$

where

(5.6) $$|K(z)| \leq C / |z|$$

for large $z \notin S$ .

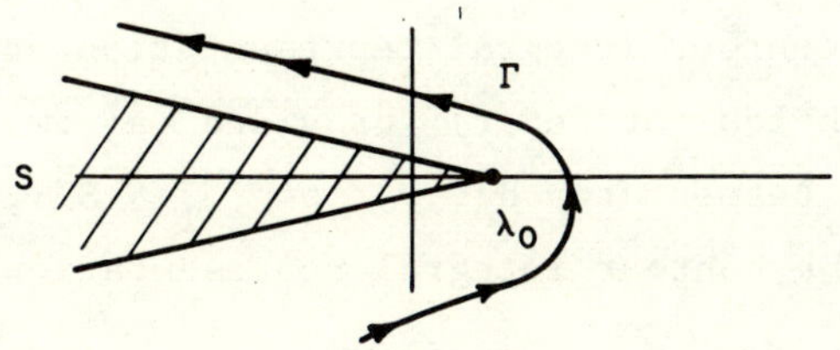


Fig. 2. The contour Γ

Let $\Gamma$ be the contour shown in Figure 5.1. Then the solution $u_0(t)$ of (5.1)-(5.2) exists and can be expressed in terms of solutions to the steady state problem (5.4) by means of contour integrals in the following way. Given $z \in \Gamma$ and the initial condition $g$ , we solve for the $w_z \in H_1$ satisfying

(5.7) $$a(w_z,v) + zb(w_z,v) = b(g,v) \quad \text{all} \quad v \in H_2 .$$

Then

(5.8) $$u_0(t) = \frac{1}{2\pi i}\int_\Gamma e^{zt} w_z dz$$

is the unique solution of (5.1)-(5.2) (see [2]).

The Galerkin approximation $\{u_h(t)\}_{t>0} \subset S_1^h$

(5.1)-(5.2) is defined by the ordinary differential equation

$$(5.9)\quad b\left(\frac{\partial u_h}{\partial t}, v^h\right) + a(u_h, v^h) = 0 \quad \text{all } v^h \in S_1^h, \ t > 0$$

and the initial condition

$$(5.10)\qquad \|u_h(t) - g_h\|_{H_1} \to 0 \quad \text{as} \quad t \downarrow 0,$$

where $g_h$ is any approximation to $g$ in $S_1^h$ (e.g., the best approximation with respect to $a(v,w)$). It is shown in [1] that a contour integral representation analogous to (5.8) also holds for $u_h(t)$ provided $a(v,w)$ is sectorial with respect to $b(v,w)$ on $S_1^h \times S_2^h$. If this is the case, the contour integral representation is

$$(5.11)\qquad u_h(t) = \frac{1}{2\pi i}\int_{\Gamma_h} e^{zt} w_z^h dz$$

where

$$(5.12)\qquad a(w_z^h, v^h) + zb(w_z^h, v^h) = b(g_h, v^h)$$

and $\Gamma_h$ is a contour like $\Gamma$ outside $S_h$.

These formulas are used to derive error bounds for $u_0(t) - u_h(t)$ in [1]. Our goal, however, is to compare $u_h(t)$ with the approximation $\tilde{u}_h(t)$ obtained from the perturbed problem

$$(5.13)\quad \tilde{b}\left(\frac{\tilde{\partial}\tilde{u}_h}{\partial t}, v^h\right) + \tilde{a}(\tilde{u}_h, v^h) = 0, \ t > 0, \ \text{all } v^h \in S_2^h$$

$$(5.14)\qquad \|\tilde{u}_h(t) - g_h\|_{H_1} \to 0 \quad \text{as} \quad t \downarrow 0,$$

which concretely is obtained from (5.9) by approximating the inner products with quadrature rules.

The first step in the comparison is to derive a

semi-group representation for $\tilde{u}_h(t)$ .

<u>Theorem 5.1</u>. Let $\tilde{a}$ be sectorial with respect to $\tilde{b}$ on $S_1^h \times S_2^h$ , and let $\tilde{S}_h$ denote the sector (5.3). Let $\tilde{\Gamma}_h$ be a contour outside $\tilde{S}_h$ such that $\mathrm{Re}(\xi) \to -\infty$ as $|\xi| \to \infty$ on $\tilde{\Gamma}_h$ . Define $\tilde{w}_z^h$ by

$$(5.15)\quad \tilde{a}(\tilde{w}_z^h, v^h) + z\tilde{b}(\tilde{w}_z^h, v^h) = \tilde{b}(g_h, v^h) \quad \text{all } v^h \in S_2^h .$$

Then if the stability condition $\eta_a^h \to 0$ as $h \to 0$ holds,

$$(5.16)\qquad \tilde{u}_h(t) = \frac{1}{2\pi i}\int_{\tilde{\Gamma}_h} e^{zt}\tilde{w}_z^h dz$$

is the unique solution of (5.13)-(5.14) for sufficiently small $h$ .

The proof of Theorem 5.1 is similar to the verification of (5.11) given in [1], however we give a short derivation here so that the reader may see the role played by our assumptions. Since the real part of $z$ goes to $-\infty$ as $|z| \to \infty$ on $\tilde{\Gamma}_h$ , the contour integral (5.16) defines a continuously differentiable function $\tilde{u}_h(t) \in S_1^h$ of $t$ for $t > 0$ . Moreover, for each fixed $t > 0$ ,

$$\tilde{b}\left(\frac{\partial \tilde{u}_h}{\partial t}, v^h\right) + \tilde{a}(\tilde{u}_h, v^h) = \frac{1}{2\pi i}\int_{\tilde{\Gamma}_h} e^{zt}\{z\tilde{b}(\tilde{u}_h, v^h) + \tilde{a}(\tilde{u}, v^h)\}$$

$$= \frac{1}{2\pi i}\, b(g, v^h) \int_{\tilde{\Gamma}_h} e^{zt} .$$

But $\int_{\tilde{\Gamma}_h} e^{zt} = 0$ by Cauchy's Theorem, thus $\tilde{u}_h(t)$ satisfies (5.13) for each $t > 0$ .

To study the behavior of $\tilde{u}_h(t)$ near $t = 0$ , we make the change of variables

$$\xi = zt$$

so that

(5.17) $$\tilde{u}_h(t) = (2\pi i t)^{-1} \int_{\tilde{\tilde{\Gamma}}_h} e^{\xi} \tilde{w}^h_{\xi/t} \, d\xi \; .$$

By Cauchy's Theorem, we may take the contour $\tilde{\tilde{\Gamma}}_h$ as shown in Figure 5-2, and independent of time $t$ .

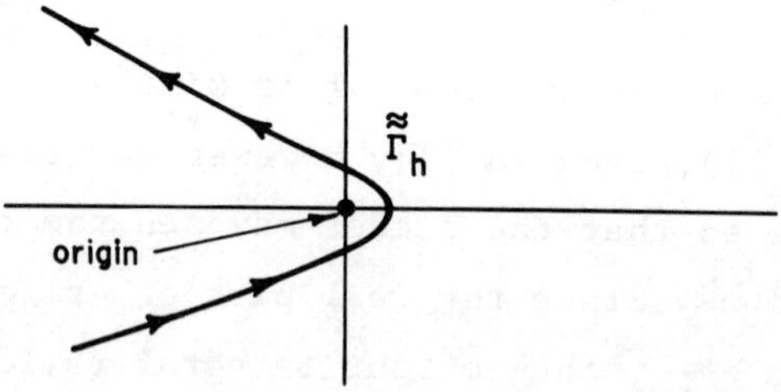


Fig. 3. The contour $\tilde{\tilde{\Gamma}}_h$

We again appeal to Cauchy's Theorem to note

$$1 = \frac{1}{2\pi i} \int_{\tilde{\tilde{\Gamma}}_h} \xi^{-1} e^{\xi} d\xi \; .$$

Hence

(5.18) $$\tilde{u}_h(t) - g_h = \frac{1}{2\pi i} \int_{\tilde{\tilde{\Gamma}}_h} e^{\xi}\{t^{-1}\tilde{w}^h_{\xi/t} - \xi^{-1} g_h\} d\xi \; .$$

But

$$(5.19)\quad \tilde{a}(t^{-1}\tilde{w}^h_{\xi/t}-\xi^{-1}g_h, v^h) = -\xi^{-1}\tilde{a}(\tilde{\tilde{w}}^h_{\xi/t}, v^h) \text{ all } v^h \in S^h_2$$

where $\tilde{\tilde{w}}^h_z$ is the unique solution of

$$(5.20)\quad \tilde{a}(\tilde{\tilde{w}}^h_z, v^h) + z\tilde{b}(\tilde{\tilde{w}}^h_z, v^h) = \tilde{a}(g, v^h) \quad \text{all } v^h \in S^h_2 \ .$$

Perhaps the best way to see (5.19) is to use the operator $\tilde{T}^h$ defined by (3.14). In particular

$$\tilde{w}^h_z = (I + z\tilde{T}^h)^{-1}\tilde{T}^h g_h$$

and

$$\tilde{\tilde{w}}^h_z = (I + z\tilde{T}^h)^{-1} g_h \ .$$

Thus

$$t^{-1}\tilde{w}^h_{\xi/t} - \xi^{-1} g_h = -\xi^{-1}(I + \frac{\xi}{t}\tilde{T}^h)^{-1} g_h$$

$$= -\xi^{-1}\tilde{\tilde{w}}^h_{\xi/t} \ .$$

Now, for $\xi \in \tilde{\Gamma}_h$ we have

$$|\tilde{\tilde{w}}^h_{\xi/t}|_{H_1} \leq C_h\left(\frac{t}{|\xi|}\right)\|g_h\|_{H_1}$$

by the sectorial property. Thus (5.18) gives

$$\sup_{\|v^h\|_{H_2}=1} |\tilde{a}(\tilde{u}_h(t) - g_h, v^h)| \leq \left(\frac{C_h t}{2\pi}\right)\|g_h\|_{H_1}\int_{\tilde{\Gamma}_h} \frac{e\xi}{|\xi|}|d\xi| \ .$$

Since $\eta^h_a \to 0$ as $h \to 0$ we conclude that

$$\|\tilde{u}_h(t) - g_h\|_{H_1} \leq C'_h t\|g_h\|_{H_1}$$

for sufficiently small $h$, and hence (5.14) holds.

Theorem 5.2. Let the conditions of Theorem 5.1 hold, and let the sectors $S_h$, $\tilde{S}_h$ be independent of $h$. Then

$$\|u_h(t) - \tilde{u}_h(t)\|_{H_1} \leq Ce^{\lambda_0 t}[\delta_a^h + \delta_b^h] \tag{5.21}$$

where $\lambda_0$ is any real number not in $\tilde{S}_h \cup S_h$.

The proof of Theorem 5.2 is quite straightforward now that the semi-group machinery has been set up. Indeed, since both of the sectors $S_h$ and $\tilde{S}_h$ are independent of $h$ we may take $\Gamma_h = \tilde{\Gamma}_h = \Gamma$. Thus

$$u_h(t) - \tilde{u}_h(t) = \frac{1}{2\pi i}\int_\Gamma e^{zt}[w_z^h - \tilde{w}_z^h] \, . \tag{5.22}$$

For fixed $t > 0$, an estimate for the errors $w_z^h - \tilde{w}_z^h$ in the steady state problem can be immediately converted into a bound on $u_h(t) - \tilde{u}_h(t)$.

So far we have considered only the homogeneous equation. However, similar techniques can also be used for the nonhomogeneous cases

$$b\left(\frac{\partial u_h}{\partial t}, v^h\right) + a(u_h, v^h) = b(f_h, v^h) \quad \text{all } v^h \in S_2^h \tag{5.23}$$

$$\tilde{b}\left(\frac{\partial \tilde{u}_h}{\partial t}, v^h\right) + \tilde{a}(\tilde{v}_h, v^h) = b(f_h, v^h) \quad \text{all } v^h \in S_2^h \, . \tag{5.24}$$

Indeed, let us denote the semi-groups $U_h(t)$ and $\tilde{U}_h(t)$

by

$$U_h(t)g_h = \frac{1}{2\pi i}\int_{\Gamma_h} e^{zt} w_z^h dz \tag{5.25}$$

$$\tilde{U}_h(t)g_h = \frac{1}{2\pi i}\int_{\tilde{\Gamma}_h} e^{zt} \tilde{w}_z^h dz , \tag{5.26}$$

where $w_z^h$ , $\tilde{w}_z^h$ are defined by (5.12) and (5.15) . Thus (5.25) and (5.26) are the solutions of (5.23) and (5.24) with $f_h = 0$ . For general $f_h$ the solutions are given by

$$U_h(t)g_h + \int_0^t \tilde{U}_h(t-\tau)f_h(\tau)d\tau$$

and

$$\tilde{U}_h(t)g_h + \int_0^t \tilde{U}_h(t-\tau)f_h(\tau)d\tau ,$$

as can be easily verified. Hence

$u_h(t) - \tilde{u}_h(t) = \mathcal{O}(\delta_a^h) + \mathcal{O}(\delta_b^h)$ in the nonhomogeneous case as well.

## REFERENCES


1. Babuška, I., "Mathematical foundations of the finite element method," Academic Press, New York, 1972, Chapter 10 -- Eigenvalue problems; Chapter 11 -- Time dependent problems -- coauthored with G. Fix.
2. Babuska, I., "Error estimates in the finite element method," Numer. Math. 16, 322-333.
3. Bramble, J. and F. Osborn, "Approximation of eigenvalues of non self-adjoint operators," Unpublished manuscript.
4. Dunford, N. and J. Schwartz, "Linear operators," Interscience 1 (1958).

5. Fix, G., "Eigenvalue approximation by the finite element method," To appear in Adv. in Math., 1972.
6. Fix, G., "Effects of quadrature error in the finite element method," Proc. of the Second Japan-U.S. Symposium on Matrix Methods in Structural Mechanics, University of Alabama Press, 1972.
7. Strang, G. and G. Fix, "An analysis of the finite element method," Prentice-Hall, New York, 1972.
8. Vianikho, G. M., "On the rate of convergence of certain approximation methods of Galerkin type in an eigenvalue problem," Amer. Soc. Transl. 86 (1970), 249-259.
9. Herbolt, R. J. and R. S. Varga, "The effect of quadrature errors in the numerical solution of two-dimensional boundary value problems by variational techniques," Aeq. Math. 7 (1972), 36-58.
10. Herbolt, R. J., M. H. Schultz, and R. S. Varga, "Quadrature schemes for the numerical solution of boundary value problems by variational techniques," Aeq. Math. 3 (1964), 96-119.
11. Yosida, K., "Functional analysis," Springer-Verlag, New York, 1965.
12. Davis, P.J., and P. Rabinowitz, "Numerical Integration," Ginn-Blaisdell, 1967.

# EXPERIENCE WITH THE PATCH TEST FOR CONVERGENCE OF FINITE ELEMENTS

Bruce M. Irons* and Abdur Razzaque**

Introduction.

The usual explanation for the astonishing popularity of finite elements is that the technique allows an engineer to specify any feasible problem geometry, material properties, etc. One seldom hears about its mathematical rigor, nor about its computational convenience -- the engineers who wrote the papers have seldom been those who wrote the programs. The program blocks are complex, but they are modular and readily checked. The storage and arithmetic requirements are nicely balanced for modern computers. Most of the innermost loops are such that the cost of the arithmetic can be halved by introducing (as options) only two or three very simple machine-language subroutines. Whatever the reasons, engineers' confidence in finite elements is expressed in perhaps hundreds of millions of dollars per annum.

This expenditure is still largely governed in detail by prejudice, precedent and intuition. One would

* Reader, University of Wales.

** Lecturer, Engineering College, Rajsahi, Bangladesh: Colombo Plan Scholar.

presume, for example, that users would welcome a simple test for convergence, probably both necessary and sufficient [34]; but until recently the patch test has been largely disregarded. The authors are not competent to discuss its mathematical validity. Instead we shall describe various useful applications extending from 1965 to mid-1972, applications which show an unexpected variety of detailed arguments.

## Origins of the patch test.

In 1965 even engineering intuition dared not predict the behavior of certain finite elements. Experience forced those engineers who doubted it to admit that inter-element continuity was important: the senior author believed that it was necessary for convergence. It is not known what ideas inspired a numerical experiment by Tocher and Kapur [1], which demonstrated convergence to within 0.3% in a biharmonic problem of plate bending, using equal rectangular elements with 1, x, y, $x^2$, xy, $y^2$, $x^3$, $x^2y$, $xy^2$, $y^3$, and $x^3y$, $xy^3$, as functional basis. The nodal variables of this Ari Adini rectangle [2] are w , $\partial w/\partial x$ and $\partial w/\partial y$ at the four corners, and this element guarantees only $C^{(0)}$ conformity.

Some months later, research at Rolls-Royce on the Zienkiewicz nonconforming triangle [3] -- a similar plate-bending element -- clarified the situation. Three elements with $C^{(1)}$ continuity were simultaneously available, and, because the shape function subroutine[†] used for numerical integration had been exhaustively tested, the results were trustworthy. It was observed (a) that every problem giving constant curvature over the whole domain was accurately solved by the conforming elements, whatever the mesh pattern, as was expected, (b) that the nonconforming element was also successful, but only for one particular mesh pattern.

Thus the patch test was born. For if the external nodes of any sub-assembly of a successful assembly of elements are given prescribed values corresponding to an arbitrary state of constant curvature, then the internal nodes must obediently take their correct values. (An internal node is defined as one completely surrounded by elements.) Conversely, if two overlapping patches can reproduce any given state of constant curvature, they should combine into a larger successful patch, provided that every external node lost is internal to one of the original patches. For such nodes are in equilibrium at their correct values, and should behave correctly as internal nodes of the extended patch.

In an unsuccessful patch test, the internal nodes take unsuitable values, which introduce inter-element discontinuities. The errors in deflection may be slight, but the errors in curvature may be ±20%. We must recognize two distinct types of error:

---

[†] A program segment which computes the values of the nodal functions $N_i$ (known as "shape functions," "interpolating functions," etc.) and their derivatives at a given point in the element.

(i) The finite element equations would not be exactly satisfied by the correct values at the internal nodes -- in structural terms, we have disequilibrium,

(ii) The answers are nonunique, because the matrix of coefficients K is semidefinite.

Note: The second is the more difficult to recognize, and is frequently overlooked -- as indeed in the preceding paragraph.

## Role of the patch test.

Clearly the patch test provides a necessary condition for convergence with fine mesh. We are less confident that it provides a sufficient condition. The argument is that if the mesh is fine, the patches are also small. Over any patch the correct solution gives almost uniform conditions, to which the patch is known to respond correctly -- provided that the small perturbations from uniform conditions do not cause a disproportionate response in the patch: we hope to prevent this by insisting that K is positive definite.

The patch test is invaluable to the research worker. Already, it has made respectable

(i) Elements that do not conform,

(ii) Elements that contain singularities,

(iii) Elements that are approximately integrated,

(iv) Elements that have no clear physical basis.

In short, the patch test will help a research worker to exploit and justify his wildest ideas. It largely restores the freedom enjoyed by the early unsophisticated experimenters.

Mixability and the primitive plate bending elements.

The experiments with the Zienkiewicz triangle showed that the diagonal stiffness coefficients $K_{ii}$ of a nonconforming element A may be only a quarter as large as those for the corresponding conforming element B . Also the forces $F_A$ given by element A in a state of constant curvature -- as in the patch test -- may be quite different from $F_B$ . We must distinguish

(i) Element pairs A and B which may be mixed indiscriminately in the same patch without causing disequilibrium.

(ii) Nonmixable element pairs.

Of the elements considered here, only the Ari Adini rectangle and the Zienkiewicz triangle are nonmixable with their conforming counterparts. Their anomalous behavior can cause embarrassment in practice: for example, the wrongful dismissal in 1965 of the technique for calculating reactions to earth by adding a large number to the appropriate diagonal stiffnesses [4]. Also -- but this could be coincidence -- they are the only elements whose patch-test convergence depends upon using a mesh of identical elements.

The analytic proof of convergence followed soon after the computer experiments. It requires symmetry, both of the formulation and of the mesh; its restricted conclusions apply to many other $C^{(1)}$ elements. We consider

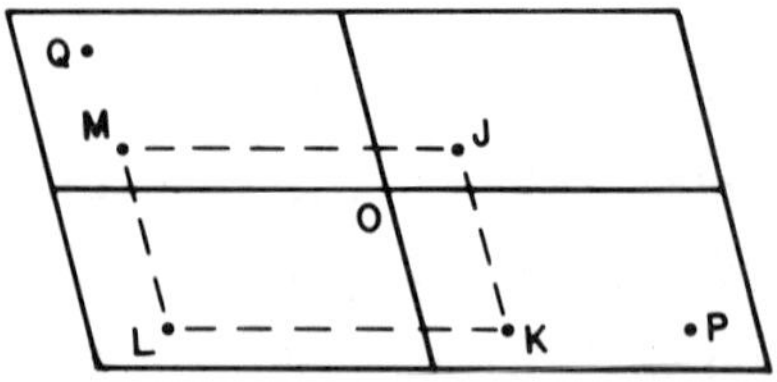


Fig. 1. Proof of patch test for primitive bending elements

first the patch of four equal parallelograms in Figure 1, whose eight external nodes are given slopes $w_x$ , $w_y$ and deflections $w$ corresponding to some state of constant curvature

(1) $$w = ax^2 + bxy + cy^2 \quad .$$

We assume tentatively that the one internal node 0 chooses $w$ , $w_x$ and $w_y$ also according to (1), and on this assumption we compute by virtual work the disequilibrium forces corresponding to $w$ , $w_x$ and $w_y$ at 0 . The strain energy is

(2) $$\frac{1}{2}[w_{xx}, w_{xy}, w_{yy}][D] \begin{Bmatrix} w_{xx} \\ w_{xy} \\ w_{yy} \end{Bmatrix} d(\text{area})$$

which gives for example the disequilibrium force at 0 due to $w_{xx} = 1$

(3) $$\int \left[ \frac{\partial w_{xx}}{\partial w_o} , \frac{\partial w_{xy}}{\partial w_o} , \frac{\partial w_{yy}}{\partial w_o} \right] [D] \begin{Bmatrix} 1 \\ 0 \\ 0 \end{Bmatrix} d(\text{area})$$

the integral being taken over the four elements. We now show that the nine integrals in such expressions are all zero.

(i) $$\int \frac{\partial w_{xx}}{\partial w_o} , \frac{\partial w_{xy}}{\partial w_o} , \frac{\partial w_{yy}}{\partial w_o} \, d(\text{area}) .$$

Here we consider small corresponding areas da at J , K , L , M in Figure 1 whose total contribution is (sum of integrands at J , K , L , M) da. But the sum of integrands $\partial w_{xx}/\partial w_o$ for example is simply the value of $w_{xx}$ caused by $w = 1$ at all four corners simultaneously of the same element, i.e., zero.

(ii) $$\int \frac{\partial w_{xx}}{(\partial w_x)_o} \, d(\text{area}) , \quad \text{etc.}$$

In these six cases the contributions from da at P and Q cancel, provided that the formulation itself is symmetric, so that $w_P = -w_Q$ .

This proves patch-test equilibrium for the Ari Adini rectangle. If we divide each parallelogram into two triangles by drawing the diagonals always from, say, the lower right-hand to the upper left-hand corner, the result holds also for the Zienkiewicz triangle, etc. Anisotropy with constant [D] is acceptable. But if the parallelograms are divided into triangles in union jack formation, for example, then argument (i) fails. The patch test also fails, rather badly [3, addendum].

Note: We are reminded here of the arguments formerly used to justify framework analogies for shells, etc. [33]. Later work shows however that the Ari Adini element passes the patch test with any mesh [10].

Eigenvalues and mixability.

Khanna [5,6] has shown how to compare the alternative formulations A and B using eigenvalues: with nodal deflection vector x ,

(4) $$F_A = K_A x \;, \quad F_B = K_B x \;, \quad F_A = \lambda F_B$$

$K_A$ and $K_B$ may be semidefinite, and we distinguish six categories of eigenvalue:

(i) $F_A = F_B = 0$ usually denotes a rigid body motion.

(ii) $F_A = 0$ , $F_B \neq 0$ means that A contains a spurious mechanism.

(iii) $F_B = 0$ , $F_A \neq 0$ means that B contains a spurious mechanism.

(iv) $F_A = F_B$ , i.e., $\lambda = 1$ , implies patch-test mixability; but it would be safer to examine the x to be certain that all the constant strain conditions are available.

As yet K is undefined: is it the matrix for an element, a patch, or the full assembly? We can for example define patch-like behavior as the case when {x} (element) are subsets of {x} (patch), with $\lambda = 1$ .

(v) $\lambda < 1$ isolates those deformations x in which A is "more flexible" than B . Conversely,

(vi) $\lambda > 1$ . In the absence of (ii) and (v) we have $K_A \geq K_B$ in the sense that $K_A - K_B$ is positive semi-definite.† Moreover, $\lambda^{(vi)}$ (assembled) are bounded by the extreme values of $\lambda^{(vi)}$ (elements).

---

†The engineer normally prefers the least positive K provided it has enough rank, and matrix inequalities are important to him. Under mixed boundary conditions, K may be part-inverted symmetrically [7] to bring the knowns to

The quartic plate bending triangle.

It was another $C^{(1)}$ plate bending triangle [8] to which (4) was first seriously applied. Its 18 degrees of freedom were:

(a) at corners, $w$ , $w_x$ , $w_y$

(b) at midsides, $w$ , $w_n$ (normal slope), $w_{ns}$ (s = along side).

The functional basis comprised:

(a) the complete quartic in $x$ , $y$ (15 terms)

(b) 3 extra terms with singularities. These existed in two forms; the first were rational functions: the second subdivided the triangle into 3 subtriangles, and specified a different quartic in each.

The second version allowed exact numerical integration, while the first did not. Reference 9 shows the convergence of the three eigenvalues in group (vi), which was rather slow, and of the twelve in group (iv), as higher integration rules were used.

The direct test of mixability.

The eigenvalue calculation (4) presents all the mixable responses automatically. Instead we can select some $\{x\}$ of class (iv) -- some arbitrarily chosen state of constant curvature -- and find the nodal loads for an isolated element

---

the left-hand side, if necessary, with a change of sign. The potential energy is given by the quadratic form of the resulting matrix. Perhaps it is not surprising then that $K_A \geq K_B$ survives the operation of part-inversion.

(5) $F_A = K_A x$ , $F_B = K_B x$ : check that $F_A = F_B$ .

Knowing as in (4) that A gives patch equilibrium, we demonstrate using (5) that B preserves equilibrium. Thus:

(i) (5) establishes mixability more cheaply than (4).

(ii) However, we lose the inequalities and the analysis of rank.

Note: Few workers measure rank, and an intelligent guess can be badly wrong. For example, our next element has a rank of only 7, not 9 as suggested by an earlier formulation.

A technique for modifying elements mixably.

The quartic triangle is too complicated for most tastes. We now describe a systematic technique for simplifying certain elements and yet improving their performance, a technique that yields a mixable replacement for the Zienkiewicz triangle. In its original form the $C^{(1)}$ element had 12 degrees of freedom, $w$ , $w_x$ , and $w_y$ at corners and $w_n$ at midsides. The basis consisted of the 10 terms of the complete cubic, {g} = good, and two rational functions, {b} = bad. The research evolved in many distinct, unplanned phases [10]:

(i) Satisfaction with 7-point integration, and admiration of results.

(ii) Dissatisfaction following the first patch test failure.

(iii) Doctoring of the array of basic integrals, to correct the patch test.

(iv) Acceptance of the artifice of drastically reducing $K$ by doctoring the "basic integrals": greatly improved results.

(v) Search for the shape function pseudo-derivatives that would produce the doctored values.

(vi) Recognition that these derivatives represented the best plane fits to the original derivatives, by equating first moments in $x$, $y$.

(vii) Discovery of the theorems relating to artificially smoothed derivatives.

(viii) Identification of the element as Allman's hybrid [11].

(ix) Continuing doubts concerning the integrability of smoothed derivatives in general, and the relation to Pian hybrids.

Clearly the patch test was administered frequently, and to good effect. Before demonstrating patch equilibrium we define the notation:

$$\{N\} = \text{nodal shape functions} = C\begin{Bmatrix} g \\ b \end{Bmatrix}$$

$$g_{sec} = \begin{Bmatrix} g_{xx} \\ g_{xy} \\ g_{yy} \end{Bmatrix} , \qquad b_{sec} = \begin{Bmatrix} b_{xx} \\ b_{xy} \\ b_{yy} \end{Bmatrix}$$

$$b^*_{sec} = \text{smoothed version of } b_{sec}$$

$$= Zg_{sec} \text{ say, where } Z \text{ is such that}$$

$$\int b^*_{sec} g^T_{sec} d(\text{area}) = \int b_{sec} g^T_{sec} d(\text{area}) \quad .$$

For the patch test we choose <u>arbitrary</u> cubic displacements from $\{g\}$.

$$w = g^T R \quad ; \text{ therefore, } x = [C^T]^{-1} \begin{Bmatrix} R \\ 0 \end{Bmatrix} \quad .$$

We prove mixability by writing the nodal forces as in (5)

$$F = \int [C] \begin{bmatrix} g_{xx} & g_{xy} & g_{yy} \\ b_{xx} & b_{xy} & b_{yy} \end{bmatrix} [D] \begin{bmatrix} g^T_{xx} & b^T_{xx} \\ g^T_{xy} & b^T_{xy} \\ g^T_{yy} & b^T_{yy} \end{bmatrix} \begin{Bmatrix} R \\ 0 \end{Bmatrix} \, d(\text{area}).$$

For $F$ is unaltered if $b$ is replaced by $b^*$ .

Another interesting theorem can be demonstrated: of all the mixable $b^*_{sec}$ , that which can be expressed as $Zg_{sec}$ gives the least positive $K$ [12,13].

Note: The most obvious practical benefit is that numerical integration becomes economical, accurate, and versatile, and the programs become short and simple -- the virtues first highlighted by the ubiquitous isoparametric elements.

Properties of the isoparametric elements [14,15].

The isoparametric and related elements use local curvilinear coordinates $\xi$ , $\eta$ , $\zeta$ with convenient limits (e.g., $-1 \leq \xi , \eta , \zeta \leq 1$) to define two sets of polynomial shape functions $N_i(\xi,\eta,\zeta)$ and $P_i(\xi,\eta,\zeta)$ and hence:

(6) coordinates $\begin{Bmatrix} x \\ y \\ z \end{Bmatrix} = \sum P_i \begin{Bmatrix} x_i \\ y_i \\ z_i \end{Bmatrix}$ and deflections $\begin{Bmatrix} u \\ v \\ w \end{Bmatrix} = \sum N_i \begin{Bmatrix} u_i \\ v_i \\ w_i \end{Bmatrix}$

where $i$ denotes nodal values. The isoparametric elements have $P_i = N_i$ and the boundary faces are usually curved.

Note: The isoparametric elements are frequently called "displacement elements." But "displacement" is a vector and $N_i$ is a scalar. It would be less pretentious to speak of "deflection elements," because an isolated nodal deflection $u_i$ causes deflections elsewhere strictly in the same direction.

The theorems, mostly trivial, depend on the functional basis $N_i$ including the $P_i$ -- hence most elements are degenerate isoparametric elements -- and on their being "deflection elements." We shall speak of the "parent element," a special case with $x = \xi$, $y = \eta$, $z = \zeta$. (If we put $u_i = x_i - \xi_i$, etc., we see the isoparametric element as a grossly distorted parent element.)

(i) If the parent gives $C^{(n)}$ continuity at any point, even across an inter-element boundary, then so will the isoparametric element provided the Jacobian is non-singular. (Usually $n = 0$. Otherwise special pseudo-coordinates $x_i, y_i, z_i$ must be introduced to ensure $C^{(n)}$ continuity of mapping between elements [35].)

(ii) Provided only that the parent provides e.g. $u = \text{constant}$, then the isoparametric element provides $u = a + bx + cy + dz$, etc., i.e., all states of rigid body motion and, often, of constant strain.

(iii) Because (6) is vectorial, the model is independent of the choice of $x$, $y$, $z$ axes.

(iv) Any symmetries involving node-swapping exhibited by the parent element survive also in the isoparametric element.

(v) If the parent admits Melosh subdivision [16] (usually on planes $\xi$ , $\eta$ or $\zeta$ = constant) then so does the isoparametric element.

Patch behavior of the isoparametric elements.

Very few workers have integrated K exactly, because the analytic integrals of rational functions are complicated and ill-conditioned. But even approximate integration typically costs about as much as reducing the equations; so we enquire what minimum accuracy gives convergence. We restrict our enquiry to $C^{(0)}$ and hence to problems involving only first derivatives; therefore (i) and (ii) ensure convergence with exact integration. To convert $\partial N_i/\partial\xi$ , etc., into $\partial N_i/\partial x$ , etc., we invert the Jacobian matrix:

(7) $$\begin{bmatrix} \partial x/\partial\xi & \partial x/\partial\eta & \partial x/\partial\zeta \\ \partial y/\partial\xi & \partial y/\partial\eta & \partial y/\partial\zeta \\ \partial z/\partial\xi & \partial z/\partial\eta & \partial z/\partial\zeta \end{bmatrix} \quad \text{where e.g. } \frac{\partial x}{\partial\xi} = \sum \frac{\partial P_i}{\partial\xi} x_i \ .$$

Then we introduce the Jacobian determinant $\partial(x,y,z)/\partial(\xi,\eta,\zeta)$ before integrating $d\xi\, d\eta\, d\zeta$ . It cancels, giving for example the load corresponding to $u_i$ due to constant stress over the element, as in (5)

$$\int \sigma_{xx} \frac{\partial N_i}{\partial x} + \sigma_{xy} \frac{\partial N_i}{\partial y} + \sigma_{xz} \frac{\partial N_i}{\partial z} \, d(\text{vol})$$

$$= - \iiint \begin{vmatrix} \frac{\partial x}{\partial\xi} & \frac{\partial x}{\partial\eta} & \frac{\partial x}{\partial\zeta} & \sigma_{xx} \\ \frac{\partial y}{\partial\xi} & \frac{\partial y}{\partial\eta} & \frac{\partial y}{\partial\zeta} & \sigma_{xy} \\ \frac{\partial z}{\partial\xi} & \frac{\partial z}{\partial\eta} & \frac{\partial z}{\partial\zeta} & \sigma_{xz} \\ \frac{\partial N_i}{\partial\xi} & \frac{\partial N_i}{\partial\eta} & \frac{\partial N_i}{\partial\zeta} & 0 \end{vmatrix} d\xi\, d\eta\, d\zeta$$

Substituting $\partial x/\partial\xi$ , etc., from (7) shows that for patch equilibrium a polynomial in $\xi$ , $\eta$ , $\zeta$ -- the $4 \times 4$ determinant -- must be integrated exactly, i.e. terms like

$$\iiint \frac{\partial N_i}{\partial \xi} \left( \frac{\partial P_j}{\partial \eta} \frac{\partial P_k}{\partial \zeta} - \frac{\partial P_j}{\partial \zeta} \frac{\partial P_k}{\partial \eta} \right) d\xi \, d\eta \, d\zeta \quad . \tag{8}$$

There is some useful cancellation of terms in the bracket. Thus:

(i) Fairly low order quadrature rules suffice, especially if the $P_i$ are a small subset of the $N_i$ -- i.e. the geometry is simple.

(ii) With fewer integrating points the structural stiffness is in general performance (but not strictly) reduced, as if $K_{approx} \lesssim K_{exact}$ . This is usually desirable.

(iii) However, the rank cannot exceed (number of integrating points) $\times$ (rank of matrix D).

<u>Note</u>: It is hotly disputed whether one should use the lowest order formulae. For example, the $2 \times 2 \times 2$ Gauss rule can generate remarkably good stresses in the 20-node brick, if the stresses are calculated at the same Gauss points. Veryard [17] described his trials with various integrating points and various stress points; the $2 \times 2 \times 2$ rule gave by far the best results -- despite the fact that his matrix K was singular. The stresses elsewhere are somewhat worse than those given by the higher rules -- in Veryard's case, much worse!

Some deviant isoparametric elements.

We now review briefly those variants of (6) which display interesting patch behavior.

The Mixoparametric elements are best illustrated by the 8 node quadrilateral:

(9a) basis of $P_i = (1,\xi,\eta,\xi^2,\xi\eta,\eta^2)$ and $\xi^2\eta$ , $\xi\eta^2$

(9b) basis of $N_i = (1,x,y,x^2,xy,y^2)$ and $\xi^2\eta$ , $\xi\eta^2$

These elements have been used sporadically for some years. The mixoparametric elements are relatively expensive, and they perform little better than the isoparametric elements, despite the extra exact solutions provided by the new complete cartesian components. Recital of their properties is therefore somewhat academic:

(i) The results do not depend on choice of (x,y) axes.

(ii) To ensure $C^{(0)}$ conformity, the inter-element faces must be plane.

(iii) High-order quadrature is needed to ensure patch equilibrium.

Note: Initially it was hoped that $K_{mixo} \leq K_{iso}$ but this is not strictly so.

Mixoparametric elements with Legendre polynomials would take, instead of (9),

(10) basis of $N_i = (1,x,y,x^2,xy,y^2)$

$$\text{and} \quad \left(\xi^2 - \frac{1}{3}\right)\eta \ , \ \xi\left(\eta^2 - \frac{1}{3}\right)$$

The two $\xi$ , $\eta$ terms are regarded as contaminants, whose importance is reduced by rendering them into Legendre polynomials. The aim is to make plane interelement faces less necessary.

The contraparametric elements of Bond and Henschel [18] would take, instead of (9),

$$\text{(11)} \quad \text{basis of } N_i = (1,x,y,x^2,xy,y^2) \text{ and } x^2y \text{ , } xy^2 \text{ .}$$

The results will depend on choice of $x$ , $y$ axes. According to (8) the patch-test forces are correctly integrated by a rule again slightly higher. But this is a misleading case: we are comparing A (exactly integrated) with B (approximately integrated) where A does not guarantee $C^{(0)}$ continuity, and so does not necessarily pass the patch test itself: thus we prove nothing.

Wilson's nonconforming brick [19] is a successful element whose patch behavior is particularly instructive. We start with the 8-node trilinear brick:

$$\text{(12)} \qquad N_i \text{ , } P_i = \frac{1}{8}(1 + \xi\xi_i)(1 + \eta\eta_i)(1 + \zeta\zeta_i)$$

where $i = 1$ to $8$ so that $\xi_i$ , $\eta_i$ , $\zeta_i$ , the nodal values, are $\pm 1$ in all combinations. Wilson adds three extra functions:

$$\text{(13)} \quad N_9 = 1 - \xi^2 \text{ , } N_{10} = 1 - \eta^2 \text{ , } N_{11} = 1 - \zeta^2 \text{ .}$$

Each responds over 4 of the 6 faces. We have 9 extra nodeless variables in $u$ , $v$ , $w$ -- for no nodal positions are specified, and it is indeed impossible to establish nodal connections simultaneously to four noncontiguous elements with the same function. Indeed they are treated as internal degrees of freedom -- Wilson invites nonconformity, on the engineering pretext that simple beam bending behavior is thereby made available.

In a patch test with constant stress, Wilson parallelopipeds mix with trilinear isoparametric elements. For the nodeless forces are shown to be zero. If the nodeless variables are zero, the corner forces are the same as those given by the trilinear elements. Therefore, the perturbating forces are zero for unequal parallelepipeds. Mixability is not proved for non-parallelepipeds, but the experiments of A. Fawkes [20] have shown unexpectedly that certain nonparallel cases also succeed, if Poisson's ratio is zero.

We now consider the complementary form of the patch test, which has implications for users of nonconforming elements -- it shows why we sometimes choose to depart from a strict energy formulation. The patch is first supported in just enough degrees of freedom to prevent rigid body motion. Then tractions consistent with some state of uniform stress are applied over its external boundaries. (We sometimes do this by numerical integration.) If the elements conform, the internal nodes automatically remain unloaded.

With the Wilson element, this process gives a boundary layer of elements whose nodeless variables are excited spuriously. Clearly to avoid this mistake we must set the nodeless forces artificially to zero.

Note: The patch test using tractions may demand higher rank.

Pian's hybrid elements [21] and the patch test.

The patch test may be one of a battery of tests used to verify a program before numerical experiments on a new element can start. A recent example is ideal in the present context, exemplifying as it does

(a) the nontrivial everyday problems of decisive debugging,

(b) the special difficulties in patch testing a hybrid element caused by its inherent contradictions, which appear to limit the stress-states available, and

(c) the challenge of translating an engineering idea from one formulation into another.

Despite its competitive performance, the Wilson nonconforming element converges only if the mesh is parallel. Our next element [22] also has the geometry of the trilinear isoparametric brick, but its internal behavior is governed by fifteen stress fields:

(i) The six fields $\sigma_{ij}$ = constant

(ii) Nine fields of pure bending. We generate the first three by considering three independent stresses in the plane $\eta\zeta$ at the central point $\xi = \eta = \zeta = 0$ . Each stress is multiplied by the distance from the plane $\eta\zeta$ .

The assumed boundary displacements are those given by the trilinear isoparametric brick. Because we have well-behaved and easily calculated functions over the volume, it is easy -- and cheap -- to compute the terms of Pian's matrix T as volume integrals. (3×3×3 Gaussian quadrature is exact for T and H .) Because the assumed boundary displacements permit any state of constant strain -- and hence constant stress -- this element should be (and was shown to be) mixable with the deflection element in any geometry. This program proceeds to calculate displacements within the element, which were also checked.

It is more difficult to check the program in the bending modes, because in general the assumed boundary displacements are grossly inconsistent with the displacements finally calculated. We had to choose a rectangular element geometry -- in the first instance oriented in the x , y , z directions -- and even then, we had to choose a special bending field (e.g., $\sigma_{xy} = z$) which was available to the deflection element. This was as far as patch testing could reasonably be expected to go.

Ahmad [23] reported a somewhat cruder hybrid 20-node brick. It satisfied constant-stress patch tests, and also an eigenvalue comparison with the deflection element: as expected, $K(\text{Pian}) \leqslant K(\text{Isoparametric})$.

Ahmad's superparametric elements.

The shell and plate bending elements for which Ahmad [24] is better known have little direct relevance to patch testing. He took e.g. a 20-node curved brick as a shell element, and attached the two or three nodes through the thickness -- say the $\zeta$ direction -- to a rigid straight line member subject to u , v , w deflections and two rotations, the rotation along the line being ineffective. This gave 8 "nodes" with 5 degrees of freedom each.

He treated the surfaces $\zeta$ = constant as membranes subject to plane stress. The shear between membranes also generated strain energy, and a factor of 1.2 was introduced to simulate parabolic shear distribution. We note:

(i) These were familiar engineering assumptions, already exploited by Winkler [25], Melosh [26], etc.

(ii) Any patch testing must reflect these assumptions rather than those of 3D elasticity.

(iii) From arguments of nodal valency [14, 15] two-thirds of the assembled variables are at midside in a large problem, although in the element only half are at midside.

(iv) Conventional shell elements make no provision for lateral shear distortion: Ahmad's make too much. (The membrane stresses vary quasilinearly, but the shear stresses -- which account for only a small proportion of the strain energy -- vary quasiquadratically.) Far from helping, this encourages spurious shear, which is often the main contaminant.

(v) Too [27] found that $2 \times 2$ Gaussian integration so obscured these shear patterns that his results improved dramatically, even for quite thin plates.

Some deviant Ahmad elements.

Ahmad plate constrained by fictitious edge-beams.

(a) The engineer surmises that in a thin plate, and approximately in a thick plate too, the deflection and the slope along the edge at midside should be governed by the slopes and deflections at the adjacent corners. These superfluous variables are now discarded, by giving them the values dictated by a beam along the edge.

If the edges are straight -- the only case clearly envisaged -- patch testing needs no computer aid. The constraint does not conflict with patch test requirements: therefore patch equilibrium is assured, for the triangle as well as the quadrilateral, in a pure bending test.

Ahmad shell constrained at Gauss points. (b) Here the 8 unwanted variables are discarded by putting the lateral shear strains $\gamma_G$ zero at the 2×2 Gauss points, for a thin shell. But it is the thick shell that clarifies our new engineering assumptions: Severn's argument for additional shear [28,29] shows, after some matrix manipulation, that even in the case of the thick shell we have merely uncoupled the unwanted variables [30]. A thin plate loses its large stiffness coefficients, and Severn's technique achieves this result in a well-conditioned manner.

Note: Uncoupling means a loss of inter-element conformity, and it is now more difficult to predict the patch-test behavior analytically. Experimentally, the thin plate passes the patch test with unequal parallelograms, as expected: otherwise, it does not fail badly, and its general performance is excellent.

Constraints on corner slopes. (c) It is certainly possible to eliminate $\theta_x$ and $\theta_y$ at the corners by considering beam action; we expect patch equilibrium with linear sides. (d) It may be possible to use $\gamma_G = 0$ at the 2×2 Gauss points, instead of imaginary beams; we expect patch equilibrium with parallelograms. These possibilities have not been tried because others are more attractive, e.g.

The use of Loof connections. H. W. Loof [31] conceived the idea of connecting nonconforming elements at Gauss points along the inter-element boundaries, to minimize the work lost in nonconformity. Postulating 8 nodes at $(\xi_i, \eta_i) = (\pm 1, \pm 1/\sqrt{3})$ and $(\pm 1/\sqrt{3}, \pm 1)$, with (9a) as basis, we cannot find shape functions because

$$N_9 = 1 - \frac{3}{4}(\xi^2 + \eta^2) \tag{14}$$

is zero at nodes 1 to 8 and lies within the basis (9a). We therefore add a corresponding node $(\xi_9, \eta_9) = (0,0)$ and we include $\xi\eta(\xi^2 - \eta^2)$ as the ninth term in the basis. Similarly for the triangle, the nodal area coordinates $(\alpha_i, \beta_i, \gamma_i)$ are $\left(\frac{1}{2} + 1/\sqrt{12}\ ,\ \frac{1}{2} - 1/\sqrt{12}\ ,\ 0\right)$, etc., (6 permutations) and $\alpha_7 = \beta_7 = \gamma_7 = \frac{1}{3}$ (centroid):

$$\text{basis of } N_1 \ldots N_7 = \text{quadratic (6 terms)}\ , \tag{15}$$

$$(\alpha - \beta)(\beta - \gamma)(\gamma - \alpha)\quad .$$

Thus, is it possible to set up an Ahmad shell or plate, and eliminate all 8 slopes $\theta_s$ along the sides? This would leave deflections and normal slopes $\theta_n$ at the 8 nodes -- a very convenient combination for multi-leaf shells, which pose a difficult problem of conformity when rotations are used as corner variables. The difficulty is most evident at the corner of the closed rectangular box of Figure 2, where three elements meet. The Kirchhoff

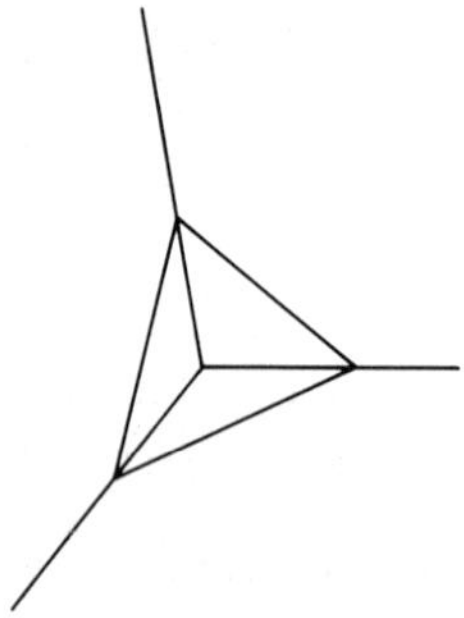

Fig. 2. The dilemma of three thin shell elements meeting at a corner

hypotheses would constrain the in-plane shear strains to zero in all three elements near the corner. This is nonsense if the elements are large compared with their thickness; it is intolerable when the corner is less pronounced, as in a nearly smooth shell, where the scissors action is inhibited. Two engineering compromises have been suggested.

The method of projected nodal rotations. (e) We find the mean normal direction at each node. We use as nodal variables the two vector rotations normal to this direction; but, for converting them into bending rotations for each element, we project the rotation into the plane of the element. This approach gives conforming normal slopes, but it sacrifices displacement continuity.

The articulated shell model. (f) We again use two nodal rotations. Imagine each node as a rigid ball drilled in the mean normal direction, as in Figure 3. The inter-element boundary between two nodes is regarded as a thin, uniform, freely extensible beam, pivoting as illustrated. This approach [10] ensures displacement conformity

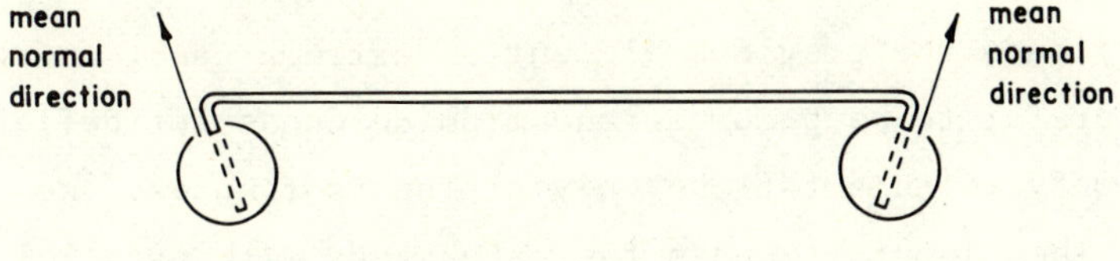


Fig. 3. The articulated shell model

but it sacrifices slope conformity.

Ahmad plate/shell with Loof nodes. (g) The problem disappears with Loof nodes, as with Visser's mixed element [32]. The elimination of $\theta_s$, the slopes along the sides, is not straightforward, however: the 8×8 matrix relating $\theta_s$ to the shears at the 2×2 Gauss points, $\gamma_G$, is singular, and the variables at the ninth node must play a part. We therefore create a 45×45 stiffness matrix in terms of $(u,v,w,\theta_n)_1 \ldots (u,v,w,\theta_n)_8$, $(\theta_s)_1 \ldots (\theta_s)_8$ $(u,v,w,\theta_x\theta_y)_9$ and introduce 8 Lagrange multipliers $\lambda_i$ to impose $\gamma_G = 0$. We can then part-invert in the natural order, from $(\theta_s)_1$ through $(\theta_y)_9$, $\lambda_1 \ldots$ to $\lambda_8$ to generate the required 32×32 stiffness matrix, and also a matrix for calculating $(\theta_s)_1 \ldots (\theta_y)_9$ subsequently [10].

We expect the element to pass the patch test if the sides are parallel.

Ahmad plate/shell with Loof rotations only. (h) The previous element had barely adequate rank. However we no longer view low valency [14,15] as altogether bad -- just as equilibrium elements have low valency, so a low valency generally implies a flexible formulation, which in itself is desirable. Although it gives more assembled variables, it seldom increases the frontwidth, so it costs little. But the rank problem associated with the low valency made the previous element an extreme case, and we might prefer to keep corner and midside nodes for deflections only, with rotations only at the Loof nodes. We expect the element to pass the patch test with parallel sides.

Directional shear constraints at the Loof nodes.

(j) An especially neat variant of (h) puts $\gamma_{ps} = 0$, $s$ = along side, $p$ = perpendicular to mid-surface, at each of the eight Loof nodes. Evidently this is always possible. Also in the patch test $\gamma_{ps} = 0$ causes similar behavior in adjacent elements, and we therefore expect the patch test to accept any quadrilateral geometry. Further, we can put both shears zero at the ninth node, $\xi = \eta = 0$, and hence eliminate the rotations there. Thus a shape function subroutine can be independent of the elastic properties.

## Patch testing of constrained isoparametric elements.

We give now a typical argument, on which the expectations for case (g) are based.

(i) We have $w$, $\theta_x$ and $\theta_y$ at all nine nodes. We assume equilibrating stresses, so we compute patch forces as $\int$ (virtual boundary displacement). d(traction). If $x$, $y$ vary linearly with $\xi$ or $\eta$, no virtual work is done on a side if neither node is displaced. Therefore the nodal force depends only on the geometry of the side on which the node lies: we thus have patch equilibrium.

(ii) Now let us free certain nodal rotations within the element, fixing the other rotations and lateral deflections -- thus virtual changes give zero shear at the Gauss points, and in a membrane we can only have

$$u \text{ and/or } v = (3\xi^2 - 1)(A + B\eta) + (3\eta^2 - 1)(C + D\xi) + (\xi^2 - \eta^2)(E + F\xi\eta) \quad . \tag{16}$$

The virtual work done by A , B , C , D is zero for linear sides. That done by E , F is zero only for a parallelogram. This is an unfortunate conclusion -- the computational procedure in (g) breaks down in just this case. Experiments have confirmed these expectations, as recorded in reference 10.

Boundary conditions for polynomial spline solutions [36].

The patch test (with a trivial redefinition of external nodes) can guide us in another current problem, namely the selection of boundary constraints when tensor products of spline functions are used with irregular boundaries. We require that the patch test would be satisfied, given suitable displacements over the entire problem boundary. We shall propose necessary and sufficient constraints -- necessary in that the patch test is satisfied, and sufficient in that any further constraint could perhaps duplicate a previous constraint, giving a singular matrix.

An "external" node is now, as before, any node

whose zone of influence trespasses outside the problem boundary. It becomes a "constrained" node if its zone covers part of the boundary where displacements are prescribed.

Rule: One and only one distinct boundary constraint must correspond to each constrained node. This constraint may be, for example, that a certain point on the boundary takes its prescribed displacement: the point must then lie inside the zone of influence of the node it specifically constrains. The rule is only slightly different with more elaborate techniques for imposing constraints.

Note: The spline is a conforming element, and this rule succeeds by analogy with orthodox elements. To have a part of one's zone outside the patch is to experience a temptation to misbehave. An internal node with conforming functions has no such temptation, and no constraint is needed. Alternatively we may argue that every node is constrained, but the constraint equations have zero coefficients for an internal node.

Conclusions.

The patch test is far more practical than the classical convergence criteria it replaces. This paper is the first to list its applications, from the dark ages of finite element history down to the present. The patch test has somehow escaped the publicity generally lavished on finite elements in recent years; yet it provides an exceptionally useful guideline in research, as workers must increasingly discover. Unfortunately it still lacks an agreed mathematical justification; but see reference 34.

REFERENCES

1. Tocher, J. L., and K. K. Kapur, Comment on ref. 16 below: JAIAA 3 (1965), 1215-1216.
2. Adini, A., and R. W. Clough, "Analysis of plate bending by the finite element method," report to National Science Foundation, U.S.A., G.7337, 1961.
3. Bazeley, G. P., Y. K. Cheung, B. M. Irons, and O. C. Zienkiewicz, "Triangular elements in bending -- conforming and nonconforming solutions," Conference on Matrix Methods in Structural Analysis, Air Force Inst. Tech., Ohio, 1965.
4. Irons, B. M., "Roundoff criteria in direct stiffness solutions," JAIAA 6 (1968), 1308-1312.
5. Khanna, J., "Criterion for selecting stiffness matrices," JAIAA 3 (1965), 1976.
6. Khanna, J., and R. F. Hooley, "Comparison and evaluation of stiffness matrices," JAIAA 4 (1966), 2105.
7. Asplund, S. O., Structural Mechanics: Classical and Matrix Methods, Prentice-Hall, Englewood Cliffs, N.J., 1966.
8. Irons, B. M., "A conforming quartic triangular element for plate bending," Int. J. Num. Meth. Eng. 1 (1969), 29-46.
9. Irons, B.M. "Testing and assessing finite elements by an eigenvalue technique," Conf. on Recent Developments in Stress Analysis, J. Brit. Soc. Stress Analysis, 1968.
10. Razzaque, A., "Finite element analysis of plates and shells," Ph.D. Thesis, University of Wales, 1972.
11. Allman, D. J., "Triangular finite element for plate bending with constant and linearly varying bending moments," IUTAM Symposium on High Speed Computing of Elastic Structures, Liège, August 1970.
12. Irons, B. M., and A. Razzaque, "Shape function formulations for elements other than displacement models," Conference on Variational Methods in Engineering,

Southampton University, September 1972.

13. Razzaque, A., "Implementation of two hybrid plate-bending elements using a shape function subroutine," Int. J. Num. Meth. in Eng. (to appear).
14. Irons, B. M., "Engineering application of numerical integration in stiffness methods," JAIAA 4 (1966), 2035-2037.
15. Irons, B. M., "Numerical integration applied to finite element methods," Conference on Use of Dig. Comp. in Struct. Eng., University of Newcastle, 1966.
16. Melosh, R. J., "Basis for derivation of matrices for the direct stiffness method," JAIAA 1 (1963), 1631-1637.
17. Veryard, D. A., "Problems associated with the convergence of isoparametric and mixoparametric finite elements," M.Sc. Thesis, University of Wales, 1971.
18. Bond, T. J., J. H. Swannell, R. D. Henshell, and G. B. Warburton, "Curved two dimensional finite elements with displacement assumptions" (to appear).
19. Wilson, E. L., and R. L. Taylor, "Incompatible Displacement models," symposium, Numerical and Computer Methods in Structural Engineering, O.N.R., University of Illinois, September 1971.
20. Fawkes, A., "Some nonconforming variants on the trilinear isoparametric brick for linear elastic problems," M.Sc. Thesis, University of Wales, 1972.
21. Pian, T. H. H., "Derivation of element stiffness matrices by assumed stress distribution," JAIAA 2 (1964), 1332-1336.
22. Irons, B. M., "An assumed stress version of the Wilson 8-node isoparametric brick," Computer Report CNME/CR/56, Civil Engineering, University, Swansea, May 1972.
23. Ahmad, S., "Curved finite elements in the analysis of solid, shell and plate structures," Ph.D. Thesis, University of Wales, 1969.
24. Ahmad, S., "Pseudo-isoparametric finite elements for shell and plate analysis," Conference on Recent Advances in Stress Analysis, Joint Brit. Committee for Stress Analysis, March 1968.
25. Winkler, memoir of 1858, discussed e.g. in chapter 25 of The Strength of Materials, J. Case, Arnold, London, 1943.
26. Melosh, R. J., "A flat triangular shell element

stiffness matrix," Conference on Matrix Methods in Structural Mechanics, Air Force Inst. Tech., Ohio, November 1968.

27. Zienkiewicz, O. C., J. Too, and R. L. Taylor, "Reduced integration technique in general analysis of plates and shells," Int. J. Num. Meth. Eng. 3, 1971.
28. Severn, R. T., "Inclusion of shear deflection in the stiffness matrix of a beam element," J. Strain Analysis 5, 1970.
29. Razzaque, A., "A conforming triangular element in the analysis of thin and semi-thick plates in bending," M.Sc. Thesis, University of Wales, 1970.
30. Irons, B. M., and A. Razzaque, "A further modification to Ahmad's shell element," letter to Int. J. Num. Meth. Eng., May 1972 (to appear).
31. Loof, H. W., "The economical computation of stiffness of large structural elements," Int. Symp. on the Use of Elect. Digital Computers in Structural Engineering, University of Newcastle-upon-Tyne, Department of Civil Engineering, 1966.
32. Visser, W., "The application of a curved, mixed-type shell element," Symp. of International Union of Theoretical and Applied Mechanics, Liège, August 1970.
33. Hrennikoff, A., "The finite element method in application to plane stress," International Association for Bridge and Structural Engineering, Vol. 28-II, Zurich, 1968.
34. Strang, G., "Variational crimes in the finite element method," article in these proceedings.
35. Dupuis, G., and J. J. Goël, "A curved finite element for thin shells," Int. J. Solids and Structures 6 (1970), 1413.
36. Babuška, I., and A. K. Aziz, Proceedings of the Symposium on the Finite Element Method with Applications to Partial Differential Equations, these proceedings.

# HIGHER ORDER SINGULARITIES FOR INTERFACE PROBLEMS*

R. B. Kellogg**

## 1. Introduction.

We consider interface problems for an elliptic partial differential equation in two independent variables. Such problems arise in engineering problems when, for example, one is studying the distribution of heat or electric potential in a composite medium consisting of several different materials. The mathematical problem (which is formulated precisely in §2) has been studied, for smooth interfaces, by many people (e.g. [1]). We consider the problem in a polygonal domain, with polygonal interfaces. Our object is to represent the solution to the problem as the sum of a finite number of special solutions, which may be considered as known, plus a remainder term whose regularity depends on the regularity of the data of the problem. These results extend the results of [2,3] by considering more regular data, and obtaining more regularity in the remainder term.

If the problem contains no interfaces, we are left with a boundary value problem for an elliptic equation

---

* This research was supported in part by the National Science Foundation under Grant GP-20555.

** Institute for Fluid Dynamics and Applied Mathematics, University of Maryland.

in a polygonal domain. The differentiability properties of the solutions of such problems have been considered, for example, in [4,5,6,7]. The paper [4] gives asymptotic properties of the solution; [5] and [6] give conditions which guarantee that the solution lies in a specified differentiability class, while [7] contains results more closely related to ours.

Our method of analysis combines techniques of [7] with a use of fractional Sobolev spaces. It is likely that similar results may be obtained for more general interface problems in two dimensions. (Higher dimensional problems contain significant difficulties; see [3,7,8]. However, because of the practical importance of interface problems for the Laplace equation, and because the analysis is simpler in certain technical points, it seems worthwhile to develop the higher order results specifically for the equations considered here.)

In obtaining our results, we require an inequality and some results on the interpolation of Hilbert spaces which may be of interest in themselves. In this note we merely state our main results; the proofs will appear elsewhere.

## 2. The principal result.

Let $\Omega$ be a bounded domain in $R^2$ whose boundary, $\Gamma = \partial\Omega$ , consists of a finite number of straight line segments. Thus, $\Omega$ is a polygon or a slit polygon. Let $\Omega$ be decomposed by a finite number of straight line segments into a finite number of subdomains $\Omega_j$ . The boundaries between the different subdomains will be called the <u>interfaces</u> of the problem. The collection of vertices of $\Gamma$, or of any of the polygons $\partial\Omega_j$, will be called <u>vertices</u> of the problem.

Let there be given functions $p(x)$, $q(x)$ in $\Omega$ such that

$$p(x) = p_j > 0 \quad , \quad x \in \Omega_j$$
$$q(x) = q_j \qquad\qquad x \in \Omega_j$$

Let the boundary $\Gamma$ be decomposed into two parts, $\Gamma = \Gamma_0 \cup \Gamma_1$, such that each part, $\Gamma_j$, is a union of a finite number of the straight line segments whose union is $\Gamma$. Let there be given a function $a(x)$ on $\Gamma_1$ such that

$$a(x) = \text{constant}$$

on each line segment of $\Gamma_1$.

We formulate our interface problem in the following weak form. Letting $H^1(\Omega)$ denote the usual Sobolev space of square integrable functions with square integrable first derivatives, we seek a function $u \in H^1(\Omega)$ such that

$$(1) \qquad \int_\Omega [p\nabla u\nabla v + quv]dx = \int_\Omega fvdx + \int_{\Gamma_1} (g_1 - au)ds \ ,$$

with $u = g_0$ on $\Gamma_0$, for all $v \in H^1(\Omega)$, $v = 0$ on $\Gamma_0$. Intuitively, the solution $u$ of (1) may be considered as a solution of the interface problem

(2) $$-p_j \Delta u + q_j u = f \quad \text{in} \quad \Omega_j ,$$

(3) $$u, p \frac{\partial u}{\partial n} \quad \text{continuous across each interface} ,$$

(4) $$u = g_0 \quad \text{on} \quad \Gamma_0 ,$$

(5) $$p \frac{\partial u}{\partial n} + au = g_1 \quad \text{on} \quad \Gamma_1 .$$

Our object is to study the differentiability properties of the solution, $u$, of this problem, if the data, $f$, $g_0$, and $g_1$, has specified differentiability properties. For this, we define the Sobolev space $H^s(\Omega)$ to be the restriction to $\Omega$ of functions in $H^s(R^2)$. If $\Omega$ contains slits, the definition of $H^s(\Omega)$ must be modified somewhat by considering $u$ as the restriction of a function which is defined in a different manner on either side of the slit.

We define $\mathcal{H}^s(\Omega)$, to be the set of functions $u \in H^0(\Omega)$ such that $u \in H^s(\Omega_j)$ for each subdomain $\Omega_j$. Similarly, letting $I$ denote a line segment in the plane, $H^s(I)$ is defined as the restriction to $I$ of functions $u$ in $H^s(\tilde{I})$, where $\tilde{I}$ is the line which is the extension of $I$. We define $\mathcal{H}^s(\Gamma)$ to be the set of functions $u \in H^0(\Gamma)$ such that $u \in H^s(I)$ for each line segment $I$ of $\Gamma$. The spaces $\mathcal{H}^s(\Gamma_j)$, $j = 0,1$, have a similar meaning. These spaces are all Hilbert spaces with the obvious norms. For properties of the fractional Sobolev spaces, see [9], Chapter I.

With these definitions our main theorem may now

be stated.

Theorem 1. Let $u \in H^1(\Omega)$ be a solution of (1). Suppose $f \in H^s(\Omega)$, $g_0 \in H^{s+3/2}(\Gamma_0)$, $g_1 \in H^{s+1/2}(\Gamma_1)$, for some $s \geq 0$. Then if $s \notin D$, a set of exceptional indices, we may write

(6) $$u = \sum_{j=1}^{J} c_j v_j + w$$

where:

a) $w \in H^{s+2}(\Omega)$, and for some $c > 0$ independent of $f$, $g_0$, and $g_1$,

(7) $$\|w\|_{H^{s+2}(\Omega)} \leq c\|f\|_{H^s(\Omega)} + c\|g_0\|_{H^{s+3/2}(\Gamma_0)} + c\|g_1\|_{H^{s+1/2}(\Gamma_1)},$$

furthermore $w$ satisfies the interface conditions (3);

b) the functions $v_j$, called the "singular functions," are independent of $f$, $g_0$, and $g_1$; each $v_j$ may be taken to vanish outside a neighborhood of one of the vertices of the problem, and depends only on the values of $p$, $q$, and $a$, and on the structure of the domain $\Omega$ near the vertex; each $v_j \notin H^{s+2}(\Omega)$, but each $v_j$ satisfies the interface conditions (3);

c) the coefficients $c_j$ are bounded linear functionals on

$$\{f, g_0, g_1\} \in H^s(\Omega) \times H^{s+3/2}(\Gamma_0) \times H^{s+1/2}(\Gamma_1);$$

d) the exceptional set, $D$, does not depend on the data, $f$, $g_0$, $g_1$; it consists of a countable sequence of positive numbers whose only limit point is $+\infty$ .

In this result, the singular functions $v_j$ may, in a certain sense, be regarded as known. In the next section we discuss some further properties of these functions and of the exceptional set D .

## 3. The singular functions.

As we have seen, in Theorem 1 the singular functions $v_j$ are each associated with a vertex of the problem. We indicate the form of the function $v_j$ , and the corresponding linear functional $c_j$ , in the case of $q(x) \equiv 0$ , $a(x) \equiv 0$ , $g_0 \equiv 0$ , $g_1 \equiv 0$ . Let there be given a vertex, which we take as the origin, and let $\Omega$ coincide with the sector $0 < \theta < \omega$ in a neighborhood of the origin. If the vertex is interior to $\Omega$ , or if it lies on the end of a slit, we have $\omega = 2\pi$ . We let the interfaces which meet at the origin be given by the lines $\theta = \theta_j$ , $1 \leq j \leq J-1$ , and we set $\theta_0 = 0$ , $\theta_J = \omega$ . If there are no interfaces at the origin, $J = 1$ . We let $p_j$ denote the value of $p$ in the sector $\theta_{j-1} < \theta < \theta_j$ , $1 \leq j \leq J$ .

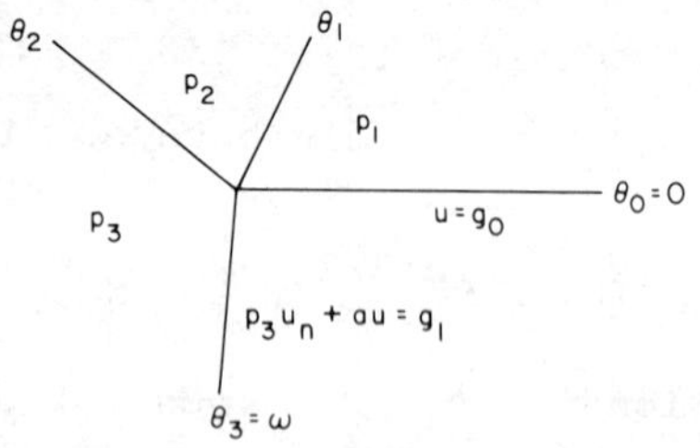

We must associate boundary conditions with the lines $\theta = 0$ , $\theta = \omega$ . If the vertex is interior to $\Omega$ , we associate the periodic boundary conditions

$$\Theta(0) = \Theta(2\pi) ,$$
$$\Theta'(0) = \Theta'(2\pi) .$$

If the vertex lies on $\Gamma$ , we use the boundary conditions on the two segments of $\Gamma$ that lead into the vertex. For definiteness, we consider the boundary conditions

$$\Theta(0) = 0 \tag{8}$$

$$p_j \Theta'(\omega) = 0 \tag{9}$$

We consider the eigenvalue problem defined by the Rayleigh quotient

$$J(\Theta) = \frac{\int_0^\omega p\Theta'^2 d\theta}{\int_0^\omega p\Theta^2 d\theta} , \quad \Theta(0) = 0 .$$

We let $\kappa_1 \leq \kappa_2 \leq \ldots$ denote the nonzero eigenvalues defined by $J$ , repeated according to multiplicity, and we let $\Theta_1 , \Theta_2 , \ldots ,$ denote the corresponding eigenfunctions, normalized according to the formula

$$\int_0^\omega p\Theta_m\Theta_n d\theta = \begin{cases} 1, & m=n \\ 0, & m\neq n \end{cases}$$

The eigenvalue $\kappa_n$ and eigenfunction $\Theta_n(\theta)$ satisfy

$$\Theta_n'' + \kappa_n \Theta = 0 \quad \text{in} \quad (\theta_{j-1}, \theta_j)$$

$$\Theta_n, \ p\Theta_n' \quad \text{continuous at} \quad \theta = \theta_j \ , \ 1 \leq j < J-1 \ .$$

$$\Theta_n(0) = 0 \ , \ p_J \Theta_n'(\omega) = 0 \ .$$

The quantities $t_n + 1$ , $t_n = \kappa_n^{1/2}$ , are members of the set D of exceptional indices. Each vertex of the problem gives rise to a set of exceptional numbers $t_n + 1$, and the set D is the union of all these sets of exceptional numbers. If $t_n \neq$ integer , and if $s > t_n + 1$ , there is a function of the form

$$v_j = \zeta(r) r^{t_n} \Theta_n(\theta)$$

occurring inthe decomposition (6), where $\zeta(r)$ is a smooth function that is $\equiv 1$ in a neighborhood of $r = 0$ , and vanishes outside a larger neighborhoood of $r = 0$ . The coefficient $c_j$ is given by the linear functional

$$c_j(f) = \int_\Omega fz dx \ ,$$

where z is a function that behaves, near $r = 0$ , like $r^{-t_n}$ . If $t_n$ = integer , the formulas are more complicated and a logarithmic dependence on r may occur in the function $v_j$ . Similar complications may occur in cases when q or a do not vanish.

## 4. Some interpolation spaces.

In the course of proving Theorem 1, we are led to

certain inequalities and results on interpolation that are interesting in themselves. We now state some of these.

Recall that, from Sobolev's inequality, a function $u \in H^s(R^2)$ is continuous if $s > 1$, and is $k$ times continuously differentiable if $s > k+1$. Let $k < s < k+1$, and let

$$\tilde{H}^s(R^2) = \{u \in H^s(R^2) : D^\alpha u(0) = 0 \ , \ |\alpha| \leq k-1\} ,$$

where $\alpha = (\alpha_1, \alpha_2)$, $|\alpha| = \alpha_1 + \alpha_2$, and

$$D^\alpha u = \frac{\partial^{|\alpha|} u}{\partial^{\alpha_1} x_1 \partial^{\alpha_2} x_2} .$$

Then $\tilde{H}^s(R^2)$ is a closed subspace of $H^s(R^2)$. The inequality in question is

$$\int r^{-2s} |u|^2 dx \leq c |u|^2_{H^s(R^2)} , \tag{10}$$

$s \neq$ integer, $u \in \tilde{H}^s(R^2)$.

Using this inequality, we may show the following results on interpolation in the Hilbert spaces $\tilde{H}^s(R^2)$. In stating these results we assume that the reader is familiar with the theory of interpolation of Hilbert spaces, as presented, for example, in [9].

<u>Theorem 2</u>. If $s_1 < s_2$, $s_1 \neq$ integer, $s_2 \neq$ integer, one has

$$[\tilde{H}^{s_1}(R^2) \ , \ \tilde{H}^{s_2}(R^2)]_\theta = \tilde{H}^{s_1\theta + s_2(1-\theta)}(R^2) ,$$

$s_1\theta + s_2(1-\theta) \neq$ integer.

For $s_1\theta + s_2(1-\theta) = k$ = integer, we define

$$\tilde{H}^k(R^2) = [\tilde{H}^{s_1}(R^2), \tilde{H}^{s_2}(R^2)]_\theta \quad .$$

Then $H^k(R^2)$ is the set of $u \in H^k(R^2)$ such that

$$\|u\|^2_{\tilde{H}^k(R^2)} = \sum_{|\alpha| \leq k} \int r^{2(k-|\alpha|)} |D^\alpha u|^2 dx < \infty \ .$$

Similar results hold in $R^n$, and for the space $\tilde{H}^s(\Omega)$ of functions in $H^s(\Omega)$ which, together with all their derivatives of the appropriate orders, vanish at the vertices of $\Omega$.

## 5. The exceptional indices.

We illustrate the way the exceptional indices enter a problem by considering, in the unit square $Q$, the problem

$$\text{(11)} \qquad \begin{aligned} \Delta u &= f \quad \text{in } Q \\ u &= 0 \quad \text{on } \Gamma = \text{boundary of } Q \ . \end{aligned}$$

We let $x^j$, $1 \leq j \leq 4$, denote the 4 vertices of $Q$. We consider Theorem 1 applied to this problem. For $s = 0$, we have $u \in H^2(\Omega)$, and there are no singular functions. The value $s = 1$ is an exceptional value of the problem. For $1 < s < 2$, we have 4 singular functions $v_j$, one corresponding to each vertex. If $r_j = |x - x_j|$ and $\theta_j$ are polar coordinates corresponding to the vertex $x^j$, we may take

$$v_j(x) = \psi(r_j)\{r_j^2[\frac{1}{4} - \frac{1}{4}\cos 2\theta_j + \frac{1}{\pi}\theta_j \cos 2\theta_j] + \tag{12}$$
$$+ \frac{1}{\pi} r_j^2 \ln r_j \sin 2\theta_j\}$$

where $\psi(t)$ is a smooth function, satisfying

$$\psi(t) \equiv 1 , \quad 0 \leq t \leq 1/3 ,$$
$$\psi(t) \equiv 0 , \quad 1/2 \leq t .$$

The decomposition (6) becomes, in this case,

$$u = \sum_{j=1}^{4} f(x^j) v_j + w , \tag{13}$$

$$w \in H^{s+2}(Q) , \ 1 < s < 2 , \ \|w\|_{H^{s+2}(Q)} \leq c\|f\|_{H^s(Q)} .$$

Since the linear functional $f \to f(x^j)$ is unbounded for $s = 1$, the decomposition (13) breaks down at $s = 1$.

In order to obtain a result for $s = 1$, we set $\phi_j = -\Delta v_j$. Thus,

$$\begin{cases} \phi_j(x) \equiv 1 , & |x - x^j| < 1/3 \\ \phi_j(x) \equiv 0 , & |x - x^j| > 1/2 \end{cases}$$

For $f \in H^s(Q)$, we set

$$f(x) = \sum_{1}^{4} f(x^j)\phi_j(x) + h(x) ,$$

where $h(x^j) = 0$, $1 \leq j \leq 4$. Thus, $h \in \tilde{H}^s(Q)$, and using this decomposition we write

$$H^s(Q) = \tilde{H}^s(Q) \times R^4 \ , \ 1 < s < 2 \ .$$

From (12) we see that the solution operator, $A$ , of our problem gives a bounded map

$$A : \tilde{H}^s(Q) \times R^4 \to H^{s+2}(Q) \times R^4 \ .$$

The same operator, $A$ , is a bounded map from

$$\tilde{H}^0(Q) \times R^4 = H^0(Q) \times R^4 \to H^2(Q) \times R^4 \ .$$

By interpolation, we find

Theorem 3. The solution $u$ of (11) admits the decomposition (13) with $w \in H^3(Q)$ provided that

(14)
$$|||f|||^2 = \sum_{j=1}^{4} \int_Q |x - x^j|^{-2} |f(x) - f(x^j)|^2 dx + \int_Q [|f|^2 + |D'f|^2] dx < \infty$$

If (14) holds, then $f$ is continuous at the vertices $x^j$ , and for some $c > 0$ ,

$$\|w\|_{H^3(Q)} \leq c|||f||| \ ,$$

$$|f(x_j)| \leq c|||f||| \ .$$

We give, finally, an example of a function $f \in H^1(Q)$ such that, if $u$ is the corresponding solution of (11), then there is no decomposition of the form

(15)
$$u = \sum c_j v_j + w \ ,$$

with $v_j$ given by (12), and with $w \in H^3(Q)$. We let the vertex $x^1 = 0$, and we let $(r,\theta)$ denote polar coordinates corresponding to $x^1 = 0$. Then we set

$$u = \psi(r)u_1 + \psi(r)u_2 ,$$

where

$$u = r^2 \ell n\, \ell n\, r \left[\frac{1}{4} - \frac{1}{4} \cos 2\theta + \frac{\theta}{\pi} \cos 2\theta\right] ,$$

and where $u_2 = R(r) \sin 2$ is the solution of the problem

$$\Delta u_2 = \frac{4}{\pi} \ell n\, \ell n\, r \sin 2\theta , \text{ in } S$$

$$u = 0 \text{ on } \partial S$$

$$u \in H^1(S) ,$$

where $S$ is the sector $0 < \theta < \frac{\pi}{2}$, $0 < r < 1$. Then one can verify with a computation that near $r = 0$,

$$\Delta u = \ell n\, \ell n\, r + \left[\frac{4}{\ell n\, r} - \frac{1}{\ell n^2 r}\right]\left[\frac{1}{4} - \frac{1}{4} \cos 2\theta + \frac{\theta}{\pi} \cos 2\theta\right] ,$$

so $u = f \in H^1(Q)$. But $u \notin H^3(Q)$, and for no constant $c$ are the third derivatives of $u - cv_1$ square integrable near $r = 0$. Hence no expansion of the form (15) can be valid.

REFERENCES


1. Schechter, M., "A generalization of the problem of transmission," Ann. Scuola Norm. Sup. Pisa, 14 (1960), 207-236.
2. Kellogg, R.B., "On the Poisson equation with intersecting interfaces," to appear in Applicable Analysis.
3. Kellogg, R.B., "Singularities in interface problems," Numerical Solution of Partial Differential Equations II, B. Hubbard, editor (1971), 351-400.
4. Wigley, N.M., "Asymptotic expansions at a corner of solutions of mixed boundary value problems," J. Math. Mech. 13 (1964), 549-576.
5. Volkov, E.A., "Differentiablity properties of solution of boundary value problems for the Laplace equation on a polygon," Proc. Steklov Inst. Math. 77 (1965), 127-159.
6. Grisvard, P., "Alternative de Fredholm relative au problème de Dirichlet dans un polygone ou un polyèdre," Boll. Un. Mat. Ital. 5 (1972), 132-164.
7. Kondrat'ev, V.A., "Boundary problems for elliptic equations with conical or angular points," Trans. Moscow Math. Soc. 16 (1967), 227-313.
8. Grisvard, P., "Problème de Dirïchlet dans un cone," Ricerche Mat. 20 (1971), 175-192.
9. Lions, J.L. and E. Magenes, Problèmes aux limites non homogènes et applications, I, Dunod, 1968.

# ON DIRICHLET PROBLEMS USING SUBSPACES WITH NEARLY ZERO BOUNDARY CONDITIONS

J. Nitsche*

## 1. Introduction.

The classical Rayleigh-Ritz methods for elliptic boundary value problems with Dirichlet data assume that the approximating functions fulfill the boundary conditions exactly. In most cases, i.e., when the boundary is curved, the construction of such subspaces is delicate and implies many difficulties in practice, see for example Miyoshi [11], Rvachev-Shklyarov [16], and Schultz [17]. In order to overcome this problem there are now known modifications of the variational principles obtained by adding appropriate boundary terms. We mention the work of Babuška [1], [2], [3], Bramble-Nitsche [6], Bramble-Schatz [7], [8], and Nitsche [15]. In these cases the approximating subspaces which are used need not fulfill any boundary conditions.

There is still another way to attack the problem, which is used in practice with good success: namely, to use subspaces which "nearly" fulfill the boundary conditions in a certain way. After a first outline in [14] this idea was further studied by Berger-Scott-Strang [4]. The variational principle used by them is not a projection

* Freiburg University, West Germany.

method and thus does not give quasi-optimal error estimates; that means the errors produced by the method are in general not as good as possible according to the approximability assumptions. The aim of the present paper is to give versions of the Ritz and least squares methods and to show them to be quasi-optimal. Besides the concept of nearly zero boundary values and a certain inverse assumption, a sharper condition of approximability than normally used is needed. In addition to the discussion of the convergence of the proposed methods, an essential part of the paper is devoted to this condition. It is proved especially for splines subordinate to a regular grid that this assumption holds true. Therefore, using such spline functions, no additional assumption is imposed. Instead we employ a seemingly up until now unrecognized property of splines. The author gratefully remembers discussions on this topic with I. Babuška and Al Schatz.

## 2. Notations, approximating subspaces.

In the following $\Omega$ denotes a bounded domain in $R^2$ with boundary $\partial\Omega$, which is for convenience of class $C^\infty$. For any domain $\tilde{\Omega} \subseteq R^2$ we denote by $W_p^k(\tilde{\Omega})$ the Sobolev-space of functions having in $\tilde{\Omega}$ generalized $L_p$-integrable derivatives up to order $k$, and by $\overset{o}{W}{}_p^k(\tilde{\Omega})$ the subspace of functions which can be approximated in the norm of $W_p^k(\tilde{\Omega})$ by $C^\infty$-functions with compact support. In the case $p = 2$ we write

$$\|u\|_{k.\tilde{\Omega}} = \|u\|_{W_2^k(\tilde{\Omega})} \tag{1}$$

and $(.,.)_{k.\tilde{\Omega}}$ for the corresponding bilinear form and omit $\tilde{\Omega}$ for $\tilde{\Omega} = \Omega$ .

If $T$ is a $C^\infty$-arc then $W_p^k(T)$ is defined similarly. For $T$ only piecewise $C^\infty$ , i.e. $T = \bigcup T_i$ and $T_i$ of class $C^\infty$ , we define

(2) $$\|u\|_{W_p^k(T)} = \left\{ \sum \|u\|^p_{W_p^k(T_i)} \right\}^{1/p} .$$

Similar to (1) we use

(3) $$|u|_{k.T} = \|u\|_{W_2^k(T)}$$

and $\langle .,. \rangle_{k.T}$ for the bilinear form omitting $T$ for $T = \partial\Omega$ . $\nabla u$ will be the gradient $(u_x^2 + u_y^2)^{1/2}$ and $D(.,.)$ the Dirichlet integral, which is the bilinear form of $\|\nabla .\|$ .

As a model equation we consider the boundary value problem

(4) $$\begin{aligned} -\Delta u &= f \quad \text{in} \quad \Omega \\ u &= 0 \quad \text{on} \quad \partial\Omega \end{aligned} .$$

By well-known a priori estimates we have $u \in \mathring{W}_2^1(\Omega) \cap W_2^{k+2}(\Omega)$ for $f \in W_2^k(\Omega)$ and $k \geq 0$ .

The solution $u$ of (4) is to be approximated by elements of subspaces which are assumed to fulfill certain conditions. We will consider a family $\{\mathring{S}_h | 0 < h \leq 1\}$ of finite-dimensional subspaces with the following properties.

Assumption 1 (approximability). There are integers $\bar{k}$ , $\bar{r}$ with $0 \leq \bar{k} \leq \bar{r}$ such that any function $u \in \mathring{W}_2^1(\Omega) \cap W_2^r(\Omega)$

with $r \leq \bar{r}$ may be approximated by elements $\chi \in \mathring{S}_h$ so that

(5) $$\|u - \chi\|_k \leq ch^{r-k}\|u\|_r \qquad (0 \leq k \leq \min(\bar{k},r)) \ ,$$

(6) $$|\chi| \leq ch^r\|u\|_r \ .$$

Assumption 2 (inverse relation). There is a number $\gamma > 0$ such that the inequality

(7) $$|\nabla\chi| \leq ch^{-\gamma}\|\chi\|_1$$

holds for $\chi \in \mathring{S}_h$ .

Assumption 3 (nearly zero boundary values). There is a number $\alpha > 0$ such that the inequality

(8) $$|\chi| \leq ch^{\alpha}\|\chi\|_1$$

holds for $\chi \in \mathring{S}_h$ .

Here and later c indicates a generic numerical constant, independent of h , which may be different at different places.

With respect to these assumptions some remarks should be added. Inequality (5) without the restriction $u \in \mathring{W}_2^1(\Omega)$ is typical for subspaces of splines or finite elements, see for instance Babuška [2], Birkhoff-Schultz-Varga [5], Bramble-Zlamal [9], Ciarlet-Raviart [10], Nitsche [12], Schultz [18], and Zlamal [19]. The additional condition (6) seems to be essentially new. If we would take the "full" spline-space $S_h$ we then would have for

$u \in W_2^r(\Omega) \cap W_2^r(\partial\Omega)$

$$(9) \qquad \|u - \chi\| + |u - \chi| \leq ch^r \{\|u\|_r + |u|_r\} \quad .$$

The inverse relation (7) was already used in Nitsche [15]; another condition which implies (7) was given in Bramble-Nitsche [6]. Spline subspaces relative to a regular grid fulfill (7) with $\gamma = 1/2$ . Since we will have optimality for the convergence of the discussed methods even for other values of $\gamma$ the restriction on the regularity of the grid may partially be lowered. For (8) see Nitsche [14] and Berger-Scott-Strang [4]. $\alpha = 3/2$ is typical, but the same arguments as for $\gamma$ apply.

In consequence of (8) $\|\nabla\chi\|$ is a norm in $\mathring{S}_h$ equivalent to $\|\chi\|_1$ with bounds independent of $h$ . With lemma 1 of [6] we have

$$\|\chi\|_1^2 \leq c\{\|\nabla\chi\|^2 + |\chi|^2\}$$

and hence

$$\|\chi\|_1^2 \leq c\|\nabla\chi\|^2 + ch^{2\alpha}\|\chi\|_1^2 \quad ,$$

or for $h$ small enough,

$$(10) \qquad \|\chi\|_1 \leq c(1 - ch^{2\alpha})^{-1/2}\|\nabla\chi\| \quad .$$

## 3. Projection methods.

The most natural way of approximating the solution $u$ of (4) by a projection is to take that element $u_h \in \mathring{S}_h$ for which the residual is orthogonal to $\mathring{S}_h$ :

$$(11) \qquad (\Delta u_h + f, \chi) = 0 \qquad \text{for } \chi \in \mathring{S}_h \quad .$$

For subspaces contained in $\overset{\circ}{W}{}_2^1(\Omega)$ this is just Ritz's method; in the following we will call this "method 1."

For the least-squares method, which we will consider here as method 2, the approximation $u_h$ is defined by

$$(12) \qquad (\Delta u_h + f, \Delta\chi) = 0 \qquad \text{for} \quad \chi \in \overset{\circ}{S}_h \quad .$$

In [14] we discussed the variational principle

$$(13) \qquad I(u - u_h) = \inf_{\chi \in \overset{\circ}{S}_h} I(u - \chi)$$

with

$$(14) \qquad I(v) = 2(v, -\Delta v) - D(v,v) \quad .$$

The advantage of using this functional is based on the duality relation

$$(15) \qquad (u,v) = I(u,w)$$

for any $u$ and any pair $v,w$ connected by

$$(16) \qquad \begin{aligned} -\Delta w &= v \quad &\text{in} \quad \Omega \\ w &= 0 \quad &\text{on} \quad \partial\Omega \quad , \end{aligned}$$

i.e., connected by the underlying boundary value problem. We will refer to this as method 3. Though we will not reproduce here the estimates explicitly for this method, the results are given in the summary at the end of this paragraph.

For completeness the approach of Berger-Scott-Strang [4], who minimize in $\overset{\circ}{S}_h$ the functional

(17) $$D(\chi,\chi) - 2(f,\chi)$$

should be mentioned.

For each of the methods $u_h$ will denote the solution and $e = e_h = u - u_h$ the corresponding error. An approximation of $u$ satisfying assumption 1 will be denoted by $U_h$, with the error $\varepsilon = \varepsilon_h = u - U_h$. Obviously we have $e - \varepsilon = U_h - u_h \in \overset{\circ}{S}_h$. If we get an error estimate for $e$ in a $W_2^k$-norm with the same power of $h$ as is possible by assumption 1 we call the method optimal in $W_2^k$. Something less than optimality in $W_2^1$ is needed in order to get an estimate similar to (6). We have -- independent of the method used --

$$\begin{aligned} |u_h| &\leq |U_h| + |e - \varepsilon| \\ &\leq ch^r\|u\|_r + ch^\alpha\|e - \varepsilon\|_1 \\ &\leq ch^r\|u\|_r + ch^\alpha\{\|e\|_1 + ch^{r-1}\|u\|_r\} \end{aligned}$$

or

(18) $$|u_h| = |e| \leq ch^r(1 + h^{\alpha-1})\|u\|_r + ch^\alpha\|e\|_1 .$$

Therefore we get

(19) $$|u_h| \leq ch^r\|u\|_r$$

already for $\alpha \geq 1$ and

(20) $$\|e\|_1 \leq h^{1-\alpha}\{ch^{r-1}\|u\|_r\} .$$

A bound for $|e_n|$, the boundary norm of the normal derivative, will be needed later. Because of the a priori estimate

$$|\nabla v|^2 \leq c\|v\|_1 \ \|v\|_2$$

we have as a consequence of (5)

$$|\nabla(u - U_h)| \leq ch^{r-3/2}\|u\|_r \quad .$$

So we get

$$|e_n| \leq |\nabla e| \leq |\nabla\varepsilon| + |\nabla(e - \varepsilon)|$$

$$\leq ch^{r-3/2}\|u\|_r + ch^{-\gamma}(\|e\|_1 + ch^{r-1}\|u\|_r)$$

or with $\gamma' = \max(1/2,\gamma)$

(21) $$|e_n| \leq ch^{r-1-\gamma'}\|u\|_r + ch^{-\gamma}\|e\|_1 \quad .$$

Now we turn to the proposed methods and discuss first the convergence properties of method 1; i.e., of

$$(\Delta e,\chi) = 0 \qquad \text{for} \quad \chi \in \overset{\circ}{S}_h$$

or after partial integration

(22) $$D(e,\chi) = \langle e_n,\chi\rangle \qquad \text{for} \quad \chi \in \overset{\circ}{S}_h \quad .$$

The choice $\chi = e - \varepsilon$ leads to

$$D(e,e) = D(e,\varepsilon) + \langle e_n, e-\varepsilon\rangle \quad .$$

Because of $D(e,\varepsilon) \leq \frac{1}{2} D(e,e) + \frac{1}{2} D(\varepsilon,\varepsilon)$ we have

(23) $$D(e,e) \leq D(\varepsilon,\varepsilon) + 2\langle e_n, e-\varepsilon\rangle \quad .$$

The first term is bounded by

(24) $$D(\varepsilon,\varepsilon) \leq \|\varepsilon\|_1^2 \leq ch^{2r-2}\|u\|_r^2 \quad .$$

Next we use

$$|e - \varepsilon| \leq ch^{\alpha}\|e - \varepsilon\|_1$$

$$\leq ch^{\alpha}\|e\|_1 + ch^{\alpha+r-1}\|u\|_r \quad .$$

Combining this inequality with (21) we get after a short computation

(25) $$|\langle e_n, e-\varepsilon\rangle| \leq ch^{2r-2+\alpha-\gamma'}\|u\|_r^2 + ch^{\alpha-\gamma'}\|e\|_1^2 \quad .$$

(24) and (25) replaced in (23) give

(26) $$D(e,e) \leq ch^{2r-2}(1 + h^{\alpha-\gamma'})\|u\|_r^2 + ch^{\alpha-\gamma'}\|e\|_1^2 \quad .$$

With (10) we can derive

$$\|e\|_1^2 \leq \|\varepsilon\|_1^2 + c\|\nabla(e - \varepsilon)\|^2$$

(27) $$\leq c\|\varepsilon\|_1^2 + cD(e,e)$$

$$\leq ch^{2r-2}\|u\|_r^2 + cD(e,e) \quad .$$

If we now have

(28) $$\gamma' = \max\left(\frac{1}{2}, \gamma\right) < \alpha$$

we find from (26) with (27)

(29) $$\|e\|_1 \leq ch^{r-1}\|u\|_r \quad ;$$

i.e., condition (28) is sufficient for optimality of method 1 in $W_2^1$. In the case $\alpha \geq 1$ [see (20)] one also has

(30) $$|e| \leq ch^r \|u\|_r \quad .$$

In order to study the convergence in $L_2(\Omega)$ we use the function v defined by

(31) $$-\Delta v = e \quad \text{in } \Omega$$
$$v = 0 \quad \text{on } \partial\Omega \quad .$$

We have $\|v\|_2 \leq c\|e\|$. Now we may write

$$\|e\|^2 = (e,-\Delta v) = D(e,v) - \langle e,v_n \rangle$$

and using the definition (22)

(32) $$\|e\|^2 = D(e,v-\chi) + \langle e_n,\chi \rangle - \langle e,v_n \rangle$$

with an arbitrary function $\chi \in \mathring{S}_h$. The last term is bounded by

$$|\langle e,v_n \rangle| \leq ch^r \|u\|_r \|v\|_2 \leq ch^r \|e\| \, \|u\|_r$$

if we assume (19) respectively (30) or $\alpha \geq 1$ and (29). Next let $\chi$ be an approximation to v ; according to our approximability assumption we will have

$$\|v - \chi\|_1 \leq ch\|v\|_2 \leq ch\|e\|$$

and

$$|\chi| \leq ch^2 \|v\|_2 \leq ch^2 \|e\| \quad .$$

Using once more (21) and the error estimate (29) already found, we get

$$|\langle e_n, \chi\rangle| \leq ch^{r+1-\gamma'}\|u\|_r\|e\| \quad .$$

Finally the first term in (32) is bounded by

$$|D(e,v-\chi)| \leq c\,\|e\|_1 \cdot h\,\|v\|_2$$

$$\leq ch^r\|e\|\,\|u\|_r \quad .$$

The estimates for the three terms in (32) give together

$$\|e\| \leq ch^r(1 + h^{1-\gamma'})\,\|u\|_r \quad .$$

In the case

(33) $$\gamma \leq 1 \,, \quad \alpha \geq 1 \,, \quad \gamma < \alpha$$

which includes (28), method 1 is optimal in $L_2$ .

Now we turn to method 2, or

(34) $$(\Delta e, \Delta\chi) = 0 \qquad \text{for} \quad \chi \in \overset{\circ}{S}_h \quad .$$

With $\chi = e - \varepsilon$ we get

$$\|\Delta e\| \leq \|\Delta\varepsilon\| \leq ch^{r-2}\|u\|_r \quad .$$

Using the auxiliary function $v$ defined by (31) and the a priori estimate $\|v\|_3 \leq c\,\|e\|_1$ , we get, with a properly chosen $\chi \in \overset{\circ}{S}_h$ ,

$$D(e,e) = \langle e,e_n\rangle - (e,\Delta e)$$

$$= \langle e,e_n\rangle + (\Delta e,\Delta v-\Delta\chi) \quad ,$$

and the estimate

$$|(\Delta e, \Delta v - \Delta\chi)| \leq \|\Delta e\| \; ch \; \|v\|_3 \leq ch^{r-1}\|u\|_r\|e\|_1$$

for the second term. Combining (18) and (21) leads after a short computation to

$$|<e,e_n>| \leq c(h^{1-\gamma'} + h^{\alpha-\gamma'})\{h^{2r-2}\|u\|_r^2 + \|e\|_1^2\} \quad .$$

Now we assume

$$(35) \qquad \gamma < 1 \; , \; \gamma' = \max\left(\frac{1}{2}, \gamma\right) < \alpha \quad ,$$

so $\beta = \min(1-\gamma', \alpha-\gamma')$ is positive, and we get

$$D(e,e) \leq ch^{2r-2}\|u\|_r^2 + ch^{r-1}\|u\|_r\|e\|_1 + ch^{\beta}\|e\|_1^2 \quad .$$

From this we derive for small $h$ similar to above (see (27))

$$\|e\|_1 \leq ch^{r-1}\|u\|_r \quad ,$$

i.e., the optimality of method 2 in $W_2^1$. In order to find the corresponding result in $L_2$ we use both the functions $v$ and $w$ defined by (31) respectively (16), and obtain

$$\|e\|^2 = (e, \Delta^2 w) = <e,v_n> + (\Delta e, \Delta w - \Delta\chi) \quad .$$

Using the a priori estimates $\|w\|_4 \leq c\,\|e\|$ and $|v|_n \leq c\,\|e\|$ we find, without additional assumptions on $\alpha$ and $\gamma$, that

$$\|e\| \leq ch^r\|u\|_r \quad ;$$

i.e., optimality in $L_2(\Omega)$ of method 2. In the case of method 3 we have the same conditions on $\alpha$ , $\gamma$ for optimality as for method 1:

Conditions for optimality

| of \ in | $W_2^1(\Omega)$ | $L_2(\Omega)$ |
|---|---|---|
| method 1 | $\max\left(\frac{1}{2},\gamma\right)<\alpha$ | $\gamma\leq 1,\ \gamma<\alpha,\ \alpha\geq 1$ |
| method 2 | $\max\left(\frac{1}{2},\gamma\right)<\alpha,\ \gamma<1$ | $\gamma<1,\ \gamma<\alpha,\ \alpha\geq 1$ |
| method 3 | $\max\left(\frac{1}{2},\gamma\right)<\alpha$ | $\gamma\leq 1,\ \gamma<\alpha,\ \alpha\geq 1$ |

## 4. Cubic spline functions.

By an $\alpha$-$\kappa$-regular triangulation $\Gamma$ we mean a triangulation of $R^2$ such that no angle of any triangle $\Delta$ is less than $\alpha > 0$ and the ratio of any two edges is bounded by $\kappa$ . The greatest length of the edges of $\Gamma$ is denoted by $|\Gamma|$ . If $\Omega$ is a domain as described in §2 then there exist regular triangulations such that any triangle having more than one point in common with $\partial\Omega$ has just two corners on $\partial\Omega$ . The triangles having all corners in $\bar{\Omega} = \Omega + \partial\Omega$ will form the triangulation $\Gamma$ and cover a polygonal domain $T$ with corners on $\partial\Omega$ . The distance between $\partial T$ and $\partial\Omega$ is of order $|\Gamma|^2$ . We will denote by $\partial\Gamma$ the collection of triangles of $\Gamma$ having two corners on $\partial\Omega$ and $\mathring{\Gamma} = \Gamma - \partial\Gamma$ . In general $T$ will differ from $\Omega$ and so we will not have a covering of $\Omega$ .

We can change the $\Delta \in \partial\Gamma$ to $\tilde{\Delta}$ in the following way: All points of $\Omega$ not in $T$ but inside the circumscribed circle around $\Delta$ are added to $\Delta$, thus giving a set $\tilde{\Delta}$, and in the union $\tilde{T}$. For small mesh sizes $|\Gamma|$ this construction is unique.

The space of (cubic) spline functions $S_\Gamma$ consists of all functions $\chi$ continuous in $\tilde{T}$ and piecewise cubic in the triangles of $\Gamma$. Such a spline is uniquely determined by its values and those of its first partial derivatives in the knots of $\Gamma$ and by its values in the centers of gravity of the $\Delta \in \Gamma$. If these values coincide with those of a function $u$, we will write $\chi = H_\Gamma u$ and call $\chi$ the Hermite-spline-interpolant of $u$. The space $\mathring{S}_\Gamma$ of splines with "nearly zero boundary values" is the subspace of $S_\Gamma$ such that for any $\chi \in \mathring{S}_\Gamma$ the values of the function and its tangential derivatives with respect to $\partial\Omega$ in the knots of $\partial T$ vanish. Obviously the Hermite-interpolation $H_\Gamma u$ of a function $u \in C^1$ vanishing on $\partial\Omega$ is in $\mathring{S}_\Gamma$.

If $u$ is defined in $\Omega$ only, then we may extend it to $\tilde{T}$ in such a way that the extension operator is bounded as mapping $W_2^r(\Omega) \to W_2^r(\tilde{T})$ for any $r$.

The main aim of this section is to show that the assumptions made in §2 are fulfilled for the constructed spline spaces $\mathring{S}_\Gamma$ with $\bar{r} = 4$ and $\alpha = 3/2$, $\gamma = 1/2$. The index $h$ is to be identified with the mesh size $|\Gamma|$; we will throughout use $h = |\Gamma|$. Only inequality (6) needs new considerations. We will first discuss the other assumptions. The estimate

(36) $$\|u - H_\Gamma u\|_k \leq ch^{r-k}\|u\|_r$$

for $0 \leq k < r \leq 4$ is well known; compare Bramble-Zlamal [9]. In the case $r < 3$ the definition of $H_\Gamma$ must be changed slightly.

In the same way as was shown for the one-dimensional case in [13] inverse relations

$$\|\chi\|_2 \leq ch^{-1}\|\chi\|_1$$

are valid. Since $\chi$ is in general not in $W_2^2(\Omega)$ the 2-norm has to be taken as

(37) $$\|\chi\|_2^2 = \sum_{\Delta\in\Gamma} \|\chi|_\Delta\|_{2.\Delta}^2 \quad .$$

It was already mentioned in Bramble-Nitsche [6] that (36) implies (7) with $\gamma = 1/2$ .

Finally as announced in [14] and then given in detail by Berger-Scott-Strang [4] the nearly zero condition (8) is correct with $\alpha = 3/2$ .

In order to come to (6) we look at one edge $t$ in the triangulation $\Gamma$ . The functions $u$ and $H_\Gamma u$ coincide as do the first derivatives in the direction of $t$ at the endpoints of $t$ which are knots of $\Gamma$ . The length of $t$ is bounded by $ch$ . Moreover $H_\Gamma u$ restricted to $t$ is a cubic polynomial in one variable. Therefore we have

$$|u - H_\Gamma u|_{L_2(t)} \leq ch^4|u|_{W_2^4(t)}$$

and so (see definition (2))

(38) $$|u - H_\Gamma u|_{0.\partial T} \leq ch^4 |u|_{4.\partial T} \ .$$

Our aim now is to pass from $\partial T$ to $\partial\Omega$. In a certain strip along $\partial\Omega$ we may introduce new variables $(\xi,\eta)$ in a diffeomorphic way such that along $\partial\Omega$ the parameter $\xi$ is the arc-length and $\eta$ is the distance from $\partial\Omega$. For any $\tilde{\Delta} \in \partial\Gamma$ we have a picture as shown in Figure 1. Let

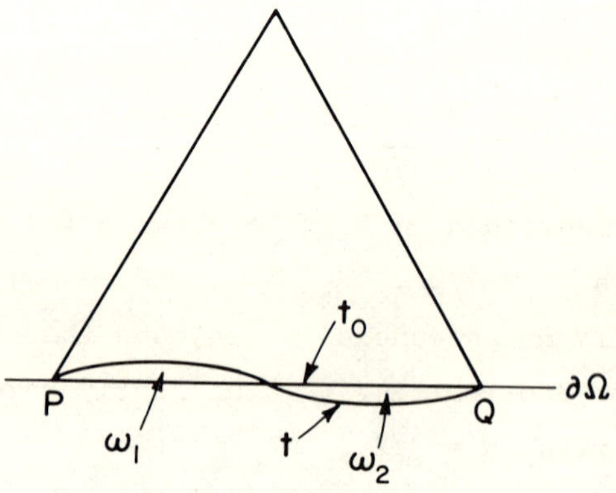


Fig. 1.

$t = \{\xi, \eta(\xi)\}$ be the boundary of $\partial T$ between P and Q and let $t_o$ be the corresponding part of $\partial\Omega$. The distance $\xi_o = PQ$ is of order h and $|\eta(\xi)|$ is bounded by $ch^2$, $|\eta'(\xi)|$ by ch. Denote further the domain between $\partial T$ and $\partial\Omega$ by $\omega$. For any function v we get

$$\left| |v|^2_{L_2(t)} - |v|^2_{L_2(t_o)} \right|$$

$$= \left| \int_o^{\xi_o} \{v^2(\xi,\eta(\xi)) \sqrt{1+\eta'^2} - v^2(\xi,0)\} d\xi \right|$$

$$(39) \qquad \leq \int_o^{\xi_o} |v^2(\xi,\eta(\xi)) - v^2(\xi,0)| d\xi +$$

$$+ \max|\eta'|^2 \int_o^{\xi_o} v^2 \, d\xi$$

$$= J_1 + J_2 \, .$$

For the second term we have at once

$$(40) \qquad J_2 \leq ch^2 |v|^2_{L_2(t)} \quad .$$

On the other hand we may write

$$v(\xi,\eta) = v(\xi,0) + \int_o^{\eta} v_\eta(\xi,\sigma) d\sigma$$

and get for any $\delta > 0$ the estimate

$$v^2(\xi,\eta) \leq (1+\delta)v^2(\xi,0) + (1+\delta^{-1}) \left\{ \int_o^{\eta} v_\eta d\sigma \right\}^2$$

$$\geq (1-\delta)v^2(\xi,0) + (1-\delta^{-1}) \left\{ \int_o^{\eta} v_\eta d\sigma \right\}^2 \quad .$$

Furthermore we have by Schwarz's inequality

$$\left\{ \int_o^{\eta(\xi)} v_\eta d\sigma \right\}^2 \leq |\eta(\xi)| \int_o^{\eta(\xi)} v_\eta^2 d\sigma \leq ch^2 \int_o^{\eta(\xi)} v_\eta^2 d\sigma \quad ,$$

and so we get

$$(41) \quad J_1 \leq \delta |v|^2_{L_2(t_o)} + c(1+\delta^{-1})h^2 \iint_\omega v_\eta^2 d\xi d\eta \quad .$$

Combining (40) and (41) we derive from (39), with $\delta = 1/2$ for example,

$$|v|^2_{L_2(t_o)} \leqslant c|v|^2_{L_2(t)} + ch^2\|v\|^2_{W_2^1(\tilde{\Delta})} .$$

By adding all the boundary terms and putting $v = u - H_\Gamma u$, we get

$$|u - H_\Gamma u|_{L_2(\partial\Omega)} \leqslant c|u - H_\Gamma u|_{L_2(\partial T)} + ch\|u - H_\Gamma u\|_{1.\Omega} .$$

To begin with, the last norm is to be taken over all $\tilde{\Delta} \in \Gamma$ only, but we may increase the region. With (38) and the already known estimate (36) in the $W_2^1(\Omega)$-norm we have, remembering $u = 0$ on $\partial\Omega$,

(42) $$|H_\Gamma u| \leqslant ch^4\{|u|_{4.\partial T} + \|u\|_4\} .$$

It remains to find a bound for $|u|_{4.\partial T}$. We look once more at Figure 1. Since the mapping $W_2^4(\tilde{\Delta}) \to W_2^3(t)$ is bounded we have

(43) $$|u|_{4.t} \leqslant |u|_{3.t} + |\nabla_t^4 u|_{o.t} \leqslant c\|u\|_{4.\tilde{\Delta}} + |\nabla_t^4 u|_{o.t} .$$

It is useful to make one more transformation in the domains $\omega_1$ respectively $\omega_2$:

$$\xi = \sigma , \quad \eta = \eta(\sigma)\tau .$$

In this way $\omega_1$ is mapped onto the square $0 < \sigma < \xi'$, $0 < \tau < 1$ and we have

$$\nabla_t^4 u = \frac{\partial^4}{\partial\sigma^4} u(\sigma,1) .$$

Since now $u(\sigma,0) = 0$ we get in quite the similar way as above

$$|\nabla_t^4 u|_{o.t} \leq ch\|u_\eta\|_{4.\omega} \quad . \tag{44}$$

Introducing (44) in (43) and summing over $\partial T$ we can replace (42) now by

$$|H_\Gamma u| \leq ch^4\|u\|_4 + ch^5\|u_\eta\|_{4.d\Omega} \tag{45}$$

with $d\Omega$ being a certain strip along $\partial\Omega$ of width $ch^2$ at most.

The next step is given by the following approximation lemma which will be proved in the last section.

<u>Lemma</u>: Let $d\Omega$ be a fixed and sufficiently small strip along $\partial\Omega$ . Any function $u \in W_2^4(\Omega) \cap \mathring{W}_2^1(\Omega)$ may be approximated for small $h$ by functions $u_h \in W_2^4(\Omega) \cap \mathring{W}_2^1(\Omega)$ such that for $0 \leq k < r \leq 4$ the inequality

$$\|u - u_h\|_k \leq ch^{r-k}\|u\|_r \tag{46}$$

is valid. Moreover $u_h$ will have a normal derivative $u_{h|\eta}$ in the boundary strip and

$$\|u_{h|\eta}\|_{4.d\Omega} \leq ch^{-1}\|u\|_4 \quad . \tag{47}$$

Let us now use as approximating elements in $\mathring{S}_\Gamma$ the Hermite-interpolation $H_\Gamma u_h$ where the index $h$ is just the mesh size of $\Gamma$ . Since $u_h$ is in $\mathring{W}_2^1(\Omega)$ the approximate element $H_\Gamma u_h$ is actually in $\mathring{S}_\Gamma$ . As a consequence of the error estimates (36), (45), and the approximation estimates (46), (47) we will have as desired

(48) $$\|u - H_\Gamma u_h\|_k \leq ch^{4-k}\|u\|_4$$
$$|H_\Gamma u_h| \leq ch^4\|u\|_4 \quad ;$$

i.e., the approximability assumption 1 is valid for $r = 4$. Repeating the arguments from (38) to (48) with the integer 4 replaced by $r$ with $1 \leq r \leq 3$ we get (6) also for the other values of $r$.

5. Proof of the lemma.

Let $\Omega_o, \Omega_i$ $(i = 1,\ldots,n)$ be a covering of $\bar{\Omega}$ such that the following conditions are fulfilled:

i. $\Omega_o \subset\subset \Omega$,

ii. $\Omega_i$ can be mapped diffeomorphically by $M_i$ onto the unit circle in the $\xi,\eta$-plane in such a way that (a) $\partial\Omega \cap \Omega_i$ is mapped on the line $\eta = 0$, $|\xi| < 1$, (b) the coordinate $\eta$ equals the distance in the origin from the boundary $\partial\Omega$.

Further, let $\Omega_i' \subset\subset \Omega_i$ such that $\Omega_o, \Omega_i'$ still is a covering of $\bar{\Omega}$. Now let $\phi_o, \phi_i$ be a partition of unity subordinate to this covering. With $\tilde{\Omega} = \bigcup_o^n \Omega_i$ any function $u \in W_2^4(\Omega)$ may be extended to $\tilde{u} \in \mathring{W}_2^4(\tilde{\Omega})$ with $\|\tilde{u}\|_{4.\tilde{\Omega}} \leq c\|u\|_{4.\Omega}$. The functions $u_i = \phi_i\tilde{u}$ have compact support in $\Omega_i$ and

(49) $$\|u_i\|_{4.\Omega_i} \leq c\|u\|_4 \quad .$$

The union $\bigcup_{i=1}^n \Omega_i'$ gives a certain strip along $\partial\Omega$ which we will identify with the one mentioned in the lemma of the preceding section. In order to get an approximation as stated in the lemma we will leave $u_o$ unchanged. Let $U(x,y)$ be one of the transformed functions $u_i : U = M_i u_i$. If $u \in \overset{\circ}{W}{}^1_2(\Omega)$ then we have

$$U(x,0) = 0 \quad . \tag{50}$$

With the help of a convolution

$$K_h U(x,y) = \int_0^1 \kappa(z)U(x,y+hz)dz \tag{51}$$

and a proper function $\eta$ we will use

$$U_h(x,y) = K_h U(x,y) - \eta(y/h)K_h U(x,0) \quad . \tag{52}$$

For $h$ small enough the function $K_h U$ will have its support in $W = \{(x,y)\,|\,x^2 + y^2 < 1\}$ since $U$ has a compact support in $W$. The same is true for the corrections $\eta(y/h)K_h U(x,0)$ if $\eta$ has a compact support in $(-1,1)$. Equation (50) is valid also for $U_h$ if we choose $\eta(0) = 1$. So the functions $u_{i|h} = M_i^{-1}U_h$ will be in $\overset{\circ}{W}{}^1_2(\Omega)$. The lemma is therefore proved provided we can guarantee the estimates

$$\begin{aligned} \|U - U_h\|_{k.W} &\leq ch^{r-k}\|U\|_{r.W} \ , \\ \|U_{h|y}\|_{4.W} &\leq ch^{-1}\|U\|_{4.W} \ . \end{aligned} \tag{53}$$

Now we consider first the convolutions (51). Evidently we have for $p = 1$ and $p = \infty$ with $I = (0,1)$

$$\|K_h U\|_{L_p(W)} \leq \|\kappa\|_{L_1(I)}\|U\|_{L_p(W)} \quad . \tag{54}$$

By interpolation this inequality is valid also for all $p$ with $1 \leq p \leq \infty$. In the case $\kappa(0) = \kappa(1) = 0$ the convolutions $K_h$ commute with differentiation and so the $L_p$-norm in (54) may be replaced by any $W_p^1$-norm.

Now we specialize the kernel of (51) by $\kappa = k^{iv}$ with $k$ chosen according to

$$k^{(\nu)}(0) = k^{(\nu)}(1) = 0 \qquad \text{for } \nu = 0,\dots,4$$

(55) with the exception of

$$k^{iii}(0) = -1 \quad .$$

Let us first put

$$V_h = K_h U \tag{56}$$

and study the approximation properties of this function. By partial integration we get

$$\partial_y V_h = -h^{-1} \int_0^1 k^v(z) U(x,y+hz)\,dz$$

and therefore because of (54),

$$\| \partial_y V_h \|_{4.W} \leq h^{-1} \| k^v \|_{L_1(I)} \| U \|_{4.W} \quad . \tag{57}$$

This implies (53.2) for $V_h$ instead of $U_h$. The conditions on $k$ lead to

$$V_h(x,y) = U(x,y) + h^4 \int_0^1 k(z)\,\partial_y^4 U(x,y+hz)\,dz \quad ,$$

and similarly for $\lambda + \mu < 4$,

$$\partial_x^\lambda \partial_y^\mu V_h = \partial_x^\lambda \partial_y^\mu U_h +$$

$$+ (-1)^{\lambda+\mu} h^{4-\lambda-\mu} \int_0^1 k^{(\lambda+\mu)}(z)\, \partial_x^\lambda \partial_y^{4-\lambda} U(x,y+hz)\,dz \quad .$$

This shows (53.1) to be valid also for $V_h$ instead of $U_h$ with $r = 4$ . The cases $r = 1,2,3$ can be handled similarly.

Since in general $V_h$ will not vanish for $y = 0$ (see (50)) we use the corrections

$$U_h = V_h - W_h \tag{58}$$

with

$$W_h(x,y) = \eta(y/h) K_h U(x,0)$$

$$= h^4 \eta(y/h) \int_0^1 k(z) U^{iv}(x,hz)\,dz \quad .$$

Here $\eta$ may be any function with $\eta \in C^5$ , $\text{supp}(\eta) \subset\subset (-1,1)$ and $\eta(0) = 1$ . We have

$$\|V_h(.,0)\|_{L_1(R)} \leq h^3 \|k\|_{L_\infty(I)} \|U^{iv}\|_{L_1(W)}$$

$$\|V_h(.,0)\|_{L_\infty(R)} \leq h^4 \|k\|_{L_1(I)} \|U^{iv}\|_{L_\infty(W)}$$

and also

$$\|\eta(./h)\|_{L_1(R)} \leq h \|\eta\|_{L_1(R)}$$

$$\|\eta(./h)\|_{L_\infty(R)} \leq \|\eta\|_{L_\infty(R)} \quad .$$

Therefore the estimate

$$\|W_h\|_{L_p(W)} \leq ch^4 \|U\|_{W_p^4(W)} \tag{59}$$

holds true for $p = 1$ and $p = \infty$ with $c$ depending only on $k$ and $\eta$, by interpolation (59) is also correct for $1 < p < \infty$.

In the same way

$$\|W_h\|_{k.W} \leq ch^{4-k} \|U\|_{4.W}$$

can be shown. This means that the function $W_h$ fulfills inequalities of the type (53). Since this is true also for the difference $V_h - U$, we have shown the error estimates (53) in the case $r = 4$ for the function $U_h$ defined by (58). The other values of $r$ can be handled in the same way. This completes the proof of the lemma stated in section 4.

## REFERENCES

1. Babuška, I., "Numerical solution of boundary value problems by the perturbated variational principle," Tech. Note BN-624, University of Maryland, 1969.
2. Babuška, I., "Error-bounds for finite element method," Numer. Math. 16 (1971), 322-333.
3. Babuška, I., "The finite element method with Lagrangian multipliers," Tech. Note BN-724, University of Maryland, 1972.
4. Berger, A., R. Scott, and G. Strang, "Approximate boundary conditions in the finite element method," Symposia Mathematica, Academic Press, New York, 1972.
5. Birkhoff, G., M. H. Schultz, and R. S. Varga, "Piecewise Hermite interpolation in one and two variables with applications to partial differential equations," Numer. Math. 11 (1968), 232-256.
6. Bramble, J. H., and J. A. Nitsche, "A generalized Ritz-least-squares method for Dirichlet problems," SIAM J.

Numer. Anal. (to appear).

7. Bramble, J. H., and A. H. Schatz, "Rayleigh-Ritz-Galerkin methods for Dirichlet's problem using subspaces without boundary conditions," Comm. Pure Appl. Math. 23 (1970), 653-675.
8. Bramble, J. H., and A. H. Schatz, "Least squares methods for 2 m th order elliptic boundary-value problems," Math. Comp. 25 (1971), 1-32.
9. Bramble, J. H., and M. Zlamal, "Triangular elements in the finite element method," Math. Comp. 24 (1970), 809-820.
10. Ciarlet, P. G., and P.-A. Raviart, "Interpolation theory over curved elements, with applications to finite element methods" (to appear).
11. Miyoshi, T., "On the convergence of Ritz-Galerkin's method," Publ. Res. Inst. Math. Sci. Ser. A. 4 (1968/69), 149-177.
12. Nitsche, J. A., "Interpolation in Sobolev'shen Funktionenräumen," Numer. Math. 13 (1969), 334-343.
13. Nitsche, J. A., "Umkehrsätze für Spline-Approximationen," Comp. Math. 21 (1969), 400-416.
14. Nitsche, J. A., "A projection method for Dirichlet-problems using subspaces with nearly zero boundary conditions" (unpublished manuscript).
15. Nitsche, J. A., "Über ein Variationsprinzip zur Lösung von Dirichlet-Problemen bei Verwendung von Teilräumen, die keinen Randbedingunger unterworfen sind," Abh. d. Hamb. Math. Sem. 36 (1971), 9-15.
16. Rvachev, B. L., and L. I. Shklyarov, "On the application of the Bubnov-Galerkin method to the solution of boundary problems for domains of complex shape," Differential Equations 1 (1965), 1211-1216.
17. Schultz, M. H., "Rayleigh-Ritz-Galerkin methods for multidimensional problems," SIAM J. Numer. Anal. 6 (1969), 523-538.
18. Schultz, M. H., "$L^2$ error bounds for the Rayleigh-Ritz-Galerkin method," SIAM J. Numer. Anal. 9 (1971), 737-748.
19. Zlamal, M., "On the finite element method," Numer. Math. 12 (1968), 394-409.

# GENERALIZED CONJUGATE FUNCTIONS FOR MIXED FINITE ELEMENT APPROXIMATIONS OF BOUNDARY VALUE PROBLEMS

J. T. Oden*

## 1. Introduction.

The idea of conjugate or dual representations for finite-element approximations was introduced in [1] in connection with purely topological properties of finite element models; i.e., in the spirit of linear graph theory, if $f(u)$ is a linear functional on a space $\mathcal{M}$ of finite element approximations $u$, then there exists a consistent (conjugate) approximation of $f$ which renders $f(u)$ invariant under transformations from the unconnected model (i.e, the system of disjoint finite elements) to the global model (the connected elements). This idea led to the theory of conjugate approximation functions introduced in [2] and applied to various problems in elasticity theory in [3-6]. A detailed summary account, together with a few generalizations is given in [6].

In the present investigation, we expand the theory of conjugate approximations to include so-called mixed formulations of linear boundary-value problems. This amounts to a natural generalization of cases studied earlier in which the subspaces $\mathcal{M}_G = \text{span}\,\{\phi_1(\underset{\sim}{x}),\ldots,\phi_G(\underset{\sim}{x})\}$

*University of Alabama.

generated by the usual displacement-type finite element formulations are self-dual. Also, all of the preliminary results in [5], which concerned situations in which different meshes are used to generate primal and dual approximations, are generalized and recast into a more meaningful framework. As a result, the concepts of "consistent" masses, forces, and even stresses [3,4] common to continuum mechanics applications can be expanded, and the notion of mixed and hybrid finite-element models (e.g. [7]) can be cast into a more general setting.

Following this introduction, we discuss several features of the theory of conjugate projections with special emphasis on discrete models of a linear operator $T$ and its adjoint $T^*$. We show that at least four sequences of projections of a class of boundary value problems involving $T^*T$ suggest themselves, and these lead directly to both the conventional and to mixed finite-element models. In Section 3 we discuss details of the finite-element formulation of mixed models, and in Section 4 we establish the relationship between our projection methods, the complementary variational principles of Noble and others [8-10], and the hypercircle method of Prager and Synge [11,12]. In terms of approximation of boundary-value problems, the notion of conjugate projections seems to suggest a broader collection of possibilities than either of the more conventional approaches. Finally, in Section 5 of the paper we address ourselves to certain preliminary questions of convergence and error estimation. There we establish convergence rates for our so-called primal, dual, and mixed finite-element models, the latter being restricted to certain special choices of test functions for dual approximations.

## 2. Conjugate Projections and Approximations.

2.1 Conjugate subspaces. We record here a number of properties of dual and conjugate subspaces. First, we consider two linear vector spaces, $\mathcal{U}$ and $\underset{\sim}{\mathcal{V}}$, defined over the same field. For the moment, $\underset{\sim}{\mathcal{V}}$ is regarded as the dual of $\mathcal{U}$ and we denote by $\langle u, \underset{\sim}{v} \rangle$ the bilinear scalar product $s: \mathcal{U} \otimes \underset{\sim}{\mathcal{V}} \to E'$, which takes ordered pairs $u, \underset{\sim}{v} (u \in \mathcal{U}, \underset{\sim}{v} \in \underset{\sim}{\mathcal{V}})$ into scalars. Further, let $\{\phi_\alpha\}$ $(\alpha = 1,2,\ldots,G)$ denote a set of $G$ linearly independent elements in $\mathcal{U}$ and $\{\underset{\sim}{\chi}^\alpha\}$ denote a set of $G$ linearly independent elements in $\underset{\sim}{\mathcal{V}}$. The sets $\{\phi_\alpha\}$ and $\{\underset{\sim}{\chi}^\alpha\}$ define $G$-dimensional subspaces $\mathcal{M}_G \subset \mathcal{U}$ and $\underset{\sim}{\mathcal{N}}_G \subset \underset{\sim}{\mathcal{V}}$, respectively. If the $G \times G$ matrix $\underset{\sim}{A}$ defined by

$$A_{\alpha}^{\cdot\beta} = \langle \phi_\alpha, \underset{\sim}{\chi}^\beta \rangle \tag{2.1}$$

$(\alpha, \beta = 1,2,\ldots,G)$ is nonsingular, then $\mathcal{M}_G$ and $\underset{\sim}{\mathcal{N}}_G$ are dual subspaces, and we use the notation $\underset{\sim}{\mathcal{N}}_G = \underset{\sim}{\mathcal{M}}_G^*$. We also note that if $\underset{\sim}{A}$ is nonsingular, we can construct sets of biorthogonal bases $(\{\phi_\alpha\}, \{\underset{\sim}{\phi}^\alpha\})$ and $(\{\chi_\alpha\}, \{\underset{\sim}{\chi}^\alpha\})$ by the transformations

$$\chi_\alpha = \sum_{\lambda}^{G} A_{\alpha}^{-1.\lambda} \phi_\lambda \quad ; \quad \underset{\sim}{\phi}^\alpha = \sum_{\lambda}^{G} \underset{\sim}{\chi}^\lambda A_{\lambda}^{-1.\alpha} \tag{2.2}$$

where $A_{\alpha}^{-1.\lambda}$ are the elements of $\underset{\sim}{A}^{-1}$. Clearly,

$$\langle \phi_\alpha , \underset{\sim}{\phi}^\beta \rangle = \langle \chi_\alpha , \underset{\sim}{\chi}^\beta \rangle = \delta_\alpha^\beta \tag{2.3}$$

where $\delta_\alpha^\beta$ is the Kronecker delta and $\alpha, \beta = 1,2,\ldots,G$ . Since transformations of the form (2.2) are always possible when $\underset{\sim}{A}$ is nonsingular, we choose either set as biorthogonal bases for $\mathcal{M}_G$ and $\underset{\sim}{\mathcal{M}}_G^*$ .

By identifying bases $(\{\phi_\alpha\},\{\underset{\sim}{\phi}^\alpha\})$ with the property (2.3), we have effectively defined projection operators $\Pi : \mathcal{U} \to \mathcal{M}_G$ and $Q : \underset{\sim}{\mathcal{V}} \to \underset{\sim}{\mathcal{M}}_G^*$ in the following sense: if $u$ is an arbitrary element in $\mathcal{U}$ and $\underset{\sim}{v}$ is an arbitrary element in $\underset{\sim}{\mathcal{V}}$ , the projection $\bar{u}$ of $u$ into $\mathcal{M}_G$ and the projection $\bar{\underset{\sim}{v}}$ of $\underset{\sim}{v}$ into $\underset{\sim}{\mathcal{M}}_G^*$ are of the form

$$\Pi u = \bar{u} = \sum_\alpha a^\alpha \phi_\alpha \quad ; \quad Q\underset{\sim}{v} = \bar{\underset{\sim}{v}} = \sum_\alpha a_\alpha \underset{\sim}{\phi}^\alpha \tag{2.4}$$

where $a^\alpha = \langle u, \underset{\sim}{\phi}^\alpha \rangle$ and $a_\alpha = \langle \phi_\alpha , \underset{\sim}{v} \rangle$ . We remark that without additional requirements, the set $\{\underset{\sim}{\phi}^\alpha\}$ with property (2.3) is, for a given $\{\phi_\alpha\}$ , not unique.

Now suppose that $\underset{\sim}{\mathcal{N}}_H$ is a subset of $\underset{\sim}{\mathcal{M}}_G^*$ spanned by a set of $H$ elements $\underset{\sim}{\omega}^\Delta (\Delta = 1,2,\ldots,H;\ H < G)$ given by

$$\underset{\sim}{\omega}^\Delta = \sum_\alpha^G R^\Delta_{\cdot\alpha} \underset{\sim}{\phi}^\alpha \tag{2.5}$$

where $R^\Delta_{\cdot\alpha}$ are the entries in an $H \times G$ matrix. If we can find a set of elements $\{\omega_\Delta\} \in \mathcal{M}_G$ biorthogonal to $\underset{\sim}{\omega}^\Delta$ , then there exists a $G \times H$ matrix $\underset{\sim}{S}$ such that

(2.6) $$\omega_\Delta = \sum_{\alpha}^{G} S_{\Delta}^{\cdot\alpha} \phi_\alpha \quad \text{and} \quad \langle \omega_\Delta, \underset{\sim}{\omega}^\Gamma \rangle = \delta_\Delta^\Gamma$$

Clearly, $\underset{\sim}{S}$ is a left inverse of $\underset{\sim}{R}^T$ (i.e. $\sum_\alpha S_\Gamma^{\cdot\alpha} R^\Delta_{\cdot\alpha} = \delta_\Gamma^\Delta$ ; $\Delta, \Gamma = 1, \ldots, H$) . If we denote by $\hat{u}$ and $\hat{\underset{\sim}{v}}$ the elements $\hat{u} = \sum_\Delta b^\Delta \omega_\Delta$ and $\hat{\underset{\sim}{v}} = \sum_\Delta b_\Delta \underset{\sim}{\omega}^\Delta$ , where $b^\Delta = \langle u, \underset{\sim}{\omega}^\Delta \rangle$ and $b_\Delta = \langle \omega_\Delta, \underset{\sim}{v} \rangle$ , then

(2.7) $$b^\Delta = \sum_\alpha R^\Delta_{\cdot\alpha} a^\alpha \quad \text{and} \quad b_\Delta = \sum_\alpha S_\Delta^{\cdot\alpha} a_\alpha$$

where $a^\alpha$ and $a_\alpha$ are the coefficients in (2.4). Should it happen that the scalar product $\langle \bar{\underset{\sim}{u}}, \bar{\underset{\sim}{v}} \rangle$ remain invariant under the transformation (2.5), then we find that

(2.8) $$\sum_\Delta b^\Delta b_\Delta = \sum_\alpha a^\alpha a_\alpha = \sum_\Delta \sum_{\alpha,\beta} R^\Delta_{\cdot\alpha} S_\Delta^{\cdot\beta} a^\alpha a_\beta$$

This means that $\sum_\Delta R^\Delta_{\cdot\alpha} S_\Delta^{\cdot\beta} = \delta_\alpha^\beta$ ; i.e. $\underset{\sim}{R}$ is a left inverse of $\underset{\sim}{S}$ .

<u>2.2 Hilbert spaces; projections of operators</u>. Often we encounter cases in which $\mathcal{U}$ and $\underset{\sim}{\mathcal{V}}$ are Hilbert spaces and, consequently, self-dual. Suppose that such is the case, and that inner products in $\mathcal{U}$ and $\underset{\sim}{\mathcal{V}}$ are denoted $\{\cdot\ ,\ \cdot\}$ and $[\cdot\ ,\ \cdot]$ , respectively. Then Gram matrices associated with the subspaces $\mathcal{M}_G = \text{span}\ \{\phi_\alpha\}_{\alpha=1}^G$ and $\underset{\sim}{\mathcal{N}}_H = \text{span}\ \{\underset{\sim}{\omega}^\Delta\}_{\Delta=1}^H$ can be constructed of the form

$$G_{\alpha\beta} = \{\phi_\alpha, \phi_\beta\} \quad ; \quad H^{\Delta\Gamma} = [\underset{\sim}{\omega}^\Delta, \underset{\sim}{\omega}^\Gamma] \tag{2.9}$$

If $\{\phi_\alpha\}$ and $\{\underset{\sim}{\omega}^\Delta\}$ are linearly independent, $G_{\alpha\beta}$ and $H^{\Delta\Gamma}$ are nonsingular, and we can compute directly biorthogonal bases

$$\phi^\alpha = \sum_\beta G^{\alpha\beta}\phi_\beta \quad ; \quad \underset{\sim}{\omega}_\Delta = \sum_\Gamma H_{\Delta\Gamma}\underset{\sim}{\omega}^\Gamma \tag{2.10}$$

where $G^{\alpha\beta}$ and $H_{\Delta\Gamma}$ are the inverses of $G_{\alpha\beta}$ and $H^{\Delta\Gamma}$. Clearly, $\{\phi_\alpha, \phi^\beta\} = \delta_\alpha^\beta$ and $[\underset{\sim}{\omega}_\Delta, \underset{\sim}{\omega}^\Gamma] = \delta_\Delta^\Gamma$ so that the necessary ingredients for defining projections of any $u \in \mathcal{U}$ into $\mathcal{M}_G$ and any $\underset{\sim}{v} \in \underset{\sim}{\mathcal{V}}$ into $\underset{\sim}{\mathcal{N}}_H$ are available. Clearly, if (2.5) holds, we could also compute the matrix

$$\underline{\underline{H}}^{\alpha\beta} = [\underset{\sim}{\phi}^\alpha, \underset{\sim}{\phi}^\beta] \quad ; \quad H^{\Delta\Gamma} = \sum_\alpha\sum_\beta R^\Delta_{\cdot\alpha}\underline{\underline{H}}^{\alpha\beta}R^\Gamma_{\cdot\beta} \tag{2.11}$$

The theory of conjugate projections described in [2.4] deals with such self-conjugate subspaces.

Now the biorthogonal bases in $\mathcal{M}_G$ and $\underset{\sim}{\mathcal{N}}_H$ are clearly independent of each other since there need now be nothing in common with the space $\mathcal{U}$ and $\underset{\sim}{\mathcal{V}}$. However, we open the door to many interesting applications if we assume that there exists a linear mapping $T:\mathcal{U} \to \underset{\sim}{\mathcal{V}}$. If $\mathcal{Q}(T)$ is the range space of the operator $T$, then the conjugate space $\mathcal{Q}(T)^* \subset \underset{\sim}{\mathcal{V}}^*$ consists of those linear functionals of the form $f(Tu) = [Tu, \underset{\sim}{v}]$, $\underset{\sim}{v} \in \underset{\sim}{\mathcal{V}} = \underset{\sim}{\mathcal{V}}^*$. To be more specific, let $\mathcal{U}$ denote a Hilbert space consisting of functions $u$ defined on a compact, convex subset $\mathcal{R}$ of $E^n$ with a smooth boundary $\partial\mathcal{R}$, and let $T$ denote a bounded linear operator mapping $\mathcal{U}$ into another Hilbert space $\underset{\sim}{V}$. If $\{\cdot,\cdot\}$ and $[\cdot,\cdot]$ are, again, the inner products defined on $\mathcal{U}$ and $\underset{\sim}{V}$, then we can identify an operator $T^*:\underset{\sim}{V} \to \mathcal{U}$ as the formal adjoint of $T$ if it satisfies the relation

$$(2.12) \qquad [\underset{\sim}{v},Tu] = \{T^*\underset{\sim}{v},u\} + [\underset{\sim}{v},Bu]_{\partial R}$$

Here $B$ is a linear operator depending on $T$ which also has an adjoint $B^*$ in the sense that

$$(2.13) \qquad [\underset{\sim}{v},Bu]_{\partial R} = \{B^*\underset{\sim}{v},u\}_{\partial R}$$

By $[\cdot,\cdot]_{\partial R}$ and $\{\cdot,\cdot\}_{\partial R}$ we mean, of course, inner products formed by restricting the domains of the entries to the boundary $\partial R$.

We recall that for any bounded linear operators $T:\mathcal{U} \to \underset{\sim}{V}$, the null spaces $N(T)$ and $N(T^*)$ are closed subspaces of $\mathcal{U}$ and $\underset{\sim}{V}$, respectively. If $N(T)^{\perp}$ and $N(T^*)^{\perp}$ denote orthogonal complements, then it can be shown that $\mathcal{U} = N(T) + N(T)^{\perp}$ and $\underset{\sim}{V} = N(T^*) + N(T^*)^{\perp}$. Moreover (see [13]), $\overline{\mathcal{Q}(T)} = N(T^*)^{\perp}$, $\overline{\mathcal{Q}(T^*)} = N(T)^{\perp}$, $\mathcal{Q}(T)^{\perp} = N(T^*)$, and $\overline{\mathcal{Q}(TT^*)} = \overline{\mathcal{Q}(T)}$ where $\overline{A}$ denotes closure of a set $A$ and $\mathcal{Q}(T)$ is the range space of $T$. If $P$ is the orthogonal projection of $\mathcal{U}$ onto $\overline{\mathcal{Q}(T^*)}$, and $Q$ is the orthogonal projection of $\underset{\sim}{V}$ onto $\overline{\mathcal{Q}(T)}$, then $Tu = TPu \;\forall\, u \in \mathcal{U}$ and $T^*\underset{\sim}{v} = T^*Q\underset{\sim}{v} \;\forall\, \underset{\sim}{v} \in \underset{\sim}{V}$. A <u>best</u> approximate solution of $Tu = \underset{\sim}{v}$ is one which minimizes $\|\underset{\sim}{w} - \underset{\sim}{v}\|$ among all $\underset{\sim}{w} \in \mathcal{Q}(T)$; such an element $\underset{\sim}{w}^*$ is such that $\underset{\sim}{v} - \underset{\sim}{w}^* \in \mathcal{Q}(T)^{\perp} = \mathcal{Q}(T^*)$ so that $T^*\underset{\sim}{v} = T^*\underset{\sim}{w}^*$. Consequently, if u is a best approximation of $Tu = \underset{\sim}{v}$ , then $T^*Tu = T^*\underset{\sim}{v} = -f$ (say). Our method of conjugate projections consists of seeking such best approximations in finite-dimensional subspaces of $\mathcal{Q}(T)$ , $\mathcal{Q}(T^*)$ , etc., and is directly aimed at the problem of finding u such that $T^*Tu + f = 0$ (see Art. 2.3).

If we consider cases in which $\mathcal{M}_G \subset \mathcal{D}_T$ and $\underset{\sim}{\mathcal{N}}_H \subset \mathcal{D}_{T^*}$ where $\mathcal{D}_T$ and $\mathcal{D}_{T^*}$ are the domains of T and T* , then a number of such approximations of T and T* suggest themselves. Since, in general, $T(\mathcal{M}_G)$ is not in $\underset{\sim}{\mathcal{N}}_H$ and $T^*(\underset{\sim}{\mathcal{N}}_H)$ is not in $\mathcal{M}_G$ , we can at least project these spaces into $\underset{\sim}{\mathcal{N}}_H$ and $\mathcal{M}_G$ . This process leads to the following rectangular matrices associated with T and T*:

(2.14) $PT(\mathcal{M}_G)$:

$$P(T\phi_\alpha) = \sum_\Delta T_{\alpha\Delta}\underset{\sim}{\omega}^\Delta = \sum_\Delta T_\alpha^{\cdot\Delta}\underset{\sim}{\omega}_\Delta$$

$$P(T\phi^\alpha) = \sum_\Delta T^\alpha_{\cdot\Delta}\underset{\sim}{\omega}^\Delta = \sum_\Delta T^{\alpha\Delta}\underset{\sim}{\omega}_\Delta$$

(2.15) $T^*(\underset{\sim}{\mathcal{N}}_H)$:

$$\Pi(T^*\omega_\Delta) = \sum_\alpha T^*_{\Delta\alpha}\phi^\alpha = \sum_\alpha T^{*\alpha}_\Delta\phi_\alpha$$

$$\Pi(T^*\underset{\sim}{\omega}^\Delta) = \sum_\alpha T^{\Delta}_{*\alpha}\phi^\alpha = \sum_\alpha T^{\Delta\alpha}_*\phi_\alpha$$

Here $P: \underset{\sim}{\mathcal{V}} \to \underset{\sim}{\mathcal{N}}_H$ and $\Pi: \mathcal{U} \to \mathcal{M}_G$ are projection operators defined by the bases $\phi_\alpha$ and $\underset{\sim}{\omega}_\Delta$ ; i.e.

$$\Pi u = \bar{u} = \sum_{\alpha}^{G} \{u, \phi_\alpha\}\phi^\alpha = \sum_{\alpha}^{G} \{u, \phi^\alpha\}\phi_\alpha \tag{2.16}$$
$$P\underset{\sim}{v} = \bar{\underset{\sim}{v}} = \sum_{\Delta}^{H} [\underset{\sim}{v}, \underset{\sim}{\omega}^\Delta]\underset{\sim}{\omega}_\Delta = \sum_{\Delta}^{H} [\underset{\sim}{v}, \omega_\Delta]\underset{\sim}{\omega}^\Delta$$

and

$$T_{\Delta\alpha} = [\underset{\sim}{\omega}_\Delta, T\phi_\alpha] \; ; \quad T^{\Delta}_{\cdot\alpha} = [\underset{\sim}{\omega}^\Delta, T\phi_\alpha] \tag{2.17}$$
$$T^{\cdot\alpha}_{\Delta} = [\underset{\sim}{\omega}_\Delta, T\phi^\alpha] \; ; \quad T^{\Delta\alpha} = [\underset{\sim}{\omega}_\Delta, T\phi^\alpha]$$

and

$$\overset{*}{T}_{\Delta\alpha} = \{T^*\underset{\sim}{\omega}_\Delta, \phi_\alpha\} \; ; \quad T^{*\alpha}_{\Delta} = \{T^*\underset{\sim}{\omega}_\Delta, \phi^\alpha\} \tag{2.18}$$
$$T^{\Delta}_{*\alpha} = \{T^*\underset{\sim}{\omega}^\Delta, \phi_\alpha\} \; ; \quad T^{\Delta\alpha}_{*} = \{T^*\underset{\sim}{\omega}^\Delta, \phi^\alpha\}$$

If we denote

$$B^{\Delta\alpha} = \{B^*\underset{\sim}{\omega}^\Delta, \phi^\alpha\}_{\partial R}, \quad B^{\cdot\alpha}_{\Delta} = \{B^*\underset{\sim}{\omega}_\Delta, \phi^\alpha\}_{\partial R} \tag{2.19}$$
$$B^{\Delta}_{\cdot\alpha} = \{B^*\underset{\sim}{\omega}^\Delta, \phi_\alpha\}_{\partial R}, \quad B_{\Delta\alpha} = \{B^*\underset{\sim}{\omega}_\Delta \; \phi_\alpha\}_{\partial R}$$

then, in view of (2.12), the matrices in (2.17) are related to those in (2.18) according to

$$T_{\Delta\alpha} = \overset{*}{T}_{\Delta\alpha} + B_{\Delta\alpha} \quad ; \quad T^{\Delta}_{\cdot\alpha} = T^{\Delta}_{*\alpha} + B^{\Delta}_{\cdot\alpha} \tag{2.20}$$
$$T^{\cdot\alpha}_{\Delta} = T^{*\alpha}_{\Delta} + B^{\cdot\alpha}_{\Delta} \quad ; \quad T^{\Delta\alpha} = T^{\Delta\alpha}_{*} + B^{\Delta\alpha}$$

Obviously, if $B^*\underset{\sim}{\omega}^\Delta$ is zero, then the $B^{\Delta\alpha}$ matrices vanish and each matrix $T^*$ of (2.18) is merely the transpose of the corresponding matrix in (2.17). Moreover, if (2.10) holds, we have

$$
(2.21)\qquad
\begin{aligned}
T_{\Delta\alpha} &= \sum_{\beta} T_{\Delta}^{\cdot\,\beta} G_{\beta\alpha} = \sum_{\Gamma} H_{\Delta\Gamma} T^{\Gamma}_{\ \cdot\alpha} = \sum_{\beta}\sum_{\Gamma} H_{\Delta\Gamma} G_{\alpha\beta} T^{\Gamma\beta} \\
T^*_{\Delta\alpha} &= \sum_{\Gamma} H_{\Delta\Gamma} T^{\Gamma}_{*\alpha} = \sum_{\beta} T_{\Delta}^{*\beta} G_{\alpha\beta} = \sum_{\beta}\sum_{\Gamma} H_{\Delta\Gamma} G_{\alpha\beta} T^{\Gamma\beta}_{*}
\end{aligned}
$$

etc. Similar transformations hold for the matrices $\underset{\sim}{B}$ of (2.19).

<u>2.3 Approximation by conjugate projections</u>. We now address the question of determining approximate solutions of mixed boundary-value problems of the type

$$
(2.22)\qquad
\begin{gathered}
T^*Tu + f = 0 \quad \text{in } \mathcal{R} \\
B(g - u) = 0 \quad \text{on } \partial\mathcal{R}_1 \quad ; \quad B^*(Tu) - S = 0 \quad \text{on } \partial\mathcal{R}_2
\end{gathered}
$$

Here $u$, $f \in \mathcal{U}$, $T$ is a linear operator from $\mathcal{U}$ into $\underset{\sim}{\mathcal{V}}$ and $T^*$ is its formal adjoint in the sense of (2.12), $Bg$ and $S$ are prescribed functions on portions $\partial\mathcal{R}_1$ and $\partial\mathcal{R}_2$ of the boundary $\partial\mathcal{R} = \partial\mathcal{R}_1 \oplus \partial\mathcal{R}_2$, and $B$ and $B^*$ satisfy (2.13). As noted earlier, the problem (2.22) can be split into the so-called canonical form by setting

$$
(2.23)\qquad Tu = \underset{\sim}{v} \quad \text{in } \mathcal{R} \quad ; \quad B(g - u) = 0 \quad \text{on } \partial\mathcal{R}_1
$$

$$
(2.24)\qquad T^*\underset{\sim}{v} = -f \quad \text{in } \mathcal{R} \; ; \quad B^*\underset{\sim}{v} - S = 0 \text{ on } \partial\mathcal{R}_2
$$

Our method of approximation consists of seeking

weak solutions of (2.23) or (2.24) and (2.25) in subspaces $\mathcal{M}_G$, $\mathcal{M}_G \times \underset{\sim}{\mathcal{N}}_H$, and $\underset{\sim}{\mathcal{N}}_H$ by use of conjugate projections. A number of possibilities suggest themselves:

1. The primal projection. This amounts to the conventional Ritz-Galerkin approximation of (2.22). We seek $\Pi u$ instead of $u$ and require that the projection of the residual into $\mathcal{M}_G$ vanish; i.e.

$$(2.25)\qquad \begin{aligned} &\Pi(T^*T(\Pi u) + f) = 0 \quad \text{in} \quad \mathcal{R} \\ &\Pi(B^*(T(\Pi u)) - S) = 0 \quad \text{on} \quad \partial\mathcal{R}_2 \end{aligned}$$

We may also require that

$$(2.26)\qquad (B(g - \Pi u)) = 0 \quad \text{on} \quad \partial\mathcal{R}_1$$

However, this requirement can be intrinsically satisfied if the basis functions $\phi_\alpha$ are suitably chosen.

Since the solution $u^*$ is unknown, we can only assert that its projection into $\mathcal{M}_G$ will be of the form $\Pi u^* = \sum_\alpha a^\alpha \phi_\alpha$ (see (2.16)). Substituting this into (2.25), we then select the coefficients $a^\alpha$ so as to reduce the residual to zero; i.e.

$$\sum_\beta \{T^*T\Big(\sum_\alpha a^\alpha \phi_\alpha\Big), \phi_\beta\}\phi^\beta + \sum_\beta \{f, \phi_\beta\}\phi^\beta = 0$$

$$\sum_\beta \{B^*T\Big(\sum_\alpha a^\alpha \phi_\alpha\Big) - S, \phi_\beta\}_{\partial\mathcal{R}_2} \phi^\beta$$

Since the $\phi^\beta$ are linearly independent, these relations lead, with the aid of (2.12), to the system of algebraic equations

$$\sum_{\beta} K_{\alpha\beta} a^{\beta} + f_{\alpha} = 0 \tag{2.27}$$

where

$$\begin{aligned} K_{\alpha\beta} &= [T\phi_{\alpha}, T\phi_{\beta}] - \{B^{*}T\phi_{\alpha}, \phi_{\beta}\}_{\partial R_1} \\ f_{\alpha} &= \{f, \phi_{\alpha}\} - \{S, \phi_{\alpha}\}_{\partial R_2} \end{aligned} \tag{2.28}$$

Solving (2.27) for $a^{\alpha}$ determines, of course the approximate solution $\sum_{\alpha} a^{\alpha}\phi_{\alpha}$ .

2. <u>The primal-dual projection</u>. We now turn to the canonical forms (2.23) and (2.24) and assume that $T^{*}\underset{\sim}{v} \in Q(\Pi)$ . We seek, instead of the solution of $\underset{\sim}{v}^{*}$ of (2.24) and element $\overline{\underset{\sim}{v}} = \sum_{\Delta} b_{\Delta}\underset{\sim}{\omega}^{\Delta}$ in $\underset{\sim}{N}_H$ such that

(2.29)

$$\Pi(T^{*}P(\underset{\sim}{v}) + f) = 0 \text{ in } R \quad ; \quad \Pi(B^{*}P(\underset{\sim}{v}) - S) = 0 \text{ on } \partial R_2$$

where P is the projection operator described in (2.16). Observing that

$$\{T^{*}P(\underset{\sim}{v}), \phi_{\alpha}\} = [\sum_{\Delta} b_{\Delta}\underset{\sim}{\omega}^{\Delta}, T\phi_{\alpha}] - \{S, \phi_{\alpha}\}_{\partial R_2} - \{B^{*}P(\underset{\sim}{v}), \phi_{\alpha}\}_{\partial R_1}$$

we obtain from (2.29) the following system of equations:

$$\sum_{\Delta} (T^{\Delta}_{\cdot\alpha} - B^{\Delta 1}_{\cdot\alpha}) b_{\Delta} + f_{\alpha} = 0 \tag{2.30}$$

Here $T^{\cdot\Delta}_{\alpha}$ is the array defined in (2.17), $B^{\Delta 1}_{\alpha} = \{B^{*}\underset{\sim}{\omega}^{\Delta}, \phi_{\alpha}\}_{\partial R_1}$ , and $f_{\alpha}$ is given by (2.28). Since $M_G$ and

$\underset{\sim}{N}_H$ are, in general, of different dimensions, $T^{\Delta}_{\cdot\alpha}$ is rectangular and no unique solution of (2.30) exists.

3. The dual-primal projection. Now we assume that $Tu \in \mathcal{Q}(P)$ and we require that the projection of the residual of (2.23) vanish in $\underset{\sim}{N}_H$ :

(2.31)

$$P(T\Pi u - P\underset{\sim}{v}) = 0 \quad \text{in} \quad \mathcal{R} \quad ; \quad P(B(g - \Pi u)) = 0 \text{ on } \partial\mathcal{R}_1$$

This leads to the system

$$\sum_{\alpha}^{G} a^{\alpha} T_{\alpha}^{\cdot\Delta} - \sum_{\Gamma}^{H} H^{\Delta\Gamma} b_{\Gamma} = 0 \tag{2.32}$$

where $T_{\alpha}^{\cdot\Delta}$ is defined by (2.17) and $H^{\Delta\Gamma}$ is given by (2.9).

While $H^{\Delta\Gamma}$ is nonsingular, insufficient information is provided by (2.32) to determine both $a^{\alpha}$ and $b_{\Delta}$. However, use of (2.30) together with (2.32) yields enough information to determine both $a^{\alpha}$ and $b_{\Delta}$, and the resulting system is said to describe a mixed discrete model of the problem (2.22). Moreover, (2.32) can be used to construct an alternate (but not equivalent) form of (2.27). Observing that

$$b_{\Gamma} = \sum_{\Delta} H_{\Gamma\Delta} T_{\alpha}^{\cdot\Delta} a^{\alpha} \tag{2.33}$$

where $H_{\Gamma\Delta}$ is the inverse of $H^{\Gamma\Delta}$, we obtain from (2.30)

$$\sum_{\beta} \tilde{K}_{\alpha\beta} a^{\beta} + f_{\alpha} = 0 \tag{2.34}$$

Here if $M^{\Delta}_{\cdot\alpha} = (T^{\Delta}_{\cdot\alpha} - B^{\Delta 1}_{\cdot\alpha})$ , then

$$\tilde{K}_{\alpha\beta} = \sum_{\Delta}\sum_{\Gamma} M^{\Delta}_{\cdot\alpha} H_{\Delta\Gamma} T^{\cdot\Delta}_{\beta} \tag{2.35}$$

A comparison of (2.34) and (2.35) with (2.27) and (2.28) illustrates clearly the difference between the fundamental matrices $K_{\alpha\beta}$ and $\tilde{K}_{\alpha\beta}$ obtained via the direct (primal) and the mixed formulations.

4. <u>The dual projection</u>. In this case, u is regarded as prescribed, $Tu \in \underset{\sim}{V}$ , and $N_H$ is a subspace of the set of elements $\underset{\sim}{v}$ such that $T^{*}\underset{\sim}{v} = 0$ in $R$ and $B^{*}\underset{\sim}{v} = 0$ on $\partial R_2$ . We consider the vanishing of the projection of the residual $Tu - P(\underset{\sim}{v})$ in $\underset{\sim}{N}_H$ :

$$P(Tu - P(\underset{\sim}{v})) = 0 \tag{2.36}$$

Noting that

$$P^2(\underset{\sim}{v}) = P(\underset{\sim}{v}) = \sum_{\Delta} b_{\Delta}\underset{\sim}{\omega}^{\Delta}$$

and

$$[Tu, \underset{\sim}{\omega}^{\Delta}] = \{B^{*}\underset{\sim}{\omega}^{\Delta}, g\}_{\partial R_1}$$

we obtain from (2.36) the system of equations

$$-\sum_{\Gamma} H^{\Delta\Gamma} b_{\Gamma} + g^{\Delta} = 0 \tag{2.37}$$

wherein

$$g^{\Delta} = \{B^{*}\underset{\sim}{\omega}^{\Delta}, g\}_{\partial R_1} \tag{2.38}$$

5. Discrete Mappings. A number of other models can be obtained by using the idea of projections. For example, consider the model obtained by successive projections of each image of the decomposed operator $T^*T$ into an appropriately defined subspace. By this we mean the following:

$$\Pi\left(T^*\left(PT(\Pi(u))\right)\right) + \Pi f = 0 \tag{2.39}$$

i.e.

$$\sum_\alpha \sum_\Delta \sum_\beta [\{T^*([Ta^\beta \phi_\beta, \underset{\sim}{\omega}^\Delta]\underset{\sim}{\omega}_\Delta, \phi_\alpha\}\phi^\alpha + f_\alpha \phi^\alpha] = 0$$

This yields a discrete model similar in form to (2.22):

$$\sum_\beta \sum_\Delta T^*_{\Delta\alpha} T^{\cdot\Delta}_{\beta} a^\beta + f_\alpha = 0 \tag{2.40}$$

Alternately, use of $(2.20)_1$ leads to the equivalent form

$$\sum_\beta \sum_\Delta (T_{\Delta\alpha} + B_{\Delta\alpha}) T^{\cdot\Delta}_{\beta} a^\beta + f_\alpha = 0 \tag{2.41}$$

Still other forms could be cited.

## 3. Finite element approximations

3.1 The finite element method. We view the finite element method as both a method of interpolation and as a method for the systematic construction of basis functions for Ritz-Galerkin approximations on irregular domains. We shall outline briefly the essential feastures of the method following the ideas in [6].

Consider an open, bounded region $\mathcal{R}$ in n-dimensional euclidean space $E^n$, with closure $\bar{\mathcal{R}} = \mathcal{R} + \partial\mathcal{R}$, and let $\underset{\sim}{x} = (x_1, x_2, \ldots, x_n)$ denote a point in $\bar{\mathcal{R}}$. A finite element model $\tilde{\mathcal{R}}$ of $\bar{\mathcal{R}}$ is the union of a finite number E of closed bounded subregions $\bar{\mathcal{R}}_e$ of $E^n$, $\bar{\mathcal{R}}_e$ being the closure of an open region $\mathcal{R}_e \subset E^n$ ($\bar{\mathcal{R}}_e = \mathcal{R}_e + \partial \mathcal{R}_e$ ; $e = 1,2,\ldots,E$) . The subregions $\bar{\mathcal{R}}_e$ are called finite elements, $\tilde{\mathcal{R}}$ is referred to as the connected model of $\bar{\mathcal{R}}$, and the open elements $\mathcal{R}_e$ are disjoint in $E^n$ :

$$\text{(3.1)} \qquad \tilde{\mathcal{R}} = \bigcup_{e=1}^{E} \bar{\mathcal{R}}_e \quad ; \quad \mathcal{R}_e \cap \mathcal{R}_f = \emptyset \ , \ e \neq f$$

In the connected model $\tilde{\mathcal{R}}$, we identify a finite number G of points, called global nodes, and we label these consecutively $\underset{\sim}{x}^1, \underset{\sim}{x}^2, \ldots, \underset{\sim}{x}^G$. Likewise, we identify within each finite element $\bar{\mathcal{R}}_e$ a number $N_e$ of points, called local nodes, and we label them consecutively $\underset{\sim}{x}_e^1, \underset{\sim}{x}_e^2, \ldots, \underset{\sim}{x}_e^{N_e}$ ; $e = 1,2,\ldots,E$. Using the notation developed in [6], we refer to a typical global node as $\underset{\sim}{x}^\alpha$ ($\alpha = 1,2,\ldots,G$) and a typical local node as $\underset{\sim}{x}_e^N$ ($N = 1,2,\ldots,N_e$; $e = 1,2,\ldots,E$). We assume, of course, that the proper correspondence between points in each $\bar{\mathcal{R}}_e$ and points in $\tilde{\mathcal{R}}$ exists, as well as a proper correspondence between nodal points $\underset{\sim}{x}_e^N$ in $\bar{\mathcal{R}}_e$ and nodal points $\underset{\sim}{x}^\alpha$ in $\tilde{\mathcal{R}}$, so that the elements fit smoothly together to form $\tilde{\mathcal{R}}$ (see the compatibility requirements, [6, p. 36]). We also remark that in some instances it is convenient to use different local coordinate systems for each finite element and a different coordinate system for the connected model (see, for example, [6, p.44]); however, for simplicity we assume here that all local and global frames are coincident.

Conceptually, either a given $\tilde{\mathcal{R}}$ is decomposed into distinct finite elements $\overline{R}_e$ , $e = 1,2,\ldots,E$ , or, as is usually done in computations, distinct elements are connected appropriately together to form $\tilde{R}$ . Mathematically, the <u>connectivity</u> and <u>decomposition</u> of a finite element model are established by the Boolean mappings

$$\text{(3.2)}\quad \underset{\sim}{x}^{\alpha} = \sum_{N=1}^{N_e} \overset{(e)}{\Lambda}{}^{\alpha}_{N}\, \underset{\sim}{x}{}^{N}_{e} \quad \text{(e fixed)}; \quad \underset{\sim}{x}{}^{N}_{e} = \sum_{\alpha=1}^{G} \overset{(e)}{\Omega}{}^{N}_{\alpha}\, \underset{\sim}{x}^{\alpha}$$

Here $\alpha = 1,2,\ldots,G$ , $N = 1,2,\ldots,N_e$ , $e = 1,2,\ldots,E$ ,

$$\text{(3.3)}\quad \overset{(e)}{\Omega}{}^{N}_{\alpha} = \begin{cases} 1 & \text{if node } \underset{\sim}{x}^{\alpha} \text{ of } \tilde{R} \text{ is coincident with node } \underset{\sim}{x}{}^{N}_{e} \text{ of } \overline{R}_e \\ 0 & \text{if otherwise,} \end{cases}$$

$\overset{(e)}{\Lambda}{}^{\alpha}_{N}$ is the transpose of $\overset{(e)}{\Omega}{}^{N}_{\alpha}$ , and the arrays $\Lambda$ and $\Omega$ have the property

$$\text{(3.4)}\quad \begin{aligned} &\sum_{\alpha=1}^{G} \overset{(e)}{\Omega}{}^{N}_{\alpha} \overset{(e)}{\Lambda}{}^{\alpha}_{M} = \delta^{N}_{M} \quad \text{(e fixed)}; \\ &\sum_{N=1}^{N_e} \overset{(e)}{\Lambda}{}^{\alpha}_{N} \overset{(e)}{\Omega}{}^{N}_{\beta} = \begin{cases} \delta^{\alpha}_{\beta} & \underset{\sim}{x}^{\alpha}, \underset{\sim}{x}^{\beta} \in \overline{R}_e \\ 0 & \underset{\sim}{x}^{\alpha}, \underset{\sim}{x}^{\beta} \notin \overline{R}_e \end{cases} \end{aligned}$$

The symbolism (3.2) effectively represents a renumbering of local and global node labels.

Now consider a function $U(\underset{\sim}{x}) \in C^m(\tilde{R})$ . The restriction $u_e(\underset{\sim}{x})$ to $\overline{R}_e \subset \tilde{R}$ is the function

$$\text{(3.5)}\quad u_e(\underset{\sim}{x}) = \begin{cases} U(\underset{\sim}{x}) & \text{if } \underset{\sim}{x} \in \overline{R}_e \\ 0 & \text{if } \underset{\sim}{x} \notin \overline{R}_e \end{cases}$$

In the finite-element method, a local representation of $u_e(x)$ of order $q$ is defined as the function

$$(3.6) \qquad u_e(\underset{\sim}{x}) = \sum_{N=1} \sum_{|\underset{\sim}{\alpha}| \leq q} a_{\underset{\sim}{\alpha}}^{N(e)} \psi_{\tilde{N}}^{\underset{\sim}{\alpha}(e)}(\underset{\sim}{x})$$

wherein $a_{\underset{\sim}{\alpha}}^{N(e)}$ are constants, $\underset{\sim}{\alpha}$ is a multi-integer (i.e., $\underset{\sim}{\alpha} = (\alpha_1, \alpha_2, \ldots \alpha_n) \in Z_+^n$, where $Z_+^n$ is the set of n-tuples of nonnegative integers), $|\underset{\sim}{\alpha}| = \alpha_1 + \alpha_2 + \ldots + \alpha_n$; $\psi_{\tilde{N}}^{\underset{\sim}{\alpha}(e)}(\underset{\sim}{x})$ are the <u>local interpolation functions</u> of order $q$, and are defined so as to have the properties

$$\psi_{\tilde{N}}^{\underset{\sim}{\alpha}(e)}(\underset{\sim}{x}) \equiv 0 \text{ if } \underset{\sim}{x} \notin R_e$$

$$(3.7) \qquad D_{\underset{\sim}{\beta}} \psi_{\tilde{N}}^{\underset{\sim}{\alpha}(e)}(\underset{\sim}{x}_e^M) = \delta_N^M \delta_{\beta_1}^{\alpha_1} \delta_{\beta_2}^{\alpha_2} \cdots \delta_{\beta_n}^{\alpha_n}$$

$$\underset{\sim}{\alpha}, \underset{\sim}{\beta} \in Z_+^n \ ; \ N, M = 1, 2, \ldots, N_e$$

Here $\delta_N^M$, $\delta_{\beta_1}^{\alpha_1}, \ldots, \delta_{\beta_n}^{\alpha_n}$ are Kronecker deltas and we have again used the multi-integer notation, $D_{\underset{\sim}{\beta}} u \equiv \partial^{|\underset{\sim}{\beta}|} u / \partial x_1^{\beta_1} \partial x_2^{\beta_2} \ldots \partial x_n^{\beta_n}$. Therefore, in view of (3.7),

$$(3.8) \qquad D_{\underset{\sim}{\beta}} \overline{u}_e(\underset{\sim}{x}_e^N) = a_{\underset{\sim}{\beta}}^{N(e)} \ ; \quad |\underset{\sim}{\beta}| \leq q$$

The <u>global finite-element representation</u> of $U(\underset{\sim}{x})$ is the function

$$(3.9) \qquad \overline{U}(\underset{\sim}{x}) = \sum_{\lambda=1}^{G} \sum_{|\underset{\sim}{\alpha}| \leq q} A_{\underset{\sim}{\alpha}}^{\lambda} \phi_{\lambda}^{\underset{\sim}{\alpha}}(\underset{\sim}{x})$$

where

(3.10)
$$A^{\lambda}_{\underset{\sim}{\alpha}} = \sum_{N=1}^{N_e} \overset{(e)}{\Lambda}{}^{\lambda}_{N} a^{N(e)}_{\underset{\sim}{\alpha}} \quad (e \text{ fixed}) \; ;$$

$$a^{N(e)}_{\underset{\sim}{\alpha}} = \sum_{\lambda=1}^{G} \overset{(e)}{\Omega}{}^{N}_{\lambda} A^{\lambda}_{\underset{\sim}{\alpha}} \; ,$$

and $\phi^{\underset{\sim}{\alpha}}_{\lambda}(x)$ are the global interpolation functions

(3.11)
$$\phi^{\underset{\sim}{\alpha}}_{\lambda}(\underset{\sim}{x}) = \bigcup_{e=1}^{E} \left\{ \sum_{N=1}^{N_e} \overset{(e)}{\Omega}{}^{N}_{\lambda} \psi^{\underset{\sim}{\alpha}(e)}_{N}(\underset{\sim}{x}) \right\}$$

It is easily verified that

(3.12)
$$D_{\underset{\sim}{\beta}} \phi^{\underset{\sim}{\alpha}}_{\lambda}(\underset{\sim}{x}^{\mu}) = \delta^{\mu}_{\lambda} \delta^{\alpha_1}_{\beta_1} \cdots \delta^{\alpha_n}_{\beta_n}$$

so that

(3.13)
$$D_{\underset{\sim}{\beta}} \overline{U}(\underset{\sim}{x}^{\mu}) = A^{\mu}_{\underset{\sim}{\beta}} \; ,$$

$|\underset{\sim}{\beta}| \leqslant q \; ; \quad \lambda, \mu = 1, 2, \ldots, G, \quad \underset{\sim}{\alpha}, \underset{\sim}{\beta} \in Z^n_+$

Hence, if we set

(3.14)
$$a^{N(e)}_{\underset{\sim}{\beta}} = D_{\underset{\sim}{\beta}} u_e(x^N_e) \quad \text{and} \quad A^{\mu}_{\underset{\sim}{\beta}} = D_{\underset{\sim}{\beta}} U(\underset{\sim}{x}^{\mu})$$

## 3.2 Finite Element Approximations by Conjugate Projections.

The beauty of the finite element method is that the essential features of an approximation can be described locally for typical disconnected elements. Corresponding global approximations are then obtained routinely by means of the transformations (3.2) and (3.10).

1. Primal finite-element model. The direct finite-element approximation of (2.22) leads to a primal model of the form (2.30). Consider $\mathcal{M}_G$ to be a subspace of $\mathcal{U}$ spanned by the functions $\phi_\lambda^\alpha(x)$ of (3.11). Then we seek coefficients $A^\lambda$ among projections of the form

$$(3.15)\quad U(\underset{\sim}{x}) = \sum_\lambda A^\lambda \phi_\lambda(\underset{\sim}{x}) = \bigcup_{e=1}^{E} \sum_\lambda \sum_N {}^{(e)}_{\;\Omega}{}^{N}_{\lambda} A^\lambda \psi_N^{(e)}(\underset{\sim}{x})$$

which satisfy

$$(3.16)\qquad \sum_\beta K_{\alpha\beta} A^\beta + f_\alpha = 0$$

Here, and in the following discussion, we have omitted the multi-indices for simplicity; $K_{\alpha\beta}$ and $f_\alpha$ are defined by (2.28) and, due to properties (3.12) of $\phi_\lambda(\underset{\sim}{x})$, the coefficients $A^\lambda_{\underset{\sim}{\alpha}}$ now can be interpreted as actual values of $D_{\underset{\sim}{\alpha}} U(\underset{\sim}{x}^\beta)$ at nodal points $\underset{\sim}{x}^\beta$.

Due to the particular choice of basis functions, we observe that

$$K_{\alpha\beta} = [T\phi_\alpha, T\phi_\beta] - \{B^* T\phi_\alpha, \phi_\beta\}_{\partial \mathcal{R}_1} =$$

$$\bigcup_{e=1}^{E} \bigcup_{f=1}^{E} \left\{ \sum_N^{N_e} \sum_M^{N_e} {}^{(e)}_{\;\Omega}{}^{N}_{\alpha}\, {}^{(f)}_{\;\Omega}{}^{M}_{\beta} [T\psi_N^{(e)}, T\psi_M^{(f)}] - \right.$$

$$\left. - \{B^* T\psi_N^{(e)}, \psi_M^{(f)}\}_{\partial \mathcal{R}_1} \right\}$$

and

$$f_\alpha = \{f, \phi_\alpha\} = \bigcup_{e=1}^{E} \sum_N \overset{(e)}{\Omega}{}^N_\alpha \{f, \psi^{(e)}_N\}$$

However, $[T\psi^{(e)}_N, T\psi^{(f)}_M] = 0$ and $\{B^* T\psi^{(e)}_N, \psi^{(f)}_M\}_{\partial R_1} = 0$ for $e \neq f$. Therefore

$$K_{\alpha\beta} = \sum_e \sum_N \sum_M \overset{(e)}{\Omega}{}^N_\alpha k^{(e)}_{NM} \overset{(e)}{\Omega}{}^M_\beta$$

(3.17)

$$f_\alpha = \sum_e \sum_N \overset{(e)}{\Omega}{}^N_\alpha f^{(e)}_N$$

where $k^{(e)}$ and $f^{(e)}_N$ are the localizations of $K_{\alpha\beta}$ and $f^{(e)}_N$:

$$k^{(e)}_{NM} = [T\psi^{(e)}_N, T\psi^{(e)}_M] - \{B^* T\psi^{(e)}_N, \psi^{(e)}_M\}_{\partial R_1}$$

(3.18)

$$f^{(e)}_N = \{f, \psi^{(e)}_N\} - \{S, \psi^{(e)}_N\}_{\partial R_2}$$

In view of the fact that $\sum_\lambda \overset{(e)}{\Omega}{}^N_\lambda A^\lambda = a^N_{(e)}$, we can see that (3.16) can be written

(3.19) $$\sum_e \sum_N \overset{(e)}{\Omega}{}^N_\alpha \left( \sum_M k^{(e)}_{NM} a^M_{(e)} + f^{(e)}_N \right) = 0$$

which implies that locally

(3.20) $$\sum_M k^{(e)}_{NM} a^M_{(e)} + f^{(e)}_N = 0$$

for each $e = 1,2,\ldots,E$ .

Equation (3.20) is the local finite-element model of (2.1). Again, (3.20) can be generated locally and the global model (3.16) can be obtained directly by using the transformations (3.17).

2.,3. Primal-dual and dual-primal (mixed) finite-element models. Since both of the approximations (2.30) and (2.32), must, in general, be used jointly to determine a unique projection into $\mathcal{M}_G$ and $\mathcal{N}_H$, we must, in general, develop local finite element approximations of these equations jointly. Then considering the restriction $\mathbf{v}_e(\mathbf{x})$ of a model $\mathbf{V}(\mathbf{x})$ of $\mathbf{v} \in \mathcal{V}$ to a typical finite element $e$ to be of the form

$$\mathbf{v}_{(e)}(\mathbf{x}) = \sum_{i=1}^{R} b_i^{(e)} \boldsymbol{\psi}_{(e)}^{i}(\mathbf{x}) \tag{3.21}$$

where $\boldsymbol{\psi}_{(e)}^{i}(\mathbf{x})$ are linearly independent functions with local support (i.e., $\boldsymbol{\psi}_{(e)}^{i}(\mathbf{x}) = 0$ if $\mathbf{x} \notin \overline{\mathcal{R}}_e$) and $R < \infty$. Following the same procedure used earlier, now on a local level, yields as the local counterparts of (2.30) and (2.32),

$$\begin{aligned} &\sum_{i=1}^{R} t_N^{(e)i} b_i^{(e)} + f_N^{(e)} = 0 ; \\ &\sum_{N}^{N_e} t_N^{(e)i} a_{(e)}^{N} - \sum_{j}^{R} F_{(e)}^{ij} b_j = 0 \end{aligned} \tag{3.22}$$

Here $a_{(e)}^{N}$ are the coefficients of the local approximation

$u_e(\underset{\sim}{x}) = \sum_N a^N_{(e)} \psi^{(e)}_N(\underset{\sim}{x})$ , $f^{(e)}_N$ are defined in (3.18), for convenience, $\partial R^{(e)}_1 = 0$ , and

$$(3.23)\quad t^{(e)i}_N = [T\psi^{(e)}_N, \underset{\sim}{\psi}^i_{(e)}] \quad ; \quad F^{ij}_{(e)} = [\underset{\sim}{\psi}^i_{(e)}, \underset{\sim}{\psi}^j_{(e)}]$$

Solving (3.22) for $b^{(e)}_j$ and reorganizing terms, we arrive at the basic equation analogous to (2.34)

$$(3.24)\qquad \sum_M \tilde{k}^{(e)}_{NM} a^M_{(e)} + f^{(e)}_N = 0$$

where

$$(3.25)\qquad \tilde{k}^{(e)}_{NM} = \sum_{i,j}^{R} t^{(e)i}_N \overset{(e)}{F}{}^{-1}_{ij} t^{(e)j}_M$$

Globally, we then obtain (2.35) with

$$(3.26)\qquad \tilde{K}_{\alpha\beta} = \sum_e \sum_{N,M} \overset{(e)}{\Omega}{}^N_\alpha \tilde{k}^{(e)}_{NM} \Omega^M_\beta$$

$\alpha,\beta = 1,2,\ldots,G$ and $f_\alpha$ given by (3.]7) .

A comparison of (3.26) with (2.34) leads to some interesting transformations. Suppose that there exist, associated with each finite element, $R_e \times H$ matrices $\underset{\sim}{R}$ with the property that

$$(3.27)\qquad T^{\cdot\Delta}_\alpha = \sum_e^E \sum_N^{N_e} \sum_i^{R_e} \overset{(e)}{\Omega}{}^N_\alpha t^{(e)i}_N \overset{(e)}{R}{}^{\cdot\Delta}_i$$

$\alpha = 1,2,\ldots,G$ ; $\Delta = 1,2,\ldots,H$ , where $T^{\cdot\Delta}_\alpha$ is defined in

(2.17). Since $\phi_\alpha(\underset{\sim}{x})$ is given by (3.11), we obtain the following relations:

$$\underset{\sim}{\omega}^{\Delta}(\underset{\sim}{x}) = \sum_{e}^{E}\sum_{i}^{R_e} \overset{(e)}{R}_{i}^{\cdot\Delta}\, \psi_{(e)}^{i}(\underset{\sim}{x}) \; ;$$

(3.28)

$$b_{i}^{(e)} = \sum_{\Delta}^{H} b_{\Delta}\, \overset{(e)}{R}_{i}^{\cdot\Delta}$$

(3.28) $$H^{\Delta\Gamma} = \sum_{e}^{E}\sum_{i,j}^{R_e} \overset{(e)}{R}_{i}^{\cdot\Delta} F_{(e)}^{ij}\, \overset{(e)}{R}_{j}^{\cdot\Gamma}$$

4. Dual finite-element model. Locally, we have

(3.29) $$\sum_{j}^{R_e} F_{(e)}^{ij} b_{j}^{(e)} + g_{(e)}^{i} = 0$$

where $g_{(e)}^{i} = \{B^{*}\underset{\sim}{\psi}^{i}, g\}_{\partial R_e}$ (g is regarded as prescribed on all of $\partial R_e$) and $F_{(e)}^{ij}$ is defined in (3.23). To provide a means for establishing the connectivity of the model, we introduce a local transformation of $b_{i}^{(e)}$ to nodal quantities $q_{N}^{(e)}$ according to

(3.30) $$q_{N}^{(e)} = \{B^{*}\underset{\sim}{v}_{(e)}, \psi_{N}^{(e)}\}_{\partial R_e} = \sum_{i}^{R_e} s_{(e)N}^{i}\, b_{i}^{(e)}$$

wherein

(3.31) $$s_{(e)N}^{i} = \{B^{*}\underset{\sim}{\psi}^{i}, \psi_N\}_{\partial R_e}$$

Introducing nodal measures $w^N_{(e)}$ such that $\sum_N w^N q_N = \sum_i b_i g^i$, we find that

$$g^i_{(e)N} = \sum_N s^i_{(e)N} w^N_{(e)} \tag{3.32}$$

Thus we have locally

$$\sum_M \hat{k}^{(e)}_{NM} w^M_{(e)} = q^{(e)}_N \tag{3.33}$$

and globally

$$\sum_\beta \hat{K}_{\alpha\beta} W^\beta = Q_\alpha \tag{3.34}$$

where

$$\begin{aligned}
\hat{k}^{(e)}_{NM} &= \sum_{i,j} s^i_{(e)N} F^{-1}_{ij} s^j_{(e)M} \\
\hat{K}_{\alpha\beta} &= \sum_e \sum_{N,M} {}^{(e)}_{\Omega}{}^{N}_{\alpha} \hat{k}^{(e)}_{NM} \Omega \\
Q_\alpha &= \sum_e \sum_N {}^{(e)}_{\Omega}{}^{N}_{\alpha} q^{(e)}_N \\
w^N_{(e)} &= \sum_\alpha {}^{(e)}_{\Omega}{}^{N}_{\alpha} W^\alpha
\end{aligned} \tag{3.35}$$

## 4. Complementary Variational Principles and the Hypercircle Method.

**4.1 Complementary variational principles.** The theory of $T^*T$ operators, initiated by Von Neumann [14] in 1932 and extended by Kato [15], Fujita [16], and others, has been revived in recent times in the study of complementary variational principles by Noble [8], Arthurs [9], and Robinson [10]. We shall summarize the essential features of this theory and show its relation to the notions of projections developed previously.

Following Noble and others (e.g. [8-10]), we begin by considering the bilinear functional

$$
\begin{aligned}
J(u,\underset{\sim}{v}) &= [\underset{\sim}{v},Tu] - \tfrac{1}{2}[\underset{\sim}{v},\underset{\sim}{v}] + \{f,u\} + \\
&\quad + \{B^*\underset{\sim}{v}, g-u\}_{\partial R_1} - \{S,u\}_{\partial R_2} \\
&= \{T^*\underset{\sim}{v},u\} - \tfrac{1}{2}[\underset{\sim}{v},\underset{\sim}{v}] + \{f,u\} + \\
&\quad + \{B^*\underset{\sim}{v} - S,u\}_{\partial R_2} + \{B^*\underset{\sim}{v},g\}_{\partial R_1}
\end{aligned}
\tag{4.1}
$$

If it is assumed that the conditions (2.23) are identically satisfied (i.e., $Tu \equiv \underset{\sim}{v}$ in $R$ and $B(g - u) = 0$ on $\partial R_1$), then we can obtain from $J(u,\underset{\sim}{v})$ the new functional

$$
\begin{aligned}
I(u) &= \tfrac{1}{2}[Tu,Tu] + \{f,u\} - \{S,u\}_{\partial R_2} + \\
&\quad + \{B^*Tu, g-u\}_{\partial R_1}
\end{aligned}
\tag{4.2}
$$

We note that the last term in this functional vanishes in the case $B = 1$ ; i.e., when $u$ takes on the prescribed variation $g$ on $\partial R_1$ .

Likewise, if it is assumed that the conditions (2.24) are identically met (i.e, if $T^*\tilde{v} + f \equiv 0$ in $R$ and $B^*\tilde{v} - S = 0$ on $\partial R_2$ ) , then we obtain from $J(u,\tilde{v})$ the new functional

$$(4.3) \qquad K(\tilde{v}) = -\frac{1}{2}[\tilde{v},\tilde{v}] + \{B^*\tilde{v},u\}_{\partial R_1}$$

Here $Bu$ is regarded as being prescribed on $\partial R_1$ .

That the functionals $I(u)$ , $J(u,\tilde{v})$ and $K(\tilde{v})$ provide a basis for complementary variational principles associated with (2.22) is made clear by the theorems we cite below. While our boundary conditions are slightly more general than those we have found elsewhere, these results are equivalent to those found in [13-15].

<u>Theorem 4.1</u>. The functional $J(u,\tilde{v})$ of (4.1) assumes a stationary value at the point $(u^*,\tilde{v}^*)$ if and only if $u^*$ is a solution of (2.23) and $\tilde{v}^*$ is a solution of (2.24)

<u>Theorem 4.2</u>. Among all the elements $u$ which satisfy the boundary conditions $B(g - u) = 0$ on $\partial R_1$ and for which there is some $\tilde{v}$ such that $\tilde{v} = Tu$ , the unique element $u^*$ such that $T(Tu^*) + f = 0$ in $R$ and $B^*(Tu^*) - S = 0$ on $\partial R_2$, makes the functional $I(u)$ of (4.2) assume its minimum value.

<u>Theorem 4.3</u>. Among all the elements $\tilde{v}$ which satisfy the boundary conditions $B^*\tilde{v} - S = 0$ on $\partial R_2$ and for which $T^*\tilde{v} + f = 0$ in $R$ , the unique element $\tilde{v}^*$

such that $Tu^* = \underset{\sim}{v}^*$ and $B(g - u^*) = 0$ on $\partial R_1$, $u^*$ being the solution of (2.22), makes the functional $K(\underset{\sim}{v})$ of (4.3) assume its maximum value.

Partial Proofs and Comments: While the proofs of these theorems are straightforward, some of the results obtained along the way are sufficiently valuable to warrant our furnishing a few details. Concerning 4.1, we fine after a little algebra that if $h$ and $\underset{\sim}{w}$ are arbitrary elements in $U$ and $\underset{\sim}{V}$, and $u = u^* + h$, $\underset{\sim}{v} = \underset{\sim}{v}^* + \underset{\sim}{w}$, then

$$J(u,\underset{\sim}{v}) - J(u^*,\underset{\sim}{v}^*) = \{T^*\underset{\sim}{v}^* + f,h\} + \{B^*\underset{\sim}{v}^* - S,h\}_{\partial R_2} + $$
$$+ [Tu^* - \underset{\sim}{v}^*,\underset{\sim}{w}] + \{B^*\underset{\sim}{w},g-u^*\}_{\partial R_1} -$$
$$- \frac{1}{2}[\underset{\sim}{w},\underset{\sim}{w}] + [Th,\underset{\sim}{w}] - \{B^*\underset{\sim}{w},h\}_{\partial R_1}$$

Clearly, if $u^*$ and $v^*$ satisfy (2.23) and (2.24), the first four terms on the right-hand side vanish and we conclude that $(u^*,\underset{\sim}{v}^*)$ is a critical point of $J(u,\underset{\sim}{v})$. Likewise, if the pair $(u^*,\underset{\sim}{v}^*)$ is a critical point of $J(u,\underset{\sim}{v})$, then $u^*$ and $\underset{\sim}{v}^*$ satisfy (2.23) and (2.24) weakly. Note that at a critical point,

(4.4)
$$J(u,\underset{\sim}{v}) - J(u^*,\underset{\sim}{v}^*) = -\frac{1}{2}\|\underset{\sim}{w}\|^2_{\underset{\sim}{V}} + [Th,\underset{\sim}{w}] - \{B^*\underset{\sim}{w},h\}_{\partial R_1}$$

where $\|\underset{\sim}{w}\|^2_{\underset{\sim}{V}} = [\underset{\sim}{w},\underset{\sim}{w}]$ and $\underset{\sim}{w} = \underset{\sim}{v} - \underset{\sim}{v}^*$, $h = u - u^*$.

Follwoing similar procedures in the case of $I(u)$ and $K(\underset{\sim}{v})$ leads to

$$I(u) = I(u^*) + \frac{1}{2}[Th,Th] \tag{4.5}$$

(4.6) $$K(\underset{\sim}{v}) = K(\underset{\sim}{v}^*) - \frac{1}{2}[\underset{\sim}{w},\underset{\sim}{w}]$$

where in (4.5) we have equated to zero terms $\{T^*Tu^* + f,h\} + \{B^*Tu^* - S,h\}_{\partial R_2}$ and in (4.6) we set equal to zero $[Tu - \underset{\sim}{v}, \underset{\sim}{w}] + \{\underset{\sim}{w},B(g - u)\}_{\partial R_1}$. In arriving at (4.6), we have used the identity, $\{B^*\underset{\sim}{w},u\}_{\partial R_1} = [\underset{\sim}{w},Tu]$. In particular, admissible variations in $u$ associated with $I(u)$ are limited to the subspace $\hat{U} \subset U$ such that $\hat{u} \in \hat{U} \rightarrow B\hat{u} = 0$ on $\partial R_1$ and $\hat{u} \in T(U)$; similarly, admissible variations in $\underset{\sim}{v}$ associated with $K(\underset{\sim}{v})$ must be such that $T^*\underset{\approx}{w} = 0$ in $R$ and $B^*\underset{\approx}{w} = 0$ on $\partial R_2$. From this latter observation we obtain the stated identity immediately.

It is clear from (4.5) and (4.6) that $I(u) \geq I(u^*)$ and $K(\underset{\sim}{v}) \leq K(\underset{\sim}{v}^*)$ since $\|Th\|^2_V \geq 0$ and $\|\underset{\sim}{w}\|^2_{\tilde{V}} \geq 0$. Consequently, $I(u)$ assumes a global minimum at $u^*$ and $K(\underset{\sim}{v})$ a global maximum at $\underset{\sim}{v}^*$. Uniqueness of $u^*$ and $\underset{\sim}{v}^*$ can be shown to follow from the fact that $-[\underset{\sim}{v},\underset{\sim}{v}]/2 + \{f,u\}$ is concave in $\underset{\sim}{v}$ and convex in $u$; i.e., the generalized Hamiltonian has saddle-point behavior (see, for example, [10]).

The relationship between conjugate projections and variational formulations is established by the following theorem.

<u>Theorem 4.4</u>. Let $\bar{u} = \Pi u = \sum_\alpha a^\alpha \phi_\alpha$ and $\bar{\underset{\sim}{v}} = P\underset{\sim}{v} = \sum_\Delta b_\Delta \underset{\sim}{\omega}^\Delta$ denote projections of $u$ and $\underset{\sim}{v}$ into the subspaces $M_G$ and $\underset{\sim}{N}_H$ described earlier. Let $I(\bar{u}) = I(a^\alpha)$, $J(\bar{u},\bar{\underset{\sim}{v}}) = J(a^\alpha,b_\Delta)$, and $K(\bar{\underset{\sim}{v}}) = K(b_\Delta)$ denote the

restriction of the functionals $I(u)$, $J(u,\underset{\sim}{v})$, and $K(\underset{\sim}{v})$ to the subspaces $M_G$ and $\underset{\sim}{N}_H$. Then

1. Among all $\bar{u} \in M_G$, that which provides a minimum for $I(\bar{u})$ is obtained by choosing the coefficients $a^\alpha$ which satisfy (2.27).

2. Among all $\bar{u} \in M_G$ and $\bar{\underset{\sim}{v}} \in \underset{\sim}{N}_H$, those which provide a stationary value to $J(\bar{u},\bar{\underset{\sim}{v}})$ are obtained by choosing the coefficients $a^\alpha$ and $b_\Delta$ which satisfy (2.30) and (2.32).

3. Among all $\bar{\underset{\sim}{v}} \in \underset{\sim}{N}_H$, that which provides a maximum for $K(\bar{\underset{\sim}{v}})$ is obtained by choosing the coefficients $b_\Delta$ which satisfy (2.37).

Proof: We simply note that $\partial I(a^\mu)/\partial a^\alpha = \sum_\mu K_{\alpha\mu} a^\mu + f_\alpha$ ; $\partial J(a^\mu,b_\Gamma)/\partial a^\alpha = \sum_\Delta T_{\alpha}^{\cdot\Delta} b_\Delta + f_\alpha$ ; $\partial J(a^\mu,b_\Gamma)/\partial b_\Delta = \sum_\alpha a^\alpha T_\alpha^{\cdot\Delta} - \sum_\Gamma H^{\Delta\Gamma} b_\Gamma$ ; and $\partial K(b_\Gamma)/\partial b_\Delta = -\sum_\Gamma H^{\Delta\Gamma} b_\Gamma + g^\Delta$ .

4.2 The hypercircle method. We shall conclude this section by establishing the relationship between the theory presented thusfar and the method of the hypercircle of Prager and Synge [11,12]. To make clear the essential differences in the two approaches, it is convenient to generalize the problem (2.1) slightly to

$$(4.7) \qquad \begin{gathered} T^* ETu + f = 0 \\ u - g = 0 \text{ on } \partial R_1 \quad ; \quad B^* ETu - S = 0 \text{ on } \partial R_2 \end{gathered}$$

where $E$ is a strictly positive nonsingular operator, self-adjoint in the sense that $[E\underset{\sim}{w}_1,\underset{\sim}{w}_2] = [\underset{\sim}{w}_1,E\underset{\sim}{w}_2]$ for every

$\underset{\sim}{w}_1, \underset{\sim}{w}_2 \in \underset{\sim}{V}$ (indeed, E may merely represent a positive variable coefficient in the governing equation and $\{B^*\underset{\sim}{v},u\}_{\partial R} = [1\cdot u,\underset{\sim}{v}]_{\partial R}$. The corresponding canonical (Euler-Hamilton) forms and functionals are easily shown to be

$$(4.8) \qquad Tu = E^{-1}\underset{\sim}{v} \quad ; \quad u - g = 0 \quad \text{on} \quad \partial R_1$$

$$(4.9) \qquad T^*\underset{\sim}{v} + f = 0 \quad ; \quad B^*\underset{\sim}{v} - S = 0 \quad \text{on} \quad \partial R_2$$

and

$$(4.10) \qquad \begin{aligned} I(u) &= \frac{1}{2}\,[ETu,Tu] + \{f,u\} - \{S,u\}_{\partial R_2} \\ J(u,\underset{\sim}{v}) &= [\underset{\sim}{v},Tu] - \frac{1}{2}\,[\underset{\sim}{v},E^{-1}\underset{\sim}{v}] + \{f,u\} + \\ &\quad + \{B^*\underset{\sim}{v},g - u\}_{\partial R_1} - \{S,u\}_{\partial R_2} \\ K(\underset{\sim}{v}) &= -\frac{1}{2}\,[\underset{\sim}{v},E^{-1}\underset{\sim}{v}] + \{B^*\underset{\sim}{v},u\}_{\partial R_1} \end{aligned}$$

Again, the last term in $I(u)$ of (4.2) now vanishes $B = 1$, $I(u)$ assumes a minimum value at the solution $u^*$ of (4.7), $K(\underset{\sim}{v})$ is a maximum at the solution $\underset{\sim}{v}^*$ of (4.9) and $J(u,\underset{\sim}{v})$ is stationary at $(u^*,\underset{\sim}{v}^*)$ .

The essense of the hypercircle approach is as follows: let $W$ denote the set of all elements u that satisfy the boundary condition $B(g - u) = 0$ on $\partial R_1$ . Each $u \in W$ is of the form $u = u_0 + \hat{u}$ , where $Bu_0 = Bg$ and $B\hat{u} = 0$ on $\partial R_1$ . Let $T(W)$ denote the image space of $W$ under the mapping T and let $X = ET(W) \times T(W)$ denote the set of ordered pairs

$$(4.11) \qquad \underset{\sim}{\Lambda} = (ETu,Tu)$$

Since each $\underset{\sim}{\Lambda} \in \mathcal{X}$ is of the form $\underset{\sim}{\Lambda} = \underset{\sim}{\Lambda}_0 + \hat{\underset{\sim}{\Lambda}}$ , where $\underset{\sim}{\Lambda}_0 = (ETu_0, Tu_0)$ and $\hat{\underset{\sim}{\Lambda}} = (ETu, Tu)$ , $\mathcal{X} = \underset{\sim}{\Lambda}_0 + \hat{\mathcal{X}}$ , where $\hat{\mathcal{X}}$ is a linear space. Similarly, let $\underset{\sim}{\mathcal{W}}$ denote the set of all $\underset{\sim}{v}$ that satisfy $T^*\underset{\sim}{v} = -f$ in $\mathcal{R}$ and $B^*\underset{\sim}{v} = S$ on $\partial\mathcal{R}_1$ . Each $\underset{\sim}{v} \in \underset{\sim}{\mathcal{W}}$ is of the form $\underset{\sim}{v} = \underset{\sim}{v}_0 + \hat{\underset{\sim}{v}}$ , where $T^*\underset{\sim}{v}_0 = -f$ , $T^*\hat{\underset{\sim}{v}} = 0$ in $\mathcal{R}$ ; $B^*\underset{\sim}{v}_0 = S$ , $B^*\hat{\underset{\sim}{v}} = 0$ on $\partial\mathcal{R}_2$. The set $\mathcal{Y} = \underset{\sim}{\mathcal{W}} \times E^{-1}\underset{\sim}{\mathcal{W}}$ consists of ordered pairs

$$\underset{\sim}{\Gamma} = (\underset{\sim}{v}, E^{-1}\underset{\sim}{v}) \tag{4.12}$$

which can be decomposed into $\underset{\sim}{\Gamma}_0 + \hat{\underset{\sim}{\Gamma}}$ , where $\underset{\sim}{\Gamma}_0 = (\underset{\sim}{v}_0, E^{-1}\underset{\sim}{v}_0)$ , $\hat{\underset{\sim}{\Gamma}} = (\hat{\underset{\sim}{v}}, E^{-1}\hat{\underset{\sim}{v}})$ . Again $\mathcal{Y} = \underset{\sim}{\Gamma}_0 + \hat{\mathcal{Y}}$ , where $\hat{\mathcal{Y}}$ is a linear space. If $u^*, \underset{\sim}{v}^*$ are the solutions of (4.8) and (4.9) there is a unique point $\underset{\sim}{X}$ of intersection of $\mathcal{X}$ and $\mathcal{Y}$ :

$$\mathcal{X} \cap \mathcal{Y} = \underset{\sim}{X} = (\underset{\sim}{v}^*, Tu^*) \tag{4.13}$$

Further, if we suppose that $\mathcal{X}$ , $\mathcal{Y} \subset \mathcal{Z}$ ; then an inner product $\langle\langle\cdot,\cdot\rangle\rangle$ can be introduced on $\mathcal{Z}$ according to rule that if $\underset{\sim}{\Lambda}_A$ is the ordered pair $(\underset{\sim}{\alpha}_A, \underset{\sim}{\beta}_A)$ , $\underset{\sim}{\alpha}, \underset{\sim}{\beta} \in \underset{\sim}{\mathcal{V}}$ , then

$$\langle\langle \underset{\sim}{\Lambda}_A, \underset{\sim}{\Lambda}_B \rangle\rangle = [\underset{\sim}{\alpha}_A, \underset{\sim}{\beta}_B] \tag{4.14}$$

Here $[\cdot,\cdot]$ is the innter product on $\underset{\sim}{\mathcal{V}}$ used previously. Symmetry of $\langle\langle \underset{\sim}{\Lambda}_A, \underset{\sim}{\Lambda}_B \rangle\rangle$ is provided through the assumption that E is self adjoint. The associated norm is $\|\underset{\sim}{\Lambda}\|^2 = \langle\langle \underset{\sim}{\Lambda}, \underset{\sim}{\Lambda} \rangle\rangle$ and the squared distances from the solution pair $\underset{\sim}{X}$ and arbitrary elements $\underset{\sim}{\Lambda} \in \mathcal{X}$ and $\underset{\sim}{\Gamma} \in \mathcal{Y}$ are

$$d^2_{x\Lambda} = \|\underset{\sim}{x} - \underset{\sim}{\Lambda}\|^2 \quad ; \quad d^2_{x\Gamma} = \|\underset{\sim}{x} - \underset{\sim}{\Gamma}\|^2 \tag{4.15}$$

The hypercircle method consists of seeking approximations to $\underset{\sim}{X}$ in $\mathcal{X}$ and $\mathcal{Y}$ by minimizing $d^2_{x\Lambda}$ and $d^2_{x\Gamma}$ among finite dimensional subspaces of $\hat{\mathcal{X}}$ and $\hat{\mathcal{Y}}$. Approximations are sought in finite dimensional subspaces of $\hat{\mathcal{X}}$ and $\hat{\mathcal{Y}}$, and the approximation is improved by computing the center of a hypercircle in $\mathcal{Z}$, the diameter of which is defined by approximate solutions in $\hat{\mathcal{X}}$ and $\hat{\mathcal{Y}}$.

The relationship between the above concepts and the theory presented earlier is made clear by the following theorems.

Theorem 4.5. Minimization of $d^2_{x\Lambda}$ of (4.16) is equivalent to minimizing $I(u)$ of (4.10). In fact,

(4.16) $$d^2_{x\Lambda} = 2I(u) + \text{constant}$$

Theorem 4.6. Minimization $d^2_{x\Gamma}$ of (4.15) is equivalent to minimizing $-K(\underset{\sim}{v})$ of (4.10). In fact

(4.17) $$d^2_{x\Gamma} = -2K(\underset{\sim}{v}) + \text{constant}$$

Proofs: Clearly,

$$\begin{aligned} d^2_{x\Lambda} = \|\underset{\sim}{X} - \underset{\sim}{\Lambda}\|^2 &= [\underset{\sim}{v}^* - ETu, Tu^* - Tu] \\ &= [\underset{\sim}{v}^*, Tu^*] + [ETu, Tu] - [ETu, Tu^*] - \\ &\quad - [\underset{\sim}{v}, Tu] \end{aligned}$$

But, owing to the self-adjoint property of $E$, $[ETu,Tu^*] = [Tu,ETu^*] = [Tu,\underset{\sim}{v}^*]$ ; moreover, $[\underset{\sim}{v}^*,Tu^*]$, for fixed $(u^*,\underset{\sim}{v}^*)$, is a constant. Thus, Theorem 4.5 follows from the relation

$$d^2_{x\Gamma} = [ETu,Tu] + 2\{f,u\} - 2\{S,u\}_{\partial \mathcal{R}_2} + \text{const.}$$

$$= 2I(u) + \text{const.}$$

Similarly, since $E^{-1}$ is self adjoint,

$$d^2_{x\Gamma} = \|\underset{\sim}{X} - \underset{\sim}{\Gamma}\|^2 = [\underset{\sim}{v}^*,Tu^*] + [\underset{\sim}{v},E^{-1}\underset{\sim}{v}] - [\underset{\sim}{v},Tu^*] - [\underset{\sim}{v}^*,E^{-1}\underset{\sim}{v}]$$

$$= [\underset{\sim}{v},E^{-1}\underset{\sim}{v}] - 2\{B^*\underset{\sim}{v},g\}_{\partial \mathcal{R}_2} + \text{const.}$$

$$= -2K(\underset{\sim}{v}) + \text{const.}$$

which proves Theorem 4.6.

## 5. Energy Error Estimates and Convergence.

While a detailed study of accuracy and convergence of mixed finite element models is, because of space limitations, postponed for a future paper, we shall outline here preliminary results obtained for certain special cases.

5.1 Primal and dual problems. Error estimates for approximations to the primal and dual problems follow the usual procedure. We denote by $h$ the maximum diameter of a finite element in the mesh $\tilde{\mathcal{R}}$ and by $u^*,\underset{\sim}{v}^*$ the solution pair of (2.23) [or (2.22)] and (2.24). If $\tilde{U} \in \mathcal{M}_G$ and $\underset{\sim}{\tilde{V}} \in \mathcal{N}_H$ are interpolants of $u^*,\underset{\sim}{v}^*$, then we can obtain the interpolation formulas [6]

(5.1) $$|D_{\underset{\sim}{\alpha}}(u^* - \tilde{U})| \leq Kh^r \quad ; \quad |D_{\underset{\sim}{\alpha}}(\underset{\sim}{v}^* - \underset{\sim}{\tilde{V}}| \leq Ch^s$$

as $h \to 0$ where $K$ and $C$ are constants independent of $h$ and $r,s > 0$. For example, if $\tilde{U}$ and $\underset{\sim}{\tilde{V}}$ are complete polynomials in $\underset{\sim}{x}$ of order $p$ and $q$ respectively, then $r = p + 1 - |\underset{\sim}{\alpha}|$ and $s = q + 1 - |\underset{\sim}{\alpha}|$. If $W \in \mathcal{M}_G$ minimizes $I(u)$ among all $U \in \mathcal{M}_G$ and if $\underset{\sim}{W} \in \underset{\sim}{\mathcal{N}}_H$ maximizes $K(\underset{\sim}{v})$ among all $\underset{\sim}{W} \in \underset{\sim}{\mathcal{N}}_H$ (i.e., if $W$ and $\underset{\sim}{W}$ are the finite element approximations of the primal and dual problems) then

$$
\begin{aligned}
|I(W) - I(u^*)| &= \frac{1}{2}[T(W - u^*),T(W - u^*)] \leq |I(\tilde{U}) - I(u^*)| \\
&= \frac{1}{2}[T(\tilde{U} - u^*),T(\tilde{U} - u^*)]
\end{aligned} \tag{5.2}
$$

and

$$
\begin{aligned}
|K(\underset{\sim}{W}) - K(\underset{\sim}{v}^*)| &= \frac{1}{2}\|\underset{\sim}{W} - \underset{\sim}{v}^*\|^2_{\underset{\sim}{V}} \\
&\leq |K(\tilde{\underset{\sim}{V}}) - K(\underset{\sim}{v}^*)| \\
&= \frac{1}{2}\|\underset{\sim}{V} - \underset{\sim}{v}^*\|^2_{\underset{\sim}{V}}
\end{aligned} \tag{5.3}
$$

Thus, if, for example, $T$ is a partial differential operator of order $m$, we have as $h \to 0$

$$
|||W-u^*||| \leq Kh^{p+1-m} \quad ; \quad \|\underset{\sim}{W} - \underset{\sim}{v}^*\| \leq Ch^{q+1} \tag{5.4}
$$

where we have used the notation $\{u,u\} \equiv |||u|||^2$ and $[\underset{\sim}{v},\underset{\sim}{v}] = \|\underset{\sim}{v}\|^2 \equiv \|\underset{\sim}{v}\|^2_{\underset{\sim}{V}}$ ; in (5.4), $|||u||| = \|u\|_{W^2_m(R)}$ .

5.2 Mixed problems. The question of convergence and accuracy of mixed finite element approximations is more difficult, and we shall be content here to establish sufficient conditions for convergence for certain special cases. In particular, we consider the class of problems in which $u$ and the finite element method approximations satisfy homogeneous boundary conditions on $\partial \mathcal{R}_1$ . Thus, if $(U^*, \underset{\sim}{V}^*)$ denote the mixed finite-element approximate solutions of the canonical problem (2.23) and (2.24), then

$$
\begin{aligned}
&\{T^*\underset{\sim}{V}^* + f, U\} = 0 \quad ; \quad [TU^* - \underset{\sim}{V}^*, \underset{\sim}{V}] = 0 \\
&\{B^*\underset{\sim}{V}^* - S, U\}_{\partial \mathcal{R}_2} = 0 \quad ; \quad \{B^*\underset{\sim}{V}, U\}_{\partial \mathcal{R}_1} = 0
\end{aligned}
\tag{5.5}
$$

$\forall U \in \mathcal{M}_G \subset \mathcal{U}$ and $\forall \underset{\sim}{V} \in \underset{\sim}{\mathcal{N}}_H \subset \underset{\sim}{\mathcal{V}}$ , with all $U = 0$ on $\partial \mathcal{R}_1$ . In most instances, $\mathcal{M}_G$ will consist of complete piecewise polynomials of degree $q$ , and $T$ is a partial differential operator of order $m$ ; then we shall also assume that there exist interpolants in $\mathcal{M}_G$ and $\underset{\sim}{\mathcal{N}}_H$ such that at each $\underset{\sim}{x} \in \overline{\mathcal{R}}$ ,

$$
\begin{aligned}
|u^* - \tilde{U}| &\leq K_1 h^{p+1} \\
|T(u^* - \tilde{U})| &\leq K_2 h^{p+1-m} \\
|\underset{\sim}{v}^* - \underset{\sim}{\tilde{V}}| &\leq K_3 h^{q+1}
\end{aligned}
\tag{5.6}
$$

If $e_u$ and $\underset{\sim}{e}_v$ denote the errors in $U^*$ and $\underset{\sim}{V}^*$ ; i.e., if

$$
e_u = u^* - U^* \quad ; \quad \underset{\sim}{e}_v = \underset{\sim}{v}^* - \underset{\sim}{V}^*
\tag{5.7}
$$

then we wish to show that each error vanishes in some normed sense as $h \to 0$ .

Our first result involves $\|\underset{\sim}{e}_v\| \equiv \|\underset{\sim}{e}_v\|_{\underset{\sim}{\mathcal{V}}} = [\underset{\sim}{e}_v, \underset{\sim}{e}_v]$ and rests on the following relations:

Lemma 5.1. Let $J(u,\underset{\sim}{v})$ denote the bilinear functional in (4.1) and let (5.5) hold. Then the following inequalities hold:

$$(5.8) \qquad \text{(i)} \quad J(U,\underset{\sim}{V}^*) - J(U^*,\underset{\sim}{v}^*) = [\underset{\sim}{e}_v, T(u^* - U)] - \frac{1}{2}\|\underset{\sim}{e}_v\|^2$$

$$(5.9) \qquad \text{(ii)} \quad J(U,\underset{\sim}{V}^*) \geq J(U^*,\underset{\sim}{V})$$

$$(5.10) \quad \text{(iii)} \quad J(U^*,\underset{\sim}{v}^*) + [\underset{\sim}{e}_v, T(u^* - U)] \geq J(U,\underset{\sim}{V}^*) \geq \geq J(U^*,\underset{\sim}{V})$$

Partial proof: These relations follow directly from the easily verified identity,

$$(5.11) \qquad \begin{aligned} J(u_1,\underset{\sim}{v}_1) - J(u_2,\underset{\sim}{v}_2) &= \{T^*\underset{\sim}{v}_2 + f, u_1 - u_2\} + \\ &+ [\underset{\sim}{v}_1 - \underset{\sim}{v}_2, Tu_1 - \underset{\sim}{v}_2] - \frac{1}{2}\|\underset{\sim}{v}_1 - \underset{\sim}{v}_2\|^2 + \\ &+ \{B^*\underset{\sim}{v}_2 - S, u_1 - u_2\}_{\partial R_2} + \\ &+ \{B^*(\underset{\sim}{v}_1 - \underset{\sim}{v}_2), g - u_1\}_{\partial R_1} \end{aligned}$$

To obtain (5.9), we use (5.11) and the fact that, generally,

$$[\underset{\sim}{V} - \underset{\sim}{V}^*, TU - \underset{\sim}{V}^*] + [\underset{\sim}{V} - \underset{\sim}{V}^*, T(U^* - U)] =$$

$$[\underset{\sim}{V} - \underset{\sim}{V}^*, TU^* - \underset{\sim}{V}^*] = 0$$

This result and (5.9) hold for the general (nonhomogeneous) boundary conditions on $\partial R_1$ and $\partial R_2$.

Collecting results (5.9) - (5.10) and using the fact that $J(U^*,\underset{\sim}{V}) - J(U^*,\underset{\sim}{v}^*) = [\underset{\sim}{V} - \underset{\sim}{v}^*, T(u^* - u^*)] - \frac{1}{2}\|\underset{\sim}{V} - \underset{\sim}{v}^*\|^2$, leads directly to the convergence theorem.

Theorem 5.1. Let (5.5) hold. Then

$$\|\underset{\sim}{e}_V\|^2 \leq \|\underset{\sim}{v}^* - \underset{\sim}{V}\|^2 + 2\|\underset{\sim}{v}^* - \underset{\sim}{V}\| \, \|T(\underset{\sim}{e}_u)\| + \tag{5.12}$$
$$+ 2\|T(u^* - U)\| \, \|\underset{\sim}{e}_V\|$$

This result establishes the fact that $\|\underset{\sim}{e}_V\| \to 0$ as $h \to 0$ if interpolation estimates of the type (5.6) hold. For example, if we set $\underset{\sim}{V} = \tilde{\underset{\sim}{V}}$ in (5.12) and make use of (5.6), then

$$\|\underset{\sim}{e}_V\| \leq K_1 h^{2(q+1)} + 2K_3 h^{2(q+1)} \|T(e_u)\| + \tag{5.13}$$
$$+ 2K_2 \|\underset{\sim}{e}_V\| h^{p+1-m}$$

It appears that nothing much sharper than (5.12) or (5.13) can be said until some estimate on $\|Te_u\|$ is established. While we shall reserve a more detailed discussion of this question for a later paper, we shall show that $Te_u \to 0$ at least weakly for a wide range of problems.

Returning to (5.5), let $W^*$ denote the usual "displacement type" finite-element approximation of the primal problem subject to the same boundary conditions on $\partial R_1$ as $U^*$, and for convenience, suppose that $U^*$ and $W^*$ belong to the same subspace $M_G$. Then we can derive the remarkably similar pair of orthogonality conditions,

$$[TW^* - V^*, TU] = 0 \quad ; \quad [TU^* - \underset{\sim}{V}^*, V] = 0 \tag{5.14}$$

The first member of the pair arises due to the fact that $\{T^*T(W^*), U\} = -\{f, U\} = \{T^*\underset{\sim}{V}, U\}$ and our assumption concerning the boundary conditions. As a measure of the distance ("angle") between the subspaces $T(\mathcal{M}_G) \subset \underset{\sim}{V}$ and $\underset{\sim}{N}_H \subset \underset{\sim}{V}$, we introduce

$$(5.15) \qquad \underset{\sim}{w} = TU - \underset{\sim}{V}$$

If we can find an interpolant $\tilde{\tilde{TU}}$ in $T(\mathcal{M}_G)$ such that $|\tilde{\tilde{TU}} - Tu^*| \leq K_4 h^n$ (generally $n = p+1-m$), then, by use of (5.6) and the triangle inequality, there is an element $\underset{\sim}{\tilde{w}} \in \underset{\sim}{V}$ such that as $h \to 0$

$$(5.16) \qquad \|\underset{\sim}{\tilde{w}}\| \leq Kh^r \quad ; \quad r = \min\{n, q+1\}$$

Now substituting (5.15) into (5.14) and use of (5.14) leads to

$$[TW^* - TU^*, \underset{\sim}{V}] \equiv [T(W^* - u^*) - Te_u, \underset{\sim}{V}] = [TW^* - TU^*, \underset{\sim}{w}]$$

Since the $\underset{\sim}{V}$ are generated by bases whose linear combinations are dense in $\underset{\sim}{V}$, and since use of the Schwarz inequality leads to

$$(5.17) \qquad [T(W^* - u^*) - Te_u, \underset{\sim}{\tilde{V}}] \leq \|TW^* - TU^*\| \cdot Kh^r$$

then $TU^*$ at least converges weakly to $Tu^*$ as $h \to 0$, assuming that $\|TW^* - \underset{\sim}{V}^*\|$ is bounded.

<u>5.3 A posteriori estimates</u>. We conclude our investigation with the observation that some of the results of Aubin and Burchard [17] can be adjusted to obtain a posteriori estimates for certain special cases of our approximations. In particular, if we replace $f(x)$ in all our results by $cu(x) + f(x)$, $c$ being a positive constant, and if we introduce the rather severe assumption that $U$ and $\underset{\sim}{V}$

satisfy the boundary conditions on $\partial R$ , and, further, if $T^*T$ is m-elliptic in the sense that

$$(5.18) \qquad a(u,u) \equiv [Tu,Tu] \geq \gamma^2 \|u\|_m^2$$

then it is easy to show that

$$(5.19) \quad K\|u^*-U\|_m^2 + c|||u^*-U|||^2 \leq K^{-1}\|\underset{\sim}{V}-Tu\|_{\underset{\sim}{V}}^2 + c^{-1}|||f-T^*\underset{\sim}{V}-cU|||$$

By setting $U = U^*$ , $\underset{\sim}{V} = \underset{\sim}{V}^*$ , we immediately obtain a bound on $\|e_u\|_m^2 + K'|||e_u|||^2$ . If $c = 0$ , the same procedure used to get (5.19) leads to

$$(5.20) \quad a(u^*-U,u^*-U) = \{T^*\underset{\sim}{v}^*-T^*\underset{\sim}{V},u^*-U\} + [\underset{\sim}{V}-TU,T(u^*-U)]$$

which leads to the less useful result,

$$(5.21) \quad K|||u^*-U|||^2 \leq \|\underset{\sim}{V}-TU\|_{\underset{\sim}{V}}^2 + \|\underset{\sim}{v}^*-\underset{\sim}{V}\|_{\underset{\sim}{V}}[\|\underset{\sim}{V}-TU\|_{\underset{\sim}{V}} + \|T(u^*-U)\|_{\underset{\sim}{V}}]$$

Acknowledgment: In recent years, my work on finite elements has been supported by the Air Force Office of Scientific Research under Contract F44620-69-C-0124. This support is gratefully acknowledged. It is also a pleasure to thank Mr. J. N. Reddy for assistance in proofreading the paper, checking certain derivations, and many helpful discussions.

## REFERENCES

1. Oden, J. T., "General theory of finite elements, I. Topological considerations," Internat. J. Numer. Methods Engrg. 1 (1969), 205-221.
2. Brauchli, H. J., and J. T. Oden, "Conjugate approximation functions in finite element analysis," Quart. Appl. Math. 29 (1971), 65-90.
3. Oden, J. T., and H. J. Brauchli, "On the calculation of consistent stress distributions in finite element applications," Internat. J. Numer. Methods Engrg. 3

(1971), 317-325.

4. Oden, J. T., "The theory of conjugate projections in finite element analysis," NATO Advanced Study Institute on Finite Element Methods in Continuum Mechanics, Lisbon, 1971.
5. Oden, J. T., and J. N. Reddy, "Mixed conjugate finite element approximations of linear operators," International Journal of Structural Mechanics 1 (1972), to appear.
6. Oden, J. T., Finite Elements of Nonlinear Continua, McGraw-Hill, New York, 1972.
7. Pian, T.H.H., "Formulations of finite element methods for solid continua," Recent Advances in Matrix Methods of Structural Analysis and Design, edited by R. H. Gallagher, Y. Yamada, and J. T. Oden, University of Alabama Press, University, 1971.
8. Noble, B., "Complementary variational principles for boundary value problems I. Basic principles," Report 473, Mathematics Research Center, University of Wisconsin, Madison, 1966.
9. Arthurs, A. M., Complementary Variational Principles, Oxford University Press, 1970.
10. Robinson, P. D., "Complementary variational principles," Nonlinear Functional Analysis and Applications, edited by L. B. Rall, publication #26 of Math. Res. Center, University of Wisconsin, Academic Press, New York (1971), 507-576.
11. Prager, W., and J. L. Synge, "Approximations in elasticity based on the concept of function space," Quart. Appl. Math. 5 (1947), 241-249.
12. Synge, J. L., The Hypercircle in Mathematical Physics, Cambridge University Press, London, 1957.
13. Yosida, K., Functional Analysis, Springer-Verlag, New York, 1965.
14. Von Neumann, J., "Uber Adjungierte Functional Operatoren," Ann. of Math. 33 (1932), 294-310.
15. Kato, T., "On some approximation methods concerning the operators T*T," Math. Ann. 126 (1953), 253-262.
16. Fujita, J., "Contribution to the theory of upper and lower bounds in boundary problems," J. Physical Soc. Japan 10 (1955), 1-8.
17. Aubin, J.P., and H.G. Burchard, "Some aspects of the method of the hypercircle applied to elliptic variational problems," SYNSPADE 1970, Academic Press, 1971.

# FINITE ELEMENT FORMULATION BY VARIATIONAL PRINCIPLES WITH RELAXED CONTINUITY REQUIREMENTS

Theodore H. H. Pian*

## Introduction.

The finite element method has long been recognized as an extension of the well-known Ritz procedure for constructing approximate solutions to the governing variational principle associated by a given boundary value problem. The method was originally used in the analysis of solid mechanics problems and several alternative variational principles in elasticity have been employed in the finite element formulation. Since in the finite element method a continuum is subdivided by finite element mesh, it is possible to modify the variational principles by allowing discontinuous fields at the interelement boundaries, and hence to create the so-called hybrid models in finite element analyses. The author has tried in several attempts to expand the classification of the finite element models in solid mechanics [5, 6, 7]. Finite element methods based on alternative and modified variational principles have also been applied to other continuum mechanics problems by Tong and his associates [12, 13, 14].

* Massachusetts Institute of Technology.

The purpose of this paper is to illustrate the various finite element formulations for two typical partial differential equations. The first one involves the harmonic equation which has its application in many continuum mechanics problems typical of which are torsion of prismatic bars, deflection of stretched membranes, heat conduction and potential flow. The second one is associated with the bi-harmonic equation which is associated, for example, with the bending of thin plates.

Finite element formulation for harmonic equation.

The governing equation for the warping displacement $\psi$ of the so-called St. Venant torsion, i.e., a prismatic bar under pure twisting moment [11] is

(1) $$\nabla^2\psi(x,y) = 0$$

in the domain $A$ which represents the cross-section of the bar. With $A$ in the x-y plane and with the origin coinciding with the center of twist the condition of vanishing shear stresses along the boundary yields

(2) $$(\psi,_x - y)\nu_x + (\psi,_y + x)\nu_y = 0$$

along $s$. Here both the shear modulus and the rate of twist are considered as unity and $\nu_x$, $\nu_y$ are the direction cosines of the normal at the boundary.

The corresponding variational principle for this problem is

(3) $$\pi_p = \int_A \left[\frac{1}{2}(\nabla\psi)^2 + (-y\psi,_x + x\psi,_y)\right] dxdy = \text{extremum.}$$

The subscript p used here has been borrowed from solid mechanics since this variational principle is the principle of minimum potential energy. It is seen that this variational principle contains only one field variable $\psi(x,y)$. By introducing the shear stresses $\tau_{xz}$ and $\tau_{yz}$

(4) $$\tau_{xz} = \psi,_x - y \quad ; \quad \tau_{yz} = \psi,_y + x$$

one can easily verify that the present problem can be defined by

(5) $$\pi_R = \int_A \left[-\frac{1}{2}(\tau_{xz}^2+\tau_{yz}^2)+\tau_{xz}(\psi,_x-y)-\tau_{yz}(\psi,_y+x)\right] dxdy$$
$$= \text{stationary} .$$

In solid mechanics this is referred to as the Hellinger-Reissner Principle. The condition $\pi_R = 0$ obviously yields

(6) $$\tau_{xz,x} + \tau_{yz,y} = 0 ,$$

which is satisfied by introducing a stress function $\phi$ defined by

(7) $$\tau_{xz} = \phi,_y \quad , \quad \tau_{yz} = -\phi,_x \quad .$$

If (6) and (2) are satisfied, the above variational principle is transformed to the Principle of minimum complementary energy

(8) $$\pi_c = \int_A \left[\frac{1}{2}(\tau_{xz}^2+\tau_{yz}^2)+\tau_{xz}y - \tau_{yz}x\right]dxdy = \text{extremum} \quad ,$$

or if (7) is introduced,

(9) $$\pi_c = \int_A \left[\frac{1}{2}(\nabla\phi)^2 + \phi,_y y + \phi,_x x\right] dxdy = \text{extremum} \quad ,$$

which yields the governing differential equation

(10) $$\nabla^2\phi = -2$$

and the condition that along the boundary $s$ , $\phi$ is equal to a constant. For a simply connected domain $\phi$ may be set to zero along $s$ .

In the conventional Ritz procedure for approximate solution of the variational problem, the assumed function $\psi$ in (3) and the function $\phi$ in (9) are to be continuous throughout the domain. In the ordinary finite element formulations piecewise interpolation functions of at least class $C^1$ are introduced for these variables and the continuity condition along the interelement boundary is of class $C^o$ . The simplest scheme is to divide the domain $A$ into triangular elements and to use linear interpolation of the field variable in terms of the values of such variable at the corner nodes of the element. The expression of $\pi_p$ , for example, can thus be expressed into terms of the nodal values of $\psi$ in the form†

(11) $$\pi_p = \sum_n \left[\frac{1}{2} \underset{\sim}{q}_n^T \underset{\sim}{k}_n \underset{\sim}{q}_n - \underset{\sim}{q}_n^T \underset{\sim}{Q}_n\right]$$

where $\underset{\sim}{q}_n$ represents a column of unknown nodal values of $\psi$ , $\underset{\sim}{k}_n$ is the so-called element stiffness matrix, and the

† The formulation based on $\pi_p$ refers to continuous warping displacement $\psi$ along the element boundary, hence it is often classified as the compatible model.

elements in $\underset{\sim}{Q}_n$ are expressed in terms of the geometry of the element and the interpolation function used for $\psi$ within the element. Equation (11) may be rearranged in a routine manner by assembling all the element matrices $\underset{\sim}{k}_n$ and $\underset{\sim}{Q}_n$ in the form of

$$\pi_p = \frac{1}{2}\, \underset{\sim}{q}^T \underset{\sim}{K} \underset{\sim}{q} - \underset{\sim}{q}^T \underset{\sim}{Q} \quad , \tag{12}$$

the stationary condition of which yields a system of algebraic equations

$$\underset{\sim}{K} \underset{\sim}{q} = \underset{\sim}{Q} \quad . \tag{13}$$

Finite element solutions of the torsion problem in terms of the warping displacement $\psi$ were obtained by Herrmann [4] and Yamada, et al. [15], while a solution in terms of the stress function was reported in the pioneering work of finite element method by Courant [2].

Modified variational principles can be obtained, however, if the continuity conditions along the inter-element boundary are introduced as conditions of constraint and the Lagrangian multiplier method is employed. For example, since in (9) only the derivatives of the stress function $\phi$ appear, it is only required that across a curve $s$ the derivative $d\phi/ds$ should be continuous. This is equivalent to the continuity of $(\phi,_y \nu_x - \phi,_x \nu_y)$. Physically, this means the continuity of the resultant shearing stress, (see (7)),

$$\tau_n = \tau_{xz}\nu_x + \tau_{yz}\nu_y \tag{14}$$

between two neighboring elements in the finite element formulation. When such condition is introduced as the condition of constraint, (9) may be written as

$$\pi_{mc} = \sum_n \left\{ \int_{A_n} \left[ \frac{1}{2}(\nabla\phi)^2 + \phi,_y y + \phi,_x x \right] dxdy - \int_{s_n} (\phi,_y \nu_x - \phi,_x \nu_y)\psi dx \right\} = \text{stationary} \tag{15}$$

where $A_n$ refers to the nth element and $s_n$ refers to the interelement boundary of the nth element. The warping displacement $\psi$ along the element boundary here is the corresponding Lagrangian multiplier. This modified variational principle now has two field variables, the stress function $\phi$ over the domain and the warping displacement $\psi$ along the element boundary.

Equation (3) may also be modified if the continuity of $\psi$ is used as a condition of constraint along the interelement boundary. The Lagrangian multiplier in this case is the resultant shear stress $\tau_n$ along the element boundary and the modified variational principle now has the warping function $\psi$ over the domain and the resultant shear stress $\tau_n$ along the element boundaries as the field variables.

The finite element formulation based on the modified variational principle with relaxed continuity condition has been called the hybrid model. The most important feature of the hybrid model based on, for example, Eq. (15) is that the variable $\phi$ may be independently assumed for each element hence both $\phi$ and its derivatives $\phi,_x$ and $\phi,_y$ may be approximated by polynomials with coefficients $\underset{\sim}{\beta}$ as undetermined parameters, i.e.,

$$\phi = \underset{\sim}{P}\underset{\sim}{\beta}$$

(16)
$$\begin{bmatrix} \phi,_x \\ \phi,_y \end{bmatrix} = \begin{bmatrix} \underset{\sim}{P}_x \\ \underset{\sim}{P}_y \end{bmatrix} \underset{\sim}{\beta} \ .$$

The functions $\underset{\sim}{P}$ must be chosen such that none of the elements in $\underset{\sim}{P}_x$ or $\underset{\sim}{P}_y$ are zero or depend on others. The variable $\psi$ along the element boundary is then interpolated in terms of the nodal values $\underset{\sim}{q}$ in the form

(17)
$$\psi = \underset{\sim}{L}\underset{\sim}{q} \ .$$

For example, if the element is a polygon, it is only necessary to express the variable $\psi$ between the ith and the (i+1)th corner as

(18)
$$\psi(s) = q_i\left(1 - \frac{s}{\ell_{i,i+1}}\right) + q_{i+1}\frac{s}{\ell_{i,i+1}}$$

where $\ell_{i,i+1}$ is the distance between the two corners and s is measured along the boundary from the ith corner.

Substituting Eqs. (16) and (17) into Eq. (15) one obtains an equation of the form

(19)
$$\pi_{mc} = \sum_n \left[\frac{1}{2}\underset{\sim}{\beta}_n^T \underset{\sim}{H}_n \underset{\sim}{\beta}_n + \underset{\sim}{\beta}_n^T \underset{\sim}{F}_n - \underset{\sim}{\beta}_n^T \underset{\sim}{G}_n \underset{\sim}{q}_n\right]$$

where

(20)
$$\underset{\sim}{H}_n = \int_{A_n} \left[\underset{\sim}{P}_x^T \underset{\sim}{P}_y^T\right] \begin{Bmatrix} \underset{\sim}{P}_x \\ \underset{\sim}{P}_y \end{Bmatrix} dA$$

$$\underset{\sim}{F}_n = \int_{A_n} \left[\underset{\sim}{P}_y^T \underset{\sim}{P}_x^T\right] \begin{Bmatrix} y \\ x \end{Bmatrix} dA$$

$$\underset{\sim}{G}_n = \int_{S_n} \left[\nu_x \underset{\sim}{P}_y^T - \nu_y \underset{\sim}{P}_x^T\right] \underset{\sim}{L} ds \ .$$

The unknowns $\underset{\sim}{q}$ and $\underset{\sim}{\beta}$ are determined from the stationary condition of $\pi_{mc}$ . Since $\phi$ for the individual elements are independently assumed, $\underset{\sim}{\beta}$ of one element are independent of that of the other elements, one can write

(21) $$\underset{\sim}{H}_n \underset{\sim}{\beta}_n + \underset{\sim}{F}_n - \underset{\sim}{G}_n \underset{\sim}{q}_n = 0$$

from which,

(22) $$\underset{\sim}{\beta}_n = \underset{\sim}{H}_n^{-1} \left[ \underset{\sim}{G}_n \underset{\sim}{q}_n - \underset{\sim}{F}_n \right]$$

for each element. Substituting Eq. (22) into Eq. (19), one obtains[†]

(23) $$\pi_{mc} = \sum_n \left[ -\frac{1}{2} \underset{\sim}{q}_n^T \underset{\sim}{G}_n^T \underset{\sim}{H}_n^{-1} \underset{\sim}{G}_n \underset{\sim}{q}_n + \underset{\sim}{q}_n^T \underset{\sim}{G}_n^T \underset{\sim}{H}_n^{-1} \underset{\sim}{F}_n + C_n \right]$$

which is of the same form as Eq. (11) and, in fact, the element stiffness matrix $\underset{\sim}{k}_n$ is defined by

(24) $$\underset{\sim}{k}_n = \underset{\sim}{G}_n^T \underset{\sim}{H}_n^{-1} \underset{\sim}{G}_n$$

and the expression for $\underset{\sim}{Q}_n$ is

(25) $$\underset{\sim}{Q}_n = \underset{\sim}{G}_n^T \underset{\sim}{H}_n^{-1} \underset{\sim}{F}_n \ .$$

Both the compatible model and the hybrid model provide directly discrete values of solutions of the partial differential equation Eq. (1). On the other hand, the finite element formulation based on Eq. (9) will provide only discrete values of $\phi$ directly and in order to evaluate the warping function $\psi$, integration process must be applied. The values of the integrals, in general, depends on the integration path; hence, no unique values of $\psi$ can be evaluated from the discrete values of $\phi$ obtained by

[†] As remarked in [8], the total number of $\beta$'s of all elements must be greater than the total number of unknown q's, in order that a solution for $\delta\pi_{mc} = 0$ is obtainable.

the finite element method. Of course, in the limit when the element size becomes vanishingly small, converging solution for $\psi$ is still obtainable.

Another feature of the finite element method based on the modified variational principle Eq. (15) is that for elements which contain the boundary $s$, it is possible to incorporate the prescribed value of $\phi$ (which is equal to zero for the present case) along that boundary. The only necessary step is to choose $\underset{\sim}{P}_x$ and $\underset{\sim}{P}_y$ so that the prescribed value for $\phi$ is satisfied. In contrast by using the finite element method based on Eq. (3), the resulting solution for $\phi$ can only satisfy the boundary condition in the integral sense.

Yamada et al. [16] have analyzed the torsion of a slender bar of equilateral triangular cross section by using element stiffness matrices obtained by the compatible model and the hybrid model. Figure 1 shows the element subdivisions. On account of the symmetry of cross-section only one sixth of the triangle was needed in the finite element solution. The numbers of elements and nodes were 123 and 82 respectively. In deriving the element stiffness matrices based on the compatible model and the hybrid model these authors used linear interpolation for $\psi$ and constant $\phi,_x$ and $\phi,_y$, respectively within the element. It turns out that when the boundary $\psi$ is assumed linear in the hybrid formulation, the resulting stiffness matrix is identical to that obtained by the compatible model. Yamada and his associate, however, realized that in deriving the stiffness matrix by the hybrid model the condition of vanishing $\phi$ can be incorporated. They obtained an alternative solution for which the stiffness matrices for these elements at the outside boundary were evaluated based on the hybrid model. The solution of the shearing stresses at the centroids of some triangular elements at the edge AB are listed in Table 1. The corresponding exact solution is also listed for comparison. It is seen that when hybrid elements are used along the outside edge, more accurate solution can be obtained for the shearing stresses. It should be pointed out that it is possible to improve the finite element solution based on the hybrid model by using higher order function for $\phi$ within each element. It turns out that the exact solutions for the equilateral triangular cross section [11] consists of a cubic polynomial for $\phi$ and linear distribution in $\psi$ along the three axes of symmetry. Thus a single hybrid element AOB, which is based on (1) cubic expansion in $\phi$, (2) vanishing $\phi$ along AB and (3) linear distributions of $\psi$ along AO and OB, will yield the exact solution of this torsion problem.

Finite element formulation for bi-harmonic equation.

For simplicity, consider the problem of a uniform thin plate which is clamped along the entire boundary and acted on by a uniformly distributed load p [10]. If the bending rigidity is unity, the governing differential equation for the lateral displacement $w(x,y)$ is

$$\nabla^2\nabla^2 w = p \tag{26}$$

in the domain A and the boundary conditions are $w = w,_\nu = 0$ along s where $\nu$ refers to the direction normal to the boundary. This partial differential equation and the corresponding boundary condition can be derived by the following variational principle, the principle of minimum potential energy

(27) $$\pi_p = \int_A \left[\frac{1}{2}(w_{,xx}^2 + 2w_{,xy}^2 + w_{,yy}^2) - pw\right] dA$$

$$= \text{extremum} \; .$$

In the finite element formulation, piecewise interpolation functions of at least class $C^2$ are introduced for w and the continuity condition along the interelement boundary is of class $C^1$. Such requirement along the element boundary creates the chief difficulty in the construction of the interpolation functions over the element. In deriving the stiffness matrix of a 12 degrees-of-freedom rectangular plate element, for example, Deák and Pian [3] applied the smooth surface interpolation which was suggested by Birkhoff and Garabedian [1] and which involves different analytical expressions in four regions of the element. To construct a 12 degrees-of-freedom element of arbitrary quadrilateral shape such smooth interpolation function would not be readily available.

The corresponding principle of minimum complementary energy is expressed in terms of the stress couples $M_{xx}$, $M_{xy}$, and $M_{yy}$ in the term of

(28) $$\pi_c = \int_A \frac{1}{2}(M_{xx}^2 + 2M_{xy}^2 + M_{yy}^2)\,dxdy = \text{extremum} \; ,$$

where

(29) $$M_{xx} = w_{,xx}$$
$$M_{yy} = w_{,yy}$$
$$M_{xy} = w_{,xy}$$

and the following equilibrium condition is satisfied

$$M_{xx,xx} + 2M_{xy,xy} + M_{yy,yy} = p \quad . \tag{30}$$

In using the variational principle for the finite element formulation, the stress couples need not be continuous at the interelement boundary, but across the boundary the resultant stresses which are represented by the following components must be continuous

$$\begin{aligned} m_x &= M_{xx}\nu_x + M_{xy}\nu_y \\ m_y &= M_{xy}\nu_x + M_{yy}\nu_y \\ Q_n &= (M_{xx,x} + M_{xy,y})\nu_x + (M_{xy,x} + M_{yy,y})\nu_y \quad . \end{aligned} \tag{31}$$

The modified variational principle with relaxed interelement continuity requirement thus becomes

$$\begin{aligned} \pi_{mc} = \sum_n \Big\{ \int_{A_n} & \tfrac{1}{2}(M_{xx}^2 + 2M_{xy}^2 + M_{yy}^2)\,dxdy \\ & - \int_{S_n} (m_x w,_x + m_y w,_y + Q_n w)\,ds \Big\} \\ & = \text{stationary} \end{aligned} \tag{32}$$

where $w,_x$ , $w,_y$ and $w$ are the corresponding Lagrangian multipliers needed to introduce the continuity of $m_x$ , $m_y$ , and $Q_n$ along the interelement boundary.

In the finite element formulation the first step is to assume the stress couples $M_{xx}$ , $M_{xy}$ , and $M_{yy}$ in terms of finite number of unknown parameters $\underset{\sim}{\beta}$ and some known functions such that (30) is satisfied. The next step is to interpolate the boundary displacement $w$ in terms of the nodal displacements $\underset{\sim}{q}$ . Since $w$ , $w,_x$ , and $w,_y$ are to be continuous across the interelement boundary it is necessary to express $w$ , at least in terms of $w$ , $w,_x$ , and $w,_y$ at the nodes. For a straight line boundary, this can be accomplished by using a cubic interpolation for $w$ in terms of $w$ and $w,_s$ at the two ends and a linear interpolation for the normal slope $w,_n$ in terms of its end values. Such assumed displacements can be easily made for elements of any polygonal shapes. The remaining derivation of the equations follows the same step as that in the previous section and the resulting expressions for $\pi_{mc}$ and $\underset{\sim}{k}_n$ are again of the form given in (24) and (25), respectively.

Numerous authors have solved the plate bending problems using the assumed stress hybrid model. Figure 2 shows the convergence behavior of the finite element solutions for the central deflection of a square plate under a concentrated load at the center [8]. The boundary conditions along the edge are $w = 0$ and $M_n = 0$ . All the finite element solutions were obtained using 12 degrees-of-freedom square elements, for which the displacement $w$ along each edge is cubic and the normal slope $w,_n$ is linear. For the hybrid elements different types of assumed stress distributions were used. They are

(a) The assumed stress is quadratic and for the elements at the plate edge $M_n = 0$

(b) The assumed stress is cubic and for the elements at the plate edge $M_n = 0$

(c) The assumed stress is linear and for all

elements the normal slopes $w,_n$ are linear along all edges

(d) The assumed stress is quadratic and for all elements the normal slopes $w,_n$ are linear along all edges

(e) The assumed stress is cubic and for all elements the normal slopes $w,_n$ are linear along all edges.

A finite element solution by the compatible displacement model obtained by Deák and Pian [3] is also plotted for comparison. As described previously the element also has 12 degrees-of-freedom and is based on a smooth surface interpolation function which must be expressed by different analytical functions in four subregions of the element. The present solution concerns the deflection under a single load, hence when the principle of minimum potential energy is used it should converge monotonically when the element size decreases gradually. Indeed, as shown in Figure 2, the compatible displacement does provide a lower bound solution. The hybrid model, however, in general, may yield the central deflection either higher or lower than the exact value. The assumed stress distribution in the interior of the element has great influence on the accuracy of the finite element solution. It appears that appropriate stress distribution can always be chosen in the hybrid model such that its solution is more accurate than that of the compatible displacement model. The use of hybrid element which satisfies the prescribed stresses along the boundary is also the source of improved accuracy.

Remarks.

The chief advantage of the formulation of the finite element method by variational principles with relaxed continuity requirement is the avoidance of interpolating the field variable to maintain its continuity along the interelement boundary. Thus, in principle, element stiffness matrix of any arbitrary shape and any number of nodes can be constructed by the hybrid model.

Another feature of the hybrid finite element method is its capability of including known types of stress singularities in the assumed stress distribution. The author and his associates [9] have demonstrated the extremely good convergence property in using such a scheme to analyze the stress intensity factors at the tips of sharp cracks in several plane stress problems.

REFERENCES

1. Birkhoff, G., and H. L. Garabedian, "Smooth surface interpolation," J. Math. and Physics 34 (1960), 258-268.
2. Courant, R., "Variational methods for the solution of problems of equilibrium and vibrations," Bull. Am. Math. Soc. 49 (1943), 1-23.
3. Deák, A. L., and T. H. H. Pian, "Application of the smooth-surface interpolation to the finite element analysis," AIAA J. 5 (1967), 187-189.
4. Herrmann, L. R., "Elastic torsional analysis of irregular shapes," J. Eng. Mech. Div. ASCE, EM6 (1965), 11-19.
5. Pian, T. H. H., "Formulations of finite element methods for solid continua," Recent Advances in Matrix Methods in Structural Analysis and Design, R. H. Gallagher, Y. Yamada, and J. T. Oden, eds., University of Alabama Press (1971), 49-83.

6. Pian, T. H. H., "Hybrid models," presented at the International Symposium on Numerical and Computer Methods in Structural Mechanics, University of Illinois at Urbana-Champaign, September 8-10, 1971 (to be published in the proceedings).
7. Pian, T. H. H., "Finite element methods by variational principles with relaxed continuity requirements," for presentation at the International Conference on Variational Methods in Engineering, Southampton, England, September 25-29, 1972.
8. Pian, T. H. H., and P. Tong, "Rationalization in deriving element stiffness matrix by assumed stress approach," Proc. Second Conf. on Matrix Methods Struct. Mech. (1969), AFFDL-TR-68-150, pp. 441-469, Wright Patterson AFB, Ohio.
9. Pian, T. H. H., P. Tong, and C. H. Luk, "Elastic crack analysis by a finite element hybrid method," presented at Third Conference on Matrix Methods in Structural Mechanics, Wright-Patterson Air Force Base, October 19-21, 1971.
10. Timoshenko, S. P., and S. Woinowsky-Krieger, Theory of Plates and Shells, Chapter 4, 2nd Ed., McGraw Hill, New York, 1959.
11. Timoshenko, S. P., and J. N. Goodier, Theory of Elasticity, Chapter 10, 3rd Ed., McGraw Hill, New York, 1970.
12. Tong, P., "The finite element method for fluid flow," Recent Advances in Matrix Methods in Structural Analysis and Design, R. H. Gallagher, Y. Yamada, and J. T. Oden, Eds., University of Alabama Press (1971), 787-808.
13. Tong, P., and Y. C. Fung, "Slow particulate flow in channel and tube - applications to biomechanics," J. Appl. Mech. (Trans ASME Series E) 38 (1971), 721-728.
14. Tong, P., and D. Vawter, "An analysis of peristaltic pumping," ASME Paper No. 72-APM-19, 1972.
15. Yamada, Y., T. Kawai, N. Yoshimura, and T. Sakurai, "Analysis of the elastic-plastic problems by the Matrix displacement method," Proc. 2nd Conf. Matrix Methods Struct. Mech., AFFDL-TR-68-150, (1969), 1271-1299.
16. Yamada, Y., S. Nakagiri, and K. Takatsuka, "Analysis of Saint-Venant torsion problem by a hybrid stress method," Seisan-Kenkyu (Monthly J. of Institute of Industrial Science, University of Tokyo) 21, No.11,1969.

TABLE 1. SHEARING STRESSES AT THE CENTROID OF TRIANGULAR ELEMENTS AT THE EDGE AB

| | ELEMENT NO. | EXACT SOLUTION | FINITE ELEMENT SOLUTION | |
|---|---|---|---|---|
| | | | ALL ELEMENTS BY COMPATIBLE MODEL | EDGE ELEMENTS BY HYBRID MODEL |
| $\tau_{xz}$ | 2 | -0.2461 | -0.2464 | -0.2472 |
| | 48 | -0.2191 | -0.2374 | -0.2160 |
| | 81 | -0.1500 | -0.1641 | -0.1420 |
| | 99 | -0.0779 | -0.0884 | -0.0776 |
| | 122 | -0.0044 | -0.0078 | -0.0043 |
| $\tau_{yz}$ | 2 | 0.4260 | 0.4325 | 0.4281 |
| | 48 | 0.3725 | 0.3625 | 0.3742 |
| | 81 | 0.2486 | 0.2294 | 0.2459 |
| | 99 | 0.1154 | 0.1251 | 0.1345 |
| | 122 | 0.0058 | 0.0024 | 0.0074 |

VARIATIONAL CRIMES IN THE FINITE ELEMENT METHOD

Gilbert Strang*

## 1. Introduction.

The finite element method is nearly a special case of the Rayleigh-Ritz technique. Both methods begin with a set of trial functions $\phi_1(x),\ldots,\phi_N(x)$ ; both work with the space of linear combinations $v^h = \Sigma\, q_j\phi_j$ ; and both choose the particular combination (we will call it $u^h$) which minimizes a given quadratic functional. In many applications this functional corresponds to potential energy. The convergence of the Ritz approximation is governed by a single fundamental theorem: if the trial functions $\phi_j$ are admissible in the original variational principle, then $u^h$ is automatically the combination which is closest (in the sense of strain energy, see equation (2)) to the true solution $u$ . Therefore convergence is proved by establishing that as $N \to \infty$ , the trial functions fill out the space of all admissible functions. This question of "completeness" dominates the classical Ritz theory.

For finite elements a new question appears. Suppose the basic rule of the Ritz method is broken, and the trial functions are not quite admissible in the true variational

*Massachusetts Institute of Technology

problem: is it still possible to prove that $u^h$ converges to $u$, and to estimate the rate of convergence? This question is inescapable, if we want to analyze the method as it is actually used. We regard the convenience and effectiveness of the finite element technique as conclusively established; it has brought a revolution in the calculations of structural mechanics, and other applications are rapidly developing. Our goal is to examine the modifications of the Ritz procedure which have been made (quite properly, in our view) in order to achieve an efficient finite element system. The hypotheses of the classical Ritz theory are violated, because of computational necessity, and we want to understand the consequences.

Suppose, in order to work with a specific example, that the given problem is to minimize the functional

$$I(v) = a(v,v) - 2b(f,v)$$
$$= \iint_\Omega (v_x^2+v_y^2)\,dxdy - 2\iint_\Omega fv\,dxdy .$$

Then there are three conditions to the basic rule:

i) the "strain energy" in the trial functions should be finite: $a(v^h,v^h) < \infty$. In our example the strain energy involves first derivatives, and <u>a piecewise polynomial</u> $v^h$ <u>has finite strain energy if and only if it is continuous</u>. A discontinuity in $v^h$ introduces $\delta$-functions into the first derivatives, and the expression $\iint (v_x^h)^2 + (v_y^h)^2$ becomes infinite: the trial function is then <u>nonconforming</u> and not admissible.

ii) <u>essential boundary conditions</u> are to be

respected. An essential conditon is one which involves only derivatives of lower order than the leading terms in the strain energy; if the true solution is required to satisfy $u = g$ on some part of the boundary, for example, then this requirement must be imposed on the functions $v$ in the variational problem. In the Ritz approximation, the trial functions need not be exactly equal to $g$, but they must share the same value $g^h$; if $u^h$ is to converge to $u$, then of course $g^h$ must converge to $g$. (Notice that since the leading terms in the strain energy need not be continuous, it would be improper to specify the values of these derivatives on the boundary; any such requirement must emerge as a natural boundary condition, satisfied by the function $u$ which actually minimizes $I(v)$ but not all the functions admissible in the minimization. The essential conditions $Bv = g$ are those which can be imposed on all functions which satisfy $a(v,v) < \infty$; $B$ is a bounded operator on this function space.)

iii) the computation of $I(v^h)$ should be exact. With $v^h = \Sigma q_j\phi_j$, the functional is a quadratic in the unknowns $q_1,\ldots,q_N$:

$$I(v^h) = a(\Sigma q_j\phi_j, \Sigma q_j\phi_j) - 2b(f, \Sigma q_j\phi_j)$$

$$= q^T K q - 2q^T F .$$

Therefore the requirement is that the stiffness matrix $K$ and the load vector $F$ should be computed by exact integration over $\Omega$.

The final step in the Ritz method is to minimize

$I(v^h)$ ; the optimal weights $Q_1,\ldots,Q_N$ satisfy $KQ = F$ , and the approximation is $u^h = \Sigma\, Q_j\phi_j$ . When the conditions (i) - (iii) are obeyed, it follows that $u^h$ not only minimizes $I(v^h)$ , but it also minimizes the error in strain energy:

$$(2) \qquad a(u-u^h, u-u^h) = \min_{v^h} a(u-v^h, u-v^h) .$$

This is the fundamental principle in proving convergence of the Ritz method; by making any convenient choice of $v^h$ we have an upper bound for the error. With finite elements this choice is often a piecewise polynomial which interpolates the true solution $u$ at the nodes, and the convergence proof becomes a standard question in approximation theory. A part of this theory, and a summary of its history, is given in [1].†

We are concerned here with the opposite problem, in which the rules (i) - (iii) are not obeyed. The property (2) is no longer valid, and another approach must be found. We had originally supposed that a useful technique would be to study the perturbations induced in the stiffness matrix $K$ . If the functional $I(v^h)$ is evaluated by numerical quadrature rather than exact integration (which

†Ciarlet and Raviart have solved the approximation problem which is induced by a piecewise polynomial (isoparametric) change of coordinates. If the derivatives of these coordinate changes are bounded, so that the elements are not very much distorted, there is no loss in the order of approximation.

has become a standard technique for complicated finite element calculations, and is even applied in some cases when the integrals involve only polynomials and could be computed analytically!) then K is replaced by a new matrix $\tilde{K}$, and F by $\tilde{F}$. These perturbations are estimated, for example, by Parlett and Johnson [2] in connection with the eigenvalue problem; the error in the eigenvalues is estimated from the size of $K - \tilde{K}$.

Surprisingly, this perturbation approach is not optimal: the error $u^h - \tilde{u}^h$ induced by numerical integration may be very small, even though the entries of K and $\tilde{K}$ are very different. This must mean that the errors in $\tilde{K}$ are too systematic to be revealed by examining each entry; translated into finite difference terms for the special case of regular meshes, the difference operators K and $\tilde{K}$ may both be highly accurate discretizations of the given equations, without resembling one another in the least.

To include also irregular meshes, our approach must be based on the variational principle. In fact it must go back to the fundamental conditions for u and $u^h$ to be minimal, namely the vanishing of the first derivative (or more properly, the first variation) of the functional to be minimized. In engineering terms this is the equation of virtual work. We denote the full admissible space in the example (1) by $\mathcal{H}_E^1$, indicating by the superscript the order of derivatives which appear in a(v,v); the subscript refers to the subspace of $\mathcal{H}^1$ which satisfies the essential boundary conditions. Then the true solution u satisfies

$$(3) \qquad a(u,v) = b(f,v) \quad \text{for all } v \text{ in } H^1_E .$$

This is the <u>weak form of the differential equation</u> $-u_{xx} - u_{yy} = f$ , which is the Euler equation for minimizing the functional $I(v)$ .

In the finite element method, we may break the Rayleigh-Ritz laws by changing to an approximate functional $\tilde{I}(v^h,v^h) = \tilde{a}(v^h,v^h) - 2\tilde{b}(f,v^h)$ . Nevertheless $\tilde{u}^h$ is a genuine minimum (barring roundoff) of <u>this</u> functional, and satisfies its own equation of virtual work:

$$(4) \qquad \tilde{a}(\tilde{u}^h,v^h) = \tilde{b}(f,v^h)$$

for all trial functions $v^h$ . The equations (3) and (4) are derived by perturbing $u$ and $\tilde{u}^h$ by $\varepsilon v$ and $\varepsilon v^h$ , respectively; the coefficient of $\varepsilon$ which appears in $I$ and $\tilde{I}$ must be zero, to have a minimum. ((2) is established in the same way, if every $v^h$ is an admissible $v$ .)

We want to use the equations (3) and (4) to estimate the error, and to find out which violations of the rules still yield a convergent approximation. In other words we want to find out whether or not crime pays; fortunately for the finite element method, it often does. The mathematical arguments vary slightly, according to which of the conditions (i) - (iii) is broken; we shall take up each case in turn, establishing the basic convergence conditions but referring for a fuller treatment to the forthcoming book [3].

## 2. Nonconforming elements.

Suppose the strain energy is found by summing the contributions from each separate polynomial piece,

$$\tilde{a}(v^h,v^h) = \Sigma \iint_{e_i} (v_x^h)^2 + (v_y^h)^2 \quad ,$$

where the domain $\Omega = \bigcup e_i$ .
Then the discontinuities on the element boundaries are ignored, and the infinite strain energy in a nonconforming element becomes finite. There are three reasons why such elements should still be used, even now that it is known how to construct piecewise polynomials with sufficient smoothness:

a) the construction may be too complicated; for $C^1$ elements on a general triangular mesh, required for plate bending (biharmonic) equations, the simplest polynomials are quintics in two variables

b) extra accuracy might be achieved by adding some nonconforming elements to an existing trial space

c) nonconforming solutions, especially with a coarse mesh, can be more accurate than conforming ones, and furthermore they are not so stiff as a true Ritz solution. The Ritz theory always leads to $a(u^h,u^h) \leq a(u,u)$ , and in most cases to an underestimate of the displacement; the structure with only N degrees of freedom is too rigid.

We must comment, however, on the risks involved in nonconforming elements. We shall show that in many cases they do not converge, so that engineers are increasingly wary of the uncertainties involved. The statement at the end of [4] would, I believe, no longer be subscribed to by all its authors.

Although a number of papers set out to study the convergence of nonconforming elements, the first successful idea came from Irons; it is now known as the <u>patch test</u>. Suppose the trial functions contain all polynomials $P_m$ of the same degree ($m=1$ in our example) as the derivatives in the strain energy; for convergence, approximation theory requires that this be so. Suppose also that, around an arbitrary patch of elements, boundary conditions are so chosen that the true solution within the patch is $u = P_m$. Then the patch test requires that <u>the approximation $\tilde{u}^h$</u>, computed in the nonconforming way by ignoring $\delta$-functions on the element boundaries, <u>must also coincide with</u> $P_m$. In a variable-coefficient problem, we must permit a small perturbation $\tilde{u}^h = P_m + O(h)$. When this test is passed, Irons believed that the trial functions would behave as if they were conforming, and expected convergence. We want to prove that he was right.

On a regular mesh we could regard the system of Ritz-finite element equations $KQ = F$ as a finite difference scheme, and then the patch test would be equivalent to <u>the formal consistency of the difference equations with the correct differential equation</u>. We shall prefer to give a variational argument, thereby admitting an irregular mesh and relating the patch test to the equation (4) of virtual work. The relation is this: the patch test is passed if and only if, for all polynomials of degree $m$ and all nonconforming trial functions $v^h$,

(5) $$a(P_m,v^h) = \tilde{a}(P_m,v^h) .$$

The left side includes $\delta$-functions at the element boundaries, but it is well defined: these mth derivatives of $v^h$ are multiplied by mth derivatives of $P_m$, in other words by constants, and integrated over the domain.

To prove the equivalence of (5) with the patch test, let $\phi_j$ be one of the basis functions for the trial space, and let the patch be the whole domain $\Omega$. Since the solution is to be $u = P_1$, the corresponding load in our example, according to the differential equation $-u_{xx} - u_{yy} = f$, will be $f = 0$. Therefore the equations of virtual work yield

(6) $$a(P_1,\phi_j) = 0 , \quad \tilde{a}(\tilde{u}^h,\phi_j) = 0 .$$

If the patch test is passed, so that $\tilde{u}^h = P_1$, then obviously (5) holds for the trial function $v^h = \phi_j$. But then it holds for all combinations $v^h = \Sigma q_j\phi_j$. Conversely, if (5) always holds then $\tilde{a}(\tilde{u}^h,\phi_j) = \tilde{a}(P_1,\phi_j)$ and it follows that $\tilde{u}^h = P_1$. So the patch test is equivalent to (5).

Remark. The first equation in (6) is, strictly speaking, not a consequence of (3): $\phi_j$ is not in $H^1_E$, and therefore the first variation of $I(v)$ at the solution $u = P_1$ is not required a priori to vanish in the direction of $v = \phi_j$. This difficulty can be circumvented by an integration by parts: since $u = P_1$ is smooth, Green's formula gives

$$a(P_1,v) = \iint_\Omega (P_1)_x v_x + (P_1)_y v_y$$
$$= -\iint_\Omega (P_1)_{xx} v + (P_1)_{yy} v \ .$$

The last integral involves no derivatives of $v$, and we can find a sequence of $v$'s which approach $\phi_j$ in the ordinary $L_2$ norm (in other words $H^0$, involving no derivative). In the limit we find

$$(7) \qquad a(P_1,\phi_j) = -\iint (P_1)_{xx}\phi_j + (P_1)_{yy}\phi_j = 0 \ .$$

Now we look at a nonconforming element which passes the test. Beginning with the standard bilinear functions on a square mesh, Wilson [5] has proposed to add two new trial functions within each square in order to improve the approximation properties of the trial space. In a typical square $x_j \leqslant x \leqslant x_{j+1}$, $y_j \leqslant y \leqslant y_{j+1}$, these two functions are simply $\theta = (x-x_j)(x-x_{j+1})$ and $\psi = (y-y_j)(y-y_{j+1})$; outside the square they are zero. Combined with the bilinear functions 1, x, y, xy which are already present, it is now possible to produce any quadratic polynomial; therefore there exists among the trial functions one which approximates the strains $u_x$ and $u_y$ to order $h^2$. However the new functions $\theta$ and $\psi$ are not continuous -- $\theta$ is not zero along the top and bottom of the square, and $\psi$ is not zero on the vertical sides. Therefore the patch test must be passed, in order to know that the particular trial function $\tilde{u}^h$ is even as close as $u_x - \tilde{u}^h_x = O(h)$, $u_y - \tilde{u}^h_y = O(h)$. (It would achieve the optimal $O(h^2)$ if the test were also passed for all polynomials $P_2$ of the next higher degree; but in our example, this doesn't happen.)

We apply the patch test to $\theta$. Suppose $P_1 = a + bx + cy$ is an arbitrary linear polynomial. Then since $\theta$ is zero outside of its square of definition,

$$\tilde{a}(P_1,\theta) = \int_{y_j}^{y_{j+1}}\int_{x_j}^{x_{j+1}} b\,\frac{\partial\theta}{\partial x}\,dxdy = 0 .$$

Since also $a(P_1,\theta) = 0$ -- as we saw in (6) and (7), by applying Green's formula over the whole of $\Omega$ -- the equality (5) holds and the test is passed. In fact the test amounts to this: the true integral of any strain ($m^{th}$ derivative $D^m\phi_j$) is zero, if the $\delta$-functions on element boundaries are included, and the test is to see whether or not the integrals of these strains are still zero if the $\delta$-functions are ignored. This will happen only if a "lucky" cancellation occurs, as it did in our example; the singularities at the top of the square are seen to cancel those at the bottom, if I apply the test in another way, by writing the difference between $\tilde{a}$ and $a$ as a line integral around the square:

$$\tilde{a}(P_1,\theta) - a(P_1,\theta) = \int \theta\, \partial P_1/\partial n ds$$

$$= \int_{x_j}^{x_{j+1}} \theta(-c)dx + \int_{x_{j+1}}^{x_j} \theta\, c(-dx) = 0 ,$$

going first along the bottom and then back along the top. $\partial P_1/\partial n$ equals $-c$ in one case and $+c$ in the other; down the sides $\theta = 0$.

Now we summarize the convergence theory for nonconforming elements. As emphasized earlier, it begins with the equations (3) and (4). Since the linear term $b(f,v^h) = \iint fv^h$ involves no singularities at the element boundaries, the sum of integrals over the separate elements gives the correct integral over $\Omega$, and $b = \tilde{b}$.

It follows from (3) and (4) that

$$\tilde{a}(u-\tilde{u}^h, v^h) = \tilde{a}(u,v^h) - a(u,v^h) \, . \tag{8}$$

This is the key identity. Ridgway Scott has observed that it leads directly to a lower bound for the error in the norm $\|u-\tilde{u}^h\| = \tilde{a}(u-\tilde{u}^h, u-\tilde{u}^h)^{1/2}$. He applies the Schwarz inequality to the left side of (8), and divides through by $\|v^h\|$ ;

$$\|u-u^h\| \geq \frac{|\tilde{a}(u,v^h) - a(u,v^h)|}{\|v^h\|} \, . \tag{9}$$

Define $\Delta$ to be the maximum of the right side of (9), taken over all $v^h$. Then the error is at least $\Delta$. Conversely, we derive in the book [3] an inequality in the reverse direction,

$$\|u-\tilde{u}^h\| \leq \Delta + \min_{v^h} \|u-v^h\| \, . \tag{10}$$

Therefore the error is governed by a nonconforming term $\Delta$ ($\Delta$ would be zero if $a = \tilde{a}$) and an approximation term. It is well-known how the latter depends on the degree of the polynomial elements; our point is that $\Delta$ depends on the patch test. In fact $\Delta \to 0$ if and only if the test is passed, and therefore the test is necessary and sufficient for convergence. The estimate $\Delta = O(h)$ goes briefly as follows: the patch test permits us to rewrite the numerator in (9) as $\tilde{a}(u-P_1,v^h) - a(u-P_1,v^h)$ . To compute the contribution from the ith square, we choose $P_1$ close to $u$ in that square and use Green's formula. Summing over all such squares (cf. [3]), the result is a discrepancy of order $h$ -- coming from the error committed in the first derivatives by linear approximation of $u$ .

It would be wrong to leave the impression that the only interesting elements are those which pass the test. There are many others for which $\Delta$ is comparable to the approximation error, at least on a coarse mesh, and the approximate solution $\tilde{u}^h$ is sufficiently accurate. Because of the simplicity, and in spite of the uncertainty, these nonconvergent elements will continue to be useful.

## 3. Violation of essential boundary conditions.

It is impossible for a polynomial to satisfy a condition like $u = 0$ on a general curved boundary. Therefore some alteration in the boundary condition will be necessary. One possibility -- perhaps the most important one -- is to change the domain. We have shown elsewhere [6] that to replace $\Omega$ by a polygon $\Omega^h$ induces an error of order $h^3$ in the strain energy in $\Omega^h$ ; this error is concentrated in a boundary layer. The error in displacement, or in strain energy over the original $\Omega$ , or in eigenvalues, is of order $h^2$ . These estimates can be derived within the classical Ritz framework, since the trial functions satisfy boundary conditions exactly on the new domain -- we speak then about change of domain, and not about violation of essential conditions.

An ingenious idea has made it possible to do still better than a polygon; the boundary can be replaced by a piecewise polynomial of higher degree, and then straightened by a change of coordinates. If the new $\xi - \eta$ coordinates are related to the old by the same class of polynomials that is used to approximate the solution itself -- so that $x(\xi,\eta)$, $y(\xi,\eta)$, and $u(\xi,\eta)$ all look alike -- then the change is called isoparametric. This idea has made it possible to overcome, in a way which finite differences never managed to do, most of the difficulties inherent in curved boundaries. Again the theoretical problem is the simpler one of change of domain.

We believe there may be other situations in which the boundary conditions are simply interpolated, for example $v^h = 0$ may be imposed only at a discrete set of boundary nodes. Between these nodes $v^h$ will be small, but it will vary from one trial function to another; this puts it outside the Ritz theory. Nitsche was the first to consider this question, and we look forward to his paper at this symposium; Blair has also contributed to the theory [7]. Our results with Berger and Scott [8] indicated the same $h^3$ strain energy error for highly refined elements as for polygonal approximation by $\Omega^h$ (presumably at least the constants are better). There was, however, a hypothetical exception in our theory: if the leading term in the integral of $v^h$ around the boundary happens to vanish, the error should decrease like $h^5$. We can now announce a case in which this exception occurs! With quadratic polynomials, the six nodes of each triangle should go as usual at the vertices and at the midpoints of the edges -- or at the middle of the arc, in the case of a curved edge at the boundary. Then, with integrations carried out over the curved triangles, the interpolation of boundary conditions will not spoil the $h^4$ strain energy convergence which is normal for quadratics. This was conjectured by Scott, easily verified from the theory in [8], and then confirmed experimentally by Berger. We don't know how useful such an alternative to isoparametrics might be; the first step will be to study quadrature formulas over curved triangles.

## 4. Numerical integration.

Numerical quadrature is used more and more frequently to compute the integrals over separate elements (triangles or rectangles) which are assembled into the final stiffness matrix $K$. In a typical program this step actually costs more than the solution of the resulting linear system $KQ = F$, but it has received much less attention from numerical analysts. We want to establish the relation between the accuracy of a numerical integration formula and the error $u^h - \tilde{u}^h$ which it induces in the finite element approximation.

Mathematically, the result of numerical integration is to replace the given potential energy $I(v)$ by a discrete sum

$$\tilde{I}(v) = \tilde{a}(v,v) - 2\tilde{b}(f,v)$$
$$= \sum w_i [v_x^2(\xi_i) + v_y^2(\xi_i) - 2f(\xi_i)v(\xi_i)].$$

In each element there are a certain number of evaluation points $\xi_i = (x_i, y_i)$, with weights $w_i$ which depend on the geometry of the element and on the particular quadrature rule. When we minimize $\tilde{I}$ over all trial functions, the result is a perturbation $\tilde{K}\tilde{Q} = \tilde{F}$ (which the computer actually solves) of the true finite element system $KQ = F$. Expressed variationally, this means that the minimizing function $\tilde{u}^h = \Sigma\, \tilde{Q}_j \phi_j$ satisfies $\tilde{a}(\tilde{u}^h, v^h) = \tilde{b}(f, v^h)$, rather than the true equation of virtual work

$$\text{(11)} \qquad a(u^h, v^h) = b(f, v^h) \qquad \text{for all } v^h .$$

The first step in the analysis is to combine these two equations into the following identity for the energy in the error:

$$\text{(12)} \quad \tilde{a}(u^h - \tilde{u}^h, u^h - \tilde{u}^h) = (\tilde{a} - a)(u^h, u^h - \tilde{u}^h) - (\tilde{b} - b)(f, u^h - \tilde{u}^h) .$$

This same identity would be used to study the change in solution due to a change of coefficients and inhomogeneous term in the differential equation; those changes can also be expressed as a move from $a$ and $b$ to $\tilde{a}$ and $\tilde{b}$. The identity follows directly from the equations of virtual work, with $v^h = u^h - \tilde{u}^h$. Notice that the energy appearing on the left side is only the approximate one, and therefore we shall need to assume that it is positive definite: $\tilde{a}(v^h, v^h) \geq \theta a(v^h, v^h)$. For numerical integration this ensures that there are enough evaluation points $\xi_i$ in each element to make $\Sigma w_i v_x^2(\xi_i)$ and $\Sigma w_i v_y^2(\xi_i)$ strictly positive. (In one dimension, if $v_x$ is a polynomial of degree $k-2$ and the weights are positive, the minimum number of points is $k-1$. It will follow from our theorem that Gauss integration on these points leads to the same $h^{2(k-1)}$ energy error in $u^h - \tilde{u}^h$ as in $u - u^h$ : the left sides of (12) and (2) are of the same order, even with variable coefficients, and the numerical integration is very efficient. This estimate was derived jointly with George Fix, and closes the gap reported in [9] between the theoretical rate of convergence, which was too low, and the rate which is observed experimentally.)

We come directly to the main result, which will be recognized as an old friend: just as for nonconforming elements, the convergence of numerically integrated finite elements depends on the patch test. In other words, the quadrature rule should yield the correct integral over each element of any of the polynomials $v_x^h$ and $v_y^h$. (In general, the test applies to the $m^{th}$ derivatives of the trial functions; $m = 1$ in our example of the Poisson equation $-\Delta u = f$ .) This test is simple to perform: for linear elements it means that constant functions must be correctly integrated, for bilinear elements $v^h = a + bx + cy + dxy$ it means that linear polynomials must be correctly integrated, and so on.† Normally the

†For rectangular elements, the trial functions $v^h$ often include extra terms of high degree -- $xy$ in the bilinear

test will be passed with something to spare; the accuracy in the quadrature rule will be more than just minimal. We can express this extra accuracy in terms of a "higher order patch test" -- the equation (13) in the following theorem, in which $n = m$ yields the basic patch test -- and deduce from it a higher rate of convergence.

Theorem. Suppose that for all polynomials $P_n$ of degree $n$, and all trial functions $v^h$,

$$a(P_n, v^h) = \tilde{a}(P_n, v^h) \tag{13}$$

Then the error due to numerical integration is of order

$$a(u^h - \tilde{u}^h, u^h - \tilde{u}^h) = \mathcal{O}(h^{2r}), \quad r = n - m + 1$$

Outline of proof. We propose to estimate the right side of the identity (12). Its first term can be rewritten, because of (13), as

$$\begin{aligned} \tilde{a}(u^h, u^h - \tilde{u}^h) - a(u^h, u^h - \tilde{u}^h) \\ = \tilde{a}(u^h - P_n, u^h - \tilde{u}^h) - a(u^h - P_n, u^h - \tilde{u}^h) . \end{aligned} \tag{14}$$

---

case and $x^3y^3$ in the bicubic. These terms have no effect on the order of the approximation error $u - u^h$, since the complete polynomials of degree two and six are not present. The integration error $u^h - \tilde{u}^h$, however, will almost certainly be decided by these terms; they must be considered equally with those of lower degree. Thus the rate of convergence is not dependent only the degree of complete polynomial which is correctly integrated; it depends on all the terms in $v^h$.

Since numerical integration is carried out an element at a time, we can concentrate on a specific element and choose $P_n$ close to $u^h$. It is known that $u^h$ will be smooth if $u$ is, and therefore there is a polynomial such that the strains (mth derivatives) of $u^h - P_n$ are of order $h^r$. Summing the results over each separate element, (14) can be bounded by $Ch^r|u^h - \tilde{u}^h|$, where single bars denote the true strain energy $|v^h|^2 = a(v^h, v^h)$.

We want to find the same $h^r$ bound for the other term in (12), the error in computing $b = \iint fv$. Suppose for simplicity that the trial strains -- the mth derivatives of the polynomials $v^h$ -- consist of all polynomials of degree $q$. Then the test (13) amounts to saying that the quadrature is correct for all polynomials of degree $n - m + q$. Therefore

$$\text{(15)} \qquad \iint fv^h - \sum w_i (fv^h)(\xi_i) = \iint (fv^h - P_{n-m+q}) - \sum w_i (fv^h - P_{n-m+q})(\xi_i)$$

With the right choice of $P$ in each element, this will be of order $h^{n-m+q+1}$, multiplied by the integral of the absolute value of derivatives up to this order of $fv^h$. But $v^h$ can be differentiated only $q+m$ times, after which (being a polynomial) it disappears. Therefore, if $f$ is smooth, the Schwarz inequality establishes that (15) is of order

$$ch^{n-m+q+1}\left[\iint |D^{q+m}v^h|^2\right]^{1/2} \leq Ch^{n-m+1}\left[\iint |D^m v^h|^2\right]^{1/2} = O(h^r|v^h|) .$$

At the last step, the removal of $q$ differentiations from $v^h$ was paid for by the factor $h^{-q}$. Thus

$$(\tilde{b}-b)(f,u^h-\tilde{u}^h) = O(h^r|u^h-\tilde{u}^h|) .$$

Substituting these bounds on the right side of (12),

$$\theta|u^h-\tilde{u}^h|^2 \leq a(u^h-\tilde{u}^h,u^h-\tilde{u}^h) = O(h^r|u^h-\tilde{u}^h|) .$$

The conclusion of the theorem follows immediately: $|u^h-\tilde{u}^h| = O(h^r)$.

We have just space enough to describe, in bare outline, the results of applying a similar argument to problems more general than $-\Delta u = f$. (We plan to publish elsewhere the details of the theory, and more examples.) The adjustments required by variable coefficients, as in $-\nabla(p(x,y)\nabla u) + q(x,y)u = f$, present no difficulty. Correction terms must be allowed in (13), but the same accuracy in the quadrature rule -- every product of an $m^{th}$ derivative of $v^h$ times a polynomial of degree $n - m$ should be integrated exactly -- still yields an error of order $h^r$. For isoparametric elements, the change of coordinates transforms

$$\iint_{e_i} \frac{\partial P_1}{\partial x}\frac{\partial v^h}{\partial x}\,dxdy$$

into

$$c\iint_{E_i} (v^h_\xi y_\eta - v^h_\eta y_\xi)\,d\xi d\eta .$$

The patch test now requires that the quadrature rule, which will be applied in $\xi - \eta$ coordinates, give the correct value for the last integral. Since $v^h$ is a polynomial of the same form as the coordinate transformation $x(\xi,\eta)$ , the integrand exactly resembles the Jacobian $J = x_\xi y_\eta - x_\eta y_\xi$ , and we have confirmed the rule given in the book the Zienkiewicz: the volume of each element (the integral of J ) should be correctly computed by the quadrature. Therefore this rule, discovered by engineering intuition, is established as the correct convergence test for numerical integration of isoparametric elements.

## REFERENCES


1. Strang, G. "Approximation in the finite element method," Numer. Math. 19 (1972), 81-98.
2. Parlett, B. and O. Johnson, "Numerical implementation of variational methods for eigenvalue problems," Mathematical Software (J. Rice, Ed.) Academic Press, 1971.
3. Strang, G. and G. Fix, An Analysis of the Finite Element Method, submitted to Prentice-Hall in May, 1972.
4. Bazeley, G., Y. Cheung, B. Irons, and O. Zienkiewicz, "Triangular elements in bending -- conforming and nonconforming solution," Conf. on Matrix Methods, AFIT, Wright-Patterson, Ohio, 1965.
5. Wilson, E., R. Taylor, W. Doherty, J. Ghaboussi, "Incompatible displacement models," Symposium on Num. Meth. in Eng., Univ. of Illinois, 1971.
6. Strang, G. and A. Berger, "The change in solution due to change in domain," Proceedings of the A.M.S. Symposium on Partial Differential Equations, Berkeley, 1971, to be published by Academic Press.
7. Blair, J., "Bounds for the change in the solution... when the boundary is perturbed," to appear.
8. Berger, A., R. Scott, and G. Strang, "Approximate boundary conditions in the finite element method," Symposium of the Istituto Nazionale, Rome, 1972 (to appear in Symposia Mathematica, Academic Press).
9. Herbold, R., M. Schultz, and R. Varga, "Quadrature schemes for the numerical solution of boundary-value

problems by variational techniques," Aequ. Math. 3 (1969), 96-119. Also Aequ. Math. 7 (1972), 36-58.

# SPLINE APPROXIMATION AND DIFFERENCE SCHEMES FOR THE HEAT EQUATION

by

Vidar Thomée*

## 1. Introduction

In this paper we shall consider the approximate solution of the initial-value problem

$$\frac{\partial u}{\partial t} = \frac{\partial^2 u}{\partial x^2}, \quad t \geq 0, \quad x \in R, \tag{1.1}$$

$$u(x,0) = v(x), \quad x \in R, \tag{1.2}$$

where $R$ denotes the real axis. It is well-known that the exact solution is given by

$$u(x,t) = (E(t)v)(x) = \frac{1}{\sqrt{4\pi t}} \int_R e^{-\frac{(x-y)^2}{4t}} v(y)dy ,$$

and that the solution operator $E(t)$ has the property that for any nonnegative integers $s_1$ and $s_2$ with $s_1 \leq s_2$ there is a constant $C$ such that

---

*Sponsored by the United States Army under Contract No.: DA-31-124-ARO-D-462.

$$(1.3) \qquad \|E(t)v\|_{s_2} \leq Ct^{-\frac{1}{2}(s_2-s_1)} \|v\|_{s_1} .$$

Here, for any nonnegative integer $s$, and $v$ in the set $C^s$ of $s$ times continuously differentiable functions on $R(D = d/dx)$,

$$\|v\|_s = \sum_{j\leq s} \|D^j v\| , \qquad \text{where} \qquad \|v\| = \sup_{x\in R} |v(x)| .$$

(For a continuous unbounded $v$ the norm is then infinite.)

We want to find by means of Galerkin's method an approximate solution of (1.1), (1.2) in terms of splines of order $\mu$ based on a regular mesh with mesh-width $h$. For this purpose let $\chi$ be the characteristic function of the interval $[-\frac{1}{2}, \frac{1}{2}]$ and consider for $\mu \geq 1$ the function $\phi = \chi^{*\mu}$ obtained by convolving $\chi$ with itself $\mu$ times. Then $\phi$ which has support in $[-\frac{1}{2}\mu, \frac{1}{2}\mu]$, is the B-spline of order $\mu$. For $h$ a (small) positive number and $\ell$ in the set $Z$ of integers, let $\phi_\ell(x) = \phi(h^{-1} x - \ell)$, with the dependence on $\mu$ and $h$ suppressed in the notation and consider the set $S_h = \{\sum_\ell c_\ell \phi_\ell\}$ where $c_\ell$ grows at most as a power of $\ell$ as $|\ell|$ tends to infinity. The elements of $S_h$, the splines of order $\mu$ of at most power growth, are piecewise polynomials of degree $\mu - 1$ and belong to $C^{\mu-2}$.

The Galerkin method consists in finding $u_h(\cdot,t) \in S_h$ for $t \geq 0$ such that

(i) $u_h(\cdot,0) = v_h$ where $v_h \in S_h$ is an approximation of $v$;

(ii) $\frac{\partial u_h}{\partial t} - D^2 u_h$ is orthogonal with respect to the $L_2$-inner product $(\cdot,\cdot)$ to $S_h$ for all $t \geq 0$.

We shall express (ii) in the weak form

$$\left(\frac{\partial u_h}{\partial t}, \phi_\ell\right) + (Du_h, D\phi_\ell) = 0 , \quad \ell \in Z .$$

In terms of the coefficients of $u_h$ with respect to the basis $\{\phi_\ell\}$ in $S_h$ this leads to an infinite system of ordinary differential equations with respect to $t$, which under suitable assumptions on $v_h$ can be solved to give a solution $u_h(\cdot,t) = G_h(t)v_h$ where $G_h(t)$ is the Galerkin solution operator at time $t$.

For the purpose of studying the Galerkin solution operator we introduce the spline interpolation operator $S_h$ which takes a continuous function $v$ of at most power growth into the element $v_h \in S_h$ which agrees with $v$ at the mesh-points $hZ$. After some preliminaries in Section 2 concerning certain trigonometric polynomials related to the B-splines, some properties of the spline interpolation operator are described in Section 3.

We prove in Section 4 that associated with the Galerkin equations there is an operator $E_h(t)$ which can be thought of as a finite difference solution operator such that $u_h$ agrees with $E_h(t)v_h$ at the mesh-points. More precisely, if we introduce an auxiliary variable $k$ such that $k/h^2 = \lambda$ is kept constant, a generalized implicit finite difference operator $F_k$ can be defined such that, for $t = nk$ with $n$ a natural number, we have the following representation of the Galerkin solution operator, namely

$$G_h(t) = S_h E_h(t) = S_h F_k^n .$$

In Section 5 we analyse the finite difference operator $F_k$ by studying its symbol. It is proved that it is parabolic and accurate of order $2\mu - 2$. Based on known results from the finite difference theory we may conclude that although the optimal error estimates in approximation using splines of order $\mu$ is $O(h^\mu)$ as $h \to 0$, the error in $u_h$ at the mesh-points is only $O(h^{2\mu-2})$ as $h \to 0$. This suggests that when the Galerkin equations are based on splines of order $\mu$ with $\mu > 2$, higher order splines could be used in the final interpolation to obtain more accurate results.

The convergence rates quoted above are valid when the initial-data are smooth enough. When this is not the case it has been shown in the finite difference situation (cf. [9]) that by applying certain smoothing operators to the initial data it is possible to retain the optimal convergence rates for $t$ bounded away from zero. In Section 6 we discuss how this applies in the present context. In particular, the orthogonal projection in $L_2$ is shown to be related to such a smoothing operator.

In the previous discussion the ordinary differential equations are assumed to be solved exactly and the finite difference operator $F_k$ based on a time step $k$ is just a construction of the proof. However, in Section 7 we shall see that it is also possible to discretize in time, using for instance the Padé approximations of $e^{-\tau}$ to obtain genuine two-level finite difference approximations $F_k$ matching the accuracy attained in the space variable.

Certain mixed initial-boundary value problems can be rephrased as pure initial-value problems by periodicity considerations. In Section 8 we give an example of such a problem and show how a specific discrete time Galerkin scheme can be applied to initial data with and without smoothing.

Since the Galerkin method is based on the inner product structure in $L_2$, error estimates in the literature have generally been derived in the $L_2$-norm. In the present paper, where the estimates are based on the existing theory for finite difference operators, the analysis could have been carried out in any metric space for which such finite difference estimates are available. For the purpose of being specific we have selected here to work only with the maximum-norm.

The program carried out here for simplicity for the heat equation can be applied to more general situations. In a subsequent paper we plan to treat general constant coefficient systems in d-dimensional space and the difference schemes resulting from Galerkin's method and multi-splines.

The presentation in this paper, particularly in Sections 2 and 3, depends heavily on Schoenberg's theory of "cardinal spline interpolation," cf. [11], [12], [13], [14]. Similar methods to his have been applied in the context of projection methods particularly for elliptic problems in e.g. Aubin [1], Babuška [2], Bramble and

Hilbert [3] and Strang [15]. For the use of Galerkin's method based on spline-spaces in connection with initial-value problems, see Douglas and Dupont [5], Dupont [6], Fix and Nassif [7], Price and Varga [10], Strang and Fix [16] and references. The material needed here from the finite difference theory can be found in Kreiss, Thomée and Widlund [9], Thomée [18], and Widlund [19].

## 2. Some Special Trigonometric Polynomials

We define, with $\chi$ the characteristic function of $[-\frac{1}{2}, \frac{1}{2}]$ and with $j$ a natural number,

$$g_j(\theta) = \sum_{\ell} \chi^{*j}(\ell) e^{-i\ell\theta} .$$

Since $\chi^{*j}$ is even and has its support in $[-\frac{1}{2}j, \frac{1}{2}j]$ we find immediately that $g_j$ is an even trigonometric polynomial of order $[\frac{1}{2}(j-1)]$.

We shall often use Fourier transforms below which we define for $v \in L_1(R)$ as

$$(Fv)(\xi) = \hat{v}(\xi) = \int_R v(x) e^{-ix\xi} dx .$$

We have the following representation for $g_j$:

Lemma 2.1. We have for $j \geq 2$,

$$g_j(\theta) = \sum_{\ell} \hat{\chi}(\theta + 2\pi\ell)^j, \quad \text{where} \quad \hat{\chi}(\theta) = \frac{2 \sin \frac{1}{2}\theta}{\theta} .$$

Proof. Noticing that

$$F(\chi^{*j}(x)e^{-ix\theta})(\xi) = \hat{\chi}(\theta+\xi)^j ,$$

we have by Poisson's summation formula,

$$\sum_{\ell} \chi^{*j}(\ell)e^{-i\ell\theta} = \sum_{\ell} F(\chi^{*j}(x)e^{-ix\theta})(2\pi\ell) = \sum_{\ell} \hat{\chi}(\theta+2\pi\ell)^j .$$

The following is a simple consequence:

Lemma 2.2. We have for $j \geq 2$,

$$g_j(\theta) = \hat{\chi}(\theta)^j + O(\theta^j) \quad \text{as} \quad \theta \to 0 .$$

Proof. By Lemma 2.1 we have

$$g_j(\theta) = \hat{\chi}(\theta)^j + \sum_{\ell\neq 0} \hat{\chi}(\theta+2\pi\ell)^j = \hat{\chi}(\theta)^j$$

$$+ \left(2 \sin \frac{1}{2}\theta\right)^j \sum_{\ell\neq 0} \frac{(-1)^{j\ell}}{(\theta+2\pi\ell)^j} ,$$

from which the result immediately follows since the last sum is bounded for small $\theta$.

The coefficients in the above trigonometric polynomials can be found easily by relating them to the following ordinary polynomials. We define recursively

$$\Pi_0(t) = P_0(t) = 1 ,$$

$$\Pi_{n+1}(t) = (1+nt)\Pi_n(t) + t(1-t)\Pi_n'(t) ,$$

$$P_{n+1}(t) = (1+(2n+1)t)P_n(t) + 2t(1-t)P_n'(t) .$$

The polynomials $\Pi_n$ of degree $n-1$ are referred to by Schoenberg [14] as the Euler-Frobenius polynomials; they can also be defined by the identities

$$\frac{\Pi_n(t)}{(1-t)^{n+1}} = \sum_{\ell=0}^{\infty} (\ell+1)^n t^\ell .$$

Similarly, the polynomials $P_n$ of degree $n$ can be defined by

$$\frac{P_n(t)}{(1-t)^{n+1}} = \sum_{\ell=0}^{\infty} (2\ell+1)^n t^\ell .$$

Both $\Pi_n$ and $P_n$ are monic reciprocal polynomials with integer coefficients and all their zeros are negative and simple (cf. [14]). They are tied to the present trigonometric polynomials in the following way:

<u>Lemma 2.3.</u> We have

$$g_j(\theta) = \begin{cases} e^{\frac{1}{2}(j-1)\theta i} 2^{-(j-1)} \dfrac{1}{(j-1)!} P_{j-1}(e^{-i\theta}) & \text{for } j \text{ odd} , \\ e^{(\frac{1}{2}j-1)\theta i} \dfrac{1}{(j-1)!} \Pi_{j-1}(e^{-i\theta}) & \text{for } j \text{ even} . \end{cases}$$

<u>Proof.</u> This is an immediate consequence of the definition of $g_j$ and the facts that with $Q_j(t) = \chi^{*j}(t-j/2)$,

$$\Pi_{j-1}(t) = (j-1)! \sum_{\ell=0}^{j-2} Q_j(\ell+1)t^\ell ,$$

$$P_{j-1}(t) = 2^{j-1}(j-1)! \sum_{\ell=0}^{j-1} Q_j(\ell+\tfrac{1}{2})t^\ell ,$$

which are proved in [14].

The following are some of the first polynomials $g_j$:

$$g_1 = g_2 = 1, \quad g_3 = \frac{1}{4}(3+\cos\theta),$$

$$g_4 = \frac{1}{3}(2+\cos\theta), \quad g_6 = \frac{2}{5!}(33+26\cos\theta+\cos 2\theta),$$

$$g_8 = \frac{2}{7!}(1208+1191\cos\theta+120\cos 2\theta+\cos 3\theta).$$

Lemma 2.4. The trigonometric polynomial $g_j(\theta)$ is positive for any $j \geq 1$.

Proof. For even $j$ this is an immediate consequence of Lemma 2.1, since $g_j$ is $2\pi$-periodic and $\hat{\chi}$ is positive for $|\theta| \leq \pi$. Consider now the case of odd $j$. Since $P_{j-1}$ is an even degree reciprocal polynomial with all its zeros simple and negative, we have that $P_{j-1}(-1) \neq 0$ and hence that $\exp(\frac{1}{2}(j-1)\theta i)P_{j-1}(e^{-i\theta}))$ is real and non-zero for all real $\theta$. Since clearly $P_{j-1}(1)$ is positive the result follows in this case by Lemma 2.3 (cf. also [3]).

We shall later have reason to consider the bilinear form $(D^j v, D^j w)$ for sufficiently smooth $v$ and $w$. We define the symbol of this form $(j < \mu)$ with respect to $S_h$ to be

$$g_{\mu,j}(\theta) = h^{2j-1} \sum_{\ell} (D^j\phi_0, D^j\phi_\ell) e^{-i\ell\theta} ,$$

where as in the introduction $\phi_\ell(x) = \phi(h^{-1}x-\ell)$, $\phi = \chi^{*\mu}$. Since $\phi$ has compact support this is a trigonometric polynomial. The factor $h^{2j-1}$ makes the symbol independent of $h$.

Lemma 2.5. We have for $j < \mu$,

$$g_{\mu,j}(\theta) = \sum_{\ell} (\theta + 2\pi\ell)^{2j}\hat{\chi}(\theta + 2\pi\ell)^{2\mu} = \theta^{2j}\hat{\chi}(\theta)^{2\mu} + O(\theta^{2\mu}) ,$$

as $\theta \to 0$.

Proof. Since $g_{\mu,j}$ is independent of $h$ we may choose $h = 1$. We then have by Parseval's relation,

$$(D^j\phi_0, D^j\phi_\ell) = \int_R D^j\phi(x+\ell)D^j\phi(x)\,dx = (2\pi)^{-1} \int_R \xi^{2j}\hat{\phi}(\xi)^2 e^{i\ell\xi}d\xi$$

$$= (2\pi)^{-1} F(\xi^{2j}\hat{\phi}(\xi)^2)(-\ell) ,$$

and hence

$$(D^j\phi_0, D^j\phi_\ell)e^{-i\ell\theta} = (2\pi)^{-1} F_\xi((\xi+\theta)^{2j}\hat{\phi}(\xi+\theta)^2)(-\ell)$$

$$= F_\xi^{-1}((\xi+\theta)^{2j}\hat{\phi}(\xi+\theta)^2)(\ell) ,$$

so that by Poisson's summation formula,

$$g_{\mu,j}(\theta) = \sum_{\ell} (\theta + 2\pi\ell)^{2j}\hat{\phi}(\theta + 2\pi\ell)^2 ,$$

which proves the first part of the result. The second part now follows as in the proof of Lemma 2.2.

The new polynomials have the following relation to the $g_j$.

Lemma 2.6. For $j < \mu$ we have

$$g_{\mu,j}(\theta) = (2 \sin \frac{1}{2} \theta)^{2j} g_{2\mu-2j}(\theta) .$$

Proof. This is an immediate consequence of Lemmas 2.1 and 2.5 and the form of $\hat{\chi}$.

## 3. Spline Interpolation

Let us first state some well-known facts about Toeplitz operators on sequences. For this purpose, consider sequences $y = (y_j)_{j=-\infty}^{\infty}$ of

$$\mathcal{S} = \{y; y_j = O(|j|^{-q}) \text{ as } |j| \to \infty \text{ for all real } q\} ,$$

$$\mathcal{S}' = \{y; y_j = O(|j|^{q}) \text{ as } |j| \to \infty \text{ for some real } q\} ,$$

$$\ell_\infty = \{y; y_j = O(1) \text{ as } |j| \to \infty\} .$$

For a sequence $y$ in $\ell_\infty$ (and hence also in $\mathcal{S}$) we define

$$\|y\| = \sup_j |y_j| .$$

Let now $\tau = (\tau_j) \in \mathcal{S}$ and consider the Toeplitz operator $T: y \to z$ defined by

$$z_j = \sum_\ell \tau_{j-\ell} y_\ell . \tag{3.1}$$

It is easy to see that $T$ maps $\mathcal{S}$, $\mathcal{S}'$ and $\ell_\infty$ into themselves. In particular,

$$\|Ty\| \leq \left(\sum_\ell |\tau_\ell|\right) \|y\| .$$

For $y \in \mathcal{S}$, let $\mathcal{F}y = \tilde{y}$ denote its Fourier transform,

$$(\mathcal{F}y)(\theta) = \tilde{y}(\theta) = \sum_j y_j e^{-ij\theta} .$$

In particular,

$$\tilde{\tau}(\theta) = \sum_j \tau_j e^{-ij\theta}$$

is the characteristic function of $T$. Notice that $\mathcal{F}\mathcal{S}$ is the set $C_{2\pi}^\infty$ of $2\pi$-periodic functions in $C^\infty$ and that for $y \in \mathcal{S}$,

$$\mathcal{F}(Ty) = \tilde{\tau}\mathcal{F}y .$$

About the invertibility of $T$ (or the solvability of (3.1)) we have the following:

Lemma 3.1. Assume $\tilde{\tau}(\theta) \neq 0$ for all real $\theta$ and set

$$\frac{1}{\tilde{\tau}(\theta)} = \tilde{\gamma}(\theta) = \sum_j \gamma_j e^{-ij\theta} \quad \text{where} \quad \gamma = (\gamma_j) \in \mathcal{S} .$$

Then $T^{-1}$ exists on $\mathcal{S}$, $\mathcal{S}'$ and $\ell_\infty$ and is the Toeplitz operator defined by

$$(T^{-1}z)_j = \sum_\ell \gamma_{j-\ell} z_\ell .$$

Now let $\mu$ be a positive integer which we shall consider fixed for the rest of the paper and let $\phi = \chi^{*\mu}$ where $\chi$ is the characteristic function of $[-\frac{1}{2}, \frac{1}{2}]$. We set, without explicitly emphasizing the dependence upon $\mu$ and $h$ in the notation, $\phi_j(x) = \phi(h^{-1}x-j)$, and

$$S_h = \{v = \sum_j c_j\phi_j, \; c = (c_j) \in \mathcal{S}'\} .$$

Thus $S_h$, the set of splines of order $\mu$ of at most power growth (cf. Schoenberg [13]) consists of functions in $C^{\mu-2}$ which reduce to polynomials of degree $\mu-1$ in each interval $(jh,(j+1)h)$, $j \in Z$, for $\mu$ even and $((j-\frac{1}{2})h,(j+\frac{1}{2})h)$, $j \in Z$, for $\mu$ odd.

Consider now functions $v$ in the set $C_*$ of continuous functions of at most power growth; i.e.

$$C_* = \{v \in C, v(x) = O(|x|^q) \text{ as } |x| \to \infty \text{ for some real } q\} .$$

We have:

<u>Lemma 3.2</u>. For $v \in C_*$ there is a unique $v_h \in S_h$ such that

$$v_h(\ell h) = v(\ell h) , \quad \ell \in Z. \tag{3.2}$$

The coefficients $(c_j)$ in $v_h = \sum_j c_j\phi_j$ can be determined from

$$g_\mu(\theta)\tilde{c}(\theta) = \tilde{v}(\theta) ,$$

where $\tilde{c} = \mathcal{F}(c_j)$, $\tilde{v} = \mathcal{F}(v(jh))$, and $g_\mu$ is defined as in Section 2.

Proof. The equations (3.2) reduce to

$$\sum_j c_j \phi(\ell - j) = v(\ell h) , \quad \ell \in Z,$$

and the result therefore follows by Lemma 3.1 in view of the fact that by Lemma 2.4,

$$\sum_\ell \phi(\ell) e^{-i\ell\theta} = g_\mu(\theta) > 0 , \quad \theta \in R .$$

Notice that without the requirement that $c \in \mathcal{S}'$ in the definition of $S_h$ we would not have uniqueness (cf. [13]).

We shall now consider the linear operator $S_h$ which takes $v$ into $v_h$ in Lemma 3.2.

Lemma 3.3. The operator $S_h : v \to v_h$ defined by (3.2) has the following properties:

(i) There is a constant $C$ independent of $v$ and $h$ such that

$$\|S_h v\| \leq C \|v\| ,$$

(ii) If $P$ is a polynomial of degree at most $\mu-1$ we have $S_h P = P$,

(iii) For any $s \in [0,\mu]$ there is a positive constant $C$ such that

$$\|(S_h - I)v\| \leq Ch^s \|v\|_s , \quad v \in C^s .$$

Proof. The first property follows easily from the fact that since $\phi$ has compact support, we have with

$$S_h v = v_h = \sum_j c_j \phi_j , \qquad c = (c_j) ,$$

$$\| S_h v \| \leq C \| c \| \leq C \| (v(jh)) \| \leq C \| v \| .$$

The property (ii) is obvious from Lemma 3.2 as $P \in S_h$ and interpolates itself. To prove (iii) finally we notice that for $s = \mu$, by a change of coordinates it is sufficient to prove that with $h = 1$,

$$\| (S_1 - I) v \| \leq C \| D^\mu v \| , \qquad v \in C^\mu . \tag{3.3}$$

When the results is thus proved for $s = \mu$ and by (i) for $s = 0$, it follows for intermediate $s$ by interpolation between $C^s$ spaces (cf. e.g. [4]). To prove (3.3), it is sufficient to prove that for $y \in (0,1)$,

$$| S_1 v(y) - v(y) | \leq C \| D^\mu v \| .$$

To prove this, we write

$$v(x) = \sum_{j=1}^{\mu-1} \frac{x^j}{j!} (D^j v)(0) + R_\mu(x) = P_\mu(x) + R_\mu(x) ,$$

where

$$| R_\mu(x) | \leq \frac{|x|^\mu}{\mu!} \| D^\mu v \| . \tag{3.4}$$

Using (ii) we have

$$S_1 v - v = S_1 P_\mu + S_1 R_\mu - P_\mu - R_\mu = S_1 R_\mu - R_\mu .$$

By (3.4), $|R_\mu(y)|$ is bounded by the right hand side of (3.3) and with the notation of Lemma 3.1, there is a $J$

such that for $y \in (0,1)$,

$$|(S_1 R_\mu)(y)| = |\sum_{|j|\leq J} c_j \phi_j(y)| \leq C \max_{|j|\leq J} |c_j| ,$$

and for these $j$ we have by (3.4),

$$|c_j| = |\sum_\ell \gamma_\ell R_\mu(j-\ell)| \leq C\left(\sum_\ell |\gamma_\ell||\ell+J|^\mu\right) \|D^\mu v\| ,$$

which completes the proof.

## 4. The Continuous Time Galerkin Method

We now return to the Galerkin equations,

$$\left(\frac{\partial u_h}{\partial t}, \phi_\ell\right) + (Du_h, D\phi_\ell) = 0 , \quad \ell \in Z , \tag{4.1}$$

$$u_h(\cdot,0) = v_h .$$

Setting

$$u_h(x,t) = \sum_j c_j(t)\phi_j(x) ,$$

this can be written

$$\sum_j \{c_j'(t)(\phi_j,\phi_\ell) + c_j(t)(D\phi_j,D\phi_\ell)\} = 0 , \quad \ell \in Z .$$

Assuming that $(c_j(t))$ and $(c_j'(t))$ are in $\mathcal{S}$ we may introduce their Fourier transforms $\tilde{c}(\theta,t)$ and $\frac{d}{dt}\tilde{c}(\theta,t)$, and since $(\phi_j,\phi_\ell)$ and $(D\phi_j,D\phi_\ell)$ only depend on $\ell-j$ we conclude that with the notation of Section 2,

$$hg_{\mu,0}(\theta)\frac{d}{dt}\tilde{c}(\theta,t) + h^{-1}g_{\mu,1}(\theta)\tilde{c}(\theta,t) = 0 .$$

This ordinary differential equation with respect to $t$ may be solved to yield

$$\tilde{c}(\theta,t) = \exp\left(-\frac{t}{h^2}\frac{g_{\mu,1}(\theta)}{g_{\mu,0}(\theta)}\right)\tilde{c}(\theta,0) .$$

It follows that

$$c_j(t) = \sum_{\ell} f_{\ell}(t/h^2)c_{j-\ell}(0) ,$$

where $f_{\ell}(t/h^2)$ are the Fourier coefficients of the exponential,

$$\exp\left(-\frac{t}{h^2}\frac{g_{\mu,1}(\theta)}{g_{\mu,0}(\theta)}\right) = \sum_{\ell} f_{\ell}(t/h^2)e^{-i\ell\theta} .$$

By Lemma 3.2 this is equivalent to

$$u_h(jh,t) = \sum_{\ell} f_{\ell}(t/h^2)v_h(jh-\ell h) . \tag{4.3}$$

Since the exponential in (4.2) belongs to $C^{\infty}_{2\pi}$ we have $(f_{\ell}(t/h^2)) \in \mathcal{S}$ so that (4.3) defines $u_h(\cdot,t)$ not only for $(v_h(jh)) \in \mathcal{S}$ but more generally for $(v_h(jh)) \in \mathcal{S}'$. In particular, for $(v_h(jh)) \in \ell_{\infty}$ we let $u_h(\cdot,t) = G_h(t)v_h \in S_h$ be the spline defined at the mesh-points by (4.3).

Now let $\lambda$ be a fixed positive number and introduce an auxiliary variable $k$ by $k = \lambda h^2$. We shall consider $t$ of the form $nk$ where $n$ is a natural number. If we are only interested in the solution for a fixed $t$ we can choose $\lambda = 1$ below without restricting the generality. However, consideration of all $t$ in an interval $[0,T]$ does require a certain uniformity in some of the estimates below with respect to $\lambda$.

Let us introduce the finite difference operator

$$F_k v(x) = \sum_j f_j v(x - jh) ,$$

with symbol

$$(4.4) \qquad \rho(\theta) = \sum_j f_j e^{-ij\theta} = \exp\left(-\lambda \frac{g_{\mu,1}(\theta)}{g_{\mu,0}(\theta)}\right) .$$

The iterated operator $F_k^n$ then has the symbol

$$\rho(\theta)^n = \exp\left(-\frac{t}{h^2}\frac{g_{\mu,1}(\theta)}{g_{\mu,0}(\theta)}\right) = \sum_j f_j(t/h^2)e^{-ij\theta} .$$

This implies that

$$u_h(jh,t) = (F_k^n v_h)(jh) , \qquad j \in Z ,$$

or

$$u_h(x,t) = (S_h F_k^n v_h)(x) .$$

We have proved the following representation result:

<u>Theorem 4.1</u>. Let $\lambda$ be a fixed positive number and consider with $k = \lambda h^2$ the finite difference operator $F_k$ with symbol defined in (4.4). Then for $t = nk$ the solution operator $G_h(t) : v_h \to u_h(\cdot,t)$ of the Galerkin equations (4.1) can be written in the form

$$G_h(t) = S_h F_k^n .$$

## 5. Application of the Finite Difference Theory

We recall that a finite difference operator

$$F_k v(x) = \sum_j f_j v(x - jh), \quad \sum_j |f_j| < \infty, \quad k/h^2 = \lambda = \text{constant} ,$$

with symbol

$$\rho(\theta) = \sum_j f_j e^{-ij\theta} ,$$

is said to be consistent with the heat equation and accurate of order $\nu$ where $\nu$ is a positive integer if

$$\rho(\theta) = \exp(-\lambda\theta^2 + O(\theta^{2+\nu})) \quad \text{as} \quad \theta \to 0 . \tag{5.1}$$

Further such a difference operator is said to be parabolic if for some positive $c$,

$$|\rho(\theta)| \leq \exp(-c\theta^2) , \quad \text{for} \quad |\theta| \leq \pi .$$

For most operators considered in the literature $\rho$ is a rational trigonometric function. In our applications below this will not necessarily be the case.

The finite difference operator

$$D_h^{(q)} v(x) = h^{-q} \sum_j d_j v(x - jh) ,$$

is said to be consistent with the differential operator $D^q$ and accurate of order $\nu$ if for $0 \leq s \leq \nu$ there is a constant $C$ such that

$$\|D_h^{(q)} v - D^q v\| \leq Ch^s \|v\|_{q+s} , \quad v \in C^{q+s} .$$

Notice that such an operator commutes with spline interpolation operators.

From the theory of difference operators we quote the following result (cf. [19] and references).

Lemma 5.1. If $F_k$ is a parabolic difference operator consistent with the heat equation and accurate of order $\nu$ and if $0 \leqslant s \leqslant \nu$ then there is a constant $C$ such that for $0 \leqslant t = nk \leqslant T$,

$$\| F_k^n v - E(nk)v \| \leqslant Ch^s \| v \|_s , \quad v \in C^s .$$

and such that if $D_h^{(q)}$ is a finite difference operator consistent with $D^q$ and also accurate of order $\nu$ then for $0 < t_0 \leqslant t = nk \leqslant T$,

$$\| D_h^{(q)} F_k^n v - D^q E(nk)v \| \leqslant Ch^s \| v \|_s , \quad v \in C^s .$$

These estimates were stated in [19] only for $v$ vanishing at infinity; it is easy to see that they are in fact valid without this restriction.

Lemma 5.2. The finite difference operator $F_k$ with symbol defined by (4.4) is parabolic, consistent with the heat equation and accurate of order $2\mu - 2$.

Proof. We have by Lemma 2.6,

$$-\lambda \frac{g_{\mu,1}(\theta)}{g_{\mu,0}(\theta)} = -\lambda (2 \sin \tfrac{1}{2}\theta)^2 \frac{g_{2\mu-2}(\theta)}{g_{2\mu}(\theta)} .$$

In view of Lemma 2.3 the parabolicity immediately follows. Also, by Lemma 2.5 we have

$$-\lambda \frac{g_{\mu,1}(\theta)}{g_{\mu,0}(\theta)} = -\lambda \frac{\theta^2 \hat{\chi}(\theta)^{2\mu} + O(\theta^{2\mu})}{\hat{\chi}(\theta)^{2\mu} + O(\theta^{2\mu})} = -\lambda\theta^2 + O(\theta^{2\mu}) \quad \text{as } \theta \to 0 ,$$

from which condition (5.1) follows with $\nu = 2\mu - 2$.

Application to the Galerkin situation gives at once:

Theorem 5.1. Let $G_h(t)$ be the Galerkin solution operator in Theorem 4.1. Then for $s$ with $0 \leqslant s \leqslant \mu$ there is a constant $C$ such that for $0 \leqslant t = nk \leqslant T$,

$$\|G_h(t)v - E(t)v\| \leqslant Ch^s \|v\|_s , \quad v \in C^s ,$$

and such that if $D_h^{(q)}$ is consistent with $D^q$ and accurate of order $\mu$ we have for $0 < t_0 \leqslant t = nk \leqslant T$,

$$\|D_h^{(q)} G_h(t)v - D^q E(t)v\| \leqslant Ch^s \|v\|_s , \quad v \in C^s .$$

Proof. By the representation of $G_h(t)$ (Theorem 4.1) we have using Lemma 3.3(i) and (iii),

$$\begin{aligned}\|G_h(t)v - E(t)v\| &\leqslant \|S_h(F_k^n - E(nk))v\| + \|(S_h - I)E(nk)v\| \\ &\leqslant C\|(F_k^n - E(nk))v\| + Ch^s\|E(t)v\|_s .\end{aligned}$$

The first result therefore follows by Lemma 5.1 and the estimate (1.3) with $s_1 = s_2 = s$. The second result follows similarly from

$$\begin{aligned}\|D_h^{(q)} G_h(t)v - D^q E(t)v\| &\leqslant \|S_h(D_h^{(q)} F_k^n - D^q E(nk))v\| \\ &\quad + \|(I - S_h)D^q E(t)v\| .\end{aligned}$$

Let us comment that if we only consider the values of the solution at the mesh-points then a direct application of Lemma 5.1 would give an $O(h^{2\mu-2})$ error estimate for smooth $v$. It is hence clear that if the interpolation was carried out instead by means of splines of order $2\mu - 2$ we would preserve this accuracy for all $x$.

## 6. Smoothing and Best $L_2$-Approximation of Initial Data

If the initial-values $v$ are not smooth enough it might be the case that the pure interpolant $S_h v$ is not the best choice of an initial-function $v_h$ for the Galerkin problem. Instead we might want to take for the set of values of $v_h$ at the mesh-points the result of some smoothing operator $M_h$ acting on $v$. More precisely, for a given $\psi \in L^1(R)$ set

$$M_h^{(\psi)} v = \psi_{(h)} * v \quad \text{where} \quad \psi_{(h)}(x) = h^{-1}\psi(h^{-1}x) ,$$

and let $v_h = P_h v$ be the spline (of order $\mu$) which agrees with $M_h^{(\psi)} v$ at the mesh-points, i.e.

$$v_h = P_h v = S_h M_h^{(\psi)} v . \tag{6.1}$$

An operator $M_h = M_h^{(\psi)}$ of this form is called a smoothing operator of order $\nu$ in [9] if certain conditions are satisfied. In particular, these conditions are met if there is a $\delta > 0$ such that

(i) $\hat{\psi}(\xi) = 1 + \xi^\nu b^{(0)}(\xi)$ for $|\xi| < 2\delta$ where $b^{(0)} \in C^\infty$,

(ii) $\hat{\psi}(\xi) = (\sin \tfrac{1}{2}\xi)^\nu b^{(1)}(\xi)$ for $|\xi| > \delta$ where $b^{(1)} \in \mathcal{F}L_1$,

where $\mathcal{F}L_1$ denotes the set of Fourier transforms of functions in $L_1(R)$.

The following result for finite difference schemes is a special case of the main result in [9]. (Again the restriction in [9] to functions which vanish at infinity is unessential.)

Lemma 6.1. Let $F_k$ be a parabolic finite difference operator consistent with the heat equation and accurate of order $\nu$ and let $M_h$ be a smoothing operator of order $\nu$. Then there is a constant $C$ such that for $0 < t = nk \leq T$,

$$\|F_k^n M_h v - E(nk)v\| \leq Ch^{\nu} t^{-\frac{1}{2}\nu} \|v\| \ .$$

Smoothing operators of the form described above were constructed in [9]. In particular, the conditions (i) and (ii) are satisfied if $\hat{\psi}$ is of the form $\hat{\psi} = \hat{\chi}^{\nu} q$ where $\chi$ is as before and $q$ is a $2\pi$-periodic $C^{\infty}$ function such that $\hat{\psi}(\xi) = 1 + O(\xi^{\nu})$ as $\xi \to 0$.

The following choices of $\psi$ correspond to smoothing operators $M_h^{(j)}$ of orders $j = 1, 2, 3$, and 4, respectively (cf. [9]):

$$\hat{\psi}_1(\xi) = \hat{\chi}(\xi) \ , \quad \hat{\psi}_2(\xi) = \hat{\chi}(\xi)^2 \ ,$$

$$\hat{\psi}_3(\xi) = \hat{\chi}(\xi)^3 \left(\frac{5}{4} - \frac{1}{4}\cos\xi\right) \ ,$$

$$\hat{\psi}_4(\xi) = \hat{\chi}(\xi)^4 \left(\frac{4}{3} - \frac{1}{3}\cos\xi\right) \ .$$

Expressed in terms of B-splines they are

$$\psi_1 = \chi, \quad \psi_2 = \chi * \chi \ ,$$

$$\psi_3(x) = \frac{5}{4}\chi^{*3}(x) - \frac{1}{8}\chi^{*3}(x+1) - \frac{1}{8}\chi^{*3}(x-1) ,$$

$$\psi_4(x) = \frac{4}{3}\chi^{*4}(x) - \frac{1}{6}\chi^{*4}(x+1) - \frac{1}{6}\chi^{*4}(x-1) . \tag{6.2}$$

Smoothing operators of the form just described can now be applied to the Galerkin situation. With the above choice of $v_h = P_h v$ the solution operator of the Galerkin problem can be represented for $t = nk$ as

$$G_h(t)P_h v = S_h F_k^n M_h v , \tag{6.3}$$

and we may state the following:

<u>Theorem 6.1</u>. Let the continuous time Galerkin solution operator $G_h(t)$ be as in Theorem 4.1 and let $M_h$ be a smoothing operator of order $\mu$. Then there is a positive constant $C$ such that for $0 < t = nk \leq T$ we have

$$\|G_h(t)P_h v - E(t)v\| \leq Ch^{\mu}t^{-\frac{1}{2}\mu}\|v\| .$$

<u>Proof</u>. We have using (6.3),

$$\begin{aligned}\|G_n(t)P_h v - E(t)v\| &\leq \|S_h(F_k^n M_h - E(t))v\| + \|(S_h - I)E(t)v\| \\ &\leq Ch^{\mu}t^{-\frac{1}{2}\mu}\|v\| + Ch^{\mu}\|E(t)v\|_{\mu} ,\end{aligned}$$

and the result follows from the estimate (1.3) with $s_1 = 0$, $s_2 = \mu$.

Again it is obvious that if $S_h$ is taken to be the spline interpolation operator of order $\nu = 2\mu - 2$ and $M_h$ is also taken to be accurate of this order $\nu$, then the result could be sharpened to read

$$\|G_h(t)v - E(t)v\| \leq Ch^{\nu} t^{-\frac{1}{2}\nu} \|v\| \quad .$$

A special initial-value operator is given for $v \in L_2$ by the $L_2$-projection into $S_h \cap L_2$. It turns out that this operator is indeed an operator of the form (6.1).

Theorem 6.2. The orthogonal projection of $L_2$ into $S_h \cap L_2$ can be represented as $P_h = S_h M_h$ where $M_h$ is a smoothing operator of order $\mu$.

Proof. The $L_2$-projection $v_h = P_h v = \sum_j c_j \phi_j$ is determined by the equations

$$\sum_j c_j (\phi_j, \phi_\ell) = (v, \phi_\ell) , \quad \ell \in Z .$$

Setting, again with $\phi = \chi^{*\mu}$,

$$\tilde{w}(\theta) = \sum_\ell w_\ell e^{-i\ell\theta} \quad \text{where} \quad w_\ell = h^{-1}(v, \phi_\ell) = (\phi_{(h)} * v)(\ell h) ,$$

we obtain as in Section 4,

$$g_{2\mu}(\theta)\tilde{c}(\theta) = \tilde{w}(\theta) , \quad \text{where} \quad \tilde{c} = \mathcal{F}(c_j) ,$$

and hence by Lemma 3.2, with $\tilde{v}_h = \mathcal{F}(v_h(jh))$,

$$\tilde{v}_h(\theta) = \tilde{w}(\theta) \frac{g_\mu(\theta)}{g_{2\mu}(\theta)} = \tilde{w}(\theta) \sum_j \beta_j e^{-ij\theta} ,$$

where the $\beta_j$ are the Fourier coefficients of $g_{2\mu}^{-1} g_\mu$. This means that $v_h$ is the spline in $S_h$ for which

$$v_h(\ell h) = \sum_j \beta_j w_{\ell - j} = \sum_j \beta_j (\phi_{(h)} * v)(\ell h - jh) = (\psi_{(h)} * v)(\ell h)$$
$$= (M_h^{(\psi)} v)(\ell h) ,$$

where $\psi(x) = \sum_j \beta_j \phi(x-j)$. The result therefore follows if we can prove that $M_h^{(\psi)}$ is a smoothing operator of order $\mu$. We have

$$\hat{\psi}(\xi) = \hat{\phi}(\xi) \frac{g_\mu(\xi)}{g_{2\mu}(\xi)} = \hat{\chi}(\xi)^\mu \frac{g_\mu(\xi)}{g_{2\mu}(\xi)} ,$$

and hence, using Lemma 2.2,

$$\hat{\psi}(\xi) = \hat{\chi}(\xi)^\mu \frac{\hat{\chi}(\xi)^\mu + O(\xi^\mu)}{\hat{\chi}(\xi)^{2\mu} + O(\xi^{2\mu})} = 1 + O(\xi^\mu) \quad \text{as} \quad \xi \to 0 ,$$

so that indeed $\psi$ generates a smoothing operator of order $\mu$.

## 7. Discrete Time Galerkin Methods

In the previous discussion it was assumed that the system of ordinary differential equations in time which resulted from the Galerkin equations (4.1) could be solved exactly. In this section we shall introduce a class of discretizations in time and analyse the corresponding difference schemes.

Let $r$ be a rational function with real coefficients,

$$r(\eta) = \frac{p(\eta)}{q(\eta)}, \quad p(\eta) = \sum_{j=0}^{M} p_j \eta^j , \quad q(\eta) = \sum_{j=0}^{N} q_j \eta^j ,$$

with the following properties:

$$q(\eta) > 0 \quad \text{for} \quad \eta \geq 0 ,$$

$$r(\eta) = e^{-\eta} + O(\eta^{\sigma+1}) \quad \text{as} \quad \eta \to 0 . \tag{7.1}$$

We introduce the bilinear forms

$$P_k(v,w) = \sum_{j=0}^{M} p_j k^j (D^j v, D^j w) ,$$

$$Q_k(v,w) = \sum_{j=0}^{N} q_j k^j (D^j v, D^j w) .$$

We assume that the degrees $M,N$ of $p,q$ are both less than $\mu$.

We can then pose the discrete Galerkin problem: let $\lambda$ be a fixed positive constant and $k = \lambda h^2$. Find $u_h^n = u_h(\cdot, nk) \in S_h$ recursively by

(i) $u_h^0 = v_h$ ,

(ii) $Q_k(u_h^{n+1}, \phi_\ell) = P_k(u_h^n, \phi_\ell)$ , $\ell \in Z$ .

Setting

$$u_h^n(x) = \sum_j c_j^n \phi_j(x) ,$$

the equations (ii) reduce to

$$\sum_j c_j^{n+1} Q_k(\phi_j, \phi_\ell) = \sum_j c_j^n P_k(\phi_j, \phi_\ell) , \quad \ell \in Z .$$

Hence, with $\tilde{c}^n(\theta) = \mathfrak{f}(c_j^n)$ we obtain in the same way as in Section 4,

$$\tilde{c}^{n+1}(\theta) q_k(\theta) = \tilde{c}^n(\theta) p_k(\theta) , \tag{7.2}$$

where $p_k$ and $q_k$ are the symbols with respect to $S_h$ of the bilinear forms $P_k$ and $Q_k$ ,

$$p_k(\theta) = \sum_\ell P_k(\phi_0,\phi_\ell)e^{-i\ell\theta}, \quad q_k(\theta) = \sum_\ell Q_k(\phi_0,\phi_\ell)e^{-i\ell\theta}.$$

By Lemma 2.5 we have

$$p_k(\theta) = \sum_{j=0}^{M} p_j k^j h^{-(2j-1)}(\theta + 2\pi\ell)^{2j}\hat{\chi}(\theta + 2\pi\ell)^{2\mu}$$

$$= h \sum_\ell p(\lambda(\theta + 2\pi\ell)^2)\hat{\chi}(\theta + 2\pi\ell)^{2\mu} \; ,$$

and similarly

$$q_k(\theta) = h \sum_\ell q(\lambda(\theta + 2\pi\ell)^2)\hat{\chi}(\theta + 2\pi\ell)^{2\mu} \; .$$

It follows in particular that $q_k(\theta)$ is positive. We may therefore solve (7.2) for $\tilde{c}^{n+1}(\theta)$, and setting

$$\rho(\theta) = \frac{p_k(\theta)}{q_k(\theta)} = \sum_j f_j e^{-ij\theta} \; ,$$

we may write

$$c_j^{n+1} = \sum_j f_\ell c_{j-\ell}^n \; ,$$

and by virtue of Lemma 3.2,

$$u_h^{n+1}(jh) = \sum_\ell f_\ell u_h^n(jh - \ell h) \; .$$

Analogously to the situation in the continuous time case we may summarize our results in the following:

<u>Theorem 7.1</u>. Let $\lambda$ be a fixed positive number and $k = \lambda h^2$. Then for $t = nk$ the solution operator $G_h(t)$ of the present Galerkin problem (i), (ii) can be written

$$G_h(t) = S_h F_k^n \; ,$$

where $F_k$ is the finite difference operator with symbol defined by

$$\rho(\theta) = \frac{\sum_{\ell} p(\lambda(\theta+2\pi\ell)^2)\hat{\chi}(\theta+2\pi\ell)^{2\mu}}{\sum_{\ell} q(\lambda(\theta+2\pi\ell)^2)\hat{\chi}(\theta+2\pi\ell)^{2\mu}} . \tag{7.3}$$

We want to analyze the properties of this difference operator $F_k$.

Lemma 7.1. The finite difference operator $F_k$ with symbol defined by (7.3) is consistent with the heat equation and accurate of order $\min(2\mu-2, 2\sigma)$.

Proof. We have by above,

$$\rho(\theta) = \frac{p(\lambda\theta^2) + O(\theta^{2\mu})}{q(\lambda\theta^2) + O(\theta^{2\mu})} = r(\lambda\theta^2) + O(\theta^{2\mu}) \quad \text{as} \quad \theta \to 0 ,$$

and hence by (7.1),

$$\rho(\theta) = \exp(-\lambda\theta^2 + O(\theta^{\min(2\mu, 2\sigma+2)})) \quad \text{as} \quad \theta \to 0 ,$$

which proves the result.

Lemma 7.2. Assume that $|r(\eta)| < 1$ for $\eta > 0$. Then the difference operator $F_k$ with symbol defined by (7.3) is parabolic.

Proof. By Lemma 7.1 we already know that $F_k$ is consistent with the heat equation. It remains only to show that $|\rho(\theta)| < 1$ for $0 < |\theta| \leq \pi$. But for all such $\theta$, $\hat{\chi}(\theta) \neq 0$ and for all $\ell \in Z$,

$$p(\lambda(\theta + 2\pi \ell)^2) < q(\lambda(\theta + 2\pi \ell)^2) \ ,$$

and hence the result follows.

The following convergence theorem is now proved exactly as in the corresponding part of Theorem 5.1.

<u>Theorem 7.2</u>. Let $G_h(t)$ be the Galerkin solution operator in Theorem 7.1. Then for $s$ with $0 \leq s \leq \min(\mu, 2\sigma)$ there is a constant $C$ such that for $0 \leq t = nk \leq T$,

$$\|G_h(t)v - E(t)v\| \leq Ch^s \|v\|_s \ , \qquad v \in C^s \ .$$

A similar comment applies as those following Theorems 5.1 and 6.1 about the higher accuracy at the mesh-points.

Special cases of operators satisfying the above conditions are obtained by choosing for $r(\eta)$ the appropriate diagonal Padé approximations of $e^{-\tau}$ (cf. e.g. [18]). Corresponding to the particular rational function

(7.4) $$r(\eta) = \frac{1 - (1-\kappa)\eta}{1 + \kappa\eta} \ , \qquad \kappa > 0 \ ,$$

we have the discrete time Galerkin equation,

(7.5)
$$(u_h^{n+1}, \phi_\ell) + \kappa k(Du_h^{n+1}, D\phi_\ell) = (u_h^n, \phi_\ell) - (1-\kappa)k(Du_h^n, D\phi_\ell) \ , \qquad \ell \in Z \ .$$

Here $\sigma = 1$ for $\kappa \neq \frac{1}{2}$, $\sigma = 2$ for $\kappa = \frac{1}{2}$ and by Lemma 7.2 the associated finite difference operator is parabolic for $\kappa \geq \frac{1}{2}$. For the case of piecewise linear functions $(\mu = 2)$ we conclude from Lemma 7.1 that the finite difference operator is accurate of order 2.

We have by Lemmas 2.5 and 2.6,

$$\rho(\theta) = \frac{g_{2,0}(\theta) - (1-\kappa)\lambda g_{2,1}(\theta)}{g_{2,0}(\theta) + \kappa\lambda g_{2,1}(\theta)} = \frac{g_4(\theta) - (1-\kappa)\lambda 2(1-\cos\theta)g_2(\theta)}{g_4(\theta) + \kappa\lambda 2(1-\cos\theta)g_2(\theta)}$$

$$= \frac{\frac{2}{3} + \frac{1}{3}\cos\theta - 2\lambda(1-\kappa)(1-\cos\theta)}{\frac{2}{3} + \frac{1}{3}\cos\theta + 2\lambda\kappa(1-\cos\theta)} .$$

A simple calculation gives

$$\rho(\theta) = \exp\left(-\lambda\theta^2 + \left[\left(\kappa - \frac{1}{2}\right)\lambda^2 - \frac{\lambda}{12}\right]\theta^4 + O(\theta^6)\right) \quad \text{as} \quad \theta \to 0 .$$

It follows that for $\kappa > \frac{1}{2}$ the choice $\lambda = (12(\kappa - \frac{1}{2}))^{-1}$ gives a difference operator of accuracy 4. The symbol of the associated difference operator before time discretization is

$$\rho_0(\theta) = \exp\left(-\lambda \frac{g_{2,1}(\theta)}{g_{2,0}(\theta)}\right) = \exp\left(-\lambda \frac{2(1-\cos\theta)}{\frac{2}{3} + \frac{1}{3}\cos\theta}\right) ,$$

for which a simple calculation shows,

$$\rho_0(\theta) = \exp(-\lambda\theta^2 - \frac{1}{12}\lambda\theta^4 + O(\theta^6)) \quad \text{as} \quad \theta \to 0 ,$$

so that the order of accuracy here is exactly 2. In this case the accuracy of the time discretized operator can therefore be higher than that of the corresponding continuous time Galerkin operator.

Let us notice the curious fact that the purely implicit method (7.5) with $\kappa = 1$, $\lambda = \frac{1}{6}$ happens to result in the fourth order accurate explicit finite difference method

$$u_h^{n+1}(jh) = \frac{1}{6} u_h^n((j+1)h) + \frac{2}{3} u_h^n(jh) + \frac{1}{6} u_h^n((j-1)h) .$$

## 8. A Numerical Example

The previous analysis may be applied to initial-boundary value problems of such nature that a corresponding pure initial-value problem exists with the same solution in the relevant domain. We shall discuss an example of this.

Consider the problem

$$\frac{\partial u}{\partial t} = \frac{\partial^2 u}{\partial x^2}, \quad 0 \leqslant x \leqslant 1, \quad t > 0,$$

$$\frac{\partial u}{\partial x}(0,t) = u(1,t) = 0, \quad t > 0,$$

$$u(x,0) = 1 - x, \quad 0 \leqslant x \leqslant 1 .$$

In order to associate with this problem a pure initial-value problem let $v$ be the 4-periodic extension of the function $1-|x|$ in $[-2,2]$. This function is even in $x$, odd in $x-1$; it can be represented as

$$v(x) = \frac{8}{\pi^2} \sum_{j=1}^{\infty} \frac{\cos \frac{1}{2}(2j-1)\, x}{(2j-1)^2} .$$

The solution of the corresponding initial-value problem will then obviously be a solution of the mixed initial-boundary value problem. Its exact solution is

(8.1)
$$u(x,t) = \frac{8}{\pi^2} \sum_{j=1}^{\infty} \frac{1}{(2j-1)^2} \cos \tfrac{1}{2}(2j-1)^2 \pi x e^{-\frac{1}{4}(2j-1)^2 \pi^2 t} .$$

Let us now consider the numerical solution based on

quadratic splines $(\mu = 3)$ with $h = J^{-1}$ and on the Crank-Nicolson discretization in time. The Galerkin equations are then with $\phi = \chi^{*3}$,

$$(u_h^{n+1},\phi_\ell) + \frac{k}{2}(Du_h^{n+1},D\phi_\ell) = (u_h^n,\phi_\ell) - \frac{k}{2}(Du_h^n,D\phi_\ell) ,$$

which corresponds to $r(\eta)$ in (7.4) with $\kappa = \frac{1}{2}$.

The symbol of the associated difference operator becomes

$$\rho(\theta) = \frac{\sum_\ell [1-\frac{1}{2}\lambda(\theta+2\pi\ell)^2]\hat{\chi}(\theta+2\pi\ell)^6}{\sum_\ell [1+\frac{1}{2}\lambda(\theta+2\pi\ell)^2]\hat{\chi}(\theta+2\pi\ell)^6}$$

$$= \frac{g_6(\theta) - \lambda(1-\cos\theta)g_4(\theta)}{g_6(\theta) + \lambda(1-\cos\theta)g_4(\theta)}$$

$$= \frac{(1+10\lambda)\cos 2\theta + (26+20\lambda)\cos\theta + (33-30\lambda)}{(1-10\lambda)\cos 2\theta + (26-20\lambda)\cos\theta + (33+30\lambda)} .$$

The discrete solution will now also be even in $x$, odd in $x-1$, and 4-periodic so that in fact with $U_j^n = u_h^n(jh)$ we only have to determine $U_j^n$ for $j = 0,\cdots,J-1$. For these quantities we obtain the following recursive system of equations for $\lambda = 1$,

$$-18\,U_2^{n+1} + 12\,U_1^{n+1} + 126\,U_0^{n+1} = 22\,U_2^n + 92\,U_1^n + 6\,U_0^n ;$$

$$-9\,U_3^{n+1} + 6\,U_2^{n+1} + 117\,U_1^{n+1} + 6\,U_0^{n+1} = 11\,U_3^n + 46\,U_2^n + 17\,U_1^n + 46\,U_0^n ;$$

$$-9\ U_{j+2}^{n+1} + 6\ U_{j+1}^{n+1} + 126\ U_j^{n+1} + 6\ U_{j-1}^{n+1} - 9\ U_{j-2}^{n+1}$$

$$= 11\ U_{j+2}^{n} + 46\ U_{j+1}^{n} + 6\ U_j^{n} + 46\ U_{j-1}^{n} + 11\ U_{j-2}^{n}\ ,$$

$$\text{for}\quad j = 2,\cdots,J-3\ ;$$

$$6\ U_{J-1}^{n+1} + 126\ U_{J-2}^{n+1} + 6\ U_{J-3}^{n+1} - 9\ U_{J-4}^{n+1}$$

$$= 46\ U_{J-1}^{n} + 6\ U_{J-2}^{n} + 46\ U_{J-3}^{n} + 11\ U_{J-4}^{n}\ ;$$

$$135\ U_{J-1}^{n+1} + 6\ U_{J-2}^{n+1} - 9\ U_{J-3}^{n+1} = -5\ U_{J-1}^{n} + 46\ U_{J-2}^{n} + 11\ U_{J-3}^{n}\ .$$

We used this scheme to compute approximately the value of $u(0,.1)$. The expected rate of convergence depends upon the smoothness of the 4-periodic extension of $v$. Since this function is only Lipschitz continuous we may only expect a $O(h)$ result as $h \to 0$. To increase the convergence rate to fourth order, matching the accuracy of the difference operator, we apply the smoothing operator $M_h^{(4)}$ defined by (6.2). This corresponds to taking

$$U_0^0 = 1 - \frac{51}{180}\,h, \quad U_1^0 = 1 - \frac{85}{90}\,h, \quad U_2^0 = 1 - \frac{719}{360}\,h,$$

$$U_j^0 = 1 - jh, \quad j = 3,\cdots,J-1\ .$$

For comparison we also computed the results by using the smoothing operator $M_h^{(1)}v = \chi_{(h)} * v$ which corresponds to taking

$$U_0^0 = 1 - \frac{1}{4}\,h, \quad U_j^0 = 1 - jh, \quad j = 1,\cdots,J-1\ .$$

The results of these computations for $J = 10,20$ are displayed in the following table together with the modulus of the errors within parentheses. The exact value can easily be computed from (8.1) to be $u(0,.1) = .64317660$.

| $v_h$ \ $J$ | 10 | 20 |
|---|---|---|
| $v$ | .64466958 (.00149298) | .64354143 (.00036483) |
| $M_h^{(1)} v$ | .64238625 (.00079035) | .64298400 (.00019260) |
| $M_h^{(4)} v$ | .64306889 (.00010771) | .64316796 (.00000864) |

REFERENCES

1. J. P. Aubin, Approximation of elliptic boundary-value problems. John Wiley & Sons, Inc. New York, London, Sydney, Toronto 1972.
2. I. Babuška, The finite element method for elliptic differential equations, Numerical solution of Partial Differential Equations II, SYNSPADE 1970, B. Hubbard ed. Academic Press, New York 1971, 69-101.
3. J. H. Bramble and S. R. Hilbert, Estimation of linear functionals on Sobolev spaces with application to Fourier transforms and spline interpolation. SIAM J. Numer. Anal. 7 (1970), 112-124.
4. P. L. Butzer and H. Berens, Semi-groups of operators and approximation. Springer-Verlag, Berlin, Heidelbert, New York, 1967.
5. J. Douglas, Jr. and T. Dupont, Galerkin methods for parabolic equations. SIAM J. Numer. Anal. 7 (1970), 575-626.

6. T. Dupont, Galerkin methods for first order hyperbolics: an example. To appear.
7. G. Fix and N. Nassif, On finite element approximations to time-dependent problems. To appear.
8. G. Fix and G. Strang, Fourier analysis of the finite element method in Ritz-Galerkin Theory. Studies in Appl. Math. 48 (1969), 265-273.
9. H. O. Kreiss, V. Thomée and O. Widlund, Smoothing of initial data and rates of convergence for parabolic difference equations. Comm. Pure Appl. Math 23 (1970), 241-259.
10. H. S. Price and R. S. Varga, Error bounds for semidiscrete Galerkin approximations of parabolic problems with application to petroleum reservoir mechanics. Numerical Solution of Field Problems in Continuum Physics. AMS Providence R.I., 1970, 74-94.
11. I. J. Schoenberg, Contributions to the problem of approximation of equidistant data by analytic functions, A and B. Quart. Appl. Math. 4 (1946), 45-99, 112-141.
12. I. J. Schoenberg, Cardinal interpolation and spline functions. J. Approximation Theory 2 (1969), 335-374.
13. I. J. Schoenberg, Cardinal interpolation and spline functions II. Interpolation of data of power growth. To appear in J. Approximation Theory.
14. I. J. Schoenberg, Cardinal interpolation and spline functions IV. The exponential Euler splines. To appear in Proceedings of the conference on "Linear operators and approximation," Oberwolfach, August 1971.
15. G. Strang, The finite element method and approximation theory, Numerical solution of Partial Differential Equations II, SYNSPADE 1970, B. Hubbard ed. Academic Press, New York 1971, 547-583.
16. G. Strang and G. Fix, A Fourier analysis of the finite element variational method. Mimeographed notes.
17. B. Swartz and B. Wendroff, Generalized finite-difference schemes. Math. Comp. 23 (1969), 37-50.
18. V. Thomée, Stability theory for partial difference operators. SIAM Rev. 11 (1969), 152-195.
19. O. B. Widlund, On the rate of convergence for parabolic difference schemes, II. Comm. Pure Appl. Math. 23 (1970), 79-96.

# PART III

# SHORT COMMUNICATIONS

# THE EXTENSION AND APPLICATION OF SARD KERNEL THEOREMS TO COMPUTE FINITE ELEMENT ERROR BOUNDS

R. E. Barnhill*, J. A. Gregory** and J. R. Whiteman**

## 1. Introduction.

Error bounds for the finite element method for elliptic boundary value problems are frequently of the form

$$\|Ru\| \leq Kh^n\|u\| , \tag{1}$$

where Ru is the remainder in approximating the solution u of the boundary value problem; h is a mesh parameter tending to zero; n is a constant depending on the smoothness of u and the polynomial precision of R ; K is a constant, usually not known, that is independent of h and u ; and the norms are appropriate Sobolev norms. In order to have a more nearly computable error bound, some value for K must be found. This paper outlines a method of calculating such constants by means of extensions of the Sard kernel theorems. Application is made to a model problem with a finite element scheme on a triangulated polygon.

## 2. Boundary value problem and interpolation remainder theory.

* University of Utah.
** Brunel University.

For simplicity of presentation, we discuss Poisson's equation, but the results can be extended to more general linear elliptic operators in divergence form and to certain monotone nonlinear operators [6].

The two dimensional Dirichlet problem with homogeneous boundary conditions is the following:

$$-\Delta[u(x,y)] = g(x,y) \quad \text{on} \quad D\,, \qquad u|_{\partial D} = 0\,. \tag{2}$$

The region $D$ must be sufficiently smooth for the <u>Sobolev imbedding theorems</u> to apply, e.g., a union of finitely many star-shaped regions. (For more generality, compare "restricted cone condition.") The relevant function space for (2) is the Sobolev space $\overset{\circ}{W}{}_2^1(D)$ with the norm

$$\|v\|_{\overset{\circ}{W}{}_2^1} = \|\nabla v\|_{L_2} = \left\{\int_D\!\!\int [(v_{1,0})^2 + (v_{0,1})^2]\,dxdy\right\}^{1/2}, \tag{3}$$

where $v_{1,0}$ is the generalized derivative $\frac{\partial v}{\partial x}$ and $v_{0,1}$ is dual. Let $a(u,v)$ be the bilinear functional

$$a(u,v) = \int_D\!\!\int [u_{1,0}v_{1,0} + u_{0,1}v_{0,1}]\,dxdy\,. \tag{4}$$

Then the weak problem corresponding to (2) is to find a function $u$ in $\overset{\circ}{W}{}_2^1$ such that

$$a(u,v) = (g,v) \quad \text{for all} \quad v \in \overset{\circ}{W}{}_2^1\,. \tag{5}$$

The Galerkin method is the solution of (5) over a finite-dimensional subspace $S^q$ of $\overset{\circ}{W}{}_2^1$. That is, $U$ is the Galerkin approximation to $u$ from $S^q$ means that

$$a(U,w^q) = (g,w^q) \quad \text{for all} \quad w^q \quad \text{in} \quad S^q\,. \tag{6}$$

We shall consider the space $S^1$ of piecewise linear functions defined on a triangulated mesh.

The relationship between interpolation remainder theory and the Galerkin method is the following:

Lemma. $U$ is a best approximation to $u$ from the subspace $S^q$ in the $\mathring{W}_2^1$ norm.

That is,

$$\|u - U\|_{\mathring{W}_2^1} \leq \|u - p\|_{\mathring{W}_2^1} \tag{7}$$

for all $p$ in $S^q$ and, in particular, for an interpolant $p$ in $S^q$. We shall let $p = p_1$, a piecewise linear interpolant.

## 3. Sard kernel theorems applied to piecewise linear interpolation.

Linear interpolation is exact for linear functions of two variables. An appropriate Sard space is $\underline{B}_{1,1}$ which consists of functions $u$ that have Taylor expansions of the form

$$\begin{aligned} u(x,y) &= u(a,b) + u_{1,0}(a,b)(x - a) + u_{0,1}(a,b)(y - b) \\ &+ \int_a^x (x - \tilde{x})u_{2,0}(\tilde{x},b)d\tilde{x} + \int_b^y (y - \tilde{y})u_{0,2}(a,\tilde{y})d\tilde{y} \\ &+ \int_b^y \int_a^x u_{1,1}(\tilde{x},\tilde{y})d\tilde{x}d\tilde{y} \ . \end{aligned} \tag{8}$$

Let $T_h$ be the right triangle with vertices at $(0,0)$, $(h,0)$, and $(0,h)$, $h > 0$. The linear interpolant $p_1(x,y)$ to $u$ at these vertices is:

$$(9) \quad p_1(x,y) = \left[1 - \left(\frac{x+y}{h}\right)\right]u(0,0) + \left(\frac{x}{h}\right)u(h,0) + \left(\frac{y}{h}\right)u(0,h) .$$

The relevant interpolation remainder functionals are

$$(10) \quad \begin{aligned} Ru(x,y) &= u(x,y) - p_1(x,y) , \\ R_{1,0}u(x,y) &= \frac{\partial}{\partial x} Ru(x,y) , \\ R_{0,1}u(x,y) &= \frac{\partial}{\partial y} Ru(x,y) . \end{aligned}$$

The Sard kernel theorems are defined for rectangular regions $D$. Lois Mansfield appears to have been the first to note that they are also valid for the triangle $T_h$. In fact, they are valid for regions $D$ such that for each $(x,y)$ in $D$, the rectangle with opposite corners at $(x,y)$ and at $(a,b)$ and with sides parallel to the coordinate axes is contained in $D$ [1]. For $T_h$ we let $(a,b) = (0,0)$. Then $R$ is a functional admissible in the sense of Sard, and so $R$ has the representation

$$(11) \quad \begin{aligned} Ru(x,y) &= \int_0^h u_{2,0}(\tilde{x},0)K^{2,0}(x,y;\tilde{x})d\tilde{x} + \text{dual term} \\ &+ \int_{T_h}\!\!\int u_{1,1}(\tilde{x},\tilde{y})K^{1,1}(x,y;\tilde{x},\tilde{y})d\tilde{x}d\tilde{y} . \end{aligned}$$

Eventually,

$$(12) \quad \begin{aligned} \|Ru(x,y)\|_{L_2(T_h)} &\leq [\|u_{2,0}(\tilde{x},0)\|_{L_2[0,h]} + \text{dual term}] \frac{h^{5/2}}{6\sqrt{5}} \\ &+ \|u_{1,1}(\tilde{x},\tilde{y})\|_{L_2(T_h)} \frac{h^2}{2\sqrt{3}} . \end{aligned}$$

Bramble and Zlámal [4] have obtained the bound:

$$\|Ru\|_{L_2(T_h)} \leqslant Kh^2 \left\{ \sum_{|i|=2} \|D^i u\|^2_{L_2(T_h)} \right\}^{1/2} . \tag{13}$$

The functionals $R_{1,0}$ and $R_{0,1}$ are not admissible in the sense of Sard [5], and so they are applied to the Taylor expansion (8) directly. The result is the following:

$$\left\{ \|R_{1,0}u\|^2_{L_2(T_h)} + \|R_{0,1}u\|^2_{L_2(T_h)} \right\}^{1/2}$$

$$\leqslant \frac{h^{3/2}}{2\sqrt{3}} \left[ \|u_{2,0}(\tilde{x},0)\|_{L_2[0,h]} + \|u_{0,2}(0,\tilde{y})\|_{L_2[0,h]} \right] \tag{14}$$

$$+ h\sqrt{2} \|u_{1,1}(\tilde{x},\tilde{y})\|_{L_2(T_h)} .$$

Birkhoff, Schultz, and Varga [3] had the idea of letting the point of Taylor expansion (a,b) also be the point of interpolation ((x,y) above). However, the analog to (14) then involves values of $u_{2,0}$ and $u_{0,2}$ outside $T_h$ . This is a disadvantage if the hypotenuse of $T_h$ is on the boundary of the overall region D .

## 4. Numerical Example.

The Galerkin approximation for nonhomogeneous Dirichlet boundary conditions has the same best approximation property as the homogeneous case. Hence we consider the square $S_\pi$ of side length $\pi$ with lower left corner at (0,0) . The boundary value problem is

$$\Delta u = 0$$

$$u|_{\partial S_\pi} = 0 \quad \text{except that} \quad u(x,0) = \sin x \ , \ 0 \leqslant x \leqslant \pi \ .$$

For linear interpolation on a triangle $T_h$ , we obtain the following bounds:

$$\|u - p_1\|_{L_2(T_h)} \leqslant \frac{h^3}{\sqrt{40}} , \tag{15}$$

$$\|u - p_1\|_{L_\infty(T_h)} \leqslant \frac{h^2}{2} , \tag{16}$$

$$\|\nabla(u - p_1)\|_{L_2(T_h)} \leqslant \sqrt{\frac{2}{5}}\, h^2 , \tag{17}$$

$$\|\nabla(u - p_1)\|^2_{L_2} = \|R_{1,0}u\|^2_{L_2} + \|R_{0,1}u\|^2_{L_2}$$

$$= \sum_{(h)} \left[\|R_{1,0}u\|^2_{L_2(T_h)} + \|R_{0,1}u\|^2_{L_2(T_h)}\right] , \tag{18}$$

where the summation is over all $T_h$ and $T_h^*$ , $T_h^*$ being the upper right triangle analogous to the lower left right triangle $T_h$ . Inequalities (15), (17) and (18) lead to the following bounds:

$$\|u - p_1\|_{L_2(S_\pi)} \leqslant .70\, h^2 , \tag{19}$$

$$\|\nabla(u - p_1)\|_{L_2(S_\pi)} \leqslant \frac{2\pi}{\sqrt{5}}\, h . \tag{20}$$

The actual errors at the midpoints of the sides of one subtriangle are .0043, - .0107, and .0127 , which compare with (16) which is .078 if $h = \frac{\pi}{8}$ .

Acknowledgments

The research of R. E. Barnhill was supported by the National Science Foundation with Grant GP 20293 to the University of Utah, by the Science Research Council with Grant B/SR/9652 at Brunel University, and by a N.A.T.O. Senior Fellowship in Science.

## REFERENCES

1. Barnhill, R. E. and J. A. Gregory, Sard kernel theorems on triangular and rectangular domains with extensions and applications to finite element error bounds. TR/11, Department of Mathematics, Brunel University, 1972.
2. Barnhill, R. E. and J. R. Whiteman, Error analysis of finite element methods with triangles for elliptic boundary value problems. In Whiteman (ed.), The Mathematics of Finite Elements and Applications, Academic Press, London, 1972.
3. Birkhoff, G., M. H. Schultz, and R. S. Varga, Piecewise Hermite interpolation in one and two variables with applications to partial differential equations. Numer. Math. 11 (1968), 232-256.
4. Bramble, J. H. and M. Zlamal, Triangular elements in the finite element method. Math. Comp. 24 (1970), 809-820.
5. Sard, A., Linear Approximation. Mathematical Survey 9 (1963), American Math. Soc., Providence, R. I.
6. Varga, R. S., The role of interpolation and approximation theory in variational and projectional methods for solving partial differential equations. IFIP Congress 71 (1971), North Holland Amsterdam, 14-19.
7. Zlamal, M., On the finite element method. Numer. Math. 12 (1968), 394-409.

# TWO TYPES OF PIECEWISE QUADRATIC SPACES AND THEIR ORDER OF ACCURACY FOR POISSON'S EQUATION*

Alan E. Berger**

We consider two types of piecewise quadratic approximate solutions for the model problem $-u_{xx} - u_{yy} = f$ in a smooth region $D$ ; $u = 0$ on the boundary $B$ of $D$, where $f \in H^1(D)$ and is continuous on $\bar{D}$. Error estimates will be described, and the results of several numerical experiments will be given.

Let $D_h$ be a triangulated polygon having its boundary vertices in $B$ where $h$ denotes the mesh size; i.e., the length of the largest side of a triangle of $D_h$. It is assumed that the ratio of the largest side to the smallest side in $D_h$ remains bounded away from zero.

For the first type of quadratic space, $S^h$, $D$ is also assumed to be convex, so that $D_h \subset D$. Elements of $S^h$ are required to be zero on the boundary of $D_h$, and are extended by zero on $D - D_h$. Let $v^h$ be the Galerkin approximate solution from $S^h$, so $B(s^h, v^h) = (s^h, f)$ for all $s^h \in S^h$, where $B(s^h, v^h) \equiv \iint_D (s^h_x v^h_x + s^h_y v^h_y)\,dxdy$ and

* Most of the work presented in this paper was done while the author was an N.S.F. Graduate Fellow at M.I.T. under the direction of Professor Gilbert Strang.

** U.S. Naval Ordnance Laboratory.

$(s^h, f) \equiv \iint_D s^h f\,dxdy$ . Then one has $E(v^h, D_h) \equiv \iint_{D_h} [(v_x^h - u_x)^2 + (v_y^h - u_y)^2]dxdy = \min_{s^h \in S^h} E(s^h, D_h)$ .
Letting $s^h$ equal $u$ at all nodes of $D_h$ except boundary chord midpoints, where $s^h$ is defined to be zero, we obtain $E(v^h, D_h) = O(h^3)$ as in [4]. A straightforward application of the Nitsche method then shows that $L(v^h, D) \equiv \|v^h - u\|^2_{L^2(D)} = O(h^4)$ .

To form the second space $T^h$ , for each boundary chord $\overline{bc}$ of $D_h$ we let $p$ be a point on $B$ within $O(h^2)$ of being midway (in terms of arclength) between $b$ and $c$ . Then $T^h$ consists of all piecewise quadratics which are zero at the boundary vertices of $D_h$ and at each point $p$ . Elements of $T^h$ are then "nearly zero" on $B$ ; such "nearly zero" spaces were originally set forth in [3] -- that "nearly zero" quadratics would have a higher order convergence was conjectured by Ridgway Scott. If $w^h \in T^h$ is the solution of $B(t^h, w^h) = (t^h, f)$ for all $t^h \in T^h$ , then from [2] and [1] we have, for some constant $C$ independent of $h$ :

$$E(w^h, D) \leq C \Big[ \operatorname{dist}_{H^1(D)} (T^h, u) + \sup_{t^h \in T^h} \|t^h\|_{H^{-1}(B)} \|u\|_{H^3(D)} \Big/ \|t^h\|_{H^1(D)} \Big]^2 \tag{1}$$

$$\leq C[O(h^2) + O(h^{5/2})]^2 = O(h^4) \quad .$$

Numerical experiments were carried out by taking $u = 1 - (x^2 + y^2)^2$ on the unit circle. As $u$ is independent of angle, the finite element method was carried out on the 60° sector whose base was on the positive x-axis. Triangulation was accomplished by dividing the sector with 1 , or 2 , or ... 5 equally spaced horizontal lines, and by placing 0 , 1 , 2 , ... , equally spaced vertices in the interior of successive lines. In this way subdivisions with 4 , 9 , 16 , 25 , 36 elements were obtained. In computing asymptotic rates of convergence, an average mesh size was used, defined to be the length of the side of the equilateral triangle whose area was the average of the areas of the triangles of $D_h$ . Numerical integration was accomplished using a composite Simpson's rule in the following manner. Dividing each element by its altitude, the iterated integral for each half was evaluated using five equally spaced points on the base, with three equally spaced points on each of the four interior vertical line segments. All calculations were done with a minimum of 14 significant digits.

In row 1 of the table below the energy error, $E(v^h,D_h)$ , and the square of the $L^2$ error, $L(v^h,D_h)$ , are given for $S^h$ with 25 elements, and the asymptotic exponents of convergence p (error $\approx$ constant$\cdot h^p$) are given based on the errors with 25 and 36 elements. The corresponding errors given for the "nearly zero" quadratics $T^h$ are $E(w^h,D)$ and $L(w^h,D)$ . The calculated exponents are seen to correspond to the theoretical estimates.

In row 3 results are given for $T^h$ when numerical integration was only carried out on $D_h$ , and $D - D_h$ was ignored. This was done to test the importance of having the boundary well approximated. The result shows that integrating only on $D_h$ negates the superior accuracy gained with $T^h$ , and suggests that for problems with a boundary whose position is somewhat uncertain, it may be just as well to use the standard Ritz quadratics $S^h$ . We note that the energy error exponent 3.01 is in excellent agreement with the theoretical exponent 3 for the effect of approximating the boundary by a polygon, as given in [2].

When using the Ritz quadratics, one obtains an approximate solution $v^h$ defined on $D^h$ (and extended by 0 in $D - D_h$). In order to obtain a better approximation to the true solution u on $D - D_h$ , we can extend the piecewise polynomial $v^h$ in the following way. Each segment G of $D - D_h$ is bounded by an arc of B and an edge of some triangle T of $D_h$ , and $v^h$ restricted to T is some quadratic polynomial $q_T$ . For any point $p \in G$ we define $v^h(p) = q_T(p)$ , and denote this extension of $v^h$ to D by $\tilde{v}^h$ . The asymptotic exponents for $E(\tilde{v}^h, D)$ and $L(\tilde{v}^h, D)$ can be shown to be the same as those for $E(v^h, D_h)$ and $L(v^h, D_h)$ , i.e., 3 and 4. The computed values for $E(\tilde{v}^h, D)$ and $L(\tilde{v}^h, D)$ are given in row 4.

In the last three rows the effect of using different integration schemes to compute $v^h \in S^h$ are given. Row 5 is the same as row 1, while for row 6 (7) the midpoint (vertex) method (which approximates the integral of a function g over a triangle T having midpoints (vertices) a , b , c by [g(a) + g(b) + g(c)] area T/3) was used in obtaining $v^h$ . Errors were calculated over $D_h$ using the composite Simpson's method. The midpoint method, exact for quadratic polynomials, gave results very close to those for Simpson's method, while taking about one-third the computer time. The results for the vertex method which is exact only for linear polynomials give an example of the consequence of using a quadrature scheme of insufficient accuracy in relation to the finite element space being used.

TABLE I

| | energy error 25 elements | energy exponent 25-36 elements | $L^2$ error squared 25 elements | $(L^2 \text{ error})^2$ exponent 25-36 elements |
|---|---|---|---|---|
| $S^h$ | .414E-2 | 2.82 | .161E-3 | 4.02 |
| $T^h$ | .572E-3 | 3.91 | .586E-6 | 5.93 |
| $T^h$, integration on $D_h$ | .794E-2 | 3.01 | .179E-3 | 4.10 |
| $S^h$, error on D for $\tilde{v}^h$ | .435E-2 | 2.87 | .163E-3 | 4.04 |
| $S^h$, Simpson | .414E-2 | 2.82 | .161E-3 | 4.02 |
| $S^h$, midpoint | .419E-2 | 2.84 | .163E-3 | 4.04 |
| $S^h$, vertex | .121E0 | 1.87 | .219E-2 | 3.86 |

REFERENCES

1. Berger, A., "Error Estimates for the Finite Element Method," Ph.D. Thesis, M.I.T., Cambridge, Mass., 1972.
2. Berger, A., R. Scott, and G. Strang, "Approximate boundary conditions in the finite element method," Proc. Symp. on Numerical Analysis of Istituto Nazionale di Alta Matematica, Rome, 1972, to appear in Symposia Matematica, Academic Press.
3. Nitsche, J., "A projection method for Dirichlet problems using subspaces with nearly zero boundary conditions," unpublished manuscript.
4. Strang, G., and A. Berger, "The change in solution due to change in domain," Proc. A.M.S. Summer Institute on Partial Differential Equations, Berkeley, 1971.

# A METHOD OF GALERKIN TYPE ACHIEVING OPTIMUM $L^2$ ACCURACY FOR FIRST ORDER HYPERBOLICS AND EQUATIONS OF SCHRÖDINGER TYPE

J. E. Dendy, Jr.*

Dupont [1] has shown that for Galerkin's method for first order hyperbolic differential equations, the $L^2$ norm of the error is not in general of the same order as the best approximation to the solution in the $L^2$ sense by functions in the finite dimensional subspace. Two methods are presented for which this defect is not present. The methods apply to the general first order linear hyperbolic system subject to a few technical constraints, that is, to systems of the form

$$\frac{\partial u}{\partial t} + \sum_{i=1}^{n} A_i \frac{\partial u}{\partial x_i} + Bu = f \;, \quad (x,t) \in \Omega \times (0,T] \;,$$

$$u(x,0) = u_0(x) \;, \quad x \in \Omega \;,$$

$$Cu(x,t) = 0 \;, \quad (x,t) \in \partial\Omega \times (0,T] \;, \quad C \in \mathcal{C} \;,$$

where $\Omega$ is a region in $R^n$ with boundary $\partial\Omega$, $u(x,t)$, $f(x,t) \in C^p$, $\forall x,t$, $A_i(x,t)$, $B(x,t)$ are $p \times p$ matrices with complete entries, and $\mathcal{C}$ is a collection of matrices describing the boundary conditions. The methods

* University of Denver.

also apply to equations of Schrödinger type, that is, to equations of the form

$$\frac{\partial u}{\partial t} - i \sum_{j,k=1}^{n} \frac{\partial}{\partial x_k}\left(a_{jk}(x,t)\frac{\partial u}{\partial x_j}\right) = f, \ (x,t) \in \Omega \times (0,T] \ ,$$

$$u(x,0) = u_0(x) \ , \ x \quad \Omega \ ,$$

with Dirichlet, Neumann, or mixed boundary conditions. For simplicity the methods are described here with respect to the simple equation

$$\frac{\partial u}{\partial t} - \frac{\partial u}{\partial x} = 0 \ .$$

Also, only the continuous-in-time schemes are described here although discrete-in-time schemes can be formulated for both methods. Let $M$ be a finite dimensional subspace of $H^1(\Omega)$ satisfying the boundary conditions. Then the first method is to find $U\colon \ [0,T] \to M$ such that $U(0)$ approximates $u_0$ well and

$$\left\langle \frac{\partial U}{\partial t} - \frac{\partial U}{\partial x} , V - \frac{\partial V}{\partial x}\right\rangle = 0 \ , \forall \ V \in M \ , \ 0 < t \leq T \ ,$$

where $\langle \cdot , \cdot \rangle$ is the $L^2(\Omega)$ inner product. The problem with this method is that its discrete-in-time analogues give rise to unsymmetric, ill-conditioned matrices to be inverted at every time step. The second scheme is to find $U\colon [0,T] \to M$ such that $U(0)$ approximates $u_0$ well and

$$\left\langle \frac{\partial^2 U}{\partial t^2} , V\right\rangle + \left\langle \frac{\partial U}{\partial x} , \frac{\partial V}{\partial x}\right\rangle - \frac{\partial U}{\partial t} V \Big|_0^1 = 0 \ , \forall \ V \in M \ , \ 0 < t \leq T \ .$$

This scheme is motivated by differentiation of the differential equation with respect to $t$, integration by parts, and substitution of the differential equation. Under reasonable assumptions on $M$, both schemes yield the same $L^2$ order of approximation as the best $L^2$ approximation in $M$ to the solution.

REFERENCES

[1] Dupont, T., "Galerkin methods for first order hyperbolics: an example," to appear.

# RICHARDSON EXTRAPOLATION FOR PARABOLIC GALERKIN METHODS

G. Fairweather* and J. P. Johnson*

Several discrete-time Galerkin methods involving two or three time levels have been proposed by various authors for the approximate solution of parabolic initial-boundary value problems. For these methods, the $L^2$ norm of the error at each time level is $O(A + (\Delta t)^p)$, where $\Delta t$ denotes the time step, $p$ is 1 or 2, and $A$ depends upon the approximating properties of the finite dimensional space containing the discrete-time Galerkin approximation. In this paper, it is shown that approximations generated by certain discrete Galerkin procedures can be modified using Richardson extrapolation to obtain approximations whose time discretisation error is $O((\Delta t)^{p(s+1)})$, where $s$ is a positive integer.

As an example, we consider the heat equation in the cylinder $D \times (0,T]$ subject to homogeneous Dirichlet boundary conditions. $D$ is a bounded domain in $R^n$. For the approximate solution of this problem we choose the family of two-level methods described by Douglas and Dupont**, which is parametrised by $\theta$, where $\theta \in [0,1]$.

---

* University of Kentucky.

** "Galerkin methods for parabolic equations," SIAM J. Numer. Anal. 7 (1970), pp. 575-626.

This family contains the backward difference Galerkin method ($\theta = 1$) and the Crank Nicolson Galerkin method ($\theta = 0$). If $W^{(0)}(\Delta t)$ denotes the approximation to the solution $u$ at time $t_m$ computed with time step $\Delta t$, we define the <u>kth extrapolate</u> $W_m^{(k)}(\Delta t)$ by

$$W_m^{(k)}(\Delta t) = \frac{2^{pk} W_m^{(k-1)}(\Delta t/2) - W_m^{(k-1)}(\Delta t)}{2^{pk} - 1}, \quad k = 1, 2, \cdots .$$

It is shown that, if $u$ is sufficiently smooth,

$$\| u(\cdot, t_m) - W_m^{(k)}(\Delta t) \|_{L^2(D)} = O(A + (\Delta t)^{p(k+1)}),$$

where $p = 1$ if $\theta \in (0,1]$ and $p = 2$ if $\theta = 0$. Thus each extrapolation improves the time discretisation error by a factor which is $O((\Delta t)^p)$.

The techniques described in this paper can be used to analyze a family of three-level methods proposed by Dupont, Fairweather and Johnson[†]. These methods are algebraically linear even for nonlinear problems. It can be shown that for a wide class of nonlinear problems one extrapolation may be performed to obtain an approximation which is fourth-order correct in time.

---

† "Three-level Galerkin methods for parabolic equations," to appear.

# GEOMETRIC ASPECTS OF THE FINITE ELEMENT METHOD

William J. Gordon* and Charles A. Hall**

## 1. Summary.

The two essential ingredients of any boundary value problem are the field equations which describe the physics and a set of relations which specify the geometry of the problem domain. Current finite element methodology is only superficially concerned with the global problem geometry. It focuses attention principally upon the discretized problem in the small, i.e., upon the individual elements and their local properties. As an alternative, this paper examines the possibility of a global treatment of the original problem domain. By exploiting the notion of vector-valued "blending function" interpolation, the authors propose to map the given problem domain one-to-one onto a canonical rectangular parameter domain. This is equivalent to the introduction of a generalized curvilinear coordinate system on the manifold representing the global geometry of the original problem. This method of treating the complex geometries of two and three dimensional boundary value problems has several significant

* General Motors Research Laboratories, Warren, Michigan.

** University of Pittsburgh.

advantages over the ad hoc methods currently employed, since it provides a systematic approach to topologically decomposing a manifold into finite elements and leads inevitably to well-structured sparse matrices.

## 2. Introduction.

This paper is concerned primarily with the geometric aspects of the finite element method. As Fix and Strang have pointed out in this volume, the first of several forms of approximation inherent in the implementation of a FEM is the approximation of the problem domain $\mathcal{R}$ by some polygonal domain $\mathcal{R}^*$. Obviously, the accuracy of the numerical solution is limited by the accuracy of the polygonal approximation. While the mainstream of the mathematical theory of the FEM focuses upon questions of approximation of the dependent variable, we shall focus upon the first phase of the method, namely, the approximation of the domain of the independent variables.

As motivation and background, let us first examine some of the deficiencies of the conventional finite element methodology which are attributable to unsystematic treatment of problem geometries. Suppose, as is typically the case in practice, that the domain of the independent variables $\mathcal{R}$ is implicitly specified by a set of engineering drawings. The task of the analyst is to extract from these drawings a sufficiently high fidelity approximation $\mathcal{R}^*$ and to select a collection of points on the boundary $\partial\mathcal{R}^*$ and the interior of $\mathcal{R}^*$ which will serve as the nodal points for a finite element analysis. Since regions of practical interest are rarely simple in shape, enormous amounts of time and effort are often involved in the manual selection of mesh nodes. These nodal points are the vertices of a graph whose topology determines the connectivity between the finite subdomains $E_j$ . Since a collection of points in Euclidean n-space ($n > 1$) is not inherently well-ordered, some ad hoc method must be used for numbering the nodes.

Next, over each $E_j$ , the dependent variable u is approximated by a low degree polynomial in the independent variables, e.g., the piecewise linear triangular approximation first suggested by Synge [10] for two-dimensional manifolds. Minimization of a quadratic functional $J[u]$ over the Ritz space of trial functions then reduces to solving a (typically) very large but sparse system of algebraic equations for the unknown nodal values of the dependent variable u . Even though the linear system is sparse, it is not systematically structured since it reflects the ad hoc nodal numbering of the unsystematically selected mesh points. In general, the nonzero entries are scattered rather haphazardly throughout the matrix. Since unstructured sparse matrices are not amenable to efficient computational solution, reordering algorithms are typically employed in an effort to isolate nonzeros within a band about the matrix diagonal. At best, however, a band reduction algorithm can only improve the gross structure of the matrix. It cannot alter the intrinsic topology of the nodal graph, nor can it recognize special block structured matrices (which are amenable to far more efficient solution) even if they do occur.

In summary, conventional finite element methodology has the following deficiencies, all of which are related to the methods used for treating the problem geometry:

a. The original problem domain $R$ is distorted.

b. The selection of nodal points in $R^*$ is unsystematic and hence they have no natural ordering.

c. The resulting sparse linear system has no discernible structure other than the gross measure of bandwidth and is thus not amenable to specialized solution techniques.

## 3. Univalent Mappings and Curvilinear Coordinate Systems.

In the recent paper [12], Zienkiewicz and Phillips have also called attention to the geometric deficiencies of conventional finite element methodology and have proposed using the local mapping methods associated with "isoparametric elements" in a zonal fashion so as to achieve an automatic mesh generation scheme. While our approach is similar in spirit to that of [12], it differs in several important respects. These differences are related to the fact that the methods developed below are based upon transfinite "blending-function methods" of bivariate and multivariate interpolation, whereas those of [12] use low order tensor product or "serendipity" methods of finite polynomial interpolation. We indicated in [8] and [9] how such finite dimensional schemes are obtained as special cases of blended interpolation formulas by discretizing the latter.

For brevity and simplicity, let us restrict attention to the case where the problem domain $R$ is an open, bounded and simply connected planar region in $E^2$ and $S$ is the open unit square $(0,1) \times (0,1)$ in $R^2$. We seek a one-to-one (univalent) mapping $\vec{U} : S \leftrightarrow R$ such that each point $(x,y) \in R$ has a unique correspondent $(s,t) \in S$. In other words, our goal is to achieve a curvilinear coordinatization of $R$ such that each point $(x,y)$ in $R$ may be uniquely referenced by its generalized coordinates $(s,t)$. As a trivial but classical example, consider the polar coordinatization of a unit disc $R$ as the graph of a map of the rectangle $(0,1) \times [0,2\pi]$ onto the circle. The polar coordinates $r$ and $\theta$ are the generalized coordinates of a point in $R$. Or, let $R$ be the triangle with vertices $(0,0)$, $(1,0)$ and $(1,1)$, and consider the univalent map of $S \leftrightarrow R$ given by $x = s$, $y = st$.

By a canonical domain (canonical geometry), we shall mean a standard domain such as a rectangle, triangle, circle, annulus, etc. for which a natural (curvilinear) coordinatization is known. If the given problem domain $R$ is reminiscent of some canonical domain, our goal will be to coordinatize $R$ with a generalized system analogous to that of the canonical region. For example, if the planar region $R$ is star-shaped with respect to a point $P$ and if $\partial R$ is smooth, then a simple radially expansive map with center $P$ serves to define an appropriate one-to-one correspondence of points $(x,y)$ in $R$ with the points $(r,\theta)$ in the unit disc. The images of the concentric circles and radial lines in the disc are "concentric contours" and radial lines in $R$. Thus, any point in $R$ may be referenced by its generalized polar coordinates $(r,\theta)$.

The most widely applicable methods for the mathematical description of free-form shapes in two and three dimensions are the so-called "blending-function" methods developed by Coons [1], Forrest [3] and Gordon [4-7]. In these, an n-dimensional manifold is expressed as weighted ("blended") combinations of manifolds of dimensionality n-1 or less. For example, a two-dimensional manifold (a planar region or surface in $E^3$) can be expressed in terms of functionally weighted combinations of curves. Here, we shall discuss the salient features of these methods as they relate to the construction of curvilinear coordinate systems.

We should remark that unlike conformal maps which could also be used to solve the same planar mapping problem, the blended mapping techniques proposed here are essentially "closed form" formulas and are easy to evaluate. Also, they are not restricted to planar regions.

To begin, let us postulate for notational simplicity the existence of a continuous vector-valued primitive function $\vec{F}$, which maps $S$ invertibly onto $R$. We emphasize that $\vec{F}$ is not known in practice (if it were, our mapping problem would be solved), but that we shall identify certain curves in $R$ as being the images under $\vec{F}$ of constant coordinate lines in $S$. We seek a simple <u>projector</u> $P$ such that the vector-valued approximation $P[\vec{F}]$ is a continuous and univalent map of $S$ onto $R$.

On $S$, consider a Cartesian product partition $\pi_s \times \pi_t$ where $\pi_s : 0 = s_0 < s_1 < \ldots < s_M = 1$ and $\pi_t : 0 = t_0 < t_1 < \ldots < t_N = 1$, and let $\{\phi_i(s)\}_{i=0}^{M}$ and $\{\psi_j(t)\}_{j=0}^{N}$ be functions satisfying the cardinality properties

(1) $$\phi_i(s_k) = \delta_{ik}, \qquad \psi_j(t_\ell) = \delta_{j\ell} .$$

Further, define the projectors $P_s$ and $P_t$ by

(2) $$P_s[\vec{F}] = \sum_{i=0}^{M} \vec{F}(s_i,t)\phi_i(s)$$
$$P_t[\vec{F}] = \sum_{j=0}^{N} \vec{F}(s,t_j)\psi_j(t) ,$$

then the <u>product</u> projection (tensor product interpolant)

(3) $$P_sP_t[\vec{F}] = \sum_{i=0}^{M}\sum_{j=0}^{N} \vec{F}(s_i,t_j)\phi_i(s)\psi_j(t)$$

interpolates $\vec{F}$ at the $(M+1)(N+1)$ points $\{(s_i,t_j)\}_{i=0,j=0}^{M,N}$. In contrast, the <u>transfinite interpolant</u> [8-9] defined by the <u>Boolean sum</u>

(4) $$P_s \oplus P_t[\vec{F}] \equiv P_s[\vec{F}] + P_t[\vec{F}] - P_sP_t[\vec{F}]$$

interpolates to $\vec{F}$ at <u>all points along the two sets of lines</u> $\{s = s_i\}_{i=0}^{M}$ and $\{t = t_j\}_{j=0}^{N}$. The functions $\phi_i$ and $\psi_j$ are called <u>blending functions</u>.

If the blending functions are chosen as the Lagrange polynomials (of minimal degree) satisfying (1), then $\mathcal{P}_s \oplus \mathcal{P}_t[\vec{F}]$ is the transfinite Lagrange interpolant studied in [8]. In particular, if $M = N = 1$ then $\mathcal{P}_s \oplus \mathcal{P}_t[\vec{F}]$ is the bilinearly blended transfinite Lagrange interpolant:

$$
\begin{aligned}
\mathcal{P}_s \oplus \mathcal{P}_t[\vec{F}] &= (1-s)\vec{F}(0,t) + s\vec{F}(1,t) + (1-t)\vec{F}(s,0) \\
&\quad + t\vec{F}(s,1) - (1-s)(1-t)\vec{F}(0,0) \\
&\quad - (1-s)t\vec{F}(0,1) - s(1-t)\vec{F}(1,0) \\
&\quad - st\vec{F}(1,1) \,.
\end{aligned} \tag{5}
$$

If the $\{\phi_i(s)\}$ and $\{\psi_j(t)\}$ are chosen as cardinal cubic splines [4,5] satisfying (1) then $\mathcal{P}_s \oplus \mathcal{P}_t[\vec{F}]$ is a cubic spline-blended transfinite mapping.

The images of the lines $s = \text{const.}$ and $t = \text{const.}$ in $\mathcal{S}$ are curves in $\mathcal{R}$. In particular, if we identify $\partial\mathcal{R}$ as being comprised of the four curve segments $\vec{F}(0,t)$, $\vec{F}(1,t)$, $\vec{F}(s,0)$ and $\vec{F}(s,1)$, we see that $\partial\mathcal{S} \to \partial\mathcal{R}$ for any choices of functions $\{\phi_i\}$ and $\{\psi_j\}$ which satisfy (1). To be more specific, consider the simplest case of the bilinearly blended map $\vec{U} = \mathcal{P}_s \oplus \mathcal{P}_t[\vec{F}]$ given by (5). It is easy to check that, as already mentioned, $\vec{U} = \vec{F}$ for all points $(s,t) \in \partial\mathcal{S}$, so that $\vec{U}$ gives an exact map of $\partial\mathcal{S}$ onto $\partial\mathcal{R}$. However, the map $\vec{U}$: $\mathcal{S} \to \mathcal{R}$ will not in general be one-to-one. To check this, one computes and graphs the images $\vec{U}(s^*,t)$ and $\vec{U}(s,t^*)$ of the lines $s = s^*$ and $t = t^*$ for various values of $s^*$ and $t^*$. The univalency requirement implies that the images of two parallel lines in $\mathcal{S}$ must not intersect in $\mathcal{R}$. If this condition is met, then the image of the Cartesian coordinatization of $\mathcal{S}$ is a legitimate curvilinear coordinatization of $\mathcal{R}$, and the mapping problem is solved. If $\vec{U}: S \to R$ is not univalent for all points in the interior of $S$, then one must define auxiliary curves $\vec{F}(s_i,t)$ and/or $\vec{F}(s,t_j)$ within $R$ and use a higher degree blending formula of the generic form (4) above. These auxiliary constraints are introduced so as to force the map to behave in the fashion dictated by the geometry of $R$. Also, some auxiliary constraints may be dictated by nonisotropic physical properties of the region $R$, e.g., material interfaces. This points up the fact that the method relies upon some form of graphical display for its practical implementation. In [8] and [9], transfinite interpolation is considered in greater detail and a heuristic strategy is given in [9] for the successful construction of univalent maps.

In passing, we note that the extensions of the methods outlined here to surfaces and to volumes in $E^3$ are straightforward ([1], [3], [4-7]).

The figures illustrate the results of applying the transfinite blending formulas as discussed above to various regions (see end of paper).

## 4. Applications to boundary value problems.

We now assume that a univalent mapping $\vec{U}: S \leftrightarrow R$ exists in the generic form of (4), and hence that $R$ has a generalized curvilinear coordinatization. Figure 5 illustrates two essentially different uses which can be made of the map $\vec{U}$.

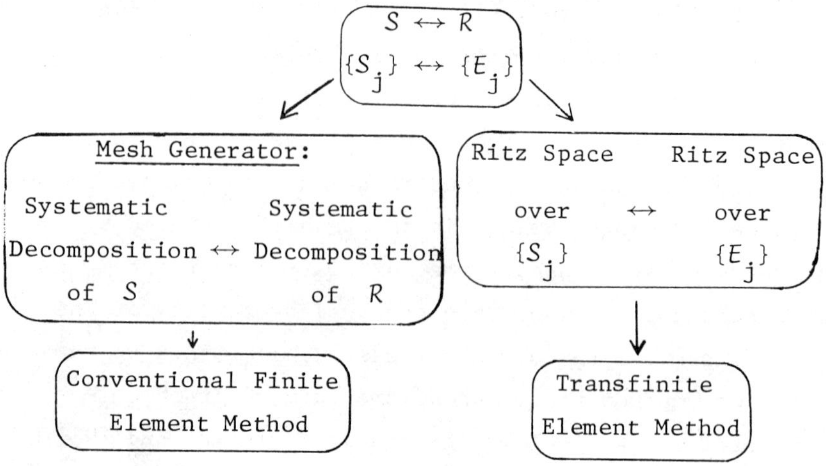


Figure 5

To explain, let us first consider the left branch of this figure. We saw earlier that the selection of a nodal topology in a general region $R$ is an arduous manual task and leads, moreover, to ill-structured sparse matrices. A quantitative relationship between $R$ and the model domain $S$, on the other hand, allows the nodal selection to be performed in the region $S$ and then mapped onto the corresponding image in $R$. In simple cases, an acceptable mesh in $R$ can be obtained from a Cartesian product grid in $S$. In other cases, the physics of the problem may dictate mesh refinements in certain subregions or may suggest a mesh other than Cartesian product. In any event, the mesh generation takes place within $S$ and the map $\vec{U}$ provides the numerical coordinates of the corresponding nodes in $R$.

One of the most significant advantages of using the mapping $\vec{U}$ as a mesh generator arises from the fact that the topology of the mesh configuration in $S$ is invariant under $\vec{U}$. In other words, a systematic mesh generation and nodal numbering in $S$ is preserved under the mapping $S \to R$. Moreover, the structure of the final sparse matrix (i.e., its binary pattern of zeros and nonzeros) is determined solely by the numbering of nodes, the Ritz space of functions and the functional $J[u]$, but is actually independent of the shape of $R$. Thus, to the extent that the mesh configuration can be standardized, the precise location of nonzeros in the sparse matrix can be predicted and the matrix can be efficiently solved by especially tailored algorithms, e.g., block direct or block iterative methods.

We now consider the right-hand branch of Figure 5 which is concerned with what we have termed <u>transfinite element methods</u> [8,9]. This class of methods differs from conventional finite element methods in that the region $R$ is not <u>a priori</u> approximated and the boundary conditions are imposed on $\partial R$, not on a polygonal approximation to $\partial R$. To explain, suppose that $\pi_S$ is some partitioning of the unit square $S$ into triangles and quadrilaterals. For concreteness, consider a linear space of piecewise analytic functions defined over $\pi_S$ such that the pieces are joined with $C^1$ continuity across the polygonal edges between adjacent subdivisions of $\pi_S$. By virtue of the map $\vec{U}$: $S \leftrightarrow R$, this Ritz space over $S$ images onto a corresponding Ritz space over the partition $\pi_R$ which is the image of $\pi_S$ in $R$. Note that since the map $\vec{U}$ is transfinite, the polygonal edges of $\pi_S$ image onto curvilinear edges of the partition $\pi_R$, so that the images of triangles and rectangles in $\mathcal{S}$ are curvilinear triangles and curvilinear quadrilaterals in $\mathcal{R}$. By the chain rule of differentiation, it is easy to see that the continuity of the transformed Ritz space over $\mathcal{R}$ is also $C^1$ provided that the univalent map $\vec{U}$ is continuously differentiable for all $(s,t) \in \mathcal{S}$.

It is very important to note that the approach outlined above completely decouples the curvilinear element generation in $\mathcal{R}$ from the definition of the Ritz space over $\pi_{\mathcal{R}}$. That is, the Ritz space used for approximating the dependent variable is totally independent of the mapping $\vec{U}$. However, one could consider transfinite schemes in which the same parametric interpolation formulas are used for both the mapping of the independent variables and the approximation of the dependent variable. Such schemes would be appropriately termed <u>transfinite isoparametric methods</u> since the term isoparametric is commonly used in the finite element literature to denote the coupling together of the independent and dependent variables via the same parametric interpolation formula. A more detailed discussion of transfinite isoparametric elements and their relation to the finite dimensional isoparametric methods of [2] and [12] is given in references [8] and [9]. Error analyses for transfinite element methods are also given in [8].

In conclusion, we mention the possibility of combining the uses of the mapping function $\vec{U}$ as both a mesh generator and a transfinite element generator on the problem domain $\mathcal{R}$. To illustrate, let $\pi_{\mathcal{S}} = \pi_s \times \pi_t$ be a Cartesian product partition of $\mathcal{S}$. Consider the subrectangle $S_{ij} = [s_i, s_{i+1}] \times [t_j, t_{j+1}]$ . For all $S_{ij}$ with no sides on $\partial S$ , use the map $\vec{U}$ to determine the images of the four corners $(s_i, t_j)$, $(s_i, t_{j+1})$, $(s_{i+1}, t_j)$, and $(s_{i+1}, t_{j+1})$ and define the element domain $E_{ij}$ to be the quadrilateral obtained by joining these four image points. For subrectangles $S_{ij}$ with one or two edges on $\partial S$ , define $E_{ij}$ to be that subregion of $R$ bounded externally by the corresponding segment of $\partial R$ and with straight sides for those edges not on $\partial R$ . In brief, the map is used as a quadrilateral mesh generator in the interior of $R$ and as a transfinite element generator for that row of elements bordering the boundary of $R$ . This approach appears to combine the best features of the two paths indicated in Figure 5.

Acknowledgments

The authors are indebted to Professor Garrett Birkhoff for valuable discussions during the preparation of this paper and to Drs. Donald H. Thomas and James A. Wixom of the General Motors Research Laboratories who generated Figures 1-4.

REFERENCES

1. Coons, S. A., "Surfaces for computer-aided design of space forms," Project MAC, MIT, 1964. Revised to MAC-TR-41, 1967. Available through CFSTI, Sills Building, 5285 Port Royal Road, Springfield, Virginia, 22151.
2. Ergatoudis, I., B. Irons, and O. Zienkiewicz, "Curved, isoparametric, 'quadrilateral' elements for finite element analysis," Int. J. Solids Structures 4 (1968), pp. 31-42.

3. Forrest, A. R., "Coons' surfaces and multivariable functional interpolation," CAD Group Doc. #38, Computer Laboratory, Cambridge University, U.K.
4. Gordon, W. J., "Free-form sufface interpolation through curve networks," General Motors Research Report, GMR-921, October 1969.
5. Gordon, W. J., "Spline-blended surface interpolation through curve networks," J. Math. Mech. 18, No. 10 (1969), pp. 931-952.
6. Gordon, W. J., "Blending-function methods of bivariate and multivariate interpolation and approximation," SIAM Numer. Anal. 8 (1971), pp. 158-177.
7. Gordon, W. J., "Distributive lattices and multivariate functions," in Approximations with Special Emphasis on Spline Functions, (I. J. Schoenberg, ed.), Academic Press, New York, 1969.
8. Gordon, W. J. and C. A. Hall, "Discretization error bounds for transfinite elements," General Motors Research Report GMR-1196, May 1972.
9. Gordon, W. J. and C. A. Hall, "Geometric aspects of the finite element method: Construction of curvilinear coordinate systems and their applications to mesh generation," General Motors Research Report, Oct. 1972.
10. Synge, J., The Hypercircle in Mathematical Physics, Cambridge Press, 1957.
11. Zienkiewicz, O. C., The Finite Element Method, McGraw-Hill, London, 1971.
12. Zienkiewicz, O. C. and D. V. Phillips, "An automatic mesh generation scheme for plane and curved surfaces by 'isoparametric' coordinates," Int. J. Num. Meth. Eng. 3 (1971), pp. 519-528.

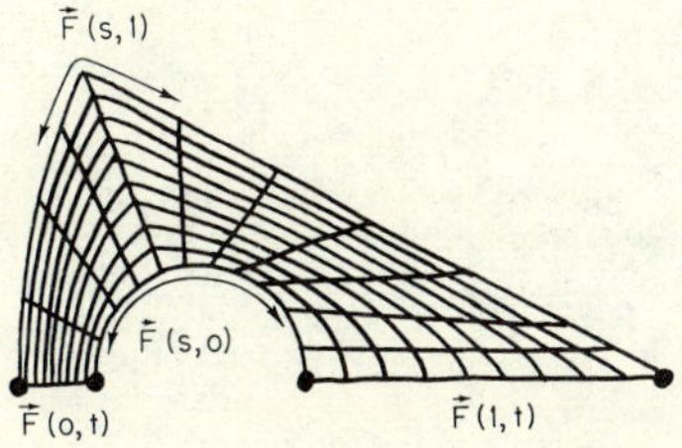


Fig. 1

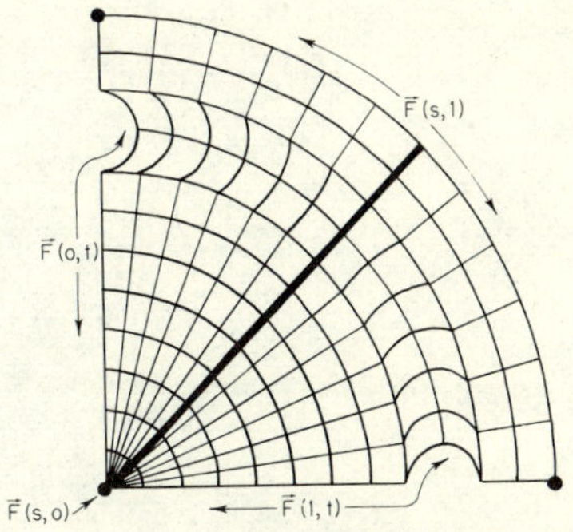


Fig. 2

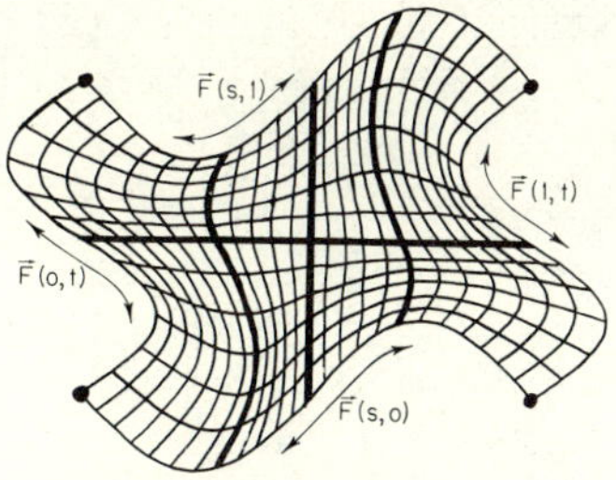


Fig. 3

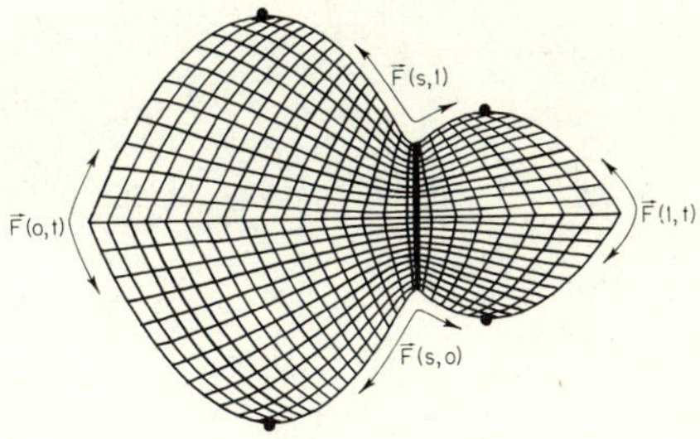


Fig. 4

# THE USE OF INTERPOLATORY POLYNOMIALS FOR A FINITE ELEMENT SOLUTION OF THE MULTIGROUP DIFFUSION EQUATION*

H. G. Kaper,** G. K. Leaf,** and A. J. Lindeman**

In this study, we investigate the use of a particular class of high-order approximation procedures for the numerical solution of the multigroup neutron diffusion equation for a realistic two-dimensional reactor configuration. The procedures are based on the use of variational principles and subspaces generated by the finite element method. In particular, we use Lagrange-type interpolatory polynomials [1,2] of degree two through five.

Earlier investigations on the application of finite element methods in reactor mathematics have been reported by Semenza, Lewis, and Rossow [3] and by Kang and Hansen [4]. The former authors confined themselves to the use of (bi)linear elements; their approximations are, therefore, basically low-order approximations. Kang and Hansen used cubic Hermite-type interpolatory elements.

The efficiency of high-order versus low-order approximation is studied for a mockup (in Cartesian geo-

---

* Work performed under the auspices of the U.S. Atomic Energy Commission.

** Argonne National Laboratory.

metry) of current designs of oxide-fueled Liquid Metal Fast Breeder Reactors. The study is based on numerical results obtained with two computer programs--DARC2D[5] (low-order finite difference approximation) and HOD (high-order approximation)--on the Argonne National Laboratory IBM System 360 Model 50/75. DARC2D is the production program for the two-dimensional multigroup diffusion module in the Argonne Reactor Computation (ARC) System, and the algorithms are similar to those used throughout the reactor community for fast reactor analyses. HOD is a prototype two-dimensional multigroup diffusion program which we developed and is based on a finite element program written by J. A. George.

The results of a four-group study are given in Table I. $\pi_p$ denotes a rectangular mesh with $p$ mesh cells for the finite difference approach and $\tau^m_{p,q}$ a triangulation with $pq^2$ triangles with a polynomial of degree $m$ for the finite element approach. The latter is obtained from a basic triangulation $\tau_{p,1}$ with $p$ triangles by subdividing the edges of each triangle in $\tau_{p,1}$ into $q$ equal parts. Relative truncation errors are measured against a benchmark value of 1.0569955, which was obtained by utilizing quintic polynomials in a sequence of mesh refinements.

The smallest relative error in $k$ for the finite difference series is 0.029% and is obtained with 1452 mesh cells at a cost of 317.4 seconds and 70,700 64-bit words. Upon comparison with the results obtained with the finite element program HOD, we observe a significant reduction of both the storage requirements and execution times for the high-order finite element approximations. HOD with quadratic elements on seventeen triangles is approximately twenty (20) times faster than DARC2D. In addition, the storage requirements are reduced from 70,700 words to 48,000 words, which represents a savings of more than 30%. If DARC2D were required to achieve a much smaller relative error (e.g., for benchmark calculations)- we expect both the timing and storage ratios to improve further in favor of the finite element program HOD.

Additional studies [6], in which we achieved execution time ratios of up to 50:1, corroborate the substantial execution time savings. However, to dwell on the quantitative aspects of these comparisons would be to miss the most essential point, which is qualitative in nature. Namely, the use of high-order polynomials in finite element methods is not just a viable alternative to the use of finite difference methods, it is a significant advancement in computational capability for the numerical solution of the multigroup diffusion equation on realistic fast reactor configurations.

## REFERENCES

1. George, J. A., "Computer implementation of the finite element method," Computer Science Department, Stanford University, STAN-CS-71-208 (1971).
2. Silvester, P., "High-order polynomial triangular finite elements for potential problems," Int. J. of Engineering Sciences 7 (1969), pp. 849-861.
3. Semenza, L. A., E. E. Lewis, and E. C. Rossow, "A finite element treatment of neutron diffusion," Trans. Am. Nucl. Soc. 14 (1971), p. 200.

4. Kang, C. M. and K. F. Hansen, "Finite element methods for the neutron diffusion equations," Trans. Am. Nucl. Soc. 14 (1971), p. 199.
5. Daly, T. A., G. K. Leaf, and A. S. Kennedy, "The ARC system two-dimensional diffusion theory capability, DARC2D," ANL-7716, Argonne National Laboratory (1972).
6. Kaper, H. G., G. K. Leaf, and A. J. Lindeman, "Applications of finite element methods in reactor mathematics--Numerical solution of the neutron diffusion equation," ANL-7925, Argonne National Latoratory (1972).

TABLE I. TIMING RESULTS AND MEMORY REQUIREMENTS FOR THE DARC2D (FINITE DIFFERENCES) AND HOD (FINITE ELEMENT) PROGRAMS

| MESH | NUMBER OF UNKNOWNS/ GROUP | k | RELATIVE ERROR (%) | EXECUTION TIME (SECONDS) | NUMBER OF 64-BIT WORDS USED |
|---|---|---|---|---|---|
| $\pi_{48}$ | 48 | 1.067176 | $9.6\cdot10^{-1}$ | 30.7 | 38,900 |
| $\pi_{140}$ | 140 | 1.064142 | $6.8\cdot10^{-1}$ | 36.4 | 41,000 |
| $\pi_{280}$ | 280 | 1.058874 | $1.8\cdot10^{-1}$ | 50.6 | 44,200 |
| $\pi_{560}$ | 560 | 1.057842 | $8.0\cdot10^{-2}$ | 87.4 | 50,500 |
| $\pi_{900}$ | 900 | 1.057838 | $5.2\cdot10^{-2}$ | 176.5 | 58,200 |
| $\pi_{1160}$ | 1160 | 1.057414 | $4.0\cdot10^{-2}$ | 284.7 | 64,100 |
| $\pi_{1452}$ | 1452 | 1.057298 | $2.9\cdot10^{-2}$ | 317.4 | 70,700 |
| $\tau^2_{17,1}$ | 46 | 1.056700 | $-2.8\cdot10^{-2}$ | 16.5 | 48,000 |
| $\tau^2_{24,1}$ | 63 | 1.056851 | $-1.3\cdot10^{-2}$ | 23.0 | 48,400 |
| $\tau^3_{17,1}$ | 94 | 1.057007 | $+1.5\cdot10^{-3}$ | 46.2 | 49,800 |
| $\tau^4_{17,1}$ | 159 | 1.056995 | $+3.3\cdot10^{-4}$ | 110.8 | 53,300 |
| $\tau^5_{17,1}$ | 241 | 1.056992 | - | 246.1 | 59,400 |
| $\tau^5_{24,1}$ | 336 | 1.056992 | - | 370.5 | 64,400 |

# A "LOCAL" BASIS OF GENERALIZED SPLINES OVER RIGHT TRIANGLES DETERMINED FROM A NONUNIFORM PARTITIONING OF THE PLANE

Lois Mansfield*

Let $\pi$ be a partition of the plane into rectangles. A basis of piecewise polynomials is constructed over right triangles determined by $\pi$ having the property that corresponding to each node of $\pi$ there is one basis function. The support of each basis function is slightly less than for the corresponding bivariate B-spline, thus providing a less dense matrix when these functions are used as basis functions in the finite element method. The linear space $T_\pi^k$ generated by these basis functions contains all polynomials of degree less than $k$. For $\Omega$ a region in $R^2$ and $f \in C^k(\Omega)$, $T_\pi^k$ also contains an element $v$ such that $\|(f - v)^{(i,j)}\|_{\infty,\Omega} = O(\bar{\pi}^{k-i-j})$, $i + j < k$.

* University of Kansas.

# LEAST SQUARE POLYNOMIAL SPLINE APPROXIMATION

Paul D. Patent

In this note we describe least square polynomial spline approximation techniques based on the classical theory of orthogonal projection in abstract Hilbert space. We confine our attention to the approximation of elements of the real Hilbert space $L^2[a,b]$ by members of finite dimensional subspaces of polynomial splines. We recall that the inner product for $f,g \in L^2[a,b]$ is defined by

$$(f,g) \equiv \int_a^b f(x)g(x)dx$$

and so the norm generated by this inner product is

$$\|f\| \equiv \left\{\int_a^b f^2(x)dx\right\}^{\frac{1}{2}} .$$

Bounds for the errors in approximating sufficiently smooth functions by such techniques can be derived using bounds for the errors in polynomial spline interpolation.

Given any $f \in L^2[a,b]$ and any polynomial spline space $Sp$ with basis $\{s_1, s_2, \ldots, s_n\}$,, minimizing $\|f - s\|^2 = (f-s,f-s)$ over all $s \in Sp$ is equivalent to solving the appropriate system of normal equations, i.e.,

$$A\underline{\alpha} = \underline{k}$$

where the entries of the positive definite, symmetric matrix $A$, denoted $a_{ij}$, and the components of the n-vector $\underline{k}$, denoted $k_i$, are defined by

$$a_{ij} \equiv (s_i, s_j) , \; 1 \leqslant i , j \leqslant n ,$$

$$k_i \equiv (f, s_i) , \; 1 \leqslant i \leqslant n .$$

Clearly, the entries of $A$ are sums of definite integrals of polynomials and can be computed "exactly" up to rounding errors. In fact, we may choose the basis functions for the spline space in such a way that $A$ is a band matrix. Consequently, once numerical values for the components of $\underline{k}$ are known, we may employ Gaussian elimination to solve our system in order to obtain the coefficients (of the basis functions) for the least square approximation in Sp to $f$. However, the components of $\underline{k}$ cannot, in general, be computed directly. In most practical applications, $f$ will be a tabulated function and we will be forced to approximate $\underline{k}$ by some n-vector $\tilde{\underline{k}}$ and solve the corresponding system, i.e.,

$$A\underline{\alpha} = \tilde{\underline{k}} .$$

Let $\hat{s} = \Sigma\alpha_i s_i$ and $\tilde{s} = \Sigma\tilde{\alpha}_i s_i$ where $\underline{\alpha} = A^{-1}\underline{k}$ and $\tilde{\underline{\alpha}} = A^{-1}\tilde{\underline{k}}$. For sufficiently smooth $f$, we can bound the quantity $\|\hat{s} - \tilde{s}\|$ for two different methods of determining $\tilde{\underline{k}}$.

One method which we consider employs a composite quadrature scheme based on n-point interpolatory formulae defined over subintervals determined by the partition $\Delta$ over which the spline space Sp is defined. In this case, the components of $\tilde{\underline{k}}$ are just the numerical values of the composite scheme applied to the components of $\underline{k}$. If the splines are of degree $d \leq n$, $\overline{\Delta}$ is the maximum mesh length for the partition $\Delta$, and $f \in C^{n+1}[a,b]$, then we can prove that

$$\|\hat{s} - \tilde{s}\| \leq K(\overline{\Delta})^{n-d+\frac{1}{2}}$$

where $K$ is a positive constant which is independent of $\overline{\Delta}$.

The other method which we consider is based on Filon type quadrature schemes, i.e., rules for integrals of product integrands in which only one of the factors requires approximation. We choose to approximate the tabulated function $f$ by a piecewise Lagrange interpolate $\tilde{f}$ and we determine $\underline{\tilde{k}}$ by applying our Filon type rule to the components of $\underline{k}$. We immediately have

$$\|\hat{s} - \tilde{s}\| \leq \|f - \tilde{f}\|$$

and, consequently, error bounds for piecewise Lagrange interpolation will provide us with bounds for the quantity $\|\hat{s} - \tilde{s}\|$. If the piecewise Lagrange interpolation is defined in terms of Lagrange interpolation of order $m$ in each subinterval of a partition $\Delta_\tau$ and $f \in C^m[a,b]$, then we can prove that

$$\|f - \tilde{f}\| \leq K(\overline{\Delta}_\tau)^m$$

where $K$ is a positive constant independent of $\overline{\Delta}_\tau$.

We remark that all these bounds give rise to asymptotic results on the convergence of suitably chosen sequences of least square polynomial spline approximations. We also note that our results can be used to examine the consistency of the orders of accuracy of errors induced by the use of these quadrature rules with the order of accuracy of the least square error itself.

In conclusion, we present plots of cubic spline approximations to data sets considered difficult to approximate using polynomials because of the "roly-poly" nature of such approximants. We employ Filon type quadrature based on piecewise linear interpolation with remarkable success.

## REFERENCE

Patent, P. D., Least Square Polynomial Spline Approximation (97 pp.). Doctoral Thesis, California Institute of Technology, 1972.

# SUBSPACES WITH ACCURATELY INTERPOLATED BOUNDARY CONDITIONS

Ridgway Scott

In a variational problem, when the minimization is done over a subspace of functions that interpolate, but do not satisfy exactly, essential boundary conditions, it is the negative Sobolev norm of the functions on the boundary that governs the error (see "Approximate Boundary Conditions in the Finite Element Method," by Alan Berger, Gilbert Strang, and the author, to appear in Symposia Mathematica, Academic Press). The present paper shows this to be a consequence of what is known in solid mechanics as Saint-Venant's principle and describes finite element subspaces that achieve optimal approximation and boundary condition satisfaction simultaneously. The latter is achieved when the shape functions vanish at certain points on the boundary, analogous to Gauss quadrature points. We also determine the boundary condition error for some subspaces that have been applied to 4th order problems; in general, the boundary condition error is larger than the approximation error with these spaces that were constructed without regard for Saint-Venant's principle. A notable exception is the Clough-Tocher cubic space, for which the boundary condition error is of the optimal order $h^{5/2}$.